Development
from Above or Below?

Development
from Above or Below?

The Dialectics of Regional Planning in Developing Countries

Edited by

WALTER B. STÖHR

Professor and Director
Interdisciplinary Institute for Urban and Regional Studies
University of Economics, Vienna

and

D.R. FRASER TAYLOR

Professor of Geography and International Affairs
Carleton University, Ottawa

JOHN WILEY AND SONS
Chichester · New York · Brisbane · Toronto

Copyright © 1981 by John Wiley & Sons Ltd.

British Library Cataloguing in Publication Data:

Development from above or below?
  1. Underdeveloped areas—Regional
  planning—Case studies
  I. Stöhr, Walter
  II. Taylor, D R
  309.2′5′091724   HT391   80-40850

ISBN 0 471 27823 8

Photosetting by Thomson Press (India) Limited, New Delhi and
printed in the United States of America.

Preface

Many of the issues related to development from above or below have a long history, but from a spatial perspective, which dominates in this book, substantive consideration of these ideas is much more recent. Considerable debate is currently taking place, some of it of a fundamental nature. Several authors who have contributed to this book are in the forefront of this debate.

The present study considers development from above or below in the Third World context. Development for all social and territorial groups is one of the greatest issues of our time and it is hoped that this book will make a contribution towards greater understanding and possibly point the way to new policy solutions.

Many people have given invaluable support and help in the production of this book. We are grateful for the cooperation and punctuality of all authors represented. Mrs. Hilde Kaufmann in Vienna and Mrs. Barbara George in Ottawa had the particularly onerous task of typing manuscripts and facilitating communication between both editors and contributors separated by thousands of miles of space, and we owe them a debt of gratitude.

Special thanks are also due to Mrs. Phyllis Kingston for editorial and typing assistance, to Christine Earl for cartographic work, and to Dr. Herwig Palme for his administrative contribution.

A task of the magnitude of this book unfortunately always demands sacrifices from wives and families, and to Erika and Monica go our thanks for their understanding and support.

Despite the help and support we have received, there are indubitably still shortcomings in this book and these, of course, are entirely the responsibility of the editors, not of the contributors and others who have helped to bring this book to fruition.

Walter B. Stöhr, Vienna
D. R. F. Taylor, Ottawa

Contents

Preface v
Walter B. Stöhr and D. R. F. Taylor

List of Figures ix

List of Tables xi

Introduction 1
Walter B. Stöhr and D. R. F. Taylor

PART I   THEORETICAL ISSUES AND CHALLENGES

1 Development from Above: The Centre-Down Development
Paradigm 15
Niles M. Hansen

2 Development from Below: The Bottom-Up and Periphery-Inward
Development Paradigm 39
Walter B. Stöhr

3 Development Theory and the Regional Question: A Critique of
Spatial Planning and its Detractors 73
Clyde Weaver

4 Basic-Needs Strategies: A Frustrated Response to Development
from Below? 107
Eddy Lee

5 Growth Poles, Agropolitan Development, and Polarization
Reversal: The Debate and Search for Alternatives 123
Fu-chen Lo and Kamal Salih

PART II   CASE STUDIES

a. ASIA

6 China: Rural Development—Alternating Combinations of Top-
Down and Bottom-Up Strategies 155
Chung-Tong Wu and David F. Ip

7 Thailand: Territorial Dissolution and Alternative Regional
Development for the Central Plains 183
Mike Douglass

8 Papua New Guinea: Decentralization and Development from the Middle 209
Diana Conyers

9 Nepal: The Crisis of Regional Planning in a Double Dependent Periphery 231
Piers Blaikie

10 India: Blending Central and Grass Roots Planning 259
R. P. Misra and V. K. Natraj

b. AFRICA

11 Nigeria: The Need to Modify Centre-Down Development Planning 283
Michael Olanrewaju Filani

12 Ivory Coast: An Adaptive Centre-Down Approach in Transition 305
M. Penouil

13 Tanzania: Socialist Ideology, Bureaucratic Reality, and Development from Below.. 329
Jan Lundqvist

14 Algeria: Centre-Down Development, State Capitalism, and Emergent Decentralization 351
Keith Sutton

c. LATIN AMERICA

15 Brazil: Economic Efficiency and the Disintegration of Peripheral Regions 379
Paulo Roberto Haddad

16 Chile: Continuity and Change—Variations of Centre-Down Strategies under Different Political Regimes 401
Sergio Boisier

17 Peru: Regional Planning 1968–77; Frustrated Bottom-Up Aspirations in a Technocratic Military Setting 427
Jos G. M. Hilhorst

PART III CONCLUSIONS

18 Development from Above or Below? Some Conclusions 453
Walter B. Stöhr and D. R. F. Taylor

Contributing Authors.. 481

Index 487

List of Figures

A   Location of case study countries   11

2.1   Historical changes in scales of spatial interaction in parts of Europe: A hypothesis   50

5.1   Trajectories of per capita income, regional disparity, urbanization, and income inequality for Japan, 1956–70..   127

5.2   Trends in income inequality and per capita GNP in selected Asian countries   129

5.3   Basic macro-spatial model   142

6.1   The provinces of China   158

6.2   Rural commune organization in China   166

7.1   Thailand: regions, provinces, and growth centres   184

7.2   Cumulative distribution of household annual income, Bangkok metropolis and central region, 1968–69 and 1972–73   191

7.3   Production and exchange relations in Thailand: circuits, regions, and the international political economy..   192

8.1   Provinces of Papua New Guinea   211

9.1   Growth axes of Nepal   233

9.2   An example of a dependent regional economy in Nepal   234

9.3   Rural development regions and zones of Nepal   236

9.4   Nepal's leaky frontier and the class interests which maintain it   ..   246

10.1   The States of India   260

10.2   India: agricultural growth rates and productivity by regions   ..   269

11.1   The States of Nigeria   284

12.1   Provinces of the Ivory Coast   306

12.2   The economy of the Ivory Coast   325

13.1   Regions of Tanzania   330

13.2   Tanzania: Imports by value, 1967–74   340

14.1   Algeria: regional development programmes   360

14.2   The economic zones of Algeria (Second Plan)   361

15.1   Regions of Brazil, 1960/70   380

16.1   Planning regions of Chile, 1970   402

17.1   Peruvian regionalization policy   432

17.2   Areas of Concentrated Action in Peru   437

List of Tables

A   Matrix of case studies   3

B   Macro-characteristics of case-study countries   4

5.1   Gini coefficients of income inequality and per capita GNP for selected Asian countries   128

6.1   Ranking of Chinese provinces by per capita gross value of industrial output (GVIO) for 1952, 1957, 1965 and 1974 (ranked in descending order)   159

6.2   Coastal provinces' share of national gross value of industrial output (GVIO) in China for selected years, in per cent   160

6.3   Percentage share of total gross value of industrial output (GVIO) for top four ranking provinces of China..   160

8.1   Socioeconomic indicators by province of Papua New Guinea ..   214

9.1   Percentage distribution of khet (irrigated) land classified by size of holding and region of Nepal 1963–69   237

9.2   Indicators of productivity and income of regions in Nepal 1970–71   238

9.3   Distribution of districts by GDP per capita as a percentage of GDP per capita for all Nepal   238

9.4   Selected development indicators for Nepal   238

9.5   Extrapolation of past trends up to 1990 for Nepal in respect of current indicators   243

10.1   Agricultural growth rates and productivity by regions for India..   268

11.1   Industrial establishments in Nigeria by type in 1945   286

11.2   Contrasts between rural and metropolitan development, Kano State, Nigeria 1971   294

11.3   Percentages of actual world prices received by Nigerian farmers for major agricultural products, 1948–57   295

11.4   Urban/rural investment allocations in selected sectors of the Nigerian economy, 1970–74 Development Plan   296

11.5   Percentage share of industrial activities by some major centres in Nigeria, 1970   298

11.6   Population growth rates of major cities in Nigeria between 1952
       and 1963   299

12.1   Annual rate of growth of the economy of the Ivory Coast..   ..   307

12.2   Over-all development of the Ivory Coast economy since 1960 ..   307

12.3   Gross domestic product, Ivory Coast   309

12.4   Total traffic for the Port of Abidjan   314

12.5   Per capita agricultural income by region of the Ivory Coast   ..   316

12.6   Number of maternity and hospital beds by administrative region of
       the Ivory Coast in 1969 and in 1975   318

13.1   The four phases of Tanzanian development since 1961   332

13.2   Concentration of Ujamaa villages by region in Tanzania   338

14.1   GDP growth, per capita GDP, and re-investment rates   354

14.2   Growth of industrial employment by sectors in Algeria 1967,
       1972, and 1976   355

14.3   Industrial location quotients for Algeria, 1974   358

15.1   Index of inequality, Brazil 1950–70   381

15.2   Spatial policies in national development plans of Brazil   382

15.3   Life-expectancy at birth by region of Brazil, 1930–40 to 1960–70   390

15.4   Life-expectancy at birth, by region and by household income for
       Brazil, 1970   391

15.5   Patterns of regional growth of industrial employment in Brazil,
       1950–70   392

15.6   Monthly income levels of employed and self-employed workers
       by region of Brazil, 1972 and 1973   395

16.1   Selected regional social indicators for Chile   408

16.2   Percentage of poor by region in Chile   410

17.1   Urban and rural population by natural zones of Peru, 1940, 1961
       and 1990   429

17.2   Income per capita and population by department and natural
       zones of Peru   430

17.3   GDP by type of expenditure in Peru, 1970–76   436

Introduction

WALTER B. STÖHR and D.R.F. TAYLOR

Spatial inequalities in the living levels of developing nations are large and, in many instances, rapidly increasing. The key issue discussed in this book is whether these inequalities can be reduced by higher or more effective functional integration, both on a national and international scale; or whether internal territorial integration (Friedmann and Weaver, 1979), and a greater degree of internal self-reliance with 'selective spatial closure' (Stöhr and Tödtling, 1978) would be more effective. The first approach we call development 'from above', and the alternate approach, development 'from below'.

Development 'from above' has its roots in neoclassical economic theory and its spatial manifestation is the growth centre concept. Until recently strategies of development 'from above' have dominated spatial planning theory and practice. The basic hypothesis is that development is driven by external demand and innovation impulses, and that from a few dynamic sectoral or geographical clusters development would, either in a spontaneous or induced way, 'trickle down' to the rest of the system. Such strategies, as well as being outward-looking or externally oriented, have tended to be urban and industrial in nature, capital-intensive, and dominated by high technology and the 'large project' approach.

Development 'from below' is a more recent strategy and is a reflection of changing ideas on the nature and purpose of development itself, as described by Seers (1977) and Goulet (1978, 1979). Development 'from below' considers development to be based primarily on maximum mobilization of each area's natural, human, and institutional resources with the primary objective being the satisfaction of the basic needs of the inhabitants of that area. In order to serve the bulk of the population broadly categorized as 'poor', or those regions described as disadvantaged, development policies must be oriented directly towards the problems of poverty, and must be motivated and initially controlled from the bottom. There is an inherent distrust of the 'trickle down' or 'spread effect' expectations of past development policies. Development 'from below' strategies are basic-needs oriented, labour-intensive, small-scale, regional-resource-based, often rural-centred, and argue for the use of 'appropriate' rather than 'highest' technology. Such strategies have received rhetorical and intellectual support but as yet have not been widely applied. Both development 'from above' and 'from below' are closely related to the two principles of societal integration described by Friedmann and Weaver (1979) in their recent book, *Territory and Function*.

The approach used in *this* book is a 'radical' one in the literal rather than political sense of that word. We are interested in examining the root (radix) causes and effects of inequality and poverty. Our concern is to appraise the results of spatial policies on the population that most of them were devised to assist—in very broad terms, 'the poor', in their respective territorial contexts. This group can be defined by social strata or by territorial unit. They can be both urban and rural, and their poverty can be measured by a variety of different indices in both an absolute and comparative way.

The first five chapters of the book examine the theoretical bases of development 'from above' and 'from below'. Niles Hansen is presenting the case for development 'from above', although not wholly uncritical of it. Walter Stöhr presents the case for development 'from below' and relates it to a number of past historical experiences. The other authors in the first, theoretical, part of the book deal with issues related to development 'from below' and the theoretical underpinnings of this strategy.

This first part is followed by a series of case studies drawn from Asia, Africa, and Latin America. The main objective of these is to examine how selected countries have used different mixes of development 'from above' and 'from below', and how effective they have been in reducing poverty and inequality. The book compares the experiences of capitalist and socialist countries in each of the three continents. Capitalist countries are usually market-oriented with externally open economic systems. Socialist countries are usually—but by no means exclusively—inward-oriented and centrally planned. The book was designed on the basis of the following conceptual matrix:

Sociopolitical Orientation	Outward-looking strategy	Inward-looking strategy
Capitalist	1	2
Socialist	3	4

Not all of the countries fit exactly into the matrix and some have oscillated over time in their sociopolitical orientation and/or their degree of external closure, making categorization difficult. But the conceptualization is useful in a comparative sense. Individual case study countries can be located in such a matrix as that shown in Table A.

The major characteristics of the countries under discussion have been compiled by Herwig Palme, and are shown in Table B.

Table A. Matrix of case studies

Continent	Capitalist/market-oriented systems		Mixed systems		Socialist systems	
	Outward-looking (1)	Inward-looking (2)	Outward-Looking (1–3)	Inward-looking (2–4)	Outward-looking (3)	Inward-looking (4)
Africa	*Ivory Coast* (M. Penouil)		*Nigeria* (M.O. Filani)		*Algeria* (K. Sutton)	*Tanzania* (J. Lundqvist)
Asia	*Thailand* (M. Douglass)		*India* (R. P. Misra and V. K. Natraj)			
			Nepal (P. Blaikie)			*China* (C. T. Wu and D. F. Ip)
			Papua New Guinea (D. Conyers)			
Latin America	*Brazil* (P. Haddad)		*Peru* (J. G. M. Hilhorst)		*Chile* (Period II) (S. Boisier)	*Cuba**
	Chile (Periods I and III) (S. Boisier)					

* A case study on Cuba was planned, but had to be dropped because of circumstances beyond the editors' control

Table B *Macro-characteristics of case-study countries*

Characteristics	Africa				Asia					Latin America		
	Algeria	Ivory Coast	Nigeria	Tanzania	China	India	Nepal	Papua New Guinea	Thailand	Brazil	Chile	Peru
Size												
Surface area[1] (1000 km²)	2381	322	923	945	9596	3287	140	461	514	8511	756	1285
Population,[1] mid-year estimates, 1976 (in millions)	17.3	5.0	64.7	15.6	852.1	610.0	12.8	2.8	42.9	109.1	10.4	16.0
Level of Development												
GNP per capita 1976[6] (US dollars)	990	610	380	180	410	150	120	490	380	1.140	1.050	800
Life expectancy at birth, 1975[6]	53	44	41	45	62	50	44	48	58	61	63	56
Food supply, calories per caput per day[2]												
Total, 1975–77	2357	2563	2291	2089	2439	1949	2070	2247	2193	2522	2644	2286
Animal products, 1975–77	228	152	82	196	256	100	138	228	135	416	429	318
Adult literacy rate [6]												
1960	—	9	25	17	—	24	10	—	68	61	84	61
1974	35	20	—	63	—	36	19	32	82	64	90	72
Hospital establishments[1] (population per bed)	365 (1969)	730 (1971)	1168 (1975)	775 (1970)	—	1590 (1969)	6630 (1974)	169 (1972)	796 (1974)	266 (1973)	362 (1975)	497 (1972)

Numbers enrolled in primary schools as percentage of age group, 1975[6]												
Total	89	86	49	57	—	65	27	59	78	90	119	111
Female	72	64	39	46	—	52	10	44	75	90	118	106
Percentage of population with access to safe water[6]	77	—	—	39	—	31	8	20	25	—	70	47
Energy consumption per capita, 1975[6]	754	366	90	70	693	221	10	278	284	670	765	682
Economic and urban structure												
Sectoral distribution of gross domestic product (percentage), 1976[6]												
Agriculture	7	25	23	45	—	47	65	28	30	8	10	16
Industry	57	20	50	16	—	23	10	—	25	39	39	31
Services	36	55	27	39	—	30	25	—	45	53	51	53
Economically active population: percentage in agriculture[2], 1978	52	81	55	82	61	65	93	83	76	40	19	39
Percentage of population in urban areas[6]												
1976	31	11	18	5	19	18	3	3	13	45	69	47
1975	50	20	29	7	24	22	5	13	17	60	83	57
Man/Land ratio												
Population density, 1976[1]	7	16	70	17	89	186	91	6	84	13	14	13
Arable land per capita in hectares[2]*												
1965	0.53	2.00	0.44	0.22	0.17	0.33	0.18	0.007	0.37	0.27	0.47	0.19
1977	0.41	1.07	0.29	0.25	0.12	0.26	0.17	0.006	0.36	0.28	0.53	0.19

Table B   *Macro-characteristics of case-study countries*

Characteristics	Africa				Asia					Latin America		
	Algeria	Ivory Coast	Nigeria	Tanzania	China	India	Nepal	Papua New Guinea	Thailand	Brazil	Chile	Peru
Development Dynamics												
Population, annual rate of increase, average, 1970–76[1]	3.2	2.6	2.7	2.7	1.7	2.1	2.3	2.2	2.8	2.8	1.8	3.0
Employment change in manufacturing 1970–76 (1970 = 100)[1]	168 (1975)	—	138 (1974)	—	—	113	—	—	—	—	—	—
Gross domestic product, average annual rates of growth 1970–1977[3]	5.3	6.5	6.2	4.5	5.8	3.0	2.8	5.0	7.1	5.3	0.1	4.6
Average annual production growth rates[6], 1970–76												
Agriculture	−8.7	3.5	−0.2	2.5	—	1.4	1.9	—	4.3	5.5	0.5	0.6
Industry	16.4	7.9	12.6	2.9	—	3.8	—	—	8.2	11.6	−2.2	6.2
Services	−4.6	7.7	9.5	2.8	—	2.4	—	—	6.9	13.1	−1.3	8.4
GNP per capita, annual growth rate, 1970–75[4]	3.6	1.3	3.8	0.6	—	−0.4	0.1	2.2	3.6	6.8	−2.6	2.7
Agricultural and food production per capita, 1978, production indices: 1969–71 = 100[2]												
Food	89	116	90	93	110	102	91	105	122	111	89	80
Agriculture	89	106	90	89	110	102	91	108	117	105	89	79

External orientation

External trade indicators.												
ratio of exports to GDP[1], 1973	25.1	33.7	26.7	18.4	—	3.9	—	22.1	14.9	7.9	—	11.2
Share of primary commodities of total imports[1] (year), percentage	20.8 (1973)	27.1 (1976)	14.0 (1973)	—	—	58.4 (1976)	—	28.3 (1973)	23.9 (1973)	29.2 (1973)	39.7 (1973)	23.1 (1973)
percentage of Total Exports	96.6 (1973)	92.4 (1976)	97.9 (1973)	—	—	43.3 (1976)	—	85.6 (1973)	73.3 (1973)	75.2 (1976)	21.1 (1973)	64.5 (1973)
Share of manufactured goods of total imports[1] (Year)	79.1 (1973)	71.6 (1976)	85.1 (1973)	—	—	41.4 (1976)	—	64.1 (1973)	72.5 (1973)	69.8 (1973)	60.1 (1973)	76.9 (1973)
Percent of total exports	3.2 (1973)	7.2 (1976)	1.1 (1973)	—	—	56.5 (1976)	—	5.9 (1973)	22.6 (1976)	23.2 (1976)	78.9 (1973)	35.2 (1973)
Percentage shares of merchandise imports[6], 1975 — Food	—	15	10	20	—	26	—	—	4	6	—	—
Fuel	—	14	3	11	—	23	—	—	22	26	—	—
Destination of merchandise exports[6] (percentage of total), 1976 — Developed countries	89	76	82	57	—	54	31	91	60	62	64	63
Developing countries	9	22	17	38	—	33	69	7	38	30	35	21
Disbursement from developed market economies and from multilateral institutions to individual developing countries or areas, of bilateral official development assistance (ODA), 1974–76 annual averages[1]. (Total per capita in US$)	7.7	19.7	1.2	15.9	—	2.2	3.4	97.7	2.0	1.4	5.3	4.9

Table B   Macro-characteristics of case-study countries

Characteristics	Africa				Asia					Latin America		
	Algeria	Ivory Coast	Nigeria	Tanzania	China	India	Nepal	Papua New Guinea	Thailand	Brazil	Chile	Peru
Debt services as percentage of exports of goods and services, 1976[6]	14.1	9.1	2.3	4.3	—	12.0	2.3	—	2.4	14.8	32.9	21.6
Net direct private foreign investment: million US$ 1976, per capita[6]*	—	7.5	6.0	—	—	—	—	—	1.8	9.2	−0.5	10.6
Development Disparities												
International:												
Relative gap—GNP per capita of individual country or region as a percentage of average GNP per capita of the OECD countries[4]												
1950	20.3	11.9	6.3	3.5	—	4.0	3.6	9.6	5.6	15.7	25.0	16.9
1975	13.7	8.8	5.5	3.1	—	2.6	1.9	7.9	6.1	17.7	13.4	14.3
Terms of trade (1970 = 100)[6]												
1960	91	81	96	97	—	77	—	—	97	90	63	68
1976	308	107	322	114	—	73	—	—	82	99	43	80

National:
Estimated distribution of income, shares of net income of population groups with different income levels[5]

The poorest, 0–20%	8	7	10	—	8	—	3	5	4
Lower middle class, 21–39%	10	7	10	—	12	—	9	10	5
Middle class, 40–60%	12	9	10	—	16	—	10	12	8
Upper middle class, 61–79%	15	16	10	—	12	—	16	21	15
Richest 20%	55	61	61	—	42	—	61	52	68
Richest 5%	29	38	43	—	20	—	38	23	48
Income share of lowest 40%, 1970[7]	20	—	7.8	—	13.1	17	10	13	9.5

Sources:

[1] *Statistical Yearbook 1977*. Department of International Economic and Social Affairs Statistical Office. Twenty-ninth issue. (United Nations, New York, 1978).

[2] *FAO Production Yearbook 1978*, Vol. 32 (FAO, Rome, 1979).

[2*] Data for 1977 computed from total arable land (*FAO Production Yearbook 1978*) and number of population in mid-1977 (*World Development Report, 1979*).

[3] *World Development Report, 1979* The World Bank; Washington, D.C., August 1979.

[4] *Twenty-five Years of Economic Development 1950 to 1975*. David Morawetz (The World Bank, Washington, 1977).

[5] An Anatomy of Income Distribution Patterns in Developing Countries, Irma Adelman and Cynthia Taft Morris. *Development Digest*, **9** (1971).

[6] *World Development Report, 1978*. Published for the World Bank. (Oxford University Press, New York, 1978).

[6*] Computed from net private investment (million US$) and number of population, from *World Development Report, 1978*.

[7] *Basic Needs Performance: An Analysis of some International Data* Glen Sheehan and Mike Hopkins. World Employment Programme Research (ILO, Geneva, January 1978).

This table was compiled by Dr. Herwig Palme, of the Interdisciplinary Institute for Urban and Regional Studies (IIR), University of Economics, Vienna.

Each case study is unique, and although the authors address the same basic questions they have chosen to do so in different ways and with differing degrees of emphasis. This is inevitable and in our view, desirable, as specific problems and their developmental context differ from country to country.

The countries analyzed give, in our view, a representative coverage, not only in terms of ideology and strategy but also in size. Although the nation-state is the dominant political unit, there is an enormous difference between an India or China on the one hand and a Nepal or Papua New Guinea on the other, in terms of both physical area and population size and density.

The countries also give a wide representation in terms of the time-frame under which they have been at least nominally responsible for their own political destiny. Papua New Guinea became independent only in 1977; whereas countries such as Brazil, Chile, and Peru gained political independence already in the first half of the past century.

Each case study has been compiled by an author who has an intimate knowledge of the country concerned and wherever possible, the author is from the country about which he or she is writing. Availability of authors in some cases was a restriction on the choice of case-study countries.

No attempt will be made at this point to summarize the essence of the country experiences, or to draw definite conclusions. The case studies present data on regional and interpersonal disparities together with an explanation of the causes of these, and from them the reader can formulate his or her own opinions. The editors' own views appear in the conclusion. A few general remarks are, however, in order at this point.

Explicit concern with regional development policies is of every recent origin, and in many countries there are no statistics available at the regional level to allow empirical analysis of trends. In others, policies on regional development exist only in rhetoric, never having been implemented. A third problem relating to data is that the methodologies for assessing social indicators relating to the fulfilment of basic needs are poorly developed even if such data were readily available. This affects the comparability of the various case studies.

Comparability is also affected by the differing analytical frameworks adopted by the authors which amongst others include institutional, functional-structural approaches. No attempt was made by the editors to standardize these approaches, although we did insist that each author address the central issue of spatial equity. The differing analyses which have emerged give a range which we hope will stimulate the reader.

Although this book is primarily about space, the time element is critical. Since Williamson's (1965) hypothesis on increasing short-term disparity, occurring as the preliminary phase of an eventual reduction of inequality, debate has raged over the time-frame which would be realistic for consideration of the success or failure of any set of policies. The studies here show trends in disparities but these can of course be analyzed in different ways depending on the time-horizon over which goals are set. In pragmatic terms, however, even if one accepts that initially increasing disparity may occur, if disparities still

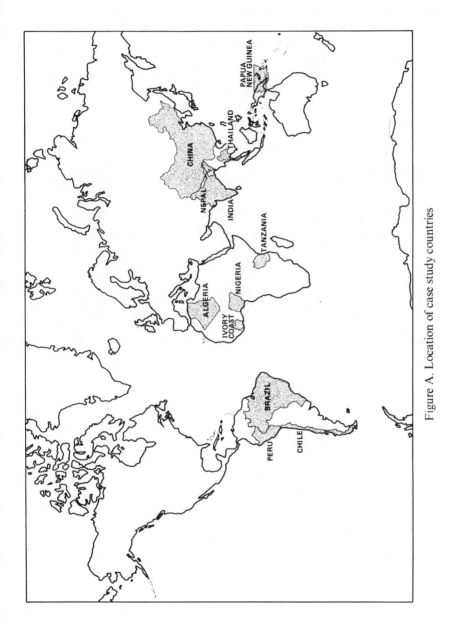

Figure A. Location of case study countries

remain or continue to widen after one or two decades, then in most Third World countries the strategies in use would have to be considered ineffective both from a political and socioeconomic standpoint.

Both editors co-operated on all chapters in this book, although primary responsibility was divided according to the focus of their personal experience. The African case studies were co-ordinated by Taylor and the Latin American case studies by Stöhr. Both the Asian case studies and the theoretical papers were jointly co-ordinated. Final editing for style was carried out by D. R. F. Taylor.

REFERENCES

Friedmann, J. and C. Weaver (1979) *Territory and Function: The Evolution of Regional Planning* (London: Edward Arnold, Berkeley and Los Angeles: University of California Press).

Goulet, D. (1978) The challenge of development economics, *Communications and Development Review*, **2** (1), 18–23.

Goulet, D. (1979) Development as liberation: policy lessons from case studies, *Dossier* International Foundation for Development Alternatives, (Geneva: January)

Seers, D. (1977) The new meaning of development, *International Development Review*, **3**, 2–7.

Stöhr, W. and F. Tödtling (1978) Spatial equity—some antitheses to current regional development doctrine, Papers of the RSA, Vol. 38, 33–53, reprinted in modified form in H. Folmer, and J. Oosterhaven (eds.), (1979) *Spatial Inequalities and Regional Development* (Leiden: Nijhoff, 1979), 133–60.

Williamson, J. G. (1965) Regional inequality and the process of national development: a description of patterns, *Economic Development and Cultural Change*, **13**, 3–45.

PART I

Theoretical Issues and Challenges

Development from Above or Below?
Edited by W. B. Stöhr and D. R. Fraser Taylor
© 1981 John Wiley & Sons Ltd.

Chapter 1

Development from Above: The Centre-Down Development Paradigm

NILES M. HANSEN

At least until recently the centre-down development paradigm has dominated spatial planning theory and practice in the developing country context. Whatever version of this approach one wishes to examine, the essential position is that development (whether spontaneous or induced) in a relatively few dynamic sectors and geographic clusters will (or hopefully should) spread over time to the rest of the spatial system.

BALANCED GROWTH THEORY AND ITS CRITICS

The centre-down paradigm has its roots in the balanced versus unbalanced growth controversy of the 1950s. It was argued that the poverty of the developing countries and of less-developed subnational areas is a result of the low productivity of labour, which is in part a function of an inadequate supply of physical capital. But the shortage of capital is attributable in large measure to the persistently low levels of saving—caused in turn by low income, thus completing the vicious circle of poverty. Because low incomes and a consequent lack of effective demand generally spell failure for any heavily concentrated investment in a single consumer-goods industry, the balanced growth advocates proposed that investment should be diversified over a broad range of such industries. Each industry, it was argued, would then generate, through its factor payments, a demand for the goods of the other industries sufficient to keep all of them viable. Investment projects that might be individually unprofitable would, taken together, be profitable (Rosenstein-Rodan, 1943; Nurkse, 1952).

The concept of balanced growth was challenged on several grounds. Fleming (1955), for example, accused the theory of overemphasis on the demand side of the problem. Inadequate supplies of factors of production in developing countries—particularly capital and skilled labour—may make diversified and simultaneous investment projects unfeasible by sharply increasing the costs of production of all industries as they compete for the limited supply of factors. Enke argued that the balanced growth approach implied a closed economy; but 'one way a country can have balanced consumption without balanced production, if it can make or grow or mine anything the world wants, is to import goods it cannot afford to produce' (Enke, 1963, p. 314). Moreover, 'in practice any implementation of the "big push" proposals would mean a large public sector. . . . Even if the government were to subsidize private firms, instead of operating public concerns, the extent of regulation would be enormous' (Enke, 1963, p. 316). Finally, Singer pointed out that the balanced-growth-via-big-push thesis lacked credibility because 'the resources required for carrying out the policy . . . are of such an order of magnitude that a country disposing of them would in fact not be underdeveloped' (Singer, 1949, p. 10).

Whatever their merit, the negative criticisms of the balanced growth thesis did not add up to a rival development theory. But such a theory clearly was 'in the air' because in the late 1950s unbalanced growth was given central importance in the highly influential and independently published works of Hirschman, Myrdal, and Perroux. The centre-down paradigm that has dominated the voluminous growth centre literature of the past two decades stems directly from these seminal contributions. Thus, they merit close attention because the inadequacies of a commonly accepted paradigm are usually not uncovered by assailing the logic of its reasoning; if it is to be dethroned it is necessary to challenge successfully its basic explicit and implicit assumptions.

UNBALANCED GROWTH AND REGIONAL DEVELOPMENT

Hirschman: Polarization and Trickle Down

The Cinderella-like transformation of unbalanced growth was largely the responsibility of Hirschman, for as will be seen his position was characterized by a certain optimism not found in the parallel writings of Myrdal and Perroux. But then American and European perspectives have often differed in this regard.

Hirschman maintained that development strategies should concentrate on a relatively few sectors rather than on widely dispersed projects; the key sectors would be determined by measuring backward-linkage and forward-linkage effects in terms of input–output maxima. He maintained that growth is communicated from the leading sectors of the economy to the followers, from one firm to another. The advantage of this approach 'over "balanced growth" where every activity expands perfectly in step with every other, is that it leaves considerable scope to *induced* investment decisions and therefore economizes our principal scarce resource, namely, genuine decision-making' (Hirschman, 1958, pp. 62–63).

Geographically unbalanced growth requires special consideration, for 'while the regional setting reveals unbalanced growth at its most obvious, it perhaps does not show it at its best' because successive growth points may all 'fall within the same privileged growth space' (Hirschman, 1958, p. 184). The principal reason for the tendency for economic activity to become overconcentrated in one or a few growth poles is that the external economies associated with the poles are consistently overrated by investment decision-makers on the ground that 'nothing succeeds like success'. Thus, whereas a clustering of investment around the initial growth poles is beneficial at the beginning of development, it may be irrational at a later period. The actual effects of the growth points on the hinterlands depend on the balance between favourable effects that trickle down to the hinterlands from the progress of the growth points and the unfavourable, or polarization, effects on the hinterlands as a consequence of the attractiveness of the growth poles.

The most important trickling down effects are generated by purchases and investments placed in the hinterlands by the growth points, though the latter may also raise the productivity of labour and per capita consumption in the hinterlands by absorbing some of their disguised unemployment. On the other hand, polarization may take place in a number of ways. Competition from the growth points may depress relatively inefficient manufacturing and export activities in the hinterlands, and the growth points may produce a 'brain drain' from the hinterlands, rather than create opportunities for their disguised unemployed.

In the long run, Hirschman argued, public investment would cease to be pulled so heavily into the developed areas, largely because of considerations of equity and national unity. Moreover, after development has proceeded for some time in the growth points 'the need for public investment in relation to private investment tends to decline and in any event an increased portion of public investment can be financed out of earnings of previous investments. This kind of change in the composition of investment is implicit in the term "social *overhead* capital"' (Hirschman, 1958, p. 194). Thus, central government funds are released for use in other regions and, in the long run, regional differences will tend to disappear. Finally, Hirschman suggests that while some investment in utilities in the hinterlands may be indispensable, the provision of infrastructure is only a permissive inducement mechanism; the essential task is to provide the hinterlands with a continuously inducing economic activity in industry, agriculture, or services.

Myrdal: Cumulative Causation

Myrdal's (1957) theory of circular causation, published at about the same time as Hirschman's analysis but developed independently, contains a number of conceptual tools that coincide with those of Hirschman. Myrdal maintained that a simple model of circular causation with cumulative effects is more consistent with actual social and economic processes than the static equilibrium analysis typical of economic theory. Myrdal found that whatever the reason for the initial expansion of a growth centre, thereafter cumulatively expanding internal and external economies would fortify its growth at the expense of other areas. These economies include not only a skilled labour force and public overhead capital, but also a positive feeling for growth and a spirit of new enterprise.

In developing his analysis, Myrdal employed the concepts of 'backwash' and 'spread' effects, which correspond closely to Hirschman's polarization and trickling down effects. The backwash effects involve the workings of population migration, trade, and capital movements. Like Hirschman, Myrdal noted the selective nature of migration from the hinterlands to the growth centre, though he emphasized the fact that the young are the most prone to move. He also dwelt on the higher fertility rates of poor areas and their impact on the working-age group to total population ratio, which is likely to be relatively unfavourable in the hinterlands. Similarly, capital tends to flow to the growth centres because of increased demand. Consequently, incomes and demand increase again, resulting in yet another round of induced investment. The tendency to increased inequality is reinforced by the flow of savings from the hinterlands, where demand for investment capital remains relatively weak, towards the centres of expansion, where returns are high and secure. In addition, Myrdal recognized the critical significance of non-economic factors to the cumulative process of maintaining poverty in the hinterlands. Their inability to support adequate health and education facilities, their generally conservative outlook—related to acceptance of the more primitive forms of tradition and religion—are all detrimental to the experimental and rational orientation of an economically progressive society.

Among the spread effects, which may counter the backwash effects, are increased outlets for the hinterland's agricultural products and raw materials and a tendency for technical advance to diffuse from the growth centres. The spread effects will be stronger the higher the level of economic development of a country. Moreover, the attractiveness of the growth centres may be weakened by increasing external technological diseconomies and high labour costs. Finally, the governments of the wealthier countries are likely to initiate policies directed towards greater regional equality.

Despite the similarities in approaches of Hirschman and Myrdal, there are considerable differences of emphasis. In particular, Hirschman was more optimistic about the long-run future of the less-developed countries; also Hirschman related his theory explicitly to less-developed areas within countries, as well as to less-developed countries; he seemed to take for granted that strong forces eventually will create a turning point, once polarization effects have proceeded for some time.

Perroux: Growth Poles

Although Perroux (1955) coined the term 'growth pole' (*pôle de croissance*) in an article published before the related works of Hirschman and Myrdal, his primary concern tended to be interactions among industrial sectors rather than spatial development processes. In contrast to the then-prevailing balanced and steady growth theories, Perroux maintained that analysis of sustained growth of total production should concentrate on the process by which various activities appear, grow in importance, and in some cases disappear; and he emphasized that growth rates vary considerably from sector to sector. Like Schumpeter, he stressed the importance of entrepreneurial innovation in the development process, which proceeds by a succession of dynamic sectors, or poles, through time. Another key element in Perroux's development theory was the concept of dominance, which consists of an irreversible or only partially reversible influence exercised by one economic unit on another because of its dimension, its negotiating strength, the nature of its activity, or because it belongs to a zone of dominant activity. As soon as any inequality appears among firms the breach is opened by which the cumulative effect of domination insinuates itself. Given these propositions, it followed that the dominant, or propulsive, firm would generally be large and oligopolistic and would exert a considerable influence on the activities of suppliers and clients. In addition, dominant and propulsive industries make the cities where they are located the development poles of their regions.

The spatial implications of Perroux's theory were elaborated by his colleague Jacques Boudeville. The highly eclectic nature of Boudeville's output precludes brief summarization, but it is pertinent to note two themes that he emphasized until his untimely death. One was the importance of big industrial complexes. Citing the Bari-Taranto complex in the Italian Mezzogiorno as a case in point, he held that a development pole is 'a complex of activities agglomerated around a propulsive activity' (Boudeville, 1972, p. 263). Another is 'that in the second half of the twentieth century, characterized by an acceleration of urbanization and accompanied by the contraction and expansion of numerous centers, the study of regional economies is crystallizing around economic polarization processes'; within the framework of technical and geographical external economies of agglomeration and large infrastructure projects, spatial planning (*aménagement du territoire*) 'has become a problem of urban growth strategy' (Boudeville, 1972, p. 263). Clearly in evidence here are basic elements of the centre-down development strategy, namely stress upon a few dynamic sectoral clusters and upon urban-industrial growth as the key to more generalized regional development.

Friedmann: Core–Periphery Interaction

The contributions discussed to this point have come entirely from the economics literature. However, the first attempt to formulate a systematic and comprehensive centre–hinterland development model was made by Friedmann, a planner. His 'core-periphery' analysis was initially set forth in a study of Venezuelan regional development policy and refined a few years later.

Another planner, Rodwin (1963), had earlier proposed a strategy of 'concentrated decentralization' for developing economically lagging peripheral regions. Although not elaborated in detail, this suggestion was consistent with the notion that induced urban growth centres should be the basis for regional development policy. However, Friedmann argued that Rodwin's approach failed

> to discriminate sufficiently among development regions and their problems. Activation of new core regions is not enough nor, indeed, always appropriate. An effective regional policy must deal as a system with the separate developments of core regions, upward- and downward-transitional areas, resource frontiers, and special problem areas. (Friedmann, 1966, p. 53)

In addressing these issues Friedmann (1972) maintained that development— that is, the unfolding of the creative potential in a society through a successive series of structural transformations—occurs through a discontinuous, but cumulative, process of innovation. Development originates in a relatively small number of 'centres of change' located at the points of highest potential inter- action within a communication field. Innovations diffuse from these centres to areas of lower potential interaction. 'Core regions' are major centres of in- novative change, while all other territory consists of 'peripheral regions', which are dependent on the core regions and whose development is largely determined by institutions of the core regions. The process by which core regions consolidate their dominance over peripheral regions tends to be self- reinforcing as a consequence of six principal feedback effects of core region growth: (1) the dominance effect, or the weakening of the periphery by resource transfers to the core: (2) the information effect, or increased interaction and innovation in the core; (3) the psychological effect, or a higher rate of innovation due to greater visibility, higher expectations, and lower risks; (4) the modern- ization effect, or social and institutional change favouring innovation; (5) linkage effects, or the tendency of innovations to induce yet other innovations; and (6) production effects, which increase scale and agglomeration economies.

Core regions are located within a nested hierarchy of spatial systems, ranging from the province to the world. A spatial system exists when a core dominates some of the vital decisions of populations in other areas, and a spatial system may have more than one core region. Peripheral regions are dependent on core regions by virtue of supply and market relations, as well as administrative organization. For any spatial system, a loose hierarchy of core regions exists in relation to 'the functional performance of each core for specified character- istics of system performance'. Innovation is diffused to peripheral regions from the core, and core-region growth will tend to promote the development process of the relevant spatial system. Eventually, however, increasing social and political tensions between the core and the periphery will tend to inhibit the development of the core, unless these tensions can be alleviated by acceleration of spread effects or decrease of the periphery's dependence on the core. The conflict between the core and the periphery may result in the repression or neu- tralization of peripheral elites, the replacement of core-region elites by peri- pheral elites, or a more equal sharing of powers between core and periphery by means of political and economic decentralization and the development of new core regions in the periphery. In the last case, authority-dependency relations between cores and their peripheries may disappear with relatively minor exceptions in remote rural areas or in urban slums.

In general, then, Friedmann's theory assigns a decisive influence to the institutional and organizational framework of society and, specifically, to the patterns of authority and dependency that result from the unusual capacity of certain areas to serve as cradles of innovation. The theory is attractive in many respects. In particular, it includes all space and it treats variables in specific areas as parts of a larger system rather than as isolated phenomena. It also integrates cultural and political processes into the process of economic development. Finally, it is sufficiently general to cover a great number of cases. Its principal weakness is that its theoretical propositions are not postulated in the form of readily testable hypotheses or mathematical statements; but then this criticism applies to much of the growth centre literature.

HIERARCHICAL DIFFUSION: A SYSTEMS APPROACH TO SPATIAL–TEMPORAL DEVELOPMENT PROCESSES

The National Setting

In the late 1960s several major geographical and economic analytical strands of thought were synthesized in a general model of hierarchical diffusion of innovation. As described by Berry (1970), this urban-oriented framework of economic activities in space has two major elements: (1) a system of cities arranged in a functional hierarchy, and (2) corresponding areas of urban influence surrounding each of the cities in the system. Given this framework, the spatial extent of developmental spread effects radiating from a given urban centre are proportional to the centre's size and functions. 'Impulses of economic change' are transmitted from higher to lower centres in the hierarchy so that continuing innovation in large cities is critical for the development of the whole system. Areas of economic backwardness are found in the most inaccessible areas, that is, between the least accessible lower-level towns in the urban hierarchy. Finally, the growth potential of an area located between any two cities is a function of the intensity of interactions between the cities.

One would conclude from this that, if metropolitan development is sustained at high levels, differences between centre and periphery should be eliminated and the space-economy should be integrated by outward flows of growth impulses through the urban hierarchy, and inward migration of labour to cities. Troughs of economic backwardness at the intermetropolitan periphery should, thereby, be eroded, and each area should find itself within the influence fields of a variety of urban centres of a variety of sizes. Continued urban-industrial expansion in major central cities should lead to catalytic impacts on surrounding regions. Growth impulses and economic advancement should 'trickle down' to smaller places and ultimately infuse dynamism into even the most tradition-bound peripheries. (Berry, 1970, pp. 45–46.)

Although static central place and urban hierarchy schemes are not in themselves adequate for analysing growth and change, the synthesis just described took these frameworks as a kind of locational landscape within which dynamic processes take place. Moreover, growth centre concepts could readily be incorporated into the general synthesis. Induced growth centres merely had to be viewed as means for linking lagging regions more closely with the national system of hierarchical filtering and spread effects from urban centres to their hinterlands.

The International Setting

José Lasuén (1973) extended the hierarchical diffusion model by examining the complex interaction between economic development and spatial business organization in an international setting. He maintained that development results from the adoption of successive packages of innovations in clusters of establishments linked to a regional export activity. These clustered sectoral sets are also clustered geographically. The diffusion and adoption of successive sets of innovations follow similar patterns, resulting in a fairly stable system of growth poles. Over time, innovations require greater scales of operation and larger markets; they also come at shorter intervals. Large cities are the earliest adopters of innovations, which then diffuse to the rest of the urban system. As a consequence of this process, the system of growth poles becomes increasingly hierarchical in nature. Lasuén placed considerable emphasis on the international generation of innovations, and he argued that if innovation diffusion is delayed because of inadequate organizational arrangements, then changes need to be made in order to minimize the costs and risks inherent in the learning process. This implies that development policies should put less emphasis on production promotion and more emphasis on marketing and technical knowhow. Moreover, permanent and long-run integrated activities (multi-city conglomerates or large single-city firms) were regarded by Lasuén as better development organizational solutions than the short-run integration of activities through contracting arrangements.

According to Lasuén, the spread of innovations in developed countries has been accelerated by the existence of multi-plant firms. In developing countries, on the other hand, the firm and the plant usually are identical, so the spatial spread of innovations cannot be carried out within firms from central plants to peripheral plants. The uncertainties and risks of adopting innovations in developing countries are less than in developed countries because the relevant products have become internationally standardized; yet the spatial spread of innovations is slower in the developing countries because they lack external complementarities (diversified skilled labour, imaginative entrepreneurship, research-oriented technical manpower, auxiliary financial and commercial services, etc.), and because the spread process must be carried out through a host of unrelated individual decisions.

Finally, high costs of production resulting from small scale are not a deterrent because innovation adoptions are forced on developing countries by balance-of-payments difficulties and consequent import-substitution policies. Significantly, innovation adopters usually are persons who previously were importers, distributors, servicers, or producers of substitution goods, familiar with the marketing, financing, and technical characteristics of the product. They adopt so that they will not be driven out of business by import substitution. As a result, the spatial structure of new productive activities is nearly identical with that of the former marketing network of the product. Innovation adoptions occur where the largest previous market areas of the product were found; normally the largest city. New producers are pressed to satisfy their own local demand and cannot supply smaller towns. Eventually, however, producers spring up in other local market areas in spite of their smaller captive markets. Thus, most innovation adoptions start in the largest cities and gradually trickle down through the urban hierarchy. This is true, argues Lasuén, whether the product in question is market-oriented or resource-oriented because at this initial stage production is tied to the location of the market.

Given Lasuén's theoretical context, what are the policy implications—both internationally and nationally—for developing countries? Lasuén emphasizes the difficulties in overcoming existing disparities between developed and developing countries, and suggests that the only possible way the latter can absorb successive innovation clusters more rapidly is to reduce technical and organizational lags by overriding national boundaries and planning multinational firms on a continental basis. Intranationally, only two basic alternatives are open to the developing countries. The first is for the largest urban centre or centres to adopt new innovations before previously adopted innovations have filtered down through the urban hierarchy. The second is to delay the adoption of innovations at the top of the hierarchy until previously adopted innovations have been adopted in turn in the rest of the country. The first choice would result in a perpetuation of economic dualism. The second would lead towards greater equality among regions but the nation as a whole would be less developed because older and less-efficient technologies would be utilized. In practice nations do not pursue either of these extreme paths but most tend (at least implicitly) to prefer a dualistic economy to one which is spatially egalitarian but generally retarded (Lasuén, 1973, p. 182).

Despite differences in emphasis among representative centre-down development theories it is evident that the relevant literature has become more complex and sophisticated over time. Certainly this is true of the hierarchical diffusion approach in relation to the earlier balanced versus unbalanced growth controversy; and it is true of Lasuén's model of the interactions through time between geographical and sectoral clusters in relation to earlier innovation diffusion models. However, increasing complexity and sophistication do not necessarily lead to improved policy guidance for spatial planning in developing countries. Indeed, many of the policy implications of the centre-down paradigm (or at least some of its more prominent versions) have been subjected to mounting criticism. The remainder of this chapter thus attempts to summarize the status of these controversial issues.

REGIONAL DISPARITIES AND NATIONAL DEVELOPMENT

So long as the developing countries continue to adopt innovations that originate in the developed countries it is difficult to imagine that disparities between them will be closed to any marked degree in the near future. However, this does not necessarily mean that per capita gross national product will not grow in developing countries, or that interregional disparities in these countries will not diminish.

It will be recalled that Hirschman believed that as developing countries matured an eventual turning point would be reached; regional convergence would replace the polarization tendencies characteristic of early development experience. His position in this regard was based rather heavily on the assumption that political motives would bring about greater spatial equity, but he did not provide much concrete supporting evidence. Lasuén's analysis of hierarchical diffusion processes led him to conclude that large cities would grow in relation to middle-sized cities and that these in turn would grow larger than smaller ones; all of them would maintain their relative places in the urban system, especially as higher levels of development are achieved. While he did not envisage a turnaround in Hirschman's sense, he held that the entire urban system would increasingly participate in the adoption of new innovations. Moreover, he maintained that if business organization policies of the kind he recommended were applied to geographical clusters at all levels of the urban hierarchy, then these clusters would become more tightly integrated in the national innovation adoption process. If Lasuén cannot be accused of excessive optimism, at least he is not pessimistic about the prospects for centre-down development.

However, in terms of generating optimism with respect to the long-run prospects for the poorer regions, probably no study has equalled the comparative investigation of Williamson (1965). Although many of the countries he examined were not in the newly developing category, he concluded that both cross-section and time-series analyses indicated that there is a systematic relation between national development levels and regional inequality. Rising regional income disparities and increasing dualism are typical of early development stages, whereas regional convergence and a disappearance of severe dualism are typical of the more mature stages of national growth and development.

Friedmann, on the other hand, countered that Williamson's thesis was only partly borne out by detailed historical studies.

Although not strictly related to our study of the developing countries, data for Canada suggest only a small amount of convergence between 1929 and 1956, while earlier periods led to a gradual widening of the income gap. And a recent study of regional income differences in Brazil, though showing a slight convergence in the decade of the fifties, is inconclusive on the long-term direction of the trend. The best that can be said from the available information is that regional income convergence is an extremely drawn-out process. It occurs, when it does, at a relatively advanced stage in the industrialization process, and it rarely changes the rank-ordering of regions by income. Core-region dominance, we may conclude, is not easily challenged. And when it is, the reversal of earlier trends may require a number of generations. (Friedmann, 1973, pp. 76–77.)

More recent research by Mera (1978) has lent some support to the Williamson thesis; indeed, his findings are, if anything, even more optimistic than those of Williamson. Mera observed that a convergence of Japanese prefectural per capita incomes started to appear around 1961 and continued in parallel with a generally high rate of national economic growth. This phenomenon appears to have been caused by a convergence of wage rates among different parts of the economy, which in turn resulted from a tightening of labour markets. It is worth noting that the Japanese government had not yet adopted a policy of decentralization during this period. Moreover, there is no evidence of decreasing economies of scale or increasing diseconomies of agglomeration in large metropolitan areas. Rather, what occurred was an upward shift of relative incomes in other areas, usually those in the immediate vicinity of metropolitan areas. In addition, Mera found that the process that produced regional per capita income convergence also helped to prevent further concentration of population in a few large metropolitan areas. Therefore, he suggested, population concentration might only be a temporary problem for developing countries. In the case of Japan the process described by Mera took about 30 years but he maintained that it should require less time in developing countries, if they can maintain a high rate of national economic growth.

In a later examination of the South Korean case, Mera found that a trend of widening regional income disparities was reversed when per capita GNP was only $240. This reversal was achieved initially by improving the terms of trade for farmers through pricing policies. When this approach was about to reach its limit (around 1967) the country's high rate of economic growth—which already had been experienced for more than 10 years—started to be effective in reducing income disparities among regions and eventually in slowing the rate of population concentration in primate urban regions. From observing the Japanese and Korean cases, Mera concluded that 'it appears quite probable that the maintenance of high growth of the economy is an effective way of reducing the rate of population concentration as well as reducing income disparities among regions once the economy reaches a certain stage' (Mera, 1977, p. 174). Mera thus clearly sides with Williamson, but emphasizes that there is 'a wide margin of flexibility' (Mera, 1977, p. 173) with respect to the divergence–convergence turnaround point.

No doubt the Williamson thesis will continue to be confirmed by some country case studies and denied on the basis of others. In light of the available evidence it would be rash to propose that developing countries merely wait patiently because there is an automatic mechanism that eventually will eliminate or significantly reduce regional disparities. With respect to non-Western economic development experience, Japan clearly is an exceptional case (setting aside Hong Kong and Singapore, where spatial development strategies obviously are no issue) and Korea is far from typical. Nevertheless, these cases illustrate what may be achievable *if* a high rate of national economic growth is sustained over a long period. In the West it has been demonstrated that macroeconomic growth at or near full employment is often needed for the effective implementation of policies aimed at the alleviation of structural problems, e.g. manpower deficiencies and regional economic disparities. Unfortunately, Williamson (and to a lesser extent Mera) ignores structural issues and implicitly adopts the view that macroeconomic policy is all that matters; regional disparities—and presumably other structural problems—will then take care of themselves. Here, as in most respects, the situation in developing countries is more complex than that in the West.

It may well be that unless and until truly basic structural problems in such domains as health, education, and even basic values are resolved, national and regional development are both threatened in developing countries. In other words, issues involving human resource development and basic values may need to receive primary attention before regional problems can be successfully attacked either directly or indirectly, through sustained national economic growth. Put yet another way, it may be necessary to deal with fundamental structural problems before national growth can proceed to a point where it positively affects remaining structural problems.

Further complication is added by the need to consider once again the notion of hierarchical diffusion. There is a considerable body of opinion that maintains that the developing countries are not really the masters of their fate because they are dependent on the developed countries, and more particularly on the large multi-national corporations based in the latter. In addition, it is often argued that both their macroeconomic and structural difficulties are in large degree related to their inability to create an urban hierarchy capable of diffusing development-inducing innovations throughout the national territory. These issues will now be examined in more detail.

SOME STRUCTURAL CONSEQUENCES: THE DEPENDENCY ISSUE

Multi-National Corporations: Pro and Con

The centre-down paradigm is not simply a matter of the national economic development of individual countries; it involves significant and highly controversial international dimensions. It will be recalled that Lasuén urged the creation of multi-national firms on a continental basis as a precondition for more rapid innovation diffusion in the developing countries. But he noted that even though the countries of the European Economic Community have been ready to pool their markets, they have had considerable difficulty in merging their firms into multi-national enterprises. 'As everyone is aware, the only really multi-national firms in the Common Market are the subsidiaries of the large multi-product multi-plant American conglomerates' (Lasuén, 1973, p. 187). If this is the case, how then, under present institutional conditions, are the developing countries supposed to do better in this regard than the nations of Western Europe?

Moreover, critics of the multi-national corporations maintain that their very efforts to extend their operations to developing countries often have negative effects from a developmental perspective. In this view these efforts are part of a process involving the accelerated growth of a subsystem oriented toward the demands of a small privileged section of the population, without changing the status of the workers and peasants. In Latin America, for example, 5 per cent of the population has a per capita income of more than $2200 but this segment generates more than half of the demand for commodities. This demand is characterized by a high level of diversification so that there is only a very small market for each line of production (Coraggio, 1975).

> The conditions required for world poles to develop their branches in backward countries restrict the possibilities of drastically changing the internal situation through coopera-tion between such countries. Such conditions therefore restrict the possibilities of allow-ing the productive apparatus to serve the population. In Latin America there is clear evidence that the slightest reformist attempt to change the 'internal' structures immediate-ly affects 'external' interests and provokes a negative reaction. This reaction can take the form of 'internal' efforts to modify the internal structures again or else it can take the form of external pressure, the effect of which is not limited to the enterprises that are directly concerned. (Coraggio, 1975, p. 369.)

Although these conclusions are stated in rather abstract terms, ample concrete examples of foreign private and governmental (not always in-dependently) meddling in the internal affairs of developing countries have been reported in the press. Moreover, these initiatives, when they have originated in the West, usually have favoured conservative social structures.

Generally speaking, the assumptions behind the notion that a strategy of economic polarization and dependent integration is beneficial for developing countries have been criticized on four grounds. First, this approach erroneously assumes that there is no structural unity among social, economic, and political phenomena. This in turn implies that a strategy aiming at social objectives can be reduced to purely economic terms; social and political considerations can be tacked on later. Second, it assumes that international relations take place in a harmonious context. Third, the state is an idealized, autonomous element in the social system; it is regarded as apart from any real power structure. Similarly, it is assumed that a neutral, rational bureaucracy exists. Thus application of the strategy would not bring about changes in the predominant political structures. Finally, the strategy assumes that polarization mechanisms can be reproduced independently at any level (Coraggio, 1975).

What these criticisms add up to is a contention that a theory of regional development must be embedded in a theory of social change; it is impossible to compose a development strategy through 'economic engineering' alone even if the main problems of development have an economic basis.

The notion that large multi-national corporations are involved in a process that exploits both the urban and rural sectors of developing-country economies is held not only by Marxist critics of the 'dependency school', but also by some Western observers with no particular ideological axe to grind. Thus, Gilbert finds that:

> Large landowners who live in the major cities transfer capital from the rural to the urban sector; trade between the rural and the urban sector tends to favour the latter; industrial concentration, fiscal transfers and the organisation of the banking system, channel the economic surplus of smaller cities into the metropolitan centres. From there, processes such as the repatriation of profits by foreign companies and the gradual deterioration of the terms of trade lead to the transfer of funds to the world metropoli. It is unnecessary to accept the full Marxist version of this argument to support the view that major transfers do take place from the rural to the urban sector. (Gilbert, 1975, p. 30.)

Of course, the multi-national corporations have quite a different view. The then President of the Ford Motor Company (now with Chrysler Corporation) argued that:

> the multinational offers less-developed nations the best available linkage with the markets, technology and knowhow of the industrialized world. In addition, through direct foreign investment in plants and equipment, it provides a ready source of scarce capital without adding to the considerable debt burdens of these countries....

> You might assume all nations would welcome investments like this. But the malevolent multinational of the exploitation myth dies hard, and many governments actually discourage such massive infusions of capital and know-how, unwittingly or otherwise. Bureaucrats dictate how much a company can export, what local materials and parts it must use, what prices it can charge and how much money it can take out of the country to finance operations elsewhere in the world. Worse yet, the uncertain investment climate is being further clouded by the proliferation of codes and guidelines now under study by various bodies in the United Nations, primarily at the urging of the developing nations. (Iacocca, 1977, p. 21.)

The role of the multi-national corporation in developing countries is no doubt more complex than one-sided defenders or detractors believe. However, the real issue for present purposes is less the multi-national corporation *per se* than the process by which the division of labour enhances productivity, and the relationship of this process to the centre-down development paradigm.

Division of Labour, Technology, and Human Resources

Economic history indicates that the process of economic development in the West since the Dark Ages has not been smooth. For example, studies of price movements and the purchasing power of a worker's wage over a thousand-year period in England (Brown and Hopkins, 1955, 1956) suggest that there have been a number of major developmental metamorphoses after long periods of relative stagnation. These periods may be termed the Commercial Revolution (1150–1325), which involved a burst of agricultural productivity, trade, and urbanization; the Capitalist Revolution (1520–1640), characterized by the rise of the nation-state and mercantilism; the Industrial Revolution (1750–1825), which saw the widespread introduction of machines and the new doctrine of *laissez-faire*; and the explosion since World War II of the post-industrial age of information and the bureaucratic welfare state. Each of these periods was marked by a sharp increase in complexity and a fresh division of labour; each also was marked by sharp increases in the level of prices because growing specialization was increasing the number of middle-men (perhaps it would be more appropriate to speak of price 'conflation' than of price inflation).

In pre-industrial, low-productivity societies factors of production, firms, and localities are relatively unspecialized as well as undifferentiated in space. When industrial urbanization takes place factors, firms, and localities become increasingly specialized and, within their respective market areas, more differentiated from each other. The progressive division of labour yields higher returns to individuals and firms, as well as to the economy as a whole. Specialized activities on a larger scale at one stage of production create opportunities for innovation and specialization at other stages, through backward and forward linkages. Firms in an expanding industry benefit from cost reductions as a result of external economies in the Marshallian sense; that is, the relevant economies (service facilities, special skills and education, etc.) are external to the firm but internal to the industry. As economic development proceeds, external economies become more general in nature. In the broader sense, then, the concept applies to all services, facilities, or activities that exist outside the firm but reduce the firm's costs. External economies so conceived are external to the firm but internal to the locality, region, or nation. Large-scale corporate enterprise may be viewed as a means to concentrate specialized but technologically interrelated processes (vertical integration) as well as to create larger and more unified systems of finance and control.

The specializations and interdependencies of towns, cities, and metropolitan areas reflect the same specialization–differentiation–reintegration tendency that yields increasing returns to scale and (internalized) external economies to firms. Thus, the vertical re-integration of specialized activities is realized at three levels of functional organization: in work processes, in business organization, and in the urban system (Lampard, 1968).

So far these remarks are largely consistent with Lasuén's view of national and international development processes. However, by concentrating on innovation diffusion and by failing to give explicit attention to the division of labour he neglects the critical importance of human resources. I have argued elsewhere that even in developed countries the greatest impediment to overcoming the structural problems of economically lagging regions is lack of attention to human resource problems (Hansen, 1968, 1970). An even stronger case can be made in this regard in the context of the developing countries.

With technological and organizational progress the quality of the labour force must also be improved so that new grades of skill can be combined profitably in novel ratios with the more complex 'roundabout' technologies. The outcome of this increasing specialization is a sophisticated division of labour between skillful workers and intricate machines.

> Division of labor was not, in short, exhausted in the form of breakdown and simplification of linear tasks, exemplified by Adam Smith's pin-making operation, or John R. Commons's slaughterhouse production line. It is precisely the long-run tendency to employ more expertise in novel combinations with more complex and sophisticated technological systems that has negated Smith's fear that simplification and monotony of job tasks in division of labour would, for all its economic promise, render a growing proportion of the town populations 'as stupid and ignorant as it is possible for a human creature to become'. Indeed, the requirement of greater knowledge-input has become a condition of rising productivity per unit of capital and labor combined in both town *and* country. It is this same requirement that transforms long-run economic growth into development. (Lampard, 1968, p. 105.)

Harbison's extended concern with manpower as the critical element in development led him to conclude that:

> human resources—not capital, nor income, nor material resources—constitute the ultimate basis for the wealth of nations. Capital and natural resources are passive factors of production; human beings are the active agents who accumulate capital, exploit natural resources, build social, economic, and political organizations, and carry forward national development. Clearly, a country which is unable to develop the skills and knowledge of its people and to utilize them effectively in the national economy will be unable to develop anything else. (Harbison, 1973, p. 3.)

Harbison may overstate his case by playing down the role of capital, but he is on firmer ground than those economists who would 'explain' economic development by means of equations containing abstract 'capital' and 'labour' variables. Both approaches, however, run the danger of oversimplification in terms of the international diffusion of innovation. Here it is necessary to return once again to the role of the multi-national corporation.

The Multi-Nationals Reconsidered

From the viewpoint of the multi-nationals, productivity consists of five basic elements: (1) capital; (2) people with the technical ability to invent and constantly improve new designs and processes; (3) a well-motivated and trained labour force; (4) an organization designed to manage people, capital, machines, and know-how; and (5) governmental co-operation (Iacocca, 1977, p. 21).

If I were committed to the dependency school, I would regard the last element as a highly convenient staging point for an attack on the multi-national corporation, though my position would be conditioned by the arguments presented earlier in this section. When the multi-national maintains that it needs 'sensible, realistic regulatory and tax policies' (Iacocca, 1977, p. 21) this is all well and good on the surface; but when it resorts to means ranging from bribes to aiding in the overthrow of unsympathetic governments in order to secure governmental co-operation, it assumes for itself a political role that is not made explicit in its demand for economic rationality. Perhaps most multi-nationals do not resort to such tactics in any significant degree; but enough have with sufficient frequency to raise doubts concerning the intentions of multi-nationals in general.

Moreover, there *is* an interdependency among the state's economic, political and social actions; and these actions *are* related to the prevailing power structure which, in many cases, may be primarily concerned with the welfare of an entrenched elite class. If a conservative power structure favours the demands of multi-national corporations for 'co-operation' with respect to regulatory and tax policies but neglects to develop the human resources of its own people, it is undercutting elements (2) and (3) of the multi-nationals' own productivity requirements. Indirectly, element (4) may also be undermined to the extent that organizational capability is a function of human talent, skill, and education.

More generally, it is not at all apparent that the multi-nationals appreciate the complexity of the interdependencies among human resource development, the division of labour, and the expansion of modern technology. Or at least they do not have this appreciation in the context of the developing countries. They are willing to 'cream' the human and natural resources of favoured segments of their economies, but this process does not lead to the larger-scale transformations required for general innovation diffusion. Moreover, the 'co-operation' of host countries only compounds the difficulties involved in achieving genuine national development.

On the other hand, the multi-nationals could argue that if they provide capital, organizational ability, and a great deal of technological expertise to the developing countries, they should not also be responsible for local human resource development. This would be a sound position. The multi-national corporation is a significant repository of power. There are those who maintain that a multitude of social and political pressures force the modern corporation to behave in a socially responsible manner (Berle, 1963); consequently, an economic system largely dominated by large corporations functions more efficiently and justly than economic theory would lead one to expect. Yet it would seem inevitable that a corporate executive who endeavours actively to further the public good is bound to impose, in greater or lesser degree, his own subjective interpretation of the public interest on relevant issues. (One of the great virtues of classical economic doctrine was the implicit notion that businessmen should not meddle in public affairs; so long as private and social costs did not deviate to any great extent, the private behaviour of businessmen would coincide with gains to the public at large and the function of businessmen needed no further justification.) One can readily imagine the uproar that would be occasioned by attempts on the part of multi-national corporations to intervene in any aspect of human resource development in a host country. The educational system, regarded in the broadest sense, is one of the surest means for attaining national cohesiveness and thus it is an essential part of the nation-state's efforts at self-perpetuation and self-determination.

In sum, then, the process of international innovation diffusion is basically consistent with the centre-down paradigm. However, this process is more complex than the arguments of either the multi-nationals or the dependency school would suggest. It is now necessary to shift attention to problems of dualistic development within countries. Although international dependency may contribute to spatial economic disparities within developing countries, in the last analysis this is an internal matter that must be confronted and dealt with internally.

THE ROLE OF GROWTH POLES

The developing countries' primate cities have enormous problems with high unemployment rates and crowded slums; but in most of these countries glaring economic and social disparities between the large cities and most of the remaining, largely rural, areas represent an even greater problem. Proponents of the center-down paradigm have addressed the latter issue in terms of one or another variant of the growth pole strategy. If the developed countries appear to have an orderly hierarchy of central places through which development-inducing innovations filter and spread, then this spatial structure must also be created in the developing countries.

The rationale for the growth pole strategy maintains that with limited resources it would be inefficient and ineffective to attempt to sprinkle developmental investments thinly over most of the national territory. Rather, key urban centres should be selected (preferably those that would help to fill out a 'rational' urban hierarchy) for concentrated investment programmes that would benefit from economies of scale and external economies of agglomeration. However, such policies usually are not justified on the basis of helping the growth poles per se, but on the ground that as a consequence of induced growth beneficial spread effects will flow to the growth poles' lagging hinterlands. It will be recalled that Hirschman's position tended to optimism in this regard (at least concerning the medium- to long-run prospect) whereas Myrdal was more cautious if not an outright pessimist.

After two decades of many experiments with growth pole strategies the evidence to date largely supports Myrdal's views; and this is the case in the context of both developed and developing countries. Gaile (1973), for example, reviewed 17 different studies of attempts to implement growth pole strategies and concluded that if a trend was discernible it was that spread effects were smaller than expected, limited in geographical extent, or less than backwash effects. Empirical studies by Nichols (1969) and Moseley (1973) indicate that if the objective of regional policy is to benefit small towns and rural areas, then it would be advisable to invest directly in these places; some 'trickle up' to larger cities will take place under such circumstances but a converse 'trickle down' situation cannot be relied upon. Moreover, Pred's (1976) empirical studies show that spread effects to the immediate hinterlands of centres of innovation are minimal in comparison with the linkages that connect these centres with numerous distant places; in other words, the innovation diffusion process is highly discontinuous in spatial terms.

It would seem that pessimism regarding growth pole strategies has become as pervasive today as enthusiasm for them was only a few years ago. Nevertheless, on the basis of Latin American experience it is argued that:

> The disenchantment with growth center policies in many countries is not evidence that the principle of polarization is wrong. On the contrary, it reflects the over-optimism and short-run time horizon of regional policy-makers, the failure of sustained political will, the use of deficient investment criteria, bad locational choices, and lack of imagination in devising appropriate policy instruments. (Richardson and Richardson, 1975, p. 169).

Richardson maintains that effective regional planning requires a 15–25-year time-horizon. He has developed a model for the analysis of spread and backwash effects over time and in this context suggests that 'a well-located growth pole, promoted with vigor in appropriate economic conditions and resistant to political trimming, should pay off as a regional planning policy instrument if the planning horizon is long enough' (Richardson, 1976, p. 8). Yet he acknowledges that the growth pole concept cannot simply be transferred to areas such as Latin America without major modifications. These would include:

> a broadening of the approach away from the functional pole concept to include political, social, and institutional changes as well as sectoral measures; a more flexible attitude to the selection, location, and size of growth centers; placing growth center policies within the broader context of a national spatial strategy and the introduction of consistent and reinforcing non-spatial policies; careful consideration of the implications of the size and shape of a country, its topography, and its climate for growth pole policies; avoiding dissipation of scarce resources by designating too many centers; taking action to reduce the 'enclave' characteristics of growth poles; and realistic expectations. (Richardson and Richardson, 1975, p. 175.)

But by being so broad it may be questioned whether this is fundamentally a growth pole strategy at all. Indeed, Richardson's approach is not unlike that of Johnson (1970), who held that while developing countries need to create functional economic areas they also need proper education and health facilities and a cultural milieu that will release the population's creative energies. Rondinelli and Ruddle (1976) likewise have stressed that developing countries need to build an articulated network of growth centres as well as linkages among them in order to encourage commercialization of agriculture, savings, and investment in productive activities. Such a system is necessary if urban goods and services are to be delivered to rural populations and vice-versa. Dispersed village economies do not permit sufficiently large concentrations to form regular institutional markets; there is little reason to save and invest, specialization and division of labour do not occur, and opportunities are few for market expansion and for non-agricultural employment. Their spatial integration strategy does not seek to substitute Western organizations, technologies, production methods, attitudes, and social relationships for local traditional institutions and practices. Rather, it calls for involving local people; building on their resources, institutions, and practices; and adapting modern technologies and services to local conditions. However, this incremental process of transformational development does require planning for the displacement of unproductive and unadaptable traditional institutions and practices as change occurs. Regional planning is necessary to this overall endeavour because central governments cannot adequately plan and implement development activities from the national capital, and local governments are simply not capable of providing the resources for area-wide development.

In sum, then, prevalent attitudes towards the growth pole approach can be said to have passed through three phases: (1) optimism with respect to possibilities for inducing growth in a few centres and to the subsequent generation of spread effects; (2) pessimism when the expectations of the early phase failed to materialize; and (3) a broader view of growth centres as one aspect of more comprehensive development planning.

SUMMARY AND CONCLUSIONS

In recent years it has become almost fashionable to deplore the use of Western theories and methods—based on an urban-industrial orientation—in dealing with problems of developing countries. Increasingly critical appraisals of the centre-down paradigm have been consistent with this tendency. It has been alleged that within developing countries this orientation has contributed to: (1) dependence on the developed countries and multi-national corporations based in these countries; (2) the persistent dominance of one or a few large cities, which have critical problems of unemployment and underemployment themselves; (3) increasing income inequalities; (4) persistent and growing food shortages; and (5) deteriorating material conditions in the countryside (Friedmann and Douglass, 1978).

The relevance of these propositions varies according to conditions in individual countries. However, while acknowledging the shortcomings of the centre-down paradigm, it may be premature to dismiss this approach as altogether useless at best and exceedingly harmful at worst.

Marxist and other critics who have emphasized the issue of economic dependency have done so largely on two grounds: (1) the political dependency that tends to accompany economic dependency; and (2) the failure of existing institutional mechanisms to alleviate mass poverty, especially in rural areas. Whatever the merits of these objections, they do not in themselves provide an alternative development strategy. Lasuén may overestimate the positive importance of multi-national corporations in the international innovation diffusion process and ignore certain pressing political and social issues, but at least implicitly he raises a question of fundamental importance; if the economic disparities between developed and developing countries are to be narrowed, what institutional alternatives are there to the multi-nationals or similar forms of organization?

Present international trade patterns indicate that the degree of economic power often attributed to US-based multi-nationals may well be exaggerated. A vast 'rationalization' process going on throughout the world recognizes that not every nation can have a stake in every industry. It is most visible in steel, which has excess capacity worldwide, but it also involves the chemical, textile, electronics, and other industries. In time, the United States could have numerous high-unemployment areas from which industry has fled to foreign locations. Perhaps slackened investment in basic industries will be taken up by investment in consumer-related industries, high-technology activities, and energy conservation. Even so, serious problems of structural change will have to be dealt with. Moreover, Marxist critics now accuse the multi-nationals of neglecting problem regions in the developed countries in favour of developing countries. Thus, Holland (1976, p. 268) argues that 'if meso-economic companies can go multi-national at different stages of the product cycle, they can also go multiregional' within developed countries.

Worldwide investment stagnation and pressures on multi-nationals to invest at home clearly are not helping the developing countries. Indeed, some critics of the centre-down paradigm maintain that rural development and bottom-up strategies are necessary in the developing countries precisely because world capitalism has entered a period of permanent crisis (Friedmann and Douglass, 1978). The implication is that centre-down development, in an international context, was not all bad but can no longer be sustained in any case. This view presents two major difficulties. First, the political structures of most developing countries are very inimical to the kind of autarchic development implied in an 'agropolitan' or similar development strategy. This approach would seem to require the explicit inclusion of a theory and programme of political revolution, the kind of revolution that does not merely re-shuffle who is doing the exploiting and who is being exploited. Second, this approach neglects the fact of international business cycles. In other words, it assumes that the recent downturn will result in permanent stagnation rather than eventual adaptation leading to a period of recovery. The trouble with this view is that while it may be correct, prophets of the doom of the West have a long and undistinguished record as predictors of the future.

Within developing countries the centre-down paradigm has leaned on two major tenets. The first is the divergence–convergence syndrome (the 'Williamson thesis'). So long as aggregate growth can be sustained (even though initially it may necessitate dependency structures within developing countries and between them and developed countries), eventually there is likely to be a convergence of levels of regional per capita income. The difficulty here is that even if the dualism problem is temporary, the transitional time-period may still be several decades; it is not easy to counsel patience under these circumstances. On the other hand, autarchic development solutions to rural development problems, even if successful, could well require at least as much time.

Attempts to accelerate the development of lagging regions have largely relied upon some kind of growth pole strategy. While some informed observers have become pessimistic about the efficacy of this approach, others maintain that it has not been implemented properly. In the latter view, investments have been too few, they have not been sufficiently concentrated, planners have worked with unrealistically short time-horizons, and such strategies need to be embedded in more comprehensive development schemes. Thus, one may say of the centre-down paradigm what Shaw said of Christianity: it is not necessarily wrong; it has simply never been tried.

Despite differences between those who espouse one form or another of the centre-down paradigm and those who advocate development from the bottom up, some areas of agreement may be found—even though emphases may vary—concerning appropriate means for alleviating poverty. These would include: (1) more attention to human resource development, (2) greater efforts to curb population growth, (3) wider and more rapid diffusion of agricultural innovations, (4) planning in terms of functional economic areas, and (5) the linking of functional economic areas by a transportation and communications policy that encourages not only more general spatial diffusion of innovations but also facilitates the movement of agricultural and light industry outputs from rural areas to large urban markets.

The centre-down paradigm suggests that, despite their faults, economic dualism and growth pole strategies will keep gross national product growing more rapidly than would be the case under alternative strategies. Moreover, it implies that so long as the divergence–convergence syndrome can be expected to operate with respect to regional and class per capita income differences, there would seem to be little point in slowing aggregate growth, that is, in making everyone more equal by giving everyone less. To what extent are these assumptions valid? The theoretical issues have been clarified considerably by the scholarly debates of the past three decades. However, a thorough empirical examination of specific cases is required not only to make the assumptions of development theory more realistic, but also to provide guidance for policies that will in fact create significantly greater economic opportunities for the developing countries' poor majority.

REFERENCES

Berle, A. A. (1963) *The American Economic Republic* (Harcourt).
Berry, B. J. L. (1970) The geography of the United States in the year 2000, *Transactions*, **51**, 21–53.
Boudeville, J. (1972) *Aménagement du territoire et polarisation* (Génin).
Brown, E. H. P. and S. V. Hopkins (1955) Seven centuries of building wages, *Economica*, **22**, 195–206.
Brown, E. H. P. and S. V. Hopkins (1956) Seven centuries of the prices of consumables, compared with builders' wage-rates, *Economica*, **23**, 296–314.

Coraggio, J. L. (1975) Polarization, development, and integration, in Kuklinski, A. (ed.), *Regional Development and Planning: International Perspectives* (Sijthoff).

Enke, S. (1963) *Economics for Development* (Prentice-Hall).

Fleming, M. (1955) External economies and the doctrine of balanced growth, *Economic Journal*, **65**, 241–56.

Friedmann, J. (1966) *Regional Development Policy: A Case Study of Venezuela* (MIT Press).

Friedmann, J. (1972) A general theory of polarized development, in Hansen, N. (ed.), *Growth Centers in Regional Economic Development*, pp. 82–107. (The Free Press).

Friedmann, J. (1973) *Urbanization, Planning, and National Development* (Sage).

Friedmann, J. and M. Douglass (1978) Agropolitan development: towards a new strategy for regional planning in Asia, in Lo, F. C. and K. Salih (eds.) *Growth Pole Strategy and Regional Development Policy* (Oxford: Pergamon Press), 163–92.

Gaile, G. L. (1973) Notes on the concept of 'spread', unpublished paper, Department of Geography, University of California, Los Angeles.

Gilbert, A. (1975) The arguments for very large cities reconsidered, *Urban Studies*, **13**, 27–34.

Hansen, N. (1968) *French Regional Planning* (Indiana University Press).

Hansen, N. (1970) *Rural Poverty and the Urban Crisis* (Indiana University Press).

Harbison, F. H. (1973) *Human Resources as the Wealth of Nations* (Oxford University Press).

Hirschman, A. O. (1958) *The Strategy of Economic Development* (Yale University Press).

Holland, S. (1976) *Capital versus the Regions* (Macmillan).

Iacocca, L. A. (1977) Myth of the big, bad multinational, *Newsweek*, 12 September, p. 21.

Johnson, E. A. J. (1970) *The Organization of Space in Developing Countries* (Harvard University Press).

Lampard, E. E. (1968) The evolving system of cities in the United States: urbanization and economic development, in Perloff, H. S. and L. Wingo, Jr. (eds.), *Issues in Urban Economics* (Johns Hopkins Press).

Lasuén J. R. (1973) Urbanisation and development—the temporal interaction between geographical and sectoral clusters, *Urban Studies*, **10**, 163–88.

Mera, K. (1978) The changing pattern of population distribution in Japan and its implications for developing countries, in Lo, F. C. and K. Salih (eds.) *Growth Pole Strategy and Regional Development Policy* (Oxford: Pergamon Press), 193–216.

Mera, K. (1977) Population concentration and regional income disparities: a comparative analysis of Japan and Korea, in N. Hansen (ed.), *Human Settlement Systems: International Perspectives on Structure, Change and Public Policy*, pp. 155–76. (Ballinger).

Moseley, M. J. (1973) The impact of growth centres in rural regions, *Regional Studies*, **7**, 80–120.

Myrdal, G. (1957) *Rich Lands and Poor* (Harper & Brothers).

Nichols, V. (1969) Growth Poles: An Evaluation of their Propulsive Effect, *Environment and Planning*, **1**, 193–208.

Nurkse, R. (1952) Some international aspects of the problem of economic development, *American Economic Review*, **42**, 571–83.

Perroux, F. (1955) Note sur la notion de pôle de croissance, *L'économie du XXème siècle*, 2d edn., pp. 142–54. (Presses Universitaires de France)

Pred, A. (1976) *The Interurban Transmission of Growth in Advanced Economies: Empirical Findings Versus Regional Planning Assumptions* (International Institute for Applied Systems Analysis, Laxenburg, Austria).

Richardson, H. W. (1976) Growth pole spillovers: the dynamics of backwash and spread, *Regional Studies*, **10**, 1–9.

Richardson, H. W. and M. Richardson (1975) The relevance of growth center strategies to Latin America, *Economic Geography*, **51**, 163–78.

Rodwin, L. (1963) Choosing regions for development, *Public Policy*, **12**, 141–62.

Rondinelli, D. A. and K. Ruddle (1976) *Urban Functions in Rural Development: An Analysis of Integrated Spatial Development Policy* (Office of Urban Development, Agency for International Development, US Department of State).

Rosenstein-Rodan, P. N. (1943) Problems of industrialization of eastern and southeastern Europe, *Economic Journal*, **53**, 202–11.

Singer, H. W. (1949) Economic progress in underdeveloped countries, *Social Research*, **16**, 1–11.

Williamson, J. (1965) Regional inequality and the process of national development: a description of the patterns, *Economic Development and Cultural Change*, **13**, 3–45.

Chapter 2

Development from Below:
The Bottom-Up and Periphery-Inward
Development Paradigm*

WALTER B. STÖHR

INTRODUCTION

The paradigm of development 'from below', like that 'from above' (Hansen, Chapter 1, of this volume), is not, as might be assumed, simply related to the level at which decisions on development are taken. A change in level of decision-making is a necessary but not a sufficient—and possibly not even the most important—condition for such a strategy. Development 'from below' implies alternative criteria for factor allocation (going from the present principle of maximizing return for selected factors to one of maximizing integral resource mobilization); different criteria for commodity exchange (going from the presently dominating principle of comparative advantage to one of equalizing benefits from trade); specific forms of social and economic organization (emphasizing territorial rather than mainly functional organization; cf. Friedmann and Weaver, 1979) and a change in the basic concept of development (going from the present monolithic concept defined by economic criteria, competitive behaviour, external motivation, and large-scale redistributive mechanisms to diversified concepts defined by broader societal goals, by collaborative behaviour and by endogenous motivation).

Development would need to be considered again as an integral process of

* An earlier draft was presented as an inaugural paper at the 18th European Regional Science Congress at Fribourg, Switzerland, in August 1978 and I want to acknowledge the valuable comments received there. I am particularly grateful for extensive comments to the collaborators of my Institute as well as to Leonhard Bauer, Vienna; Vera Cao-Pinna, Rome; John Friedmann, Los Angeles; Paul Hanappe, Paris; Kalman Kadas, Budapest; Jan Lundqvist, Bergen (Norway); and D. R. F. Taylor, Ottawa. All remaining shortcomings are as always my own responsibility. I am grateful to Hildegard Kaufmann for patiently typing and re-typing the text from often terribly untidy manuscripts, and to P. Fritz for drawing Figure 2.1.

widening opportunities for individuals, social groups, and territorially organized communities at small and intermediate scale, and mobilizing the full range of their capabilities and resources for the common benefit in social, economic, and political terms. This means a clear departure from the primarily economic concept of development held in the 1950s and 1960s with its ensuing pressure on individuals, social groups, and territorially organized communities to develop only a narrow segment of their own capabilities and resources as determined 'from above' by the world system, and neglecting other capabilities and self-determined objective in order to retain a competitive position in economic and political terms *vis-à-vis* the rest of the world.

Unlike development 'from above', which was nurtured by the economic theories of the past three decades (particularly the neoclassical one), there seems to be no well-structured theory available as yet for an alternative paradigm of development 'from below'. Some efforts in this direction have been undertaken: at international level by the search for a New International Economic Order (Tinbergen *et al.*, 1976), or for 'another development' (Nerfin, 1977); at the subnational level by such concepts as Agropolitan Development (Friedmann and Douglass, 1978), Ecodevelopment (Sachs, 1976), and the search for a 'theory of rural development' (Haque *et al.*, 1977). But there is still a pressing need for a coherent and systematic framework for an alternative approach. One reason for the lack of such a coherent framework may be that it would need to be supported by a variety of disciplines and not primarily by economics—as our present theory suggests—and the cumulative co-operation between different disciplines is apparently very difficult to achieve. In addition there may not be only one strategy of development 'from below'—as has been the case for the predominantly monolithic industrialization-urbanization strategy 'from above' to date. Beyond some basic common features, different cultural areas will need to construct their own development strategies which will require compatibility only at certain points of mutual contact. Alternatively, the contact points may need to be restricted to those types of interaction where compatibility is feasible and desired by each of them. Alternative strategies of development 'from below' need to emerge from, and be adapted to, the requirements of different cultural areas. Such strategies may change over time, possibly alternating with phases of development 'from above'. The traditional development paradigm 'from above', a centre-down-and-outward paradigm (Hansen, this volume, Chapter 1) provides a starting point in the search for alternatives.

The past three decades, dominated by development strategies 'from above', have not led to decreased disparities in living levels. Disparities have in general increased as many of the case studies in this book show. This applies both to disparities between social strata (Adelman and Morris, 1973) and between geographical areas (Stöhr and Tödtling, 1977).

Both paradigms are conceptual constructs which in practice rarely occur in pure form. Real situations will always (especially in today's highly interactive national, continental, and worldwide systems) consist of a mix of both these elements. In different national or regional situations there is undoubtedly considerable variation in the make-up of these two elements as well as in their temporal sequence.

Let me briefly contrast the two concepts with each other in more or less pure form.

A REAPPRAISAL OF THE CENTRE-DOWN PARADIGM

As Hansen has shown in Chapter 1, the concept of development 'from above' assumes that development, whether spontaneous or induced, starts only in a relatively few dynamic sectors and geographical clusters, from where it will (or hopefully should) spread to the remaining sectors and geographical areas. This 'trickle down' process is essentially supposed to start at the global level (from worldwide demand, or world innovation centres), and than filter down and outward to national or regional units, either through the urban hierarchy (Berry, 1972), through input–output relations (Perroux, 1964), or through the internal channels of multi-plant business organizations (Lasuén, 1973; Pred, 1977), or large-scale government organizations.

Its emphasis therefore has been on urban and industrial, capital-intensive development, the highest available technology, and maximum use of external and scale economies. This usually involves: large investment projects; increasing units of functional and territorial integration; increasing scale of the private and public organizations required to transmit development through these integrated units; large redistributive mechanisms; and the reduction of economic, social, cultural, political and institutional 'barriers' (including distance friction and institutional differentiation), which might hinder transmission effects within and between these units.

The hypotheses upon which the paradigm of development 'from above' is based, explicitly or implicitly, are: (1) development in its economic, social, cultural, and political dimensions can be generated only by some very few select agents such as Schumpeter's (1934) entrepreneurial pioneer, the white, the urbanite, or the intellectual; (2) the rest of the population are considered 'incapable of initiatives in making improvements, consequently everything must be done for them . . . ' (Uphoff and Esman, 1974, pp. 28–29); (3) these few agents are able and willing to allow all others to participate in this development within a reasonable time-span and on a reasonably equal basis; (4) these other groups are able and willing to adopt the same type of development; (5) the specific type of development (economic, social, cultural, and political) initiated by the few select agents is the most suitable one for all the other members of the increasingly interactive system and should therefore replace other existing notions of development; (6) the (socially and culturally) new and the (economically and politically) more powerful notions of development are also the 'better' ones and therefore the ones which the rest of society strives for. Essentially it therefore presumes an eventually monolithic and uniform concept of development, value systems, and human happiness, which automatically or by policy intervention will spread over the entire world.

These notions are related very closely to the interests of the large-scale (private or government) organizations which were installed to serve as the 'motor' of development. In many cases, however, these have in turn come to dominate the system, very often overruling the interests of those—e.g. local and regional communities—they were meant to serve. This has led to a high concentration of power in a few private or governmental organizations which now dominate the greater part of the world system. At the same time such a strategy ignores or overrules: (1) the great diversity of value systems and aspirations created by the differences—often historically grown and territorially defined—between cultural systems; (2) the great variations in natural conditions (which in part have brought about the latter); (3) the fact that, with different aspirations and cultural and natural preconditions, the imposition of a uniform concept and measurement of development is bound to relegate some groups to what is today called 'underdevelopment', leaving them further away from those set standards and plunging them still further into the role of the disadvantaged in any measurement of standards as defined by the dominant culture or group— as has been shown in another context (Stöhr and Tödtling, 1978); (4) that this subsequently leads to differing levels of 'dependence' as has also been shown in the latter paper. Once such economically disadvantaged groups start to interact more intensively with the more developed ones on the latters' terms, they are increasingly forced to adopt the same social, cultural, political, and institutional norms in order to try to compete with them in economic terms, economics being the medium through which most large-scale interactions are regulated; (5) that this entails the subordination of broader societal and cultural values to economic determinants, a fact that can be observed in most highly developed countries and those directly influenced by them today.

The alternative would be for economically less-developed social groups and areas to give clear priority to their own self-determined societal standards, and to subordinate external economic and other interactions to these standards. This increasingly is being demanded in fact by developing countries which are beginning to demand some form of collective self-reliance and which view increased nationalism as an important component of a 'new meaning of development' (Seers, 1977).

Normally, large-scale organizational linkages between areas of greatly differing levels of development lead (due to factors such as unequal distribution of power, selective factor withdrawals, unequal terms of trade, unequal distribution of scale economies) to increasing spatial divergence rather than a convergence of living levels. In other words, even with explicit regional development policies operating through large-scale private or public organizations, the sum of backwash effects in most cases still seems to exceed spread effects.

With a 'centre-down' development strategy this can only be avoided if at the national level there is both a strong control mechanism avoiding leakages to the exterior (a control on commodity and factor flows) and a strong internal redistributive mechanism with broad public participation. In large countries the same mechanisms will also be necessary at subnational levels, to avoid major concentrations of power. Few developing countries seem to have been willing (or able) to do this so far, and it will be interesting to see the evaluation of respective experiences, particularly those of some of the more inward-looking developing countries of a socialist type to be presented in this book.

DEVELOPMENT 'FROM BELOW'

Development 'from below' would require the control of the backwash effects of development 'from above' mentioned before, and the creation of dynamic development impulses within less-developed areas. The crucial question is how these two requirements can be fulfilled. The first requires changes in the interaction between different regions and countries and will be treated here; the second requires the creation of endogenous factors of change for increased equity and developmental dynamics; this will be considered in a later section of this chapter (see pp. 62–3).

Instead of optimizing selected factor components in a 'centre-down' fashion on a large (national or international) scale, thus 'creaming' the human and natural resources of favoured segments of (national or regional) economies (Hansen, Chapter 1 of this volume), the basic objective of development from below is the full development of a region's natural resources and human skills (what Richardson, 1973, calls 'generative' growth), initially for the satisfaction in equal measure of the basic needs of all strata of the regional or national population, and subsequently for developmental objectives beyond this. Most basic-needs services are territorially organized, and manifest themselves most intensely at the level of small-scale social groups and local or regional communities. Development 'from below' therefore would require that the greater part of any surplus (created through production specialization within an area) should be invested regionally for the diversification of the regional economy. By 'region' here we mean the smallest territorial unit above the rural village where such activities are still feasible (cf. also Friedmann and Douglass, 1978 on 'agropolitan districts'), and which comprise commuting and service provision areas of acceptable internal accessibility. This process is then envisaged to occur at successively higher scales. Through retention of at least part of the regional surplus, integrated economic circuits within less-developed regions would be promoted (Santos, 1975; Senghaas, 1977) and development impulses would be expected to successively pass 'upward' from the local through regional to national level. Policy emphasis therefore will need to be oriented towards: territorially organized basic-needs services; rural and village development; labour-intensive activities; small and medium-sized projects; technology permitting the full employment of regional human, natural, and institutional resources on a territorially integrated basis.

The basic hypotheses underlying a strategy of development 'from below' are the following:

(1) Major regional disparities in living levels have emerged as the negative

consequences of insufficiently prepared large-scale economic integration, through the withdrawal effects mentioned above. Previously regional civilizations developed independently, achieving broadly similar levels of material progress (Abdalla, 1978, p. 19). Population distribution adjusted to the long-term resource potential of individual regions (except for unforeseen natural disasters) and the social mechanism on the whole retained a man–resource balance.

(2) There are many concepts of development depending on the natural and social environment of different communities and the development over time of specific cultural and institutional conditions. In fact, these represent major factors of development potential and should not be subordinated to the short-term pressures of any externally dominated or 'anonymous' market mechanism.

(3) The basic impulse for formulation and implementation of such differentiated concepts of development must come from within the respective communities. This requires discarding the presently dominant hypothesis that small-scale (local, regional, or national) communities can develop only through the intermediary of other more highly developed communities or countries, and must do this by applying the latter's—usually materially defined—concept of development. Discarded also must be the presently dominant hypothesis that, in order to achieve this, poor communities must produce more commodities for demand of the rich ones (the export-base concept) at reduced cost and return to their own factors (mainly labour and natural resources) and in return, must receive transfers of capital, technology, and organizational skill ('development aid') from the more developed countries. Many of these factor transfers, along with withdrawal of natural resources and unequal terms of trade, actually weaken rather than strengthen the comparative development potential of less-developed areas. They also reduce the respective communities' capacity to mobilize their own capital, technology, and organizational skills, thereby making them increasingly dependent on more developed areas.

(4) There should be greater national and regional self-determination on the degree and type of interaction needed in these territorial units. Most 'less-developed' countries (as defined by developed countries' criteria) are underdeveloped mainly because of their reduced capability for formal large-scale interaction to provide goods and production factors for large-scale (international) interchange systems. Many such communities, however, have a much higher potential for informal small-scale interaction (interpersonal social relations, group identity, small-scale solidarity, active cultural participation) than have materially highly developed areas. Such small-scale potential (related to Allardt's (1973) conditions of 'loving' and 'being') is important to human beings but is usually not interchangeable at larger scales. It has use-value, not exchange-value, and therefore does not enter into the calculations of 'comparative advantage' on a large (worldwide) scale, and cannot be measured by our usual indicators of development. These

potentials are, however, greatly affected—usually negatively—by the economic, social, and political transformations caused by a rapid integration into
the world economy. This is visible in virtually all formerly colonial countries
(Abdalla, 1978).

Since such small-scale potentials, often operating on an informal basis, are
also important for such necessities as social security, health care, environmental
protection, and education, quality of life in these sectors is often detrimentally
affected by factor withdrawals, or by rapid social transformations brought
about by large-scale economic interchange between areas of greatly differing
levels of economic development. Usually the state has to intervene through
large-scale cost-intensive institutions to compensate for the informal social
mechanisms which had served such purposes before. But in less-developed
countries this is usually not feasible.

Development 'from below' therefore may require a certain degree of 'selective spatial closure' (Stöhr and Tödtling, 1978) to inhibit transfers to and from
regions or countries which reduce their potential for self-reliant development.
This could be done by control of raw material or commodity transfers which
contribute to negative terms of trade and/or by control of factor transfers
(capital, technology), and by the retention of decision making powers on commodity and factor transfers in order to avoid the underemployment or idleness
of other regional production factors, or major external dependence.

Instead of maximizing return of selected production factors on an international scale, the objective would be to increase the over-all efficiency of all
production factors of the economically less-developed region in an integrated
fashion. This integration of territorially available resources and social structures
should form a basis for more internally initiated development impulses.

Development of large-scale activities and centres will then primarily be
based on territorially defined local and regional inputs and demand, and will
correspond to their requirements rather than the reverse. Large cities would
not be able to grow as fast as they have in the past—which would help solve a
key problem of spatial development in most developing countries. Benefits
accruing from activities exceeding this scale would be subject to spatial redistribution, Hoselitz's (1954) dichotomy between 'generative' and 'parasitic' cities
would be rephrased: cities would primarily generate activities for their immediate hinterland rather than for an abstract inter-urban system. In this way
the hierarchical urban-industrial system would essentially be sustained 'from
below' by the relatively stable (human, social, political, and environmental)
needs and potential of their territorial hinterland and its population, rather
than by the fortuitous and uncontrollable trickling down of impulses 'from
above'. Urban centres would develop primarily as a supportive component
of their respective hinterlands, rather than the hinterland developing (as in the
centre-down strategy), as a function of the selective requirements of the urban
system.

The spatial pattern of urban-industrial development resulting from such strategies will be comparatively decentralized and regionalized. As industrialization and service growth are based primarily on demand and resources within their respective regions and are not (primarily) export-base activities (which should be promoted later at successive stages), the urban system will tend to be more inward-oriented than the existing coastal patterns of developing countries which have resulted from the 'centre-down' process. Urban centres will tend to maximize accessibility within their regions (on which their development primarily depends) rather than maximizing accessibility to the outside. Inward-oriented urban systems which maximize internal accessibility provide the best conditions for equal provision of basic-needs services to all parts of the population. This corresponds to the urban pattern of those countries which were able to establish an urban system and an institutionally disaggregated (e.g. multi-tier federal) nation-state before large-scale transport integration and industrialization took place.

A strategy of development 'from below' would need to combine selected elements of what Hirschman (1958) called advantages of integration (or 'surrender of sovereignty') with selected elements of advantages of separatism or 'equivalents of sovereignty'. 'If only we could in some respects treat a region as though it were a country and in some others treat a country as though it were a region, we would indeed get the best of both worlds and be able to create situations particularly favorable to development' (Hirschman, 1958, p. 199). The question is how and whether this is possible.

The 'advantages of integration' for less-developed areas would seem to lie particularly in free and self-determined access to technological and organizational innovation as well as in access to potential markets for products surplus to their own need (Furtado, 1978).

The advantages of 'equivalents of sovereignty' would seem to lie in cultural and institutional semi-autonomy of less-developed areas and in the possibility of a certain degree of 'spatial closure' (Stöhr and Tödtling, 1978) or a 'cellular economy' (Donnithorne, 1972), in which barriers to free resource withdrawals, to negative terms of trade, and to a further concentration of power would be drawn along territorial lines at different hierarchical levels.

At least for a 'cellular' socialist country such as China it is maintained that self-reliance together with decentralized planning (as introduced in 1975) did not reduce the development potential of the poorest regions, but '..seems to have narrowed ... interprovincial differentials ... even before the important state interprovincial transfer of funds has been taken into account' (Paine, 1976). It may well be that the mobilization of additional resources in the regions—beyond increasing the production and income of even the poorer regions—may in fact increase aggregate growth and thereby make more funds centrally available for additional redistribution.

Apart from these rather fragmentary strategy elements, is there a more coherent theoretical or practical experience available to development processes 'from below'?

Dudley Seers, in his paper 'The new meaning of development' (1977) maintains that we still lack a theory for an alternative development paradigm:

. . . we do not yet understand much about what self-reliance implies for development strategies, but some of the economic aspects are obvious enough. They include reducing dependence on imported necessities, especially basic foods, petroleum and its products, capital equipment and expertise. This would involve changing consumption patterns as well as increasing the relevant productive capacity. Redistribution of income would help, but policies would also be needed to change living styles at given income levels—using taxes, price policies, advertising and perhaps rationing. In many countries, self-reliance would also involve increasing national ownership and control, especially of sub-soil assets, and improving national capacity for negotiating with transnational corporations. . .

There are other implications as well, especially in cultural policy. These are more country-specific, but as a general rule, let us say that 'development' now implies, *inter alia*, reducing cultural dependence on one or more of the great powers—i.e. increasing the use of national languages in schools, allotting more television time to programmes produced locally (or in neighbouring countries), raising the proportion of higher degrees obtained at home, etc. . . .

On this approach, development plans would henceforward not put the main emphasis on overall growth rates, or even on new patterns of distribution. The crucial targets would be for (i) ownership as well as output in the leading economic sectors; (ii) consumption patterns that economised on foreign exchange (including imports, such as cereals and oil); (iii) institutional capacity for research and negotiation; (iv) cultural goals like those suggested above, depending on the country concerned. . . .

Of course, an emphasis on reducing dependence does not necessarily mean aiming at autarchy. How far it is desirable, or even possible to go in that direction, depends on a country's size, location and natural resources; on its cultural homogeneity and the depth of its traditions; on the extent to which its economy needs imported inputs to satisfy consumption patterns which have to be taken—at least in the short term—as political minima. The key to a development strategy of the type suggested is not to break all links, which would almost anywhere be socially damaging and politically unworkable, but to adopt a selective approach to external influences of all types. (Seers, 1977, pp. 5–6.)

IN SEARCH OF A THEORY OF DEVELOPMENT "FROM BELOW": SOME HISTORICAL PARALLELS

Niles Hansen has shown in Chapter 1 that the theoretical foundations of development strategies 'from above' as practised during the past three decades stem mainly from the economic theory dominant during the last quarter of a century, strongly influenced by neoclassical economic thought. Such strategies are still widely considered a prerequisite for further economic progress, although in view of the results produced, increasing doubts are being raised.

The lack of an alternative theory and practice for development 'from below' which Seers alludes to, refers in fact only to the period of the last 30 years, however. It is true that during this period development 'from above' has been the dominant paradigm in the theory and practice of economic policy in particular, and in the evolution of the two systems of super-powers in general. This was the case with respect to the Western hemisphere, in spite of the fact that one of the major assumptions of classical and neo-classical economic thought is individual and independent private decision-making—seemingly the extreme opposite of the mechanism of development 'from above'. This contradiction becomes clear in hindsight, however, as classical or 'positivistic' economics was concerned essentially with individual resource allocation for the production of 'private goods' (Machlup, 1977); the increasingly important production of public goods for collective objectives, however, needs to focus essentially on external effects and therefore requires different allocation mechanisms. In the eastern hemisphere, the Soviet model is by definition based on the operation of a monolithic central planning and redistributive mechanism 'from above' which has increasingly been challenged by peripheral countries in the Socialist sphere of influence.

There have been many examples in previous periods however, under changing cultural and political conditions, of periods dominated by development 'from below' intermittent with periods of development 'from above', as we shall try to show has been the case in central Europe. This has been done in spite of the fact that the present book deals mainly with Third World countries, in order to show that the presently dominant paradigm of development 'from above' may not be as irreversible as it might seem in short-term perspective and from an isolated economic point of view.

The swing between these two approaches in the past seems among other factors, to have been related to changing scales of societal interaction, to a changing subordination of rationalizing economic activity under broader societal norms (philosophical, religious, political, social, etc.), and more recently, to changing rates of economic expansion. Let me give a few short—and necessarily overly simplified—examples. A concise and very pointed description of these changing trends, including their spatial dimension for earlier periods, is contained in Heinrich (1964), for more recent periods in Krondratieff (1926), Schumpeter (1939), Lewis (1978), and Rostow (1978). For recent periods the specific interrelations between long waves of economic growth and scales of spatial industrial interaction have been analysed by Hanappe and Savy (1978), and their relations with socio-political integration by Siegenthaler (1978).

For the present purpose I shall go back into the history of economic and societal thought much further than Niles Hansen has done in describing the centre-down paradigm dominant during the past 25 years:

Over the past 2500 years the scale of societal and spatial interaction in specific cultural areas has changed at various times from periods dominated by small-scale societal interaction (from 'below') to others dominated by large-scale interaction (from 'above'). Along with these changes have gone changes in sociopolitical institutions from small- to large-scale and vice-versa. Furthermore, they were usually associated with changes in the dominant power structure between different social, political, and religious groups within the respective societies. More recently these changes have also been associated with major technological innovations (Schumpeter, 1939; Lewis, 1978). Periods of dominating small-scale spatial interaction ('from below') tended to be irrationally and metaphysically dominated; they focused on what Touraine (1976) calls 'ideological' accumulation. Firmly subordinated under such explicit societal norms (metaphysical, institutional, social, cultural, ritual, etc.) of the respective territorial community were specific functional economic, technological, and other rational activities or what Touraine has called processes of 'functional' accumulation. These small-scale economic and societal interaction patterns (small-scale compared to those of preceding or following periods) might be explained by the fact that irrationally based (emotionally, sensually based) human interaction is limited to a much smaller scale than functional (abstract, rationally, or economically based) human interaction (Hall, 1966; Greenbie, 1976; Laszlo, 1977). Whereas the first kind of interaction takes place mainly in concrete (spatially delimited) territorial space, the latter takes place primarily in abstract (functionally delimited) space (Friedmann and Weaver, 1979) and can therefore bridge distance much more easily. To put it in Allardt's terms, human conditions of 'loving' and 'being' depend much more on small-scale interaction; whereas the human condition of 'having' benefits considerably from large-scale interaction.

Laszlo (1974) assumes a continuous increase in scales of spatial interaction throughout history, whereas according to our hypothesis spatial expansion has oscillated considerably between rationally and metaphysically dominated historical periods.

In periods of dominating large-scale interaction 'from above', such as during the past 25 years, functional (rational, including economic and technological) activities gained considerable autonomy from broader societal norms (metaphysical, institutional, social, ritual, etc.), and often even became supra-ordinate over them. This usually was accompanied by a rapid acceleration in the rate of innovation in many fields including production, transport, and communications technology. These innovations promoted the rapid expansion of interaction scale which only functional processes could readily bridge. These periods of development 'from above', possibly due to the destabilizing effects of accelerated innovation (Laszlo), in general tended to be much shorter than the preceding or subsequent periods of development 'from below'. These changes seem to have oscillated around a long-term trend towards increasing scales of interaction which, however, were periodically reduced once more to smaller scale, and thereby prevented from moving ahead at a continually fast rate.

Figure 2.1 illustrates such periods of development 'from above' and 'from below'. Samples are taken from specific periods in time and only from a relatively small part of the world: the Mediterranean and central Europe. More profound analysis is required on these issues, particularly on the impact each of these periods has had on the living levels of the poorest population strata. It is hoped that the present hypotheses will arouse the interest of historians in undertaking more systematic studies of the interrelations between scale of socioeconomic interaction, rate of economic growth, and spatial disparities in living levels.

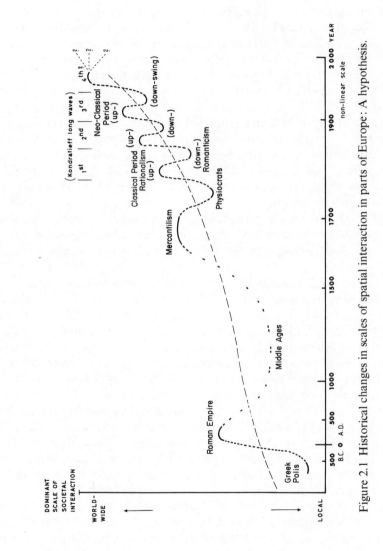

Figure 2.1 Historical changes in scales of spatial interaction in parts of Europe: A hypothesis.

Following are some of the examples (see also Figure 2.1):

Greek Polis:
In the Greek Polis, urban and economic development, particularly in its initial phase, was *essentially self-reliant* (sometimes even autarchic within city-states) and rather rigidly subordinated to the philosophical framework of Nicomachean ethics and their norms of justice rooted in the works of Plato and Aristotle. The Polis was a narrowly delimited territorial unit usually without tendency towards contiguous expansion. Instead, more or less rigidly organized city leagues were established. This period lasted for about half a millennium from the seventh century BC onwards.

Roman Empire:
During the first century and a half AD. (Caesar to Trajan), the Roman Empire (see Figure 2.1) created the *first 'world economy'*, with a comparatively well organized monetary, credit, and transport system, and relatively little subordination to metaphysical norms.

Middle Ages:
In the third and fourth centuries AD there followed a *reversal* to a large number of *small-scale economic areas*, a receding of the monetary and credit economy, and an increasing orientation towards agricultural and rural forms. During the Middle Ages (Figure 2.1) economic activities for over four centuries were subjected to a multiplicity of religious and social restrictions, with emphasis on the concept of individual and common 'need' determining a just price (*'justum precium'*), prohibition on the taking of interest on capital, etc. Evidently this long period was—both temporally as well as spatially—much more diversified than Fig. 2.1 indicates.

The subjection of economic activities to philosophical, religious, and social norms introduced important elements of solidarity (against those of competition), provided stability (against the dynamics of an autonomous economic process), and in general reduced the rate of societal change: the relatively small-scale framework of the Greek Polis and the Medieval system lasted four or five centuries each; that of the Roman Empire only for one and a half centuries.

Mercantilism:
Towards the end of the Middle Ages, there occurred in central Europe a change from the relatively self-contained, small-scale agricultural and rurally dominated interaction patterns of the twelfth and thirteenth centuries to the *increasing interaction radii* of Mercantilism (Figure 2.1). The latter was intended to serve 'as a unifying system' attempting to overcome the 'economic disintegration' caused by feudal powers, by the system of river tolls and road tolls, and by 'local disintegration in other spheres' (Heckscher, 1955). It was based on state-guided promotion of handicrafts and urban-based

trade which created decentralized urban-centred regional economies and in which rural areas became strictly subservient to their urban centres. In the Mercantilist period, for the first time, explicit economic policy by the state existed and was scarcely embedded in metaphysical norms. The major objective was to increase the wealth of the state as a whole by producing sufficient amounts of the goods required for nourishment, material needs, and comfort of its inhabitants. The (aggregate) positive balance of trade became an overriding objective, oriented towards the development of the state's economic potential and reduction of foreign economic influence. Major instruments were the prohibition on import of final products, and export of raw materials and food products. All foreign commodities, particularly those serving 'unnecessary luxury' rather than real needs, were banned from import and later were subjected to high tariffs. It was for the first time, though on a small scale, a centralist state-run policy to develop to the maximum the full economic potential of individual states. It therefore combined elements of development 'from above' with those of development 'from below'. Emphasis was on urban activities in a larger national context to safeguard political independence. At the same time it was the period when the first overseas colonial acquisitions were made by European countries.

Physiocrats:
As a reaction to the previous urban and handicraft-oriented state-run economic policy there emerged in the second half of the eighteenth century, particularly in central Europe, the first *partial laissez-faire* policy of the 'Physiocrats' (Figure 2.1). *Laissez-faire* principles, however, were mainly applied to commerce as their application in agriculture (the Physiocrats' key sector, according to Quesnay's Tableau Economique) would have made large rural regions idle. Free trade thereby essentially fortified the power of the agricultural and administrative sectors. The directly productive employment potential of cities was minimal as industrialization had barely started. *Handicrafts and trade were considered 'sterile'* activities. At the same time the rural unrest of the Peasant Wars was still remembered. *Tariff protection for agriculture*, the key sector, was therefore a necessary exception to general *laissez-faire* attitudes applied to the 'sterile' sectors, a restriction aiming at full employment of natural resources, especially of agricultural land.

Classical Free-trade Era and First Kondratieff Up-swing:
Finally, towards the end of the eighteenth century the Enlightenment (Figure 2.1), classical economic theory and the free-trade philosophy came into domination. By this time the chief country of origin of this doctrine, England, had acquired sufficient initial advantage to benefit from such an economic arrangement. This coincided with the upswing of Kondratieff's (1926) first 'long economic wave'. Although Kondratieff's waves are mainly based on variables indicating economic expansion (prices, production, wages, etc.) they are indirectly also related to scales of spatial interaction (foreign trade).

The initiation of Kondratieff's 'long up-swings' has been associated with the enlargement of world markets by assimilation of new (and especially of colonial) countries, with the application of important discoveries and inventions in the technique of production and communication, and by extensive wars and revolutions (Samuelson, 1978).

Along with it in many European countries came the elimination of rigid institutional restrictions in the fields of artisanry and commerce (guilds), and in rural activities (feudalism). Previous small-scale interaction units were transformed into larger, internally increasingly uniform political entities: the nation-states of France, England, Sweden, and Russia in the eighteenth century; Germany, Italy, and the USA in the nineteenth century, etc. These nation-states increasingly *enlarged their areas of influence* on non-European continents as well.

This period saw a major extension of European forms of development to other continents. In the previous Mercantilist and Physiocratic periods, colonies had mainly been considered as sources of prestige, monetary wealth (precious metals), and luxury goods (spices and tea, etc.), while their social and cultural systems on the whole were left intact with the exception of slave-trade areas. With the emerging technological and industrial revolution in Europe in the latter part of the eighteenth century, however, the role of colonies was increasingly transformed into their becoming sources of raw materials for industry (cotton, wool, vegetable oils, jute, dyestuffs, etc.), and into potential markets for industrial products. In order to attain these objectives major social, cultural, and legal changes were imposed upon the colonies by the motherland: (1) changes in existing land and property arrangements in the direction of private property, and the expropriation of land for plantation agriculture; (2) creation of a labour supply for commercial agriculture and mining, by means of forced labour, and by indirect measures aimed at generating a body of wage-seeking labourers; (3) spread of the use of money and exchange of commodities, by imposing money payments for taxes and land rent and by inducing a decline in home industry; and (4) where the pre-colonial society already had a developed industry (e.g., cotton in India), a curtailment of production and export by native producers. These changes were perpetuated by the introduction of legal codes, administrative techniques and the culture and language of the dominant power, and the promotion of a local élite willing to co-operate with the colonial power (Magdoff, 1978). Industrialization had thus led to a sociopolitical and cultural penetration (i.e. a territorial disintegration) of the colonies. A similar process to this international one also occurred within most of these countries, in what has recently been termed 'internal colonialism' (Hechter, 1975).

The economic sphere had finally liberated itself from centuries-old religious, ethical, social, and institutional restrictions, and began to evolve as an autonomous field of activity. With rapid technological innovation and capital accumulation it turned from being an object into becoming a driving force for social, political, religious, and environmental transformation. The broader societal and environmental spheres became relegated to the status of 'externalities' of economic processes, both in the original motherland and successively also in the dependent colonies. Substituted for former principles of solidarity, satisfaction of basic needs, justice, and integrated resource mobilization, were new criteria of competition, efficiency, selective worldwide resource optimization, and survival of the fittest. Along with successive elimination of obstacles to increasing efficiency such as religious, ethnic, and institutional barriers came the reduction of state barriers and the introduction of a nationless and spaceless economic doctrine. Intranational and international trade became institutionally less and less differentiated. Tariffs were discarded as distorting comparative advantage and discriminating between social classes (by presenting gifts to the landowners!). Other territorially manifested 'externalities' in spheres such as environment, employment of human and natural resources, social conditions, etc., were increasingly neglected. It was assumed (as later also in neoclassical economics) that under conditions of perfect competition and full mobility of factors, private efficiency and social welfare would converge. Foreign competition would lead to modernization of national manufacturing, and the principle of comparative advantage would increase efficiency, reduce prices, and as a consequence also wages; increased capital accumulation and increasing radii of spatial interaction would further raise over-all efficiency. The transformation of economic structures (once embedded in stabilizing complex societal norms) had taken place, into economic processes now autonomous and spaceless. Devoid of institutional or other societal constraints, a process of quantitative growth, but also of qualitative deterioration and instability at a rate never known before, had started.

In economic theory and policy, criteria for resource allocation such as 'need' or 'absolute cost' (locally or regionally manifested), was substituted by the concept of (worldwide) comparative advantage. There were a number of important consequences:

(1) This concept cemented the dominance of economic criteria over what now became 'external' (social, cultural or environmental) criteria as described above.

(2) It promoted the systematic over- or under-exploitation of territorially organized or less mobile resources, especially land, human, and institutional resources.

(3) It initiated an increasing exposure of local and regional communities to external (national and international economic, and other) influences. The level of development and the wealth of a regional or national community was no longer determined by the amount of resources (material, cultural, institutional, etc.) which this community was able to mobilize for the satisfaction of its own needs, but rather by the value attached (from outside) to exportable segments of its resources in exchange for imported goods. The value of an imported commodity, on the other hand, was determined not by

its local or regional production cost, but by the cost of export goods exchanged for it, and by the elasticity of national demand for foreign goods, compared to that of foreign demand for national goods (the equation of international demand; Mill, 1848). This meant that territorially defined communities were not masters of their destiny any more, but had become subservient to the 'objective' mechanism of worldwide commodity and factor markets and those national and international functional groups able to control them.

(4) It created and perpetuated situations of 'underdevelopment' along territorial lines, where following the previous argument, 'less-developed' regions or countries in order to develop had to reduce the cost (and price) of their export commodities (based mainly on an abundant supply of natural resources and cheap labour) in order to pay for imported finished goods and capital equipment (based mainly on inputs of scarce capital and technology); the unequal distribution of surpluses reaped from these processes increasingly widened territorial disparities in living levels.

(5) It promoted inequalities between social strata, as the required reduction in cost of export commodities had to be achieved mainly via the reduction of real wage inputs in order to pay for imports of consumption or capital goods required by a small regional or national élite.

(6) It promoted the extension of private or state capitalism throughout major parts of the world, as the 'less-developed' regions and countries, in order to close their capital and balance-of-payments gap, had to import capital or take on long-term loans and other (including political) commitments.

Economic interaction in fact seemed to have overcome distance friction as commodity and capital flows were able to reach the most remote corners of the globe. Aggregate economic growth and efficient use of scarce factors (particularly capital) advanced at a rate unheard of before. But territorially articulated economic 'externalities' and others distributionally and structurally articulated (such as economic inequality; forceful transformations of political, cultural, and social structures; fluctuations and disparities in level of employment of human and natural resources at national, regional, and local levels) also exceeded any measure known before. A strong disintegration of territorially organized communities and their environment had taken place. Müller (1809) in fact had called Adam Smith's classical economics a theory of the successive and radical dissolution of the state.

Romanticism and the first Kondratieff down-swing

It is therefore not surprising that the pendulum soon swung back again in the early nineteenth century. This swing coincided with what Kondratieff (1926) considered the first counter-wave after the up-swing 1780–1810 (Figure 2.1). Parallel to this down-swing and a reduction of interaction scales there occurred a change from the previous rationalistic conception of the economy and society (enlightenment, classical economics), to the more irrational and metaphysical conceptions of Romanticism (Figure 2.1) in Europe. There was substituted for the timeless and ubiquitous rationality of the previous period, a *renewed emphasis on historical and territorial specificity* (for example the German 'historical' school of economics). The economy ceased to be considered as an autonomous field of activity, and was once more subordinated to normative societal and institutional constraints (e.g. by the 'institutionalists' such as Thornstein Veblen in the USA). In contrast to the previous rationally oriented period of the Enlightenment and of functionally oriented classical economics, aesthetic, metaphysical, qualitative, and structural elements—often defined in territorial terms—came to the foreground, such as man's relation to nature (Rousseau), Fichte's 'Closed Commercial State' aiming at maximum resource mobilization for equal supply to all population strata; List's theory of broad intersectoral 'mutually supporting' national economic development which, particularly for a country's intermediate agricultural–industrial transition phase, demanded the introduction of educational tariffs (List, 1840). There are many connections, by the way, between List's economic theory and self-reliant or 'dissociative' development (Senghaas, 1977). Free trade was considered admissible only between countries at equal stages of economic development, and only once they had articulated internal structures which could retain the vital elements required for development, letting dissipate to the outside only the *superavit*. Otherwise, it was pointed out, free trade would cause external dependency and instability of production (Hildebrand, 1848; Lehr, 1877). Beyond monetary price and market conditions, the state has to consider the national future and long-range development of an integral societal system (Schmoller, 1873).

In contrast to the individualism dominant in the preceding period of classical economics, group solidarity again became an important objective and in this sense constituted a transition to subsequent Christian social (and also Communist social) theories of the second half of the nineteenth century and subsequent periods.

Christian—especially Catholic—social theory, in fairly consistent form over time starting from the encyclica 'rerum novarum' of 1891 onwards, has emphasized principles of solidarity and subsidiarity, both in spatial and social terms. In spatial terms this has manifested itself in a call for federalism; in social and economic terms in the preference for a 'corporate' structure of society based on natural law (but not however dominated by the state, in contrast to Fascism). Communist social theory was much more exposed to changes over time. From a Socialism and Communism very much 'from below' in its early anarchic, utopian, and co-operative forms, it has moved increasingly to a socialism 'from above' in the form of state Communism and Leninism, intended to operate not only from the national but also from the international level downwards. It would be going far beyond the scope of this chapter to discuss the complex development of theory and practice of these doctrines in more detail. However we have all more recently observed a tendency towards the lowering of the scale of societal reform to national (Euro-Communism) and even small-group citizens' levels, both in the East and West. Citizens' initiatives in East and West, and the New Philosophers (particularly in France) rejecting the dominance of large-scale power, are symptomatic of this.

Kondratieff's second long wave
Returning to the earlier historical description, between *1850 and 1873* there occurred a rapid *expansion* both in spatial interaction scales and in economic growth along with Kondratieff's second up-swing (Figure 2.1), followed by the down-swing of *1873 to 1895*, and a concurrent *contraction* of international trade. Increasingly these changes in spatial interaction scales expressed themselves also in changes of economic growth rates and of capital investment.

Neoclassical era and Kondratieff's third long wave
From about 1895 up until the *First World War* came Kondratieff's third up-swing, coinciding with the neoclassical economic era (Figure 2.1).

World War I and the *inter-war period* of economic crises then led to an abrupt *breakdown of large-scale economic interactions*, and to a trend towards national economic self-sufficiency (Hilgerdt, 1942), as well as towards increased intranational political solidarity. This coincides with Kondratieff's (1926) third counterwave which in the event lasted until World War II (Figure 2.1). For many developing countries the breakdown of the civilian overseas shipping network and the changeover of most industrialized countries to a war or crisis economy triggered import substitution and initial industrialization, thereby reinforcing intranational economic circuits and intranational interaction patterns.

Post-World War II *Expansion*
In the *late 1950s* there occurred, particularly in the capitalist hemisphere, *a successive replacement of the foregoing dominantly territorial approach, again by a functional one*, both in economic and spatial development. This can be related to the efforts towards reconstruction of Europe after World War II, and to the emergence of neo-positivistic scientific thinking. It coincides with the projection of what would have been the up-swing in a new Kondratieff cycle, to begin around 1946 and last for about 25 years until about 1973 (Figure 2.1), at which time a new important turning-point would have had to be expected (Kaldor, 1977). It actually occurred in reality.

In economic development theory and practice this trend was underpinned by Perroux (1964) in an attempt to show that the economies of nations (in this case the Germany of post-World War II) could grow not only by expanding their territorial borders but also by functional input–output relations across them with external dynamic sectors. Hansen in Chapter 1 has excellently analysed this evolution of functional thinking in economic theory which constituted the basis for centre-down development strategies. In the field of spatial development theory and practice, Friedmann and Weaver (1979) have related this transformation of theory and practice to the introduction of the concept of the urban system as a national network of interaction nodes (rather than the previous concern for individual cities), and to the definition of systems of growth centres through which development was expected to 'filter down' to all parts of a national territory (Berry, 1972). Friedmann and Weaver (1979) have also demonstrated in concrete terms how, in the USA, regional planning in the late 1950s shifted its major concern from contiguous territorial space to discrete functional space, and how 'although surely nothing more than sheer coincidence, in exactly fifty years we have passed through a complete cycle of territorial dominance' (p. 7), from a territorially to a functionally oriented one and back again. In the last two chapters of their book Friedmann and Weaver present a similar finding for international development policy which, as they try to show, in the past decades has predominantly attempted to solve national and regional development problems and inequalities with what we here call strategies of development 'from above'.

This is where things stood around the *middle of the 1970s*. Since then the *call for a reversal of development theory and practice* which—as Friedmann and Weaver (1979) have shown—had already started at the end of the 1960s during the 'second development decade', became stronger and manifested in such thinking as the new periodical *Development Dialogue* (Dag Hammarskjöld Foundation, Uppsala, since 1975), in publications of the International Foundation for Development Alternatives (Geneva), in a great number of so-called 'basic-needs' strategies (however fuzzy the term may be) initiated several years ago by the International Labour Office, and successively introduced into the programmes of most international development organizations (such as the World Bank, the Continental Development Banks, the UN Regional Commissions, and UN affiliated planning institutes—particularly those for Asia and Africa).

Large-scale organizations such as the World Bank and the International Institute of Applied Systems Analysis (IIASA, Laxenburg, Austria) are trying to develop methods for a more balanced integration and maintenance of these complex systems by modelling approaches towards 'Redistribution with Growth' (Chenery *et al.*, 1974 and various successive applications of this model) and by the use of systems analysis (IIASA). A number of applied studies based on the 'Redistribution with Growth' model come to the conclusion that the ultimate requirements for the successful application of such large-scale balancing strategies are basic structural reforms within respective societies, accompanied by a redistribution of power, wealth, and income (Adelman and Robinson, 1978)—but none of these analyses indicates how such crucial structural transformations are to be brought about.

There is at the same time an increasing call for more 'self-reliant' development (however fuzzy this term also remains), initially for national units, but more recently for subnational regional and even rural communities as well (Haque *et al.*, 1977).

These efforts are all rather diffuse still, and it is impossible to do them justice in the present context, but their features are becoming consistently clearer and more coherent, and may well lead to what Friedmann and Weaver (1979) have called the imminent paradigm shift in development thinking and the doctrine of spatial development. This shift in development thinking is most clearly and radically manifested in the Third World by the present strong Islamization of some Arab countries and their repudiation of both Western and Eastern rational value systems. It is symbolized in Europe by symptoms of a 'new irrationalism', manifested for example by the New Philosophers who—comprising both former Marxist and Catholic personalities—oppose any kind of large-scale power structures, including the dominance of the state or of a positivistic science. These tie in with the European Romantic tradition of the nineteenth century and its protest against the state and a science made to dominate man (Coletti *et al.*, 1978).

We have thus shown that development doctrine in the past has alternated between different combinations of development 'from below' and 'from above'. Individual stages have at times gone to extremes on either side, and swung back again. This description does not intimate any historical determinism or any automatic (inescapable) historical periodicity; but there appears to be some association—though with varying intensity—between on the one hand predominantly rationally guided periods, rapid technological innovation, large-scale societal interaction patterns, large-scale formal organizations, emphasis on urban activities, a neglect of man-environment relations, a sub- or over-utilization of natural resources, and the decline of rural activities. These periods have many features of what we here call development 'from above'. On the other hand, predominantly metaphysically guided periods seem to be associated with social control of technological innovation, a narrowing of societal interaction scales, a preference for informal organization, an emphasis on rural activities, love of landscape and nature, and a general emphasis on man–environment relations. These periods have many features in common with what is here called development 'from below'.

The swing from periods of development 'from above' to those 'from below' in many cases seems to have been prompted by one or several of the following phenomena:

(1) the extension of interaction and dominance scales beyond the conflict-resolving capacity of the respective society;
(2) the emergence of few centres exerting decision-making powers over increasingly large areas pushed into dependency roles;
(3) resistance against pressure by the centres towards institutional and cultural penetration and uniformity;
(4) rapid increases in application of new technology to production processes and the transport media (Schumpeter, 1939), and the destabilizing effects these exert on societal structures (Laszlo, 1977);

(5) increasing disparities in living levels and wealth, in part objectively brought about by large-scale resource withdrawals and unequal benefits from commodity exchange, and in part subjectively caused by the increasing scale of communication and the ensuing trend towards uniformity of aspirations and perference patterns;

(6) changes in terms of trade between industry and agriculture to the disadvantage of the latter (explicitly shown for the Kondratieff down-swing 1873–96—cf. Lewis, 1978, p. 27);

(7) unequal or foreign utilization and the partial idleness of human, institutional, and natural resources (resulting from over-utilization in some areas, and under-utilization in others)—Axinn (1977), for example, considers the degree of integrated resource utilization as the major criterion for the under- or over-development of social systems);

(8) neglect of provision of basic needs for the whole population, and lack of broader systems-wide solidarity;

(9) increasing instability of systems due to lack of broader common societal norms.

Many of these points in fact are critical issues in today's worldwide problems. It is true that reversal of these trends in the past has often been accompanied by violence and bloodshed. This makes the early initiation of such changes all the more necessary, particularly if they are in the interests of the poor who make up the majority, before too great tensions have built up and while a smooth transition is still feasible.

We have dealt mainly in our analysis with the evolution of development doctrine in Europe. For long periods (at least until the end of the Middle Ages), the evolution of a comparable doctrine in other parts of the world may have followed different paths, and with varying cyclical time-sequences, such as Silberman (1978) describes for Judaism, and Axinn (1977) for ancient Hindu and Buddhist thinking. For the post-Colombian period in Mexico, such relations between economic political variables and scales of territorial integration have been shown by Friedmann and Gardels (1979). From the time of the European Enlightenment, industrialization, and capitalist expansion, however, the development doctrine of Europe (and later of North America and the Soviet Union) penetrated more and more into other parts of the world and has led to a relatively small number of worldwide development paradigms, all essentially operating with mechanisms 'from above'.

The present demand of many materially less-developed countries (as well as less-developed subnational areas and social groups) for more self-reliant development may in fact be an effort not only to escape from, or to change, the established economic and institutional channels of centre-down development, but also to change the present predominant value system oriented towards rationalistic and material objectives and steered by large-scale interaction and organizational systems.

This paradigm shift may be starting from the functional and geographical peripheries—far from the major nodes of the present global interaction system—which have been increasingly frustrated by this value system and the institutional structures sustaining it. Or is this just an idealistic notion of a few intellectuals both in industrialized and developing countries? The broad pressure in this direction on the part of many subnational, territorially organized ethnic groups on a concerted multi-national basis even in the European periphery (Seers *et al.*, 1979) seems a symptom of broad political relevance. Such an alternative strategy of development 'from below' might—at least for certain parts of the world (and possibly for a transitional period)—be the only way to escape from a downward spiral of increasing disparities and underdevelopment.

PROBLEMS OF TRANSITION FROM DEVELOPMENT 'FROM ABOVE' TO DEVELOPMENT 'FROM BELOW'

Most developing countries have historically been exposed to external interaction and colonial dependence, for varying lengths of time. The question therefore in most cases (except possibly for some isolated pre-capitalist societies) is not how to start a new development 'from below', but rather whether and how is it possible for developing countries and regions today to transform the past sectoral and spatial patterns of development 'from above' to incorporate more elements of development 'from below', and thereby reduce existing social and spatial disparities in levels of development. One of the problems of development 'from below' is that there is neither a uniform concept to strive for nor a uniform transitional process to follow. Each country and region may therefore have to devise its own strategy, although some basic characteristics will be the same.

Recent development 'from above' assumed that the ultimate objective for each country and region should be to reach a high degree of industrialization and urbanization resembling the structures of the most developed countries today, by a unilineal process of increasing the use of capital, technology, and energy, and by utilizing ever-increasing agglomeration and scale economies in order to participate with increasing specialization in the worldwide market according to their comparative advantages in factor endowment, which in fact rarely occurs precisely in this fashion as Streeten (1974, p. 262) and Stöhr and Tödtling (1978) indicate.

Alternatively, development 'from below' means the instituting of a diversity of structural objectives and transition paths—as discussed earlier, with the broadest possible participation of the respective communities. Development 'from below', however, does not mean a negation of growth objectives which, in view of the great material needs particularly of poor population groups and regions in developing countries, would be irresponsible. It means instead basing growth on increased and integrated resource mobilization in a regional context rather than on selective resource withdrawal under optimizing criteria derived from the world market.

The possibilities of changing formerly externally oriented economic functional (e.g. trade) and geographic structures (e.g. urban and transport networks) to more region-serving and equitable ones will vary.

Developing regions or countries with a high degree of external interaction (and possibly dependence), with small internal markets but high-value saleable resources, may in fact need to proceed at an accelerated pace through the raw-material export–capital goods import–export diversification sequence (described by Paauw and Fei, 1973, for 'small open economies' such as Taiwan and Thailand). Other larger regions or countries with sizeable potential internal markets, few saleable resources and strong autochthonous cultural traits, might find the change to a more self-reliant development strategy not only easier but also more useful in solving their own problems (various regions of India are cases in point). Institutional constraints unfortunately may be inverse. In large countries it may be more difficult to change existing social and economic structures unless heavily populated regions are still not integrated into the national economic and political system (Goulet, 1979).

Equally, countries with strongly outward-oriented urban and transport systems may have more difficulty moving towards self-reliant development than countries with internally oriented urban and transport systems. There seems to be a relatively strong relationship between the direction (inward or outward) of the dominant functional relations of an economic and social system and the spatial pattern of its urban and transport network (Appalraju and Safier, 1976; Taafe et al., 1973). Some countries have already reacted, perhaps intuitively, to this fact and have taken drastic steps towards changing the spatial structure of their urban and transport network, e.g. by transferring their national capital from an outward-oriented (coastal) location to an inward-oriented, central one (e.g. Brazil, Nigeria, Pakistan, Tanzania), although this in itself does not mean that the functional system has also been transformed. It will be interesting to see from case studies what an impact the re-location of the national capitals in these countries has on changing the entire spatial structure of the urban system, economic relations, and the spatial disparities in living levels.

Finally, doubt is often expressed that societal change towards more egalitarian social, economic, and political structures can be initiated from within local, regional, or even national societies. External influence transmitted through contact with agents of external change is frequently considered a prerequisite for bringing about more egalitarian social, economic, and political structures.

Contrarily, evidence that such change can also be initiated by internal forces is perhaps not so spectacular and therefore not as widely publicized as external intervention (an extreme of external influence); but such evidence exists. A few examples since World War II have been Yugoslavia, Cuba, Portugal, and Spain. Transformation in the first two has taken place despite the opposition of neighbouring external powers, while in Iberia only some complementary support from outside functional units (national or international political parties, trade unions, church organizations) was received. In the many cases, however, where societal change has been induced by outside agents (mainly supported by one or other of the two 'super-powers') it has not in fact led to more egalitarian structures, but rather to external imposition of a leading stratum as the link between the country and the external change agents in an effort to retain their influence. Such leading anti-egalitarian strata have been externally established both in capitalist and in socialist spheres of influence.

At the subnational level similar conditions apply. National liberation movements have often gained decisive momentum from peripheral subnational areas, and from there have reformed the entire national societal system. On the other hand many 'revolutions' initiated in the national capital have soon led to petrified national power structures, extended over peripheral regions by nationally appointed governors (even in formally democratic federal systems as found in various Latin American countries), by central planners or by centrally responsible regional administrators. It is therefore assumed that to operate in a sustained way, societal change towards more egalitarian structures, as well as towards broad economic, social, and political development, has to be supported from within the social system. The fact that inward-motivated, 'cellular' (Paine, 1976) or 'segmentary' (Renfrew, 1973b) societies can in fact have non-hierarchical, relatively egalitarian structures was demonstrated by the aforementioned authors for quite different time-periods and geographical areas, as well as for earlier (pre-colonial) social patterns on the continents analysed here, many of which are today called 'underdeveloped'. Examples are pre-capitalist agricultural societies in Asia (Haque *et al.*, 1977, pp. 16 and 51), in Africa (Brooks, 1971) and in some parts of pre-colonial Latin American societies (Clastres, 1974). Various studies in this book also refer to this fact (e.g. Chapter 8, Conyers on Papua New Guinea, and Chapter 9, Blaikie on Nepal).

A STRATEGY FOR DEVELOPMENT 'FROM BELOW' FOR LEAST-DEVELOPED REGIONS IN THIRD WORLD COUNTRIES

The available evidence as quoted above indicates that traditional spatial development policies (predominantly of the centre-down-and-outward type) in most cases have not been able—at least within a socially or politically tolerable time-span—to improve or even stabilize living levels in the least-developed areas of Third World countries. Alternative spatial development strategies therefore should be urgently considered. The idea of solving problems of spatial disparities in living levels in Third World countries by increased urbanward migration from peripheral areas has meanwhile been widely discarded in view of the utter futility of trying to absorb such masses of migrants into restricted urban labour markets, and supplying basic urban infrastructure and services for them.

As an alternative, the possibility of a development strategy 'from below' is discussed in this chapter. It might be particularly suitable for the many areas which in the near future cannot expect to benefit from traditional 'centre-down' development strategies. At the very least it can be used as a transitional strategy while competitiveness with the world economic system is improved, and it might be especially useful for subnational peripheral (predominantly rural) areas with the following characteristics (Stöhr and Palme, 1977, see also Chapter 5, p. 143):

(1) contiguous less-developed (predominantly rural) areas with relatively large population size providing a potential internal market for basic services and commodities;
(2) a low per capita resource base (natural resources or human skills) for worldwide demand;
(3) low living levels compared with other regions and distant from developed core regions;
(4) few internal dynamic urban centres able to absorb large rural populations into their labour market, infrastructure or service system;
(5) areas sufficiently different in sociocultural aspects from their neighbours to have a regional identity of their own.

Development 'from below' needs to be closely related to specific socio-cultural, historical and institutional conditions of the country and regions concerned. No uniform patent recipe for such strategies can be offered as is often done for strategies of development 'from above'. The guiding principle is that development of territorial units should be primarily based on full mobilization of their natural, human and institutional resources. In this context, the following elements are seen as essential components of development strategies 'from below':

(1) *Provision of broad access to land* and other territorially available natural resources as the key production factors in most less-developed areas; in many pre-capitalist African, Asian, and Latin American societies this was the case and sometimes still is, but it has been superseded since colonization by European systems of private landownership. Retention or introduction of fairly equal access to land and natural resources (e.g. by land reform) seems an essential prerequisite for equalizing income, achieving broad effective demand for basic services, and creating broad rural decision-making structures, essential for development 'from below'.

(2) *The introduction of new, or revival of the old, territorially organized structures for equitable communal decision-making* on the integrated allocation of regional natural and human resources. These territorially organized communal decision-making processes would have to extend also to the processing of regional resources, to regional allocation of surpluses generated from these activities, and the introduction of regionally adequate technologies (Goulet, 1979). Such structures for territorial governance should be established at different territorial levels and should operate on a mutual contracting basis (Friedmann, 1978, 1979), in which the decision-making potential of lower territorial units should be exhausted before higher levels become involved.

(3) *Granting a higher degree of self-determination to rural and other peripheral areas* in the utilization or transformation of existing (or the creation of new) peripheral institutions to promote diversified peripheral development in line with self-determined objectives-instead of utilizing primarily external institutions to promote development by externally defined (mainly core-region) needs and standards. This should lead to a greater balance between peripheral and core-region population groups in decision-making powers at all government levels. This should facilitate the retention in peripheral areas of a greater share of the surplus generated there, for the mobilization of own resources. An important role in this context could be played by 'agropolitan districts' suggested by Friedmann and Douglass (1978).

(4) *Choice of regionally adequate technology* oriented towards minimizing waste of scarce, and maximizing use of regionally abundant, resources. For most less-developed areas such technology would therefore have to save capital and facilitate full employment of the region's human, institutional and natural resources, as well as taking careful note of regional cultural patterns and value systems.

It should furthermore contribute to the recuperation of renewable, and the preservation of non-renewable, natural resources in the region, which Furtado (1978) sees as a starting-point in the struggle against dependence. Idle regional resources should be used, with the priority being to improve local and regional infrastructure, possibly by community self-help. Regionally adequate technology should furthermore, according to the region's conditions, 'be simple to learn and quickly implemented, and produce something for which there is a genuine demand' (Vacca, 1978). The application of labour-intensive technology and the improvement of basic infrastructure will at the same time promote the creation of broad effective demand, as Lele (1975, p. 189) suggests.

(5) *Assignment of priority to projects which serve the satisfaction of basic needs* of the population (food, shelter, basic services, etc.), using to a maximum regional resources and existing formal or informal societal structures. Such a priority—rather than the one given presently to export-base production—would reduce dependence on outside inputs and the backwash effects caused by them discussed earlier in this chapter. Thus the strengthening of local and regional economic circuits is emphasized (Santos, 1975) and the promotion of local and regional trade and service facilities for consumer goods and agricultural inputs, and regionally based activities are stressed such as food processing, animal breeding, power supply, repair of agricultural machinery, etc., preferably by co-operative local or regional groups (Haque *et al.*, 1977, p. 64).

(6) *The introduction of national pricing policies* which offer terms of trade more favourable to agricultural and other typically peripheral products.

(7) In case of peripheral resources being insufficient for satisfaction of peripheral (e.g. rural) basic needs, *external* (national or international) *assistance* would be solicited but should be mainly considered as compensation for the eroding effects of previously emerging dependencies; this external assistance should be oriented mainly towards (a) the full productive utilization of peripheral human and natural resources, (b) the satisfaction of basic needs of the peripheral population, (c) the improvement of intra-peripheral transport and communication facilities, and (d) the formulation and implementation of locally initiated social and economic projects for the further processing of peripheral resources and the satisfaction of basic needs; the allocation of such external assistance should be made with majority participation of local and regional decision-making bodies;

(8) *Development of productive activities exceeding regional demand* (so-called export-based activities) should be promoted only to the extent that they lead to a broad increase in living levels of the population of the territorial unit. Otherwise, 'the monetary wealth it can get in excess of the needs of national (regional, local) development stimulates conspicuous consumption, investment in ... prestige projects, ... moreover the surpluses are deposited or invested mainly in industrialized nations, regions or centres' (Abdalla, 1978). Priority should be given to activities which promote (a) full employment of regional labour and natural resources; (b) the application of technology to safeguard full employment of regional resources; (c) competitiveness in extra-regional markets by qualitative product differentiation, rather than by purely quantitative price competition in standardized mass production; (d) particular promotion should be given to small-scale labour-intensive activities in rural areas, building on autochthonous technologies or developing them further, as these usually permit the most efficient use of all locally available resources, a maximization of output with available resources, the greatest adaptability to specific regional demand, and broad distribution of the income generated; (e) if competitiveness on external markets seems to require introduction of large-scale technology, the adaptation of locally or regionally available technology and improvement of local production conditions by local resource inputs (including communal self-help) should be attempted first.

(9) *Restructuring of urban and transport systems* to improve and equalize access of the population in all parts of the country to them, rather than strengthening such systems oriented to the outside. If both these systems are heavily oriented externally, this may require a relocation of major urban functions (including national government) from peripheral (e.g. coastal) locations to the interior of the country. This should also apply to intermediate-level functions and the improvement of intraregional accessibility within peripheral areas in order to improve their relative starting position. In case of competing demands between such functions at national and subnational levels, preference would have to be given to the improvement of intermediate

urban functions in peripheral areas (e.g. educational facilities oriented to specific regional development potentials) able to retain development impulses within the region, rather than giving priority to national functions and national transport integration which might drain development potential into the core regions.

(10) *Improvement of rural-to-rural and rural-to-village transport and communications facilities* (rather than the present priority given to rural-to-large-urban communications) should have preference, in order to (a) increase commodity and service markets within peripheral areas—thus permitting realization of economies of scale and a better supply of basic services to all parts of the population; (b) increase the scale and diversity of factor markets within peripheral areas in order to reduce production costs; and thereby (c) facilitate increased processing, purchasing, and marketing activities in peripheral areas sustained by peripheral groups, and a more decentralized national pattern of capital accumulation with greater participation of peripheral areas.

(11) *Egalitarian societal structures and a collective consciousness* are important prerequisites for a strategy for development 'from below'. They should by preference be retained or initiated through internal (local or regional) initiative. Where this is not possible, external support may be necessary. This applies at different levels of society, such as the level of the small group (basic rights of children, women, etc.), or that of local or regional units (reduction of the power of vested territorial interest groups). Where such external support 'from above' is necessary it should be facilitated, both between and within countries, preferably on a non-government basis, through ideological or religious groups, organizations of committed intellectuals, through 'link cadres' (Haque *et al.*, 1977), or Independent Volunteer Cadres (Monlik, 1979)—as local or regional government representatives will tend to feel responsible to their hierarchical superiors rather than to their target groups (Haque *et al.*, 1977).

ON THE FEASIBILITY OF DEVELOPMENT 'FROM BELOW'

The most frequent reaction to proposals like the above, even from professionals open to change, is one of 'scepticism about the concrete possibility of implementing such an attractive change of world conditions' (Vera Cao–Pinna, personal communication, November 1978).

There is nothing more convincing than reality. Let me therefore point to some instances where approaches similar to those proposed above have actually been implemented with some success. Most of these examples are on a relatively small regional scale at subnational level. Experiences on such a small scale are usually not recorded centrally—much in line with the macrostructure of our dominant information systems—and in fact many more probably exist all over the world. Only recently, institutions such as the International Foundation for

Development Alternatives (IFDA), Geneva, or the International Research Centre on Environment and Development (CIRED) in Paris, have started to collect and report systematically on such efforts of small-scale alternative development. Further examples are given by Lo and Salih (Chapter 5 in this volume, p. 136).

Here are some examples which have been more systematically reported upon:

(1) In Asia the Santhal Movement in Bihar, India (Haque *et al.*, 1977, pp. 20 ff.), based mainly on local common use of implements and labour and on common local contributions in kind for night school, political activities, etc.

(2) The Bhoomi Sena (Land Army) movement by some groups of the Adivasi aboriginal inhabitants of India who in total comprise about 50 million people (Silva *et al.*, 1979), based mainly on regaining access to land, the abolition of bonded labour and creation of village forums.

(3) The Rangpur Self-reliant Movement in Bangladesh (Haque *et al.*, 1977, pp. 32 ff.), based mainly on communal decision-making on the use of land and its proceeds, on the communal purchase of agricultural inputs, the reduction of dependence from external loans, and communal labor for economic infrastructure.

(4) The Sarvodaya Sharamadana Movement in Sri Lanka (Goulet, 1979, p. 6), based on student volunteer participation in local infrastructure work and on motivating local community deliberation and action.

(5) The Quechua Alto Valle co-operatives in central Bolivia (Goulet, 1979), oriented mainly towards the co-operative production of woollen products and ceramics; the decisions of the co-operative are subordinated to the broader interests and values of the entire territorial community: this restricts the introduction of new technologies to those 'which are in harmony with ancient Quechua rural values of mutual help and sharing the benefits in all improvements', and to 'assigning a share of the surplus to all members of the village, whether they belong to the co-operative or not' (Goulet, 1979, p. 13).

(6) An Agricultural Development Project for the Altiplano in the Peruvian province of Puno (Morlon, 1978), oriented towards mobilizing the endogenous development potential of large highland areas and their agricultural and forest resources (including agricultural infrastructure such as extensive terraces, etc.) which have become depleted by changes in the natural and in the social environment (including the take-over by large sheep-breeding 'haciendas').

(7) Different programmes for the evolution of rural space in mountain areas and unproductive Mediterranean regions of the Pyrenees and various other French regions (Poly, 1977) involving the integrated transformation of agrarian structures, land-use policies and modifications in production systems oriented mainly towards sylvo-pastoral activities; similar programmes are being undertaken by Longo Mai co-operatives in Costa Rica based on ex-

periences of this international group of co-operatives in mountain areas of the French Provence, of Switzerland and southern Austria.

Many of these examples are explicitly built on autochthonous value systems and principles of community collaboration, decision-making and the sharing of proceeds. In a number of them, for example in the Quechua co-operatives of Bolivia, rational decisions—such as the introduction of new technologies— are explicitly subordinated to the community's wider and more basic cultural needs (Goulet, 1979, p. 13).

Most of these examples are relatively small in scale, although some of them are spreading to neighbouring areas or to other communities suffering from similar phenomena of underdevelopment in their present relations with the outside economy. Such movements 'generated internally from within . . . their own experience (of the daily life of its villagers) are like bubbles that will disappear and tend to be forgotten unless external forces . . . sustain them' (Haque *et al.*, 1977, p. 35). Lack of support from higher levels (particularly national) may be a major reason why such examples have hardly developed on a larger scale (except temporarily in China). A recent case where autochthonous local development appears to be successfully supported by the state is in West Bengal in India, via the reformed revival of traditional local *panchayat* institutions (Marshall, 1979). While general support needs to come from national governments, concrete action should preferably come from independent groups such as 'committed link-cadres' (Haque *et al.*, 1977, p. 61) as mentioned above. The major limiting factor for expansion of such initiatives today seems to be fear by central (national or international) decision-making centres that they might lose control and their competitive efficiency *vis-à-vis* other states, enterprises, or power blocks.

Does this mean that the success of development 'from below' will to a great extent hinge—even if more equitable social structures can be achieved at different subnational and national levels—on the reduction of international tension and competition, particularly between super-powers, and successively also between nation-states, regions, and ultimately between individuals?

REFERENCES

Abdalla, J. S. (1978) Heterogeneity or differentiation—the end of the Third World?, *Development Dialogue*, **2**, 3–21.

Adelman, I. and C. T. Morris (1973) *Economic Growth and Social Equity in Developing Countries* (Stanford: Stanford University Press).

Adelman, I. and S. Robinson (1978) *Income Distribution Policy in Developing Countries*, Published for the World Bank (Oxford: Oxford University Press).

Allardt, E. (1973) *About Dimensions of Welfare*, Research Report No. 1 (Helsinki).

Appalraju, J. and M. Safier (1976) Growth centre strategies in less-developed countries, in A. Gilbert (ed.), *Development Planning and Spatial Structure* (London: Wiley & Sons).

Axinn, G. H. (1977) The development cycle: new strategies from ancient concepts, *International Development Review*, **XIX** (4), 9–15.

Berry, B. J. L. (1972) Hierarchical diffusion: the basis of development filtering and spread in a system of cities, in N. M. Hansen (ed.), *Growth Centres in Regional Economic Development* (New York: The Free Press).

Brooks, L. (1971) *Great Civilizations of Ancient Africa* (New York: Four Winds Press).

Chenery, H. H. S. Ahluwalia, C. L. G. Bell, J. H. Duloy, and R. Jolly, (1974) *Redistribution with Growth* (London: Oxford University Press).

Clastres, P. (1974) *La Societé contre l'Etat* (Paris: Les Editions de Minuit).

Coletti, L., A. Glucksmann, and B. H. Lévy (1978) Ich bin für das Licht—aber nicht für die Festbeleuchtung des Staates, *Der Neue Irrationalismus* (Reinbeck bei Hamburg: Literaturmagazin 9, Rowolt), 13–36.

Donnithorne, A. (1972) China's cellular economy: some trends since the cultural revolution, *China Quarterly*, No. 52 (Oct.–Dec. 1972), 605–19.

Friedmann, J. (1978) Concerning contract programs, *Ecodevelopment News*, No. 7 (December), 3–4.

Friedmann, J. (1979) The active community: towards a political–territorial framework for rural development in Asia, *Comparative Urbanization Studies* (Los Angeles: University of California), mimeo.

Friedmann, J. and M. Douglass (1978) Agropolitan development: towards a new strategy for regional planning in Asia, in Lo, F. C. and K. Salih (eds.), *Growth Pole Strategy and Regional Development Policy* (Oxford: Pergamon Press), 163–92.

Friedmann, J., N. Gardels, and A. Pennink (1979) The politics of space: regional development and planning in Mexico. Paper presented at the International Seminar on National Development Strategies for Regional Development in Latin America, Bogota, September 1979.

Friedmann, J. and C. Weaver (1979) *Territory and Function: The Evolution of Regional Planning* (London: Edward Arnold).

Furtado, C. (1978) *Criatividade e Dependencia na Civilização Industrial*, (Rio de Janeiro: Paz e Terra)

Goulet, D. (1979) Development as liberation: policy lessons from case studies, International Foundation for Development Alternatives, *Dossier* 3, January (Geneva).

Greenbie, B. B. (1976) *Design for Diversity* (Amsterdam: Elsevier Scientific Publishing Company).

Hall, E. T. (1966) *The Hidden Dimension* (New York: Doubleday).

Hanappe, P. and M. Savy (1978) Transports industriels, internationalisation de l'economie et mouvements de Kondratieff, Prospective et Amenagement, Montrouge (mimeogr.)

Haque, W., N. Mehta, A. Rahman and P. Wignaraja (1977) Towards a theory of rural development, *Development Dialogue* 2.

Hechter, M. (1975) *Internal Colonialism, the Celtic Fringe in British National Development* (London: Routledge & Kegan Paul).

Heckscher, E. F. (1955) *Mercantilism* (London: Allen & Unwin, 1st. Swedish edn., Stockholm, 1931).

Heinrich, W. (1964) *Wirtschaftspolitik* (Berlin: Duncker & Humblot).

Hildebrand, B. (1848) *Nationalökonomie der Gegenwart und Zukunft* (Frankfurt/Main).

Hilgerdt, F. (1942) *The Network of World Trade* (Geneva: League of Nations).

Hirschman, A. (1958) *The Strategy of Economic Development* (New Haven and London: Yale University Press).

Hoselitz, B. F. (1954) Generative and parasitic cities, *Economic Development and Cultural Change*, **3**, 278–94.

Kaldor, N. (1977) Structural causes of the world economic recession. (mimeo.).

Kondratieff, N. D. (1926) Die langen Wellen der Konjunktur, *Archiv für Sozialwissenschaft und Sozialpolitik*, December.

Lasuén, J. R. (1973) Urbanization and development—the temporal interaction between geographical and sectoral clusters, *Urban Studies*, **10**, 163–88.

Laszlo, E. (1974) *A Strategy for the Future* (New York: Braziller)

Laszlo, E. (1977) *Goals for Mankind. A Report to the Club of Rome* (New York: E. P. Dutton).

Lehr, J. (1877) *Zollschutz und Freihandel* (Berlin).

Lele, U. (1975) *The Design of Rural Development: Lessons from Africa*, published for the World Bank by Johns Hopkins University Press, Baltimore.

Lewis, W. A. (1978) *Growth and Fluctuations 1870–1913* (London: George Allen & Unwin).

List, F. (1840) *Das Nationale System der Politischen Ökonomie* (Berlin).

Machlup, F. (1977) *A History of Thought on Economic Integration* (London: Macmillan).

Magdoff, H. (1978) Colonialism, European expansion since 1763, *Encyclopaedia Britannica*, Macropaedia, vol. 4, 890.

Marshall, T. (1979) Power to rural poor: local councils revived in India grass roots move, *Los Angeles Times*, September 15

Mill, J. S. (1848) *Principles of Political Economy* (London).

Montelius, O. (1899) *Der Orient und Europa* (Berlin: Asher & Co.)

Monlik, T. K. (1979) Strategies of implementation of rural development in India. International Foundation for Development Alternatives, *Dossier* 5, March (Geneva).

Morlon, P. (1978) Agricultural development project for the Altiplano in the Peruvian province of Puno, *Ecodevelopment News*, CIRED., International Research Center on Environment and Development (Paris), No. 5 (June), 3–19.

Müller, A. (1809) *Elemente der Staatskunst* (Berlin).

Nerfin, M. (ed.) (1977) *Another Development: Approaches and Strategies.* (Uppsala: Dag Hammarskjöld Foundation)

Paauw, D. S. and J. C. H. Fei (1973) *The Transition in Open Dualistic Economies* (New Haven: Yale University Press)

Paine, S. (1976) Balanced development: Maoist conception and Chinese practice, *World Development*, **4**, (4), 277–304.

Perroux, F. (1964) *L'économie du XXème siécle* 2nd ed. (Paris: Presses Universitaires de France).

Poly, J. (1977) Recherche agronomique. Réalités et perspectives, Institut National de Recherche Agronomique, 27–28 June (Paris) (mimeo). Excerpt translated under title: Ecodevelopment for underdeveloped regions, in *Ecodevelopment News*, CIRED (Paris), No. 5 (June 1978), 38.

Pred, A. (1977) *City Systems in Advanced Economies* (London: Hutchinson).

Renfrew, C. (1973a) *Before Civilization. The Radio-Carbon Revolution In Prehistoric Europe* (London: Jonathan Cape).

Renfrew, C. (ed.) (1973b), *The Explanation of Cultural Change, Models in Prehistory* (London: Duckworth).

Renfrew, C. (1976) Megaliths, territories and population, *Acculturation and Continuity in Atlantic Europe*, Dissertationes Archaelogicae Godenses (ed. Laet), de Tempel (Belgium: Brugge), **XVI**.

Richardson, H. W. (1973) *Regional Growth Theory* (London: Macmillan).

Rostow, W. W. (1978), *The World Economy, History and Prospect* (Austin: University of Texas Press).

Sachs, J. (1976) *Ecodevelopment* (Rome: Ceres) **42**, 8–12.

Samuelson, R. J. (1978) The long waves in economic life, *Economic Impact*, A Quarterly Review of World Economics; Washington), No. 26, 60–64.

Santos, M. (1975) *L'Espace Partagé—les deux circuits de l'économie urbaine des pays sous-développés*. Editions M.-Th. Génin—Libraires Techniques (Paris).

Schmoller, G. (1873) Verhandlungen der Eisenacher Versammlung zur Besprechung der sozialen Frage, Opening Assembly of the *Verein für Socialpolitik* (Leipzig).

Schumpeter, J. (1934) *The Theory of Economic Development* (Cambridge, Mass. : Harvard University Press).

Schumpeter, J. (1939) *Business Cycles* 2 vols. (New York: McGraw-Hill).

Seers, D. (1977) The new meaning of development, *International Development Review*, 3, 2–7.

Seers, D., B. Schaffer, and M. L. Kiljunen (1979) *Underdeveloped Europe: Studies in Core-Periphery Relations* (Hassocks, Sussex: The Harvester Press).

Senghaas, D. (1977) *Weltwirtschaftsordnung und Entwicklungspolitik, Plädoyer für Dissoziation* (Frankfurt/Main: Suhrkamp).

Siegenthaler, H. (1978) Capital Formation and Social Change in Switzerland 1850 to 1914, *Jahrbücher für Nationalökonomie und Statistik*, Stuttgart, 193, (1); 1–29.

Silberman, L. H. (1978) Judaism, *Encyclopaedia Britannica*, Macropedia, vol. 10 (Chicago), 288 ff.

Silva, G. V. S. de, N. Metha, A. Rahman, and P. Wignaraja (1979) Bhoomi Sena—a struggle for people's power. International Foundation for Development Alternatives, *Dossier 5*, March (Geneva).

Snead, W. (1975) Self-reliance, internal trade and China's economic structure, *China Quarterly*, No. 62 (June), 302–07.

Stöhr, W. and H. Palme (1977) Centre–periphery development alternatives and their applicability to rural areas in developing countries. Paper presented at the Joint LASA/ ASA Meeting, Houston, November.

Stöhr, W. and Tödling, F. (1977) Evaluation of regional policies—experiences in market and mixed economies, in Hansen, N. M. (ed.), *Human Settlement Systems*, International Institute of Applied Systems Analysis (New York: Ballinger).

Stöhr, W. and F. Tödtling (1978) 'Spatial equity—some antitheses to current regional development strategy. Papers of the Regional Science Association, vol. 38, pp. 33–53. Reprinted in modified form in H. Folmer and J. Oosterhaven (eds.) 1979. *Spatial Inequalities and Regional Development* (Leiden: Nijhoff).

Streeten, P. (1974), On the theory of development policy, in Dunning, J. (ed.), *Economic Analysis and the Multinational Enterprise*, pp. 252–79. (London: George Allan & Unwin Ltd.)

Taafe, E. J., R. Morrill, and P. R. Gould (1973) Transport expansion in the underdeveloped countries: a comparative analysis, in Hoyle, B. S. (ed.), *Transport and Development*, (London: Macmillan).

Tinbergen, J. (coord.) *et al.* (1976) *Reshaping the International Order*, A Report to the Club of Rome (New York: E. P. Dutton).

Touraine, A. (1976) *Was nützt die Soziologie* (Frankfurt/Main: Suhrkamp).

Uphoff, N. T. and M. J. Esman (1974) *Local Organization for Rural Development: Analysis of Asian Experience.* Rural Development Committee, Center for International Studies, Cornell University.

Vacca, R. (1978) *Tecniche Modeste per un Mondo Complicato* (Milano: Rizzeli).

Chapter 3

Development Theory and the Regional Question: A Critique of Spatial Planning and its Detractors*

CLYDE WEAVER

The centre-down model of development was born in the period immediately following World War II, in part as a response to the imminent collapse of Europe's crumbling world empire. The break-up of traditional colonialism, with its dependence on finance capital, the Royal Navy, and various trading companies and colonists from Britain, France, Belgium, Holland, and Portugal required a new mechanism to integrate the southern and eastern hemispheres into the international economy. South-east Asians were understandably loath to welcome back their old taskmasters after the ebb of Japan's 'Co-prosperity Sphere'. British India fell to Gandhi's gentle revolution; and Africa seethed with growing discontent. Working through the newly formed United Nations, the *Pax Americana* brought in its wake the Myth of Modernization: urban industrialization through direct foreign investment; the development of indigenous entrepreneurs and domestic capital accumulation; the creation of manufacturing jobs; and eventually, the spread of prosperity and a dawning age of high mass consumption.

This image of following in the footsteps of the industrial West was very powerful, especially for the new national leaders of Africa and Asia, trained as they were at 'Oxbridge', the Sorbonne, and Harvard. With their pride soothed

*This chapter was written in part while the author was a Visiting Fulbright Lecturer at the Centre d'Economie Régionale, Université d'Aix-Marseille. I would like to thank John Friedmann, Derek Diamond, J. C. Perrin, François Perroux, Bernard Planque, Walter Stöhr, and D. R. F. Taylor for their comments on various earlier drafts. Thanks are also due Dudley Seers for acquainting me with recent core–periphery research in Europe, and to Elizabeth Lebas and Robert Kraushaar for trying to help me unravel current regional science developments in the United Kingdom.

and the fires of nationalism partially dampened, the doctrine of polarized development spread the good news that through hard work the benefits of Western civilization could be had at home. Furthermore, the new economic theories clearly justified the key role of these selfsame personalities in national development. Modern men—touched by the Enlightenment and the revolutionary ideas of nineteenth-century Europe—would be needed to point out the path, to help forge new ways of life, and start the difficult task of capitalist savings and investment which would lay the foundations for universal betterment.

This ideology also suited businessmen and government officials in the West. To Washington and the American corporate leadership it clarified and sanctioned their new role in fuelling world economic growth. (The over-stoked American economy had emerged from World War II not to the much-feared return of depression, but rather to a clamouring world market—without competitors.) To the reeling heads of European élites the torch of world domination had passed west-ward all to quickly: physical destruction; humiliation in war; insurrection abroad; a new, unfamiliar dependence: the new economic theories soothed their pride as well, promising a continuing if different role in world affairs.

Spatial planning arrived late to this scenario (Gilbert, 1976, pp. 1–19). Although national planning had been part and parcel of the centre-down model of development from the outset, recognition of the relationship between sectoral planning and the geography of modernization was long in coming. The spatial dimension of development was not an important theme for neoclassical economists, nor were regional differences in the outcomes of development popular topics among politicians in the emerging nations, bent on national integration and political control. It was not until around 1960 that the regional aspects of economic planning attracted significant attention. This came about largely in reaction to the practical problems of project planning (Hirschman, 1958, 1963, 1967) and the political need for some visible concern about geographically balanced national growth (Friedmann, 1977). Latin America provided some of the best-documented examples of this movement, and 'Northern' ideas provided the appropriate spatial planning model: a model founded on the theory of polarized economic development, the idea of regions as open systems within the broader national economy, and the concept of spatial planning from above.

In this chapter, first, I will trace the origins of the top-down approach to regional development and show how it has come under attack today, predominantly through ideas which had their origin in the underdeveloped world. Then I will argue for an alternative bottom-up approach which emphasizes territorial development, a strategy aimed towards meeting the needs of regional population groups, as opposed to hastening their functional integration into the global economic system. This perspective carries on the best historical traditions of regional planning, complementing Stöhr's analysis in Chapter 2, although it is in evident conflict with the synthesis presented by Hansen in Chapter 1, as well as the predominant themes of recent neo-Marxist criticism.

DEVELOPMENT FROM ABOVE: LIBERAL THEORIES OF ECONOMIC DEVELOPMENT

Before we can examine the spatial component of the centre-down paradigm of development, we must have a clear idea of its theoretical context (i.e. the modern theory of capitalist economic growth) and how it was applied historically to the Third World. In this discussion I draw heavily from the analyses presented by Friedmann (1977) and Friedmann and Weaver (1979, pp. 89–113).

The major concepts in economic growth theory had been accumulating since the 1930s. The most important of these were: (1) the crucial role of innovation and the private entrepreneur in development (Schumpeter, 1934, orig. 1912); (2) the idea that government could and should intervene in the market economy to achieve full employment (Keynes, 1936); and (3) the concept of national accounts as a tool for measuring a country's economic performance (Clark, 1938; Kuznets, 1941). These ideas were synthesized and applied to the newly emerging nations in 1951 by an international panel of economists called together by the United Nations. The UN experts made the now-common distinction between developed and underdeveloped countries, and argued that there was only one course open to former colonial areas if they were to achieve the high standards of living of the Western world: they must attempt to replicate the recent economic history of the already industrialized nations. They must 'modernize' their social and economic structures so as to achieve a high rate of capital accumulation and industrialization. The appropriate indicators of progress in this battle were said to be a savings rate of 12 per cent per annum and a steady increase in GNP per capita.

Four keys to success were identified. First, national economies must be open to trade and economic stimulation from the world economy. This meant that the developing countries would continue providing raw materials for the core economies of the industrial world, and that in return they would be the beneficiaries of foreign investment, grants-in-aid, and technological assistance. Second, modernization could only be accomplished through a process of urban industrialization. It was only through the development of urban-based manufacturing industries that the creation of new jobs could keep pace with population growth, increasing foreign earnings and the per capita growth rate of GNP. Agriculture would be modernized and made more efficient as rural/urban migration absorbed surplus labour in the growing cities and the rural economy could be restructured along efficient, capital-intensive lines. Third, going back to Joseph Schumpeter's arguments about the propulsive effect of entrepreneurship, capital must be concentrated in the hands of a naturally small number of progressive-minded élites, who would have a high propensity to save their earnings and reinvest in further industrial development. For economic growth to occur, development, by definition, must be 'unequal'; the question of 'efficiency' was all-important and any consideration of economic equity must wait until modernization had been achieved. Finally, national economic planning was seen as a concomitant to economic growth. Only through a rational process of allocating scarce resources could the necessary concentration of productive forces be assured, with economic entrepreneurs providing the energizing force behind industrial expansion, and government making the required supporting investments in public services and infrastructure.

This 'success' model of unequal development was elaborated by Lewis (1955), but it was formulated most succinctly, if simplistically, in 1961 by Walt Rostow as *The Stages of Economic Growth: A Non-Communist Manifesto*. In Rostow's much-criticized but influential essay, published from a series of lectures he gave at Cambridge University in the autumn of 1958, he laid out a simplified step-model of economic modernization, arguing that development was a unidirectional process, and that underdeveloped countries could follow in the path of the industrialized world, eventually reaching the post-maturity stage of high mass consumption. After a period of transition, this goal would be reached through what Rostow called 'take-off' into self sustaining economic growth.

Thus, for economists, modernization was a process of unequal development in which productive wealth and capital were to be concentrated in the hands of a small urban élite. People were to be pushed into the cities to provide the necessary labour force, and agriculture was to be restructured in an efficient corporate pattern.

REGIONAL PLANNING AS SPATIAL DEVELOPMENT

The theory of economic development and modernization was an urban-dominated paradigm, although early writers, clinging to the traditions of neoclassical economics, failed to elaborate the implications of this fact. The locational aspects of development, as well as an explicitly subnational adaptation of economic growth theory, came into the picture through regional planning.

The lineage of regional theory can be traced back directly to the work of the classic utopian socialists and anarchists, as well as civic reformers and early city planners. The first regional planners, writing in the 1920s and 1930s—men like Lewis Mumford, Benton MacKaye, and Howard Odum—were deeply influenced by the ideas of Patrick Geddes, Ebenezer Howard, Peter Kropotkin, Pierre-Joseph Proudhon, and, of course, Charles Fourier and Robert Owen. But the first paradigm of regional planning was radically transformed in the postwar era, and the contemporary doctrine of regional development only dates back a little over 20 years. (See Friedmann and Weaver, 1979 and Weaver, 1979 for the evolution of early regional planning ideas.)

It was in the decades following World War II that a direct concern for economic development was combined with theories purporting to explain the location of economic activities, giving birth to regional science and spatial development planning. On the American side, the main architects of this synthesis were Douglass North, Walter Isard, and John Friedmann (Friedmann,

1955; North, 1955; Isard, 1956). Working independently in Europe, François Perroux, Jacques Boudeville, and Jean Paelinck developed what was to become a complementary line of reasoning (Perroux, 1950, 1955; Boudeville, 1961, 1966; Paelinck, 1965).

Douglass North, an economic historian at the University of Washington, proposed that regional economic growth takes place in response to exogenous demand for regional resources. Although Alvin H. Hansen and Harvey S. Perloff had presented most of the fundamental components of North's argument over a decade earlier, their discussion of natural resources and regional development was not presented within the emerging economic development paradigm (see Hansen and Perloff, 1942). North divided the regional economy into two sectors: basic and residentiary. The basic sector was said to be the propulsive force in economic growth, supplying needed inputs to the national economy and thus bringing outside wealth into the local economic system. Multiplier effects from the expansion of basic activities would lead to the development of local service industries, nourished by linkages to various stages of production for export and final demand for consumer goods. The similarities between these ideas and the general theory of economic growth are self-evident. The basic assumptions of North's model were twofold. First, growth is a function of external capital accumulation and subsequent interregional flows of revenue and capital, and second, trade in a market economy proceeds on the basis of comparative advantage and equal exchange. Although North's theory was stated in terms of trade relations between discrete regional units, the concepts of regional complementarity and increasing regional interdependency clearly pointed to a continuing process of functional integration of the national space economy. It was posited that the prosperity and well-being of people living in one region were, in fact, dependent upon the conscious elaboration of economic ties with people living in other areas.

Even given the conceptual simplicity of North's export-base notion, it has provided the touchstone for most subsequent theorizing about regional growth (Perloff *et al.*, 1960; Borts and Stein, 1962; Thompson, 1968; Richardson, 1969, 1973a, 1978; Perrin, 1974). *Increasing the territorial division of labour, decreasing the friction of distance, and augmenting the level of interregional trade are still usually said to be the keys to local economic growth and development.* Stated another way, it is argued—or perhaps more correctly it is taken as a fundamental truth—that people cannot satisfy their own human needs and provide themselves with an increasing level of prosperity through their own labour and the use of their own resources.

Walter Isard, the founder of the Regional Science Association, accomplished the feat of integrating the German tradition of location studies into the framework of neoclassical economics. Notwithstanding prior efforts by Edgar Hoover (1937, 1948), it was not until the appearance of Isard's *Location and Space Economy* in 1956 that the English-speaking world began to consider seriously the spatial dimension of economic activities. Following the lead of Alfred Weber (1928), Isard placed the emphasis of his work on what he called the *transport inputs* to production. He argued that the cost of overcoming the friction of distance should be considered of equal importance with the traditionally recognized production factors: labour, resources, and capital. Applying the equilibrium doctrine of neoclassicism to the space economy, Isard argued that market mechanisms would arrange economic activities in their optimal, profit-maximizing locations, creating an hierarchical economic landscape based largely on substituting transport costs for other production inputs.

This was an abstract, open-system model of the space economy. Ideographic problems such as natural resource locations and political boundaries were largely assumed away, leaving the two primal economic forces—production and consumption—to be balanced off in locational terms. All other things being equal, this logic suggested that eventually, all economic activities should gravitate towards the same selected set of locations. Ultimately the locational problem would be solved through development of an urban network of nodes and linkages, closely resembling August Lösch's earlier formulation (1954).

Cities became point locations exerting varying amounts of economic attraction, depending primarily on their size. Their main locational characteristic was their relative situation *vis-à-vis* other places in a theoretically unlimited urban system. All connections with the concrete world of cities and regions with proper names and individual identities were lost, etheralized into the *n*-dimensional realm of economic space.

Two critical assumptions lay at the heart of Isard's theory. First, to create such a neat functional ordering of economic activities it was necessary to postulate that all the factors of production are more or less foot-loose, moving around freely in response to private economic decisions. Secondly, these micro-scale decisions, meant to maximize profit at the level of the individual firm, must also contribute to achievement of aggregate scale economic efficiency. It is clear that such an economy would be predominantly urban-centred. What is more, by definition, this would mean a high degree of polarization in the location of people, resources, and capital, as well as a marked dependence on relatively cheap transportation. Overcoming the friction of distance became the major imperative in an Isardian world.

John Friedmann was among the first to become interested in the policy implications of such a theory for regional economic development. If aggregate economic efficiency were dependent on a highly integrated urban economy, then the best way to achieve regional economic growth was through encouraging development of the urban system. This meant that regional planning must become spatial systems planning, and that the main concern of planners should be optimizing the location of economic activities (Friedmann, 1955, 1963).

In emphasizing the importance of locational decision-making to regional development Friedmann had cast his lot with the budding multi-disciplinary field of regional science. In doing so he unavoidably borrowed its central intellectual traditions. Regional planning as a field of study and professional practice was to interest itself in economic location theory, central place studies, urbanization, and regional economic development. Its methods were to be rigorously scientific and its goal was to be the functional integration of the space economy, concentrating people, resources, and economic activities in a tightly woven network of cities and their adjoining regions. As we will see below, this notion of spatial systems planning, when combined with corollary ideas about scale economies, polarization, and unequal development, has provided the dominant framework for debate about regional growth until the present day.

As should be apparent from the foregoing discussion, there was a basic internal contradiction among the several components of regional development theory. The line of reasoning began with North's proposition that economic growth takes place through the stimulus of trade between different regions. What North had in mind were discrete, bounded territorial units which would set in motion a process of 'spatial interaction', based on complementary economic resources. Isard's theory of location and the space economy focused on the flows and linkages between these different places, recognizing the overwhelming importance of cities in the industrial economy but abstracting them out of recognizable geographical space into a mathematical world of nodes and networks. Friedmann grafted the North/Isard models together to forge an eclectic strategy for regional development: capturing the manifold benefits of trade and regional complementarity by expanding and tightening the bonds of the urban spatial system. In the several steps of this process, territorial regions were overlaid with a functional network of structural economic relationships, while the original building blocks—the regions—faded from view. The problem, of course, is that with them went the initial logic of the whole edifice: the export-base concept of exogenously induced growth was replaced by an 'integrated system' which feeds on its own dynamism. In retrospect, development either became a 'zero sum game,' with some places and therefore some people benefiting from other people's labour and resources, or it was, in fact, a self-initiating process which could just as well have taken place within the confines of the originally postulated regions.

At the time, neither of these alternative interpretations was given centre stage, but the subsequent history of regional studies demonstrates the discrepancy between the territorial and functional elements of the theory. In practice, analysts either had to choose to deal with one part or the other; they either had to concentrate on regional-territorial characteristics (Hirschman, 1958; Perloff et al., 1960; Borts and Stein, 1962; Klaassen, 1965, Alonso, 1968; Siebert, 1969), or opt for the spatial-functional approach (Friedmann, 1955; Ducan et al., 1960; Berry, 1967, 1971, 1973; Hansen, 1970; Hermansen, 1971; Richardson, 1973a; Pedersen, 1974). During the decade of the 1960s the spatial-functional school came into increasing ascendance, but both sides tended to look upon their differences as basically methodological problems, problems which could be worked out as regional science techniques became more sophisticated. It was not until the fundamental ideological assumptions of top-down development came to be challenged during the late 1960s that the theoretical nature of the conflict began to surface.

POLARIZED DEVELOPMENT AND GROWTH POLE DOCTRINE

Regional development theory did not stop with the idea of integrating the space economy, and regional planning needed more specific objectives if it were to transcend the level of arcane philosophizing. It was from an interest in polarized development and growth pole theory that conceptual evolution continued. In 1957 Gunnar Myrdal wrote a modest-sized book entitled *Economic Theory and Underdeveloped Regions*. Myrdal's well-known theme was that economic development, having started in certain favoured locations, would continue through a process of 'circular and cumulative causation'. From a geographical standpoint, growth would be transmitted through a network of spread and backwash effects, meaning, respectively, the positive and negative impacts of continued growth in the original areas on other regions. Myrdal displayed a strong concern that the cumulative advantages experienced in the initial growth areas would cause backwash effects to prevail in most other places and that conscious policy intervention would be necessary to prevent a truly explosive global political situation. This was one of the first troubled references to ethics and politics which entered post World War II economic theorizing, but it was largely ignored.

In the following year, Albert Hirschman published an independently conceived essay covering many of the same ideas. Hirschman's views were more in line though with the equilibrium arguments which dominated the contemporary intellectual scene. In *The Strategy of Economic Development* (1958), he spoke of 'trickle down' and 'polarization' instead of spread and backwash, but the concepts were fundamentally the same as Myrdal's. Hirschman, however, had much greater faith in the efficiency of market forces for allocating the factors of production, and believed that while development might tend to polarize around certain initial growth centres, eventually trickle down effects would become predominant in the never-ending search for resources and new markets. This difference of interpretation between Myrdal and Hirschman became a classic point of contention among planners.

Hirschman's argument fitted neatly into the emerging paradigm of regional studies, and set the tone for regional development and planning for more than a decade. It contained three important contentions. First, because of the friction of distance, agglomeration economies would cause growth to be polarized in certain existing locations at the outset. Second, although by its very nature growth must be unequal or unbalanced, it also contains within itself the imperatives for geographical expansion. And third, such expansion will take place through the emergence of subsequent growth points or growth poles. Because of their importance, I will briefly trace the evolution of each of these ideas as they pertain to regional planning.

To begin with the agglomeration economies concept, Hirschman equated it with overcoming the friction of distance: getting everything close together to create a common pool of skilled labour, services, information, and infrastructure. In later years, economists and geographers have subdivided external economies into a bewildering host of categories, but the fundamental idea is that it is cheaper to be together. Sometimes this principle has been set at odds with the idea of scale economies internal to the firm, but more typically it is argued that large plants require large urban markets, meaning big cities and an integrated space economy. No one has ever been able to measure scale economies very precisely, especially as they apply to the external environment. But this line of reasoning led directly to the consideration of city size as a measure of scale advantages, and eventually to theories about optimum city size and the optimum shape of city-size distributions (Berry, 1961, 1971; Neutze, 1965; El-Shakhs, 1965, 1972; Thompson, 1965; Alonso, 1971, 1972; Richardson, 1972, 1973b; Mera, 1973). These two conclusions are not necessarily mutually supportive, but they have managed to co-exist very fruitfully, providing a partial rationale for growth pole doctrine.

Turning to growth poles, as Hirschman had been aware, this idea originated with the celebrated French economist François Perroux. Perroux's argument was that leading industrial sectors could act as strategic *pôles de croissance* within interindustrial economic space, starting a process of self-sustaining economic growth which would radiate throughout the economy. Perroux's colleague, Boudeville, helped to transform the growth pole notion into a concept applicable in geographical space. Hirschman had already made this connection, and with later contributions by Lloyd Rodwin (1963) and Friedmann (1966), among others, growth centres provided the necessary link between theories of unequal development and the idea of inducing regional growth through integration of the space economy. The eventual logic of growth pole doctrine ran something like the following: disparities in welfare between different regions may be overcome by extending the polarized development process into depressed areas, through establishment of growth centres which link such areas to the economic growth impulses generated within the broader urban system.

Without becoming mired in the subsequent debates among growth pole theorists (see, Hansen, 1967, 1972; Darwent, 1969; Kuklinski, 1972; Kuklinski and Petrella, 1972; Mosely, 1974; and Friedmann and Weaver, 1979), the key questions here are: (1) 'What is the true nature of polarized development?' and (2) 'Are growth pole policies really a means of transferring economic growth to new areas (spread effects), or are they, in fact, misdirected, reinforcing the cumulative effects of polarized development (backwash effects)?' Most analysts at first tended to side with Hirschman's more optimistic view. Two of the most important studies by Borts and Stein (1962) and Williamson (1965) presented strong positive arguments. Borts and Stein attempted to show that eventual equilibrium in a space economy was inevitable, because capital would necessarily move outward from core areas with a high K/L ratio in search of higher marginal returns on investments. And conversely, labour would tend to migrate from low productivity areas in response to higher wage incentives. In the end everyone would get their fair share.

Williamson's argument was that interregional revenue disparities could be presented on a graph in the form of a bell-shaped curve. Polarization during the initial stages of the growth process would cause increasing regional inequalities, but at some unspecified point continuing polarization would naturally push things over the top of the curve and regional inequalities would again decrease. The secular trend would eventually point to an equalization of regional incomes.

Interestingly, Perloff *et al.* had taken a very different view of similar empirical evidence in their earlier 1960 study. While this classic work stood outside the prevailing imagery of polarized development, its conclusions are very pertinent: the relative decline of certain areas, in the *volume* of economic activities, is an inevitable feature of a rapidly changing economy. This conclusion seemed to suggest that national economic growth would lead to two separate processes at the subnational or regional level. Some areas, perhaps changing over time in response to changes in the structure of the national economy, would grow and become increasingly wealthy, while other areas would experience an absolute decline in their volume of economic activities. This could throw an altogether different light on the apparent convergence in regional income noted by Williamson!

After lengthy practical experience in Brazil, Venezuela, and Chile, Friedmann re-emphasized this less optimistic view. In his 'A General Theory of Polarized Development' (1972, original 1967) Friedmann attempted to conceptualize all the various factors—cultural, political, and economic—which enter into regional growth, and tried to anticipate the results of a policy of induced urbanization. His position was equivocal but somewhat disturbing. Polarized development appeared as a predominantly political process, with a dominant core area systematically exploiting its surrounding periphery through a monopoly of information and political power. He argued that eventually a crisis of transition would occur, either leading to a diffusion of political power and economic opportunity, or ending in continued exploitation and possible political revolution. It seemed that at last Myrdal's original fears were gaining wider currency: regional development might indeed be a fundamentally ethical/political process and the ideology of polarized development might be a dangerous ploy. This was only the beginning, however, of a growing critique of unequal development.

UNDERDEVELOPMENT AND THE TRANSNATIONALS

In the same year that Friedmann first wrote his 'General Theory', a former University of Chicago colleague, André Gunder Frank, published a forceful ideological critique of polarized development (Frank, 1967). For Frank, Myrdal's notion of cumulative causation and backwash effects was far too benevolent; he felt foreign-dominated industrialization in Latin America could only lead to a process of underdevelopment. In redefining the term 'under-development', Frank turned the whole theory of regional growth on its head. He argued that economic development could only be understood in terms of a global economic system, dominated by American capital. All appearance of development in Third World countries was a ruse. Foreign exploitation of natural resources and foreign industrial investment—the heart of the export-base concept—only led to dependency, alienation, social disintegration, and suppression; presided over by local élites paid by the international power structure. Picking up on Lenin's 50-year-old argument: international economic relations became a form of neocolonialism, a zero sum game in which for some places to develop they would necessarily have to exploit the potential surplus value of other areas (Lenin, 1917). Development and underdevelopment became two sides of the same coin, a double-edged sword. Any hope for underdeveloped regions to follow in the path of international core areas was definitionally impossible.

The Chilean economist, Osvaldo Sunkel (1969, 1970, 1973), made a comprehensive application of dependency theory to spatial development, concentrating on the role of transnational corporations, which he referred to as Trancos. The world capitalist economy, according to Sunkel, is being reorganized into a new institutional structure. The Trancos, backed by the governments of the industrialized countries, are concentrating capital and brain-power within their own organizations. Third World countries are being increasingly forced into a position of dependent underdevelopment. People that do not serve the needs of transnational capital are excluded from the modern sector of the economy; 'marginalized', in Sunkel's terminology. This marginalized population 'either becomes a landless rural proletariat or moves into the cities where it is absorbed into the "lower circuits" of the urban economy at levels of productivity that remain close to physical survival' (Friedmann and Weaver, 1979, p. 177).

José Luis Coraggio (1972, 1975), trained in Walter Isard's programme at the University of Pennsylvania, applied the neo-Marxist critique to growth pole theory. Coraggio, concentrating on the work of Perroux, Boisier, and Lasuén, followed Frank's example, turning the whole idea upside-down. Growth centre policies not only could do little to spread economic growth and development, they actually thwarted it, implanting new points of capital expropriation and dominance in the dependent space economy. Growth centres merely extend the spatial pattern of underdevelopment. (Ironically, perhaps, in writing his own analysis of national economic independence François Perroux, the father of growth pole theory, cited Sunkel approvingly; see Perroux, 1969, p. 34. In fact, Perroux's continuing elaboration of the essential role of power relationships in the international economy (e.g. Perroux, 1973), while still formulated within the misleading equilibrium model, suggests that his view of polarized development was significantly more sophisticated than its various adaptations and caricatures.)

There is one further explicitly spatial aspect of underdevelopment thinking at the supra-urban level; this deals with the crippling legacy of inherited colonial transportation patterns. The standard liberal model of transport network development was presented in Taaffe, Morrill, and Gould's classic 1963 article, 'Transport expansion in underdeveloped countries'. Their argument was basically an application of spatial equilibrium theory to the 'linkage component' of urban systems growth. Briefly stated: route development in colonial countries would begin by penetration of the interior by a simple transport corridor, linking indigenous collection points with a coastal port tied into the imperial trading network. This simple linear pattern would be expanded later to connect natural resource centres into the system, forming a dendritic pattern, similar in appearance to a natural drainage basin. Further elaboration of the national settlement hierarchy would eventually create an increasing number of horizontal linkages within the structure, finally providing a mature transport network and the backbone of an integrated space economy. Taaffe *et al.* were forced to admit, however, that this final stage was largely hypothetical, since it had not been attained as yet in actual Third World settings. As with regional income equality and city size distributions, the final equilibrium state was an assumption based on belief in the universal, uni-dimensional character of the development process.

Underdevelopment theorists have found the transport network argument easy game (e.g. see McNulty, 1976). A mere glance at the map of transportation routes in almost any Third World country will suffice to demonstrate the continuing dendritic nature of circulation structures. More formalized connectivity analyses yield the same results.

Defrocking underdevelopment theory of its typically accusative tone, the central argument is relatively simple. *Functional economic power, removed from the control of territorial authority, can only exacerbate the social and geographical inequities inherent in polarized development.* Transnational capital, working partially through local élites, will create an ever-increasing dependency on outside economic interests. For poor countries and poor regions, labour, resources, and capital will be exploited by unavoidably unequal terms of exchange. Small-scale local production and traditional social relationships will be destroyed by a flood of overly competitive international products and the imperatives of uncontrolled urban-industrial development. The hypothesized spread effects of economic growth will suffer fatal leakages outside the national economy, as multipliers are captured by faraway industries and financial institutions. The rich will get richer and the poor will become even poorer. Only the lucky few will win: people and places which serve the ends of transnational power.

By the early 1970s Latin American ideas had taken root across the Atlantic in Africa. Some authors, because of contact with European Marxist thinking, presented analyses of an increasingly orthodox nature (Amin, 1974; Stuckey, 1975). Others investigated the economic heritage of colonialism and neocolonialism (Rodney, 1972; Brett, 1973; Leys, 1974) and the role of urbanization in underdevelopment (De Souza and Porter, 1974; Soja and Weaver, 1976). In an interesting twist of hypothetical innovation diffusion patterns, several variants of underdevelopment theory had sprung up on the north side of the Mediterranean by the mid-1970s. How close their ties are with Afro-American theorizing is hard to say: several of the Latin American writers are now living in exile in England and France. In any case, radical approaches to regional problems have become an idea in good currency. Their influence has already reached such proportions that it can be traced in official government policy and in major political platforms.

Looking first at Great Britain, Stuart Holland in the European School at the University of Sussex presented an early Western formulation of the under-development/dependency hypothesis in his analysis of 'Regional Under-Development in a Developed Economy: The Italian Case' (1971). As I will show directly below, Holland later went on to elaborate several of the central compo-nents of underdeveloping thinking, based at least in part on his experience with the Wilson government. Focusing on the UK, others have used this reasoning to account for industrial decline and separatist movements in Scotland and Wales. (The 'touchy' case of Northern Ireland has been handled in a distinct body of literature; see Perrons, 1978, for an excellent introduction.) Carter (1974) and the American sociologist Hechter (1975) identified underdevelop-ment with internal colonialism and argued that Scotland had been exploited through the core – periphery nature of economic relations between England and its Celtic fringe. This latter core – periphery concept has been used more recently to analyse peripheral development in a variety of European situations (Seccchi, 1977; Lee, 1977; Tarrow, 1977; Seccchi, 1978; Seers *et al.*, 1978; Tarrow *et al.*, 1978).

Holland continued his analysis of underdevelopment with an empirical study of the role of transnational corporations in the European economy (Holland, 1974, 1976a, 1976b). In an analysis reminiscent of Sunkel's work in Latin America, Holland argued that the meso-sector—the Trancos—are beyond the control of territorial authority. Government investment in retarded areas has resulted in few, if any, local multiplier effects through linked industries, and transnational capital would much sooner locate in Portugal or the Third World than in relatively expensive labour areas such as Scotland. He argued that in the international economy the friction of distance has proven less important than high labour costs. It is Holland's view that regional development can only succeed through subordinating functional economic power to territorial interests. In his best-known work (1976a), he contends that the answer to regional inequalities lies in widespread nationalization, even given the disap-pointing results of historical experience to date with such policies in both Western and Eastern Europe. More recently (1978), Holland has helped out-line a socialist strategy for regaining control of the meso-economic sector, suggesting planning reforms at the national level and at the level of the EEC.

Although the underdevelopment/internal colonialism argument continues to be used (see, for example, Williams' 1977 discussion of Wales), most critical theorists find it increasingly unacceptable. This seems to stem primarily from two things: first, the fact that, historically, socialism has had a strong 'inter-nationalist' bias, posing global working-class solidarity as the appropriate response to labour exploitation by national groupings of entrepreneurs, aided by the national state. (During World War II this even took the form of the Comintern actively working against anti-colonial national liberation movements in North Africa, which it branded as fascist. See for instance Lottman, 1979 as well as Debray's 1977 assessment of Marxism and nationalism.) Secondly, Marxist theory, which provides the conceptual framework for an increasing proportion of critical regional scientists, has no appropriate causal niche for the geographical or 'spatial' component of social relationships. Like so many liberal economists and sociologists, Marxists have little but ridicule for the importance of such notions as the 'friction of distance', 'areal differentiation' and 'economic space', the geographical imagination being most frequently attacked as a form of 'fetishism'.

The first Marxist critics of the underdevelopment hypothesis in the UK appear to be Lebas (1974) and Davies (1974), both working at the time on a research project dealing with the sociology of regional planning at the Polytechnic of Central London. These papers, while partially ambiguous about the real role of polarization, core–periphery relations and imperialism in development, set out most of the subsequent elements of the Marxist analysis of regional decline. The earliest comprehensive revision of underdevelopment theory came out of work centred largely at the University of Durham, like Lebas and Davies financed by a Ford Foundation grant administered by the Centre for Environmental Studies. In a path-breaking paper given at the Urban Change and Conflict Conference at the University of York, Carney et al. (1975) presented a thorough-going attack on the fetish of space and 'area'. Analysing the 'underdevelopment' of north-east England, they emphasized the theoretical importance of the organic composition and circulation of capital, as well as the crucial role of the state in capitalist development, setting the tone for the even stronger critique which was soon to follow. The most recent contributions are best discussed as part of the 'regional question', in the British context meaning primarily the devolution and separatist movements in Scotland and Wales.

THE REGIONAL QUESTION

Political movements for the independence of historically identifiable cultural regions have gained widespread attention in Europe during the last few years, and are largely responsible for the new-found visibility of regional development problems, which had languished in an atmosphere of benign neglect during much of the 1970s. Varying theoretical explanations have been put forward in the UK (and in France), assigning differing degrees of importance to imperialism, economic decline, nationalism, and underdevelopment. As mentioned earlier, Carter (1974) and Hechter (1975) emphasized the dependency/underdevelopment argument in their analyses of internal colonialism in Scotland, as did Williams (1977) for Wales. More recently, though, strenuous opposition has arisen within the ranks of the Left.

The most outspoken critic has been Lovering (1978), writing about the case of Wales. Referring to the 'model of Internal Colonialism', he argues that '. . . the conceptions of "economic exploitation" and the State employed in the model derive from an eclectic, ahistorical approach with no firm material basis' (p. 55). And that: 'the empirical predictions are not borne out by an inspection of reality, and they actually obsure important aspects of the present conjuncture' (p. 55). For Lovering, underdevelopment theory does not reveal the contemporary role of the state in holding together the Welsh economy, and even worse, it obscures the true nature of class relations. In less categorical analysis of Scottish oil development, Moore (1979) and Shapiro (1979) come nonetheless to similar conclusions, arguing that the underdevelopment notion is of only marginal utility in understanding the present situation in Scotland and that the idea of 'exploitation under conditions of dependent development' adds no theoretical insight to the broader Marxist concept of labour exploitation.

Adding another dimension to the debate, Harvie (1977) in his discussion of *Scotland and Nationalism* comes into a head-on collision with Hechter's 'magnificently wrong-headed' (p. 72) notion of internal colonialism. Reaching back to Gramsci's idea of the pivotal role of intellectuals for a legitimate entree into the Marxist camp, Harvie argues the importance of Scottish nationalism as an ironical force for 'Union' with England, obscuring conflicting class interests and, through assimilation, allowing Scotland to share in the capitalist spoils of empire. In a stimulating variation on this theme, Rees and Lambert (1979) detail the formulation of a 'regionalist consensus' in Wales; a form of outward-reaching nationalism which has legitimated the role of the British state in rationalizing the southern Welsh economy into the evolving national structure.

The strongest confrontation over nationalism and the regional question, however, has taken place within the ethereal realm of the *New Left Review*. Tom Nairn (1977), the proponent of a 'Marxist theory of nationalism', argues that an adequate materialist explanation of 'the break-up of Britain' must take into account the central role of nationalism in Northern Ireland, Scotland, and Wales, as well as England. After a lengthy account of UK nationalisms, he goes on to criticize the 'anti-imperialist' underdevelopment theories of Emmanuel (1972) and Amin—their 'necessary resort to populism'—and then identifies absense of an adequate comprehension of the nationalist phenomenon as the weakest component of socialist theory, calling for an urgent attempt to reach a Marxist understanding of nationalism and the 'irrational'.

Nairn's protagonist is the renowned Marxist historian Eric Hobsbawm. Hobsbawm's (1977) sharp reply asks: 'Does the present phase of nationalism require any change in the attitude of Marxists to this phenomenon?' (p. 8) His answer: 'If Nairn's book is anything to go by, it certainly appears to require, rather than the by now ritual breast-beating about theoretical deficiencies in this field, a reminder of the basic fact that Marxists as such are not nationalists' (p. 8). After rejecting the temptations of separatism as a 'detour on the road to revolution', recalling Lenin's critique of Stalin's 'chauvinist behaviour', Hobsbawm concludes that the traditional Marxist position on nationalism provides adequate theoretical guidelines for contemporary political practice.

It must be noted in passing, whatever the *theoretical* merits of Nairn's contestations or Hobsbawm's rebuttal, that the 'regional question', in the form of the failure of the Scottish and Welsh devolution referendums, and their aftermath, has served to break up the Callaghan government's ruling coalition, and, thus, caused Labour to fall from power. Mrs. Thatcher's new Tory regime, by all accounts, promises to be the most conservative and 'nationalistic'—meaning English nationalism—of the post-war era.

THE DEATH OF REGIONAL POLICY

This brings us to the last two major sub-fields in Britain's radical regional science: the 'death' of regional policy writings, and critical location studies. These two issues are interrelated and their key elements are (1) the (somewhat eclectic) role of multinational monopoly capital (i.e. the Trancos à la Sunkel or the meso-sector à la Holland) in controlling national economic decisions, and (2) the role of the capitalist state in rationalizing and legitimating the accumulation and circulation of capital. Both these concepts relate to the work of Lebas, Davies, and Carney *et al.*, and ultimately draw part of their inspiration from French writers such as Palloix, Dulong, Lipietz, Bleitrach and Chenu, Castells, Poulantzas, Lojkine, and Lefebvre. They provide the strongest link between regional analysis and the more elaborate French/Italian school of Marxist urban sociology, and in so doing carry on its all but overwhelming internal contradiction; they organize themselves explicitly to exercize the 'fetish of space' (regionalism), while at the same time acting as almost the sole credible force for introducing 'spatial relations' into the Marxist discourse.

Doreen Massey (1976, 1978a, 1978b), one of the leading writers in both policy criticism and location theory, framed the problematique most succinctly when she asked: 'In what sense a regional problem?' Her answer takes the form of a carefully organized, thoughtful review of the various trends among radical analysts in Great Britain, France, and Italy. Her argument, along with Mellor (1975), Mingione (1977), and Secchi (1977), is that regional underdevelopment should be viewed as a normal facet of capitalism rather than a special attribute of imperialism, applying the classic Marxist concept of uneven development in a spatial sense. This geographically uneven development is the product of an international spatial division of labour based on the requirements of capital accumulation and circulation. And today, the nineteenth-century regional specialization of labour by industrial sector has been replaced in advanced industrial countries by a technical division of labour, founded upon the stages of the production process. Massey divides these into three for the United Kingdom: the first stage specializing in highly automated work, requiring only low-paid unskilled labour, and locating in areas with little experience in manufacturing or labour organization; stage two represents production processes which still require a more skilled labour input and remain located in traditional industrial areas (although these activities are distinctly in decline); and the third stage, management and R&D activities, locate in the most dynamic urban centres, i.e. south-east England, to take advantage of its numerous urban economies. Thus, as Firn (1975) and McDermott (1976) showed, peripheral regional economies, specialized in first and second stage industrial processes, are dependent in the sense that they are controlled in their managerial and production decisions from the outside. Massey's explicit location studies (Massey, 1974, 1978c; and Massey and Meegan, 1978, and forthcoming), an outcome of the Centre for Environmental Studies' Industrial Location Project, lend significant empirical and theoretical weight to her arguments.

Because of the conceptual importance of Massey's work, it is only fair to point out several of its shortcomings. First, while her reformulation of location principles is much more intellectually satisfying than its immediate predecessors, fundamentally its explanatory powers lie in the fact that the whole edifice is taken in tandem by a theoretical analysis of capitalist economics which (once again) imputes a rigid rationality to economic decision-making. Second, in Massey's attempt to place location theory within a Marxist framework, she fails to identify the historical sources of the two most central features of her model: the idea of 'product cycle' (e.g. Vernon, 1959, 1966), and Swedish research into the behaviour of 'contact systems' (Tornqvist, 1970; Thorngren, 1970). The importance of this omission is that ideas about the technical–geographical division of labour can be, and were originally, founded in entirely different intellectual milieux, and lend themselves to various sorts of syntheses. Third, with this retreat into rationalism, Massey—like the early location theorists—creates a determined world: an economic landscape which in large measure disregards the many 'minority locations' which do not at all fit the more general pattern. And like David Smith's (1971) geographical margins of profitability, these minority locations represent a mighty challenge to the idea of the materialist determination—liberal or Marxist—of industrial location. Finally, fourth, like so many cultural, social, and economic geographers before her, Massey has disavowed the importance of natural site characteristics, arguing that these hold little value in explaining location. She even goes on explicitly to invoke the argument that local culture and socioeconomic characteristics of the specific places involved offer few positive determinants, at least, of location. Both these contentions are undeniably fashionable, but seem extremely tenuous if pushed very far.

Significant complementary work has been done in related areas by Michael Dunford (1977, 1979) at the University of Sussex. Confronting the idea of a 'regional question', like Massey, he finds it at best mis-framed, and goes on to reformulate it within a Marxist context. Dunford's major theoretical contributions have been: (1) an elaboration of Fine's (1975) model of the circuit of an individual industrial capital, in combination with some of Palloix's (1970, 1975b, 1976) arguments about the tendencies of capitalist accumulation to increase concentration, centralization, and international specialization, and the combination of the forces and relations of production; as well as (2) an elaboration of the role of the state in regional development, drawing heavily on the ideas of Castells (1976), Poulantzas (1976) and O'Connor (1973); and finally

(3) a discussion of the Rey–Lipietz model of the transition from feudalism to capitalism, agricultural change, and class alliances in France (Rey, 1973; Lipietz, 1977), and an introduction to English readers of Dulong's related work. More than most continuing writers in the regional studies area, Dunford has served admirably as what Lebas refers to as a 'culture broker', bringing many important continental concepts to a fuller understanding in the UK through his analyses of regional development in Italy and France.

LA CONTRIBUTION FRANÇAISE

The marked impact of French ideas on British work logically carriers us to a brief overview of critical regional thinking in France. Radical analysis of urban and regional problems in France can be traced, in large measure, directly to the *événements* of 1968, which set in motion what has become a veritable school of applied critical studies. After a short gestation period, an impressive number of publications began to appear in the early and mid-1970s. In what seems an ironic train of events, however, the 'Revolution of May '68'—so clearly Anarchist, Maoist, and Trotskyist in origins and intent, and so clearly betrayed by the French Communist Party (Singer, 1970)—left as one of its major intellectual legacies an academic research paradigm which has tended increasingly towards Marxist orthodoxy. The strong heritage of libertarion and syndicalist thought, as well as the continuing regionalist tradition in France (see Gras and Livet, 1977), have modified these tendencies somewhat, however.

It is not my intention here even to attempt a review of the massive French urban sociology literature. Besides the enormity of the task, although its conceptual ties with critical regional analysis are numerous (e.g. the theory of the capitalist state, the means of collective consumption, the role of the built environment in the circulation of capital), its most immediate impact on regional studies has probably come in a roundabout fashion, through the work of British and American writers. (Pickvance, 1975, 1976, 1977; Harloe, 1978; and Lebas, 1979, give a summary of important aspects of the French urban debate.)

One of the longest-sustained contributions to critical 'spatial' theory in France has been the work of Palloix (1970, 1972, 1975a, 1975b, 1975c, 1976, 1977). Palloix began with a sceptical examination of the question of unequal exchange, which as we have already seen was an important component of the international debate about dependency and underdevelopment, and then proceeded to elaborate on the notion of an 'internationalization' of capital—the mandatory restructuring of capitalist production through the transnational corporation. He went on to argue that this self-expansion of capital on a world scale has been accompanied by new forms of labour exploitation, which he identified as *Neo-Fordism*. Palloix's ideas have had an influence on the conceptual formulations of several other writers, including Dunford and Massey in Great Britain.

The most systematic treatment of the locational strategies of transnational corporations and their impact upon the new international division of labour was presented by Charles-Albert Michalet of the University of Paris X at Nanterre. Writing for the OECD and the International Labour Organization, Michalet (1975a, 1975b) surveyed the world economic order under oligopoly capitalism and reviewed corporate investment and location strategies. His theoretical interpretation, while taking its global outlines from Marxism, drew very heavily upon the work of liberal economists. Michalet's summary of the changing geographical structure of industrial production emphasized the falling rate of profit hypothesis and the 'product cycle' notion mentioned above. In an important addition to his discussion of capital circulation, he also underscored the impact of internalization of technology by the transnationals. Adding a more detailed organizational analysis to Sunkel's argument, cited earlier, Michalet contended that the new international division of labour is exacerbated by the control of information flows. Thus, with the explicit limitation of innovation diffusion—limited in most underdeveloped areas only to the know-how necessary for consumption—regional specialization is the product not only of ownership of the means of production, but also 'ownership' of the techniques of production.

In the same year, Michon–Savarit (1975) proposed two interesting scenarios which, it was said, could lead to entirely different patterns of regional specialization in France. Like Michalet, pointing to the importance of developing endogenous national management and R&D capabilities, it was argued that in a world dominated by the Cold War US/USSR power blocs, France would remain in the background of industrial advance, continuing to specialize in the production of less innovative products, fitting roughly into what Massey called the second stage of development. In an alternative world order, based on the mutation of five new power configurations, France would play a more innovative role, leading to an emphasis on 'third stage' industries (i.e. growth industries based on the development of new technologies), a decline in regions specializing in traditional manufacturing activities, and the rise of a glamour region along the Mediterranean periphery. This latter phenomenon is an historical reality in the United States: the Sunbelt.

The trends towards a decline of the North and industrialization of the French Midi were, in fact, already in evidence. In 1973, Yves Durrieu wrote an analysis of the new industrial centre of Fos, near Marseille, employing fundamentally a Marxist critique. He saw the 'failure' of development planning in Languedoc as primarily a function of centralization and the ties between state power and the capitalist ruling class. Durrieu's conclusions called for a new socialist regionalization of France and a territorially responsive planning process. Other writers also criticized the new patterns of capitalist growth in Provence; most notably perhaps are Bleitrach and Chenu (1971, 1975), Cultiaux (1975), and Robert (1975).

Renaud Dulong and Louis Quere, working at the Centre d'Etude des Mouvements Sociaux in Paris have begun a less eclectic Marxist analysis of French regional problems (Dulong, 1976a, 1976b, 1978; Dulong and Quere, 1976). Dulong, concentrating on the fiery question of Brittany, argues that regional underdevelopment is caused by fragmentation of the ruling class and new class alliances. National élites and the capitalist state have lost their common interests with regional rulers, turning their face toward the transnationals and the world economy. This split causes decisions to be made in favour of international capital, causing deterioration of the regional superstructure and regional exploitation. Dulong's solution is establishment of a leftist government and nationalization of key industries, rather like Holland's ideas for Great Britain.

Because of the entirely different conceptual basis of Dulong's approach it is very difficult to juxtapose against liberal theories of regional development. It is closely related, though, to another school of Marxist thinking which directly challenges traditional location economics and regional theory. This approach is best represented in Alain Lipietz's recent book *Le Capital et son Espace* (1977). After an historical analysis of the French transition to industrial capitalism, Lipietz gives a critical review of economic-base theory, Borts and Stein, and François Perroux's ideas. He then proposes the beginnings of a Marxist theory of economic space centered around monopoly capital and spatial reproduction of the mode of production. Lipietz's theorizing acknowledges the reality of polarized development but offers a very different interpretation of the mechanisms behind it. Beyond a call for immediate class struggle, his work remains exclusively critical, offering little insight as to an eventual socialist solution to regional inequalities.

A final line of regional thinking stems directly from resurgence of the 'Regional Question'. From Brittany to Occitania, Pays Basque, and Corsica, cultural regionalism has once again become a salient political issue, as in the United Kingdom (and Spain and Canada). What begins as poetry ends in politics.

This struggle of territorial values against functional power has been described vividly in France by Robert Lafont. In a decade-long series of publications (e.g. 1967, 1971, 1976), Lafont analyses the fate of territorial communities under the centralized capitalist state. He argues that peripheral culture areas have been underdeveloped by a process of internal colonization directed by the national core area. In a mixture of tradition and socialism, he calls for a new marriage of regional consciousness and class consciousness, leading to an electoral revolution, followed by a genuine decentralization of political and economic power and regional self-management. Lafont himself ran as a presidential candidate to dramatize the seriousness of his position, and today many of his ideas find expression in an alliance between progressive and regionalist forces in France (Lagarde, 1977).

An important symbol of this union was its prominence in the 1978 French parliamentary campaign and the ill-fated Common Programme of the Left (Parti Socialiste Français, 1978). From a regional perspective, the *Programme Commun* presented four important themes: nationalization, self-management, decentralized planning, and local political power. Nationalization of key industries was the centrepiece of the left's political strategy and the supposed point of contention between the Socialist and Communist parties. Public ownership of the means of production was combined with a commitment to self-management of the workplace, placing major decision-making responsibility in the hands of workers and their institutional representatives. Economic planning and physical development decisions were to be democratized through a process of dialogue and compromise among various sectoral and territorial groups. Especially in hypercentralized France, this requires vesting real authority in local units of government and thus a restructuring of French municipal administration. To this end, it was proposed that regions become a new unit of territorial government, with real budgetary and decision-making powers, controlled by an elected regional council. It was evident, however, that nationalization and worker participation in management were the *idées fixes* of the *Programme Commun*, and that proposals for territorial decentralization played a decidedly secondary role.

DEVELOPMENT FROM BELOW: TOWARDS A DOCTRINE OF TERRITORIAL DEVELOPMENT

Sorting through the vast array of conflicting regional theories, it becomes evident that from North and Isard to Holland, Lipietz, and Massey, regional inequities have been recognized as a fact of contemporary economic life. Economic orthodoxy, on both right and left, clings to various notions of polarized development and centralized ownership. Their planning strategies typically call for either modified forms of transnational hegemony or domination by a new class of national political managers. It is clear, however, that during the last 20 years, theories of regional development and planning have come full circle. Arguments for functional integration of the space economy are being displaced by calls for decentralization and regional autonomy.

What are we to make of this complete reversal of professional opinion? To me it suggests that, quintessentially, the world is our own representation, a matter of will (Schopenhauer, 1966). Abandoning the determinisms of economic thinking, regional development is above all an ethical/political question. If development is to occur, what is needed is a doctrine of territorial development: negating the bonds of unequal exchange by an explicit theory of wilful community action, selective regional closure, and strategic regional advantage.

Territorial development simply refers to the use of an area's resources by its residents to meet their own needs. The main definitives of these needs are regional culture, political power, and economic resources. Territorial development can be compared and contrasted to the idea of functional development, i.e., the narrow exploitation of a region's potentials only because of the role these play in the larger international economy. (For further discussion of the differences between territorial and functional development, see Stöhr, in Chapter 2 of this volume, and Friedmann and Weaver, 1979.) The animating force behind territorial development—the thing that makes it possible—I have called wilful community action. This means clarifying and transforming the shared heritage within regional communities in regard to the things they value and want to accomplish, and using this new feeling of purpose and unity as the basis for bold new initiatives and action (Weaver, 1978). Two of the major substantive components of territorial development are selective regional closure and strategic regional advantage. These concepts are in direct opposition to the traditional notions of free trade within an open economy based on comparative advantage. Trade theory, one of the foundations of regional science, makes the presumption that transactions between economic actors take place on the basis of equal exchange. Even the briefest reflection upon the daily newspaper will quickly dispel such quaint notions, however, and their acceptance as the basis for self-interested economic policies is only in the advantage of those in favoured positions of economic and political power. Regions cannot specialize in 'what they do best' and expect to get equal returns for their efforts, because the 'resources' at the disposal of some other areas and the transnational corporations give them an undeniable bargaining edge. So for the weak, the 'lagging', the disinherited, the only feasible solution is to refuse to play the game by rules that, by definition, will beat them (Perroux, 1969).

Obviously there are practical limits to such a strategy, but these may well be broader than often imagined. Selective regional closure emphasizes doing as many as possible of the things necessary to meet a region's needs within the area itself, and taking whatever political measures are necessary and feasible to see that this is accomplished. As part of an important elaboration of this idea, Stöhr and Tödtling (1977) specify a number of preconditions:

(1)   the broadening of explicit spatial development policy beyond economic to a more explicit consideration of social and political processes;

(2)   the reformulation of distance friction from a negative concept (to be diminished as an obstacle to large-scale integration and spatial equilibrium) to a positive one for the structuring of a spatially disaggregate interaction and decision system (Isard);

(3)   greater attention to be paid to non-market and non-institution-based activities and to the requirements of small-scale human and man–environment relations;

(4)   a shift of decision-making powers from today's mainly functional or vertical (sectoral) increasingly to horizontal (territorial) units at various levels. The scale of the territorial decision-making level should ideally be the one within which a maximum of the repercussions or external effects of the respective decision can be internalized. This means to short-circuit decision-making scales with spatial impact scales to the maximum degree possible. In case of doubt the lower level should be given preference. (Stöhr and Tödtling, 1977, p. 47.)

As suggested below, strategic regional advantage means limiting development of resources for export within a carefully regulated parallel export sector, to those situations where to the greatest possible extent the region enjoys a favoured bargaining position—for whatever political or economic reasons.

It is no use trying to dream away the power of transnational corporations. They exist. It is no less futile attempting to deny the realities of the centralizing nation-state. But accepting their ideologies is a matter of belief. There are many worlds and many realities—it is for us to choose—and this choice is the paramount fact of regional life. Functional power is dependent upon suppression of regional consciousness and the will of territorial communities. The fate of a community of destiny lies in the common beliefs and values of its people. To mention only a few striking examples, think of the Irish in 1920, India in 1945, Algeria in 1960. Who would have predicted their success? Their political victories were a matter of will. What has not been sufficiently recognized is that the economic world is analogous. All technocratic propaganda notwithstanding, economy is, indeed, political economy. Power and control are the definitives of human society, but at all scales the interface between one level and the next is a function of consensus: shared beliefs. If the will to power can be spread throughout a regional society, there will be development. From the Massif Central in France to the High Plateaux of Algeria, people can satisfy their legitimate needs by the work of their own hands and the use of their own resources.

There are several important considerations in building an internally oriented regional economy, but at the broadest level, the exploitative nature of core—periphery relations must be ended by a conscious choice to use regional resources for regional purposes. To accomplish this goal, sympathy on the part of central governments will be useful, but initiative rests squarely with the regional communities themselves. Regional development is a regional project!

As applied to regional planning, a doctrine of territorial development could include the following components:

Job Creation through Meeting Regional Needs

High unemployment rates in underdeveloped regions are one of the most glaring of ironies. Where there is much to be done there is much potential work. Jobs can be created through local initiative, using local skills and resources to meet local needs. These needs may be basic, social, or personal: the secret is drawing on regional ideas and manpower for their fulfilment. Such solutions will frequently be resource-conserving, labour-intensive and energy-cheap.

Residentiary Activities as the Key to Growth

It has been recognized in recent years that the service or residentiary sector has become the leading generator of economic growth in metropolitan areas. This recognition underlies the contradiction in spatial development theory identified earlier. Economic expansion is dependent on the creation of value through the use of human labour and natural resources, and while this can be done through the export market, it can also be done *in situ*. A space economy at almost any scale is capable of generating its own growth. This is as true of the single region as it is for the world economy. At the regional scale, however, there is less chance of leakages, backwash effects and outside domination.

Regional Infrastructure and Community Facilities

A major area for selfhelp in meeting regional needs is the provision of modern, region-serving infrastructure and community facilities. Non-dendritic transportation systems providing horizontal linkages between existing population centres are essential. Multi-purpose public buildings at convenient locations are also a must. Many people can contribute to such ventures through regional bond sales, in-kind donations and their own labour.

A Crucial Framework of Regional Institutions

Institution-building is of prime importance. Besides the basic skeleton of physical structures, a largely self-sufficient region must have a well-developed network of services and community support institutions. The most important are probably basic-level health and dental services, co-operative savings institutions which invest exclusively in regional projects, producer and consumer cooperatives for the collection and distribution of regional products, democratic planning mechanisms within the framework of regional government, practically oriented institutes of technology doing research on concrete regional problems, co-operative child-care centres—to free women to become active participants in regional life, and adult educational extension services.

Community Education and Territorial Values

The substantive components of regional education are of particular significance. They must first of all promote a consciousness of regional identity, regional problems and regional interests. If this is to be done, there must be an emphasis on concrete problems which immediately affect the community and its members. People should develop their learning capabilities by working together in their own communities; only in this way can regional productive forces be mobilized to provide labour and capital for regional development. To escape mystification and eventual alienation within the regional population, education and other components of territorial development must be based on a concept of equal opportunity and equal development.

Two of the most important components of such a doctrine are the encouragement of historical regional values and a sensitive ideology of frugality and ecological awareness. Patrick Geddes' ideas and experiments are very instructive in this respect, as are the writings of the Regional Planning Association of America and the great American regionalist Howard Odum (Mumford, 1925, 1938; MacKaye, 1928; Odum, 1934; Odum and Moore, 1938; Sussman, 1976; Boardman, 1978; Friedmann and Weaver, 1979).

Decentralization, Small Scale, and Local Control

For wilful community action to be a reality, territorial control of the regional economy is imperative. This means small-scale production units, preferably in decentralized locations, which cater to the needs of the regional population. In scope, this can include consumer goods—wage goods, perishables, and even consumer durables—as well as service activities and intermediate inputs to other regional industries. The form of ownership of the means of production is less important than the residential status of the owner, the scale of the operation, and the adoption of self-management procedures which can be meshed with territorial planning processes. It is essential that multiplier effects are not allowed to leak from the regional economy. Capital accumulated in the region must be kept there. Extraregional industrial linkages must be developed very selectively. For these reasons branch plants and proposals for subassembly operations should be analysed with exceeding care. Transnational investment and national public investment should be reserved for a very special category of economic activity, a parallel export sector, designed to earn needed 'foreign exchange'.

Regional service industries need protection from unfair competition if they are to survive. Keynes himself recognized this fact at the level of the newly industrializing nation-state, and it is doubly true for the underdeveloped region in an integrated space economy. This presents a real problem at the subnational scale, but policies such as promoting community education and adopting discriminatory sales and use taxes could offer viable approaches.

Another form of decentralization can be achieved through a spatial mix of rural and urban activities. This means ruralizing the city as well as locating industrial pursuits in the countryside. Kropotkin, Odum, and Goodman all elaborated on this theme (Kropotkin, 1899; Odum, 1939; Goodman, 1947). Regional planning must be conceptualized as a process of comprehensive resource development rather than a quest for urban industrialization. Growth centres doctrine must be replaced by a model of over-all territorial development.

Natural Resource Development

The role of natural resources in regional economic development needs to be reconsidered. Reliance on raw materials as an export base is both risky and probably short-sighted. The export of petroleum and other energy resources provides an important example. Even when man/resources ratios are extremely high, as they are in the Middle East, the long-run effects of using energy resources as an entrée into the international economy are uncertain. Massive foreign exchange earnings may provide the basis for later economic diversification, but the inability of underdeveloped economies to absorb large infusions of capital forces much of this revenue back into the global financial circuit, further fuelling polarized development. Using North Sea oil and gas as another example, the meagre income multipliers of highly mechanized supertanker ports, petroleum refineries and related chemical industries will offer little stimulus to the Scottish economy. Most benefits will be captured by the transnationals and, secondarily, by the central government, through various royalties and export duties. Only an unforeseeably generous allocation of national earnings to regional development programmes could be hoped to achieve any other outcome. For natural resources to contribute to territorial development, three essential conditions must be met. First the proportion of resources allocated to earning foreign exchange should be critically scrutinized and regulated by the criterion of strategic regional advantage. Second, the bulk of resources should be used in meeting regional production needs, within a strategy of decentralized, equal development. And third, restraint and conservation should be the by-words. Equality and stability seem very difficult to attain in an extravagant consumer economy (Illich, 1973, 1977).

CONCLUSION

By way of conclusion, I want to state in the strongest possible terms that territorial development through selective regional closure is not a blanket replacement for growth pole doctrine. Regional problems vary from time to time and place to place. The generalized prescription of growth centre policies for all nature of regional ills was one of the most unrealistic aspects of the last 20 years of regional studies. As Benton MacKaye wrote some 50 years ago, territorial development is a philosophy of regional planning, its world-view makes it unavoidably egalitarian and decentralizing. Wilful community action is a method for bringing it to earth. Its transactive approach explains its epistemological commitment to social learning and self-management. But within these necessary bounds, procedures, strategies, and plans have to be responsive to particular historical and geographical situations. These can be defined and interpreted in myriad ways, but each region presents a different setting for territorial development.

Traditional theories of regional growth called for economic polarization based on overcoming the friction of distance. Further polarization was seen as the key to eventually overcoming regional inequalities. Recent critics of this doctrine of unequal development point to its inconsistent internal logic and its anomalous historical results. In this chapter I have presented a review of regional theory and have argued for a new territorial approach to regional development. In its broad outlines territorial development through wilful community action cannot help but appear less 'practical and realistic' than the established orthodoxies. (In a recent sympathetic analysis Planque, 1979, has pointed up some of the very real empirical difficulties which must be met head-on by territorial development strategies.) But this apparent utopianism will quickly fade as more information comes to light on the history of regionalist thought and the numerous regional movements now in full swing in many parts of the world come to fruition. (Encouraging examples of territorial development strategies applied to widely differing circumstances can be found in Friedmann and Douglass, 1975, Friedmann, 1978, Sullam et al., 1978, and Weaver, 1979.) As it becomes more fully elaborated the territorialist doctrine can provide a vital response to the regional question and a workable strategy for development from below.

REFERENCES

Alonso, W. (1968) Urban and regional Inbalances in economic development, *Economic Development and Cultural Change*, **17**, 1–14.

Alonso, W. (1971) The economics of urban size. *Papers of the Regional Science Association*, **26**, 67–83.

Alonso, W. (1972) The question of city size and national policy, in R. Funck (ed.) *Recent Developments in Regional Science* (London: Pion).

Amin, S. (1974) *Accumulation on a World Scale: A Critique of the Theory of Underdevelopment*, 2 vols. (New York: Monthly Review Press).

Berry, B. J. L. (1961) City size distributions and economic development, *Economic Development and Cultural Change*, **9**, 573–87.

Berry, B. J. L. (1967) *Spatial Organization and Levels of Welfare: Degree of Metropolitan Labor Market Participation as a Variable in Economic Development*, (Washington, DC: US Economic Development Administration).

Berry, B. J. L. (1971) City size and economic development: conceptual synthesis and policy problems, in Jacobson, L. and V. Prakash (eds.), *South and Southeast Asia Urban Affairs Annual*, vol. 3, (Beverley Hills: Sage Press).

Berry, B. J. L. (1973) *Growth Centers in the American Urban System*, 2 vols. (Cambridges Mass.: Ballinger).

Bleitrach, D. and A. Chenu (1971) Le rôle idéologique des actions d'aménagement du territoire—l'exemple de l'aire métropolitaine marseillaise, *Espaces et Sociétés* (Dec.), 43–55.

Bleitrach, D. and A. Chenu (1975) L'aménagement: régulation ou approfondissement des contradictions sociales? un exemple: Fos-sur-Mer et l'aire métropolitaine marseillaise, *Environment and Planning A*, **7**, 367–91.

Boardman, P. (1978) *The Worlds of Patrick Geddes* (London: Routledge & Kegan Paul).

Borts, G. H. and J. L. Stein (1962) *Economic Growth in a Free Economy* (New York: Columbia University Press).

Boudeville, J. R. (1961) *Les Espaces Economiques* (Paris: Presses Universitaires de France).

Boudeville, J. R. (1966) *Problems of Regional Economic Planning* (Edinburgh: Edinburgh University Press).

Brett, E. A. (1973) *Colonialism and Underdevelopment in East Africa: The Politics of Change, 1919–1939* (New York: NOK Publications).

Carney, J., R. Hudson, G. Ive, and J. Lewis (1975) Regional underdevelopment in late capitalism: a study of the north east of England, in *Proceedings of the Conference on Urban Change and Conflict*, University of York, January 1975. (London: Centre for Environmental Studies).

Carter, I. (1974) The Highlands of Scotland as an underdeveloped region, in de Kadt, E. and G. Williams (eds.), *Sociology and Development*, pp. 279–311 (London: Tavistock).

Castells, M. (1976) Crise de l'état, consommation collective et contradictions urbaines, in Poulantzas, N. (ed.), *La Crise de l'Etat*, (Paris: PUF).

Clark, C. (1938) *National Income and Outlay* (London: Macmillan).

Coraggio, J. L. (1972) Hacia una Revision de la Teoria de los Polos de Desarrollo, *Revista Latinoamericana de Estudios Urbanos y Regionales*, **2**, 25–40 (English trans., *Vierteljahres Berichte* (Bonn), **53**, September 1973).

Coraggio, J. L. (1975) Polarization, development and integration, in Kuklinski, A. (ed.), *Regional Development and Planning: International Perspectives*, (Leyden: Sijthoff International Pub. Co.) (Original 1973.)

Cultiaux, D. (1975) L'aménagement de la region Fos-Etang de Berre, *Notes et Etudes Documentaires*, Nos. 4164–66 (February).

Darwent, D. F. (1969) Growth poles and growth centers in regional planning: a review, *Environment and Planning*, **1**, 5–31.

Debray, R. (1977) Marxism and nationalism, *New Left Review*, **105**, 25–41.

De Souza, A. R. and P. W. Porter, (1974) *The Underdevelopment and Modernization of the Third World*, Commission on College Geography, Resource Paper No. 28 (Washington, DC: Association of American Geographers)

Dulong, R. (1976a) *La Question de Bretagne*, Ecole des Hautes Etudes en Sciences Sociales (Paris: Centre d'Etude des Mouvements Sociaux).

Dulong, R. (1976b) La crise du rapport etat/societe locale vue au travers de la politique régionale, in Poulantzas, N. (ed.), *La Crise de l'Etat* (Paris: PUF).

Dulong R. (1978) *Les regions, l'Etat et la Societe Locale*, (Paris: PUF).

Dulong, R. and Quere, L. (1976). *La Question Regionale en France*, Ecole des Hautes Etudes en Sciences Sociales (Paris: Centre d'Etude des Mouvements Sociaux).

Duncan, O. W. R. Scott, S. Lieberson, B. Duncan, and H. H. Winsborough (1960) *Metropolis and Region* (Baltimore: Johns Hopkins University Press).

Dunford, M. (1977) *Regional Policy and the Restructuring of Capital*, Urban and Regional Studies, Working Paper 4. University of Sussex.

Dunford, M. (1979) *Capital Accumulation and Regional Development in France*, Urban and Regional Studies, Working Paper 12, University of Sussex.

Durrieu, Y. (1973) *L'Impossible Régionalisation Capitaliste: Temoignages de Fos et du Languedoc* (Paris: Editions Anthropos).

El-Shakhs, S. (1965) 'Development, Primacy and the Structure of Cities'. Unpublished Ph.D. dissertation, Harvard University.

El-Shakhs, S. (1972) Development, primacy, and systems of cities, *Journal of Developing Areas*, **7**, 11–36.

Emmanuel, A. (1972) *Unequal Exchange; A Study of the Imperialism of Trade* (London: New Left Books).

Fine, B. (1975) The circulation of capital, ideology and crisis, *Bulletin of the Conference of Socialist Economists* (October).

Firn, J. (1975) External control and regional development: the case of Scotland, *Environment and Planning*, **7**, 393–414.

Frank, A. G. (1967) *Capitalism and Underdevelopment in Latin America* (New York: Monthly Review Press).

Friedmann, J. (1955) *The Spatial Structure of Economic Development in the Tennessee Valley* (Department of Geography, University of Chicago).

Friedmann, J. (1963) Regional planning as a field of study, *Journal of the American Institute of Planners*, **29**, 168–74. Quotations are taken from the reprint in Friedmann J. and W. Alons. (eds.), *Regional Development and Planning: A Reader* (Cambridge, Mass.: MIT Press, 1974).

Friedmann, J. (1965) *Venezuela: From Doctrine to Dialogue* (New York: Syracuse University Press).

Friedmann, J. (1966) *Regional Development Policy: A Case Study of Venezuela*. (Cambridge, Mass.: MIT Press)

Friedmann, J. (1972) A general theory of polarized development, in Hansen, N. (ed.), *Growth Centers in Regional Economic Development* (New York: Free Press, New York —original 1967; revised 1969).

Friedmann, J. (1977) *The Crisis of Transition: A Critique of Strategies of Crisis Management*. Comparative Urbanization Studies, School of Architecture and Urban Planning, University of California, Los Angeles.

Friedmann, J. (1978) *The Active Community: Towards a Political–Territorial Framework for Rural Development in Asia*. UN Centre for Regional Development, Nagoya, Ms.

Friedmann, J. and M. Douglass (1975) Agropolitan development: towards a new strategy for regional development in Asia, in Lo, F. C. and K. Salih (eds.) *Growth Pole Strategy and Regional Development Planning in Asia*. Proceedings of a Seminar, UN Centre for Regional Development, Nagoya, (Printed version: 1978, Oxford: Pergamon Press).

Friedmann, J. and C. Weaver (1979) *Territory and Function: The Evolution of Regional Planning* (London: Edward Arnold; Berkeley and Los Angeles: University of California Press).

Gilbert, A. (ed.) (1976) *Development Planning and Spatial Structure* (London: John Wiley).

Goodman, P. and P. (1947) *Communitas: Means of Livelihood and Ways of Life*, (New York: Random House).

Gras, C. and G. Livet (1977) *Régions et Régionalisme en France du XVIII à Nos Jours* (Paris: Presses Universitaires de France).

Hansen, A. H. and H. S. Perloff (1942) *Regional Resource Development* (Washington, DC: National Planning Association).

Hansen, N. (1967) Development pole theory in a regional context, *Kyklos*, **20**, 709–25.

Hansen, N. (1970) *Rural Poverty and the Urban Crisis: A Strategy for Regional Development*, (Bloomington: Indiana University Press).

Hansen, N. (ed.) (1972) *Growth Centers in Regional Economic Development* (New York: Free Press).

Harloe, M. (1978) Marxism, the state and the urban question: critical notes on two recent French theories, in Crouch, C. (ed.), *British Political Sociology Year-book* (London: Croom Helm).

Harvie, C. (1977) *Scotland and Nationalism; Scottish Society and Politics, 1707–1977*, (London: George Allen & Unwin).

Hechter, M. (1975) *Internal Colonialism: The Celtic Fringe in British National Development, 1536–1966* (London: Routledge & Kegan Paul).

Hermansen, T. (1971) *Spatial Organization and Economic Development*, Institute of Development Studies, University of Mysore.

Hirschman, A. O. (1958) *The Strategy of Economic Development* (New Haven: Yale University Press).

Hirschman, A. O. (1963) *Journeys Toward Progress: Studies of Economic Policy-Making in Latin America* (New York: The Twentieth Century Fund).

Hirschman, A. O. (1967) *Development Projects Observed* (Washington, DC: The Brookings Institution).

Hobsbawm, E. (1977) Some reflections on 'The Breakup of Britain', *New Left Review*, **105**, 3–23.

Holland, S. (1971) Regional under-development in a developed economy: the Italian case, *Regional Studies*, **5**, 71–90.

Holland, S. (1974) Multinational companies and a selective regional policy, in Expenditure Committee, Trade and Industry Sub-Committee, *Minutes of Evidence Session 1973/74 on Regional Development Incentives*.

Holland, S. (1976a) *Capital Versus the Regions* (London: Macmillan).

Holland, S. (1976b) *The Regional Problem* (London: Macmillan).

Holland, S. (ed.) (1978) *Beyond Capitalist Planning*, esp. pp. 137–202. (New York, St. Martin's Press).

Hoover, E. M. (1937) *Location Theory and the Shoe and Leather Industries* (Cambridge, Mass.: Harvard University Press).

Hoover, E. M. (1948) *The Location of Economic Activity* (New York: McGraw-Hill).

Illich, I. (1973) *La Convivialité* (Paris: Editions du Seuil).

Illich, I. (1977) *Le Chômage Créateur* (Paris: Editions du Seuil).

Isard, W. (1956) *Location and Space Economy*, (Cambridge, Mass.: MIT Press).

Keynes, J. M. (1936) *The General Theory of Employment, Interest and Money* (London: Macmillan).

Klaassen, L. (1965) *Area Economic and Social Redevelopment. Guidelines and Programs* (Paris: OECD).

Kropotkin, P. (1899) *Fields, Factories and Workshops* (London: Hutchinson; current edition: London: George Allen & Unwin, 1974).

Kuklinski, A. (ed.) (1972) *Growth Poles and Growth Centers in Regional Planning* (Paris: Mouton).

Kuklinski, A. and Petrella, R. (eds.) (1972) *Growth Poles and Regional Policies*, (Paris: Mouton).

Kuznets, S. (1941) *National Income and Its Composition* (New York: National Bureau of Economic Research).

Lafont, R. (1967) *La Révolution Régionaliste* (Paris: Gallimard).

Lafont, R. (1971) *Décoloniser en France*, (Paris: Gallimard).

Lafont, R. (1976) *Autonomie de la Région à l'Autogestion* (Paris: Gallimard).

Lagarde, P. (1977) *La Régionalisation* (Paris: Seghers).

Lebas, E. (1977) Regional policy research: some theoretical and methodological problems, in Harloe, M. (ed.), *Captive Cities: Studies in the Political Economy of Cities and Regions* (London: John Wiley)

Lebas, E. (1979) The evolution of the state monopoly capital thesis in French urban research—or the crisis of the urban question. Centre for Environmental Studies, London (MS).

Lee, R. (1977) Regional relations and economic structure in the E.E.C., *London Papers in Regional Science*, 7, 19–35.

Lenin, V. I. (1917) *Imperialism, The Highest Stage of Capitalism* (Petrograd; current edition: Peking: Foreign Languages Press, 1970).

Lewis, W. A. (1955) *Theory of Economic Growth* (London: George Allen & Unwin).

Leys, C. (1974) *Underdevelopment in Kenya: The Political Economy of Neo-Colonialism, 1964–1971* (Berkeley: University of California Press).

Lipietz, A. (1977) *Le Capital et son Espace* (Paris: François Maspero).

Lösch, A. (1954) *The Economics of Location (trans. of Die Räumliche Ordnung der Wirtschaft*, 2nd edn., 1944) (New Haven: Yale University Press).

Lottman, H. R. (1979) *Albert Camus: A Biography* esp. pp. 147–60 (New York: Doubleday, Garden City, New York).

Lovering, J. (1978) The theory of the 'internal colony' and the political economy of Wales, *Review of Radical Political Economics*, 10, 55–67.

MacKaye, B. (1928) *The New Exploration: A Philosophy of Regional Planning.* (New York: Harcourt, Brace).

Massey, D. B. (1974) *Towards a Critique of Industrial Location Theory* (London: Centre for Environmental Studies).

Massey, D. B. (1976) *Restructuring and Regionalism: Some Spatial Effects of the Crisis.* Paper presented to the American Regional Science Association. Centre for Environmental Studies, Working Note 449.

Massey, D. B. (1978a) *Regionalism: Some Current Issues. Capital and Class*, 6, (Autumn).

Massey, D. B. (1978b) In what sense a regional problem? Paper presented to the Regional Studies Association Conference: The Death of Regional Policy, Glasgow, 31 January, 1978. Centre for Environmental Studies, Working Note 479.

Massey, D. B. (1978c) Capital and locational change: the UK electrical engineering and electronics industries, *Review of Radical Political Economics*, 10, 39–54.

Massey, D. B. and Meegan, R. A. (1978) Industrial restructuring versus the cities, *Urban Studies*, 15 (3), p. 273–288.

Massey, D. B. (forthcoming) *The Industrial Location Project*, Final Report. (London: Centre for Environmental Studies).

McDermott, O. (1976) Ownership, organisation and regional dependence in the Scottish electronics industry, *Regional Studies*, 10, 319–35.

McNulty, M. L. (1976) West African urbanization, in Berry, B. J. L. (ed.), *Urbanization and Counter-Urbanization.* (Beverley Hills: Sage Press).

Mellor, J. R. (1975) The British experiment: combined and uneven. *Proceedings of the Conference on Urban Change and Conflict*, University of York, January 1975 (London: Centre for Environmental Studies).

Mera, K. (1973) On urban agglomeration and economic efficiency, *Economic Development and Cultural Change*, **21**, 309–24.

Michalet, C. A. (1975a) *Les Firmes Multinationales et la Nouvelle Division Internationale du Travail.* Document de Travail, Recherches pour le Programme Mondial de l'Emploi (Geneva: Bureau International du Travail, ILO).

Michalet, C. A. (1975b) *Transfert technoloqique par les FMN et Capacité d'absorption des Pays en Voie de Développement.* (Paris: OCDE).

Michon-Savarit, C. (1975) La place des régions françaises dans la division internationale du travail: deux scenarios contrastés, *Environment and Planning, A*, **4**, 449–54.

Mingione, E. (1977) Theoretical elements for a Marxist analysis of urban development, in Harloe, (ed.), *Captive Cities* (London: John Wiley).

Moore, R. (1979) Urban development in the periphery of industrialised societies. Paper presented to the Urban Change and Conflict Conference, Nottingham University, 5–8 January 1979.

Moseley, M. J. (1974) *Growth Centres in Spatial Planning* (Oxford: Pergamon Press).

Mumford, L. (ed.) (1925) Regional plan number. *Survey Graphic*, **54**.

Mumford, L. (1938) *The Culture of Cities* (New York: Harcourt, Brace).

Myrdal, G. (1957) *Economic Theory and Underdeveloped Regions* (London: Duckworth).

Nairn, T. (1977) *The Break-up of Britain* (London: New Left Books).

Neutze, G. M. (1965) *Economic Policy and the Size of Cities*, (Canberra: Australia National University Press).

North, D. C. (1955) Location theory and regional economic growth, *Journal of Political Economy*, **63**, 243–58.

O'Connor, J. (1973) *The Fiscal Crisis of the State* (New York: St. Martin's Press).

Odum, H. W. (1934) The case for regional–national social planning, *Social Forces*, **10**, 164–75.

Odum, H. W. (1939) *American Social Problems: An Introduction to the Study of the People and their Dilemmas* (New York: Henry Holt).

Odum, H. W. and H. E. Moore (1938) *American Regionalism: A Cultural-Historical Approach to National Integration* (New York: Henry Holt).

Paelinck, J. (1965) La théorie de développement régional polarisé, *Cahiers de l'Institut de Science Economique Applquée*, Series L, **15**, 5–48.

Palloix, C. (1970) La question de l'échange inégal, *Table Ronde du C.E.R.M.*, Paris.

Palloix, C. (1972) The question of unequal exchange: a critique of political economy, *Bulletin of the Conference of Socialist Economists*, **2**.

Palloix, C. (1975a) *L'Internationalisation du Capital: Eléments Critiques* (Paris: Maspero).

Palloix, C. (1975b) *La Crise Organique du Capitalisme: Essai sur la Crise de l'Organisation Capitaliste de la Production et du Process de Travail*, (Grenoble: IREP).

Palloix, C. (1975c) *L'Economie mondiale capitaliste et les firmes multinationales* (Paris: Maspero).

Palloix, C. (1976) Labour process: from Fordism to neo-Fordism, in *The Labour Process and Class Strategies*. (London: Conference of Socialist Economists).

Palloix, C. (1977) The self-expansion of capital on a world scale, *Review of Radical Political Economics*, **9** (Summer), 1–29.

Parti Socialiste Français (1978) *Le Programme Commun de Gouvernment de la Gauche: Propositions Socialistes pour l'Actualisation* (Paris: Flammarion).

Pedersen, P. O. (1974) *Urban-Regional Development in South America: A Process of Diffusion and Integration* (Paris: Mouton).

Perloff, H. S., E. S. Dunn, Jr., E. E. Lampard, and R. F. Muth (1960) *Regions, Resources and Economic Growth*, Printed for *Resources for the Future*, (Baltimore: Johns Hopkins University Press).

Perrin, J. C. (1974) *Le Développement Regional* (Paris: Presses Universitaires de France).

Perrons, D. (1978) *The Dialectic of Region and Class in Ireland*, Working Paper 8, Urban and Regional Studies, University of Sussex.

Perroux, F. (1950) Economic space: theory and applications, *Quarterly Journal of Economics*, **64**, 90–7.

Perroux, E. (1955) Note sur la notion de pôle de croissance, *Economie Appliqué*, **1**, and **2**, 307–20.

Perroux, E. (1969) *Indépendance de l'Economie Nationale et Interdépendance des Nations* (Paris: Union Generale d'Editions).

Perroux, E. (1973) *Pouvoir et Economie*, (Paris: Dunod).

Pickvance, C. G. (1975) On the study of urban social movements, *Sociological Review*, **23** (1), 29–49.

Pickvance, C. G. (1976) *Urban Sociology: Critical Essays* (London: Tavistock).

Pickvance, C. G. (1977) Marxist approaches to the Study of urban politics: divergences among some recent French studies, *International Journal of Urban and Regional Research*, **1**, (2), p, 219–55.

Planque, B. (1979) *Specialisation fonctionnelle des espaces et capacites locales de développement*. Paper presented to the Colloquim on Redeploiment Industriel et Developpement Regional, Clermont-Ferrand (France), 4 May 1979.

Poulantzas, N. (ed.) (1976) *La Crise de l'Etat* (Paris: PUF).

Rees, G. and J. Lambert (1979) *Urban Development in a Peripheral Region: Some Issues from South Wales*. Paper presented to the Conference on Urban Change and Conflict, University of Nottingham, 5–8 January 1979.

Rey, P. P. (1973) *Les Alliances de Classes* (Paris: Maspero).

Richardson, H. W. (1969) *Regional Economics: Location Theory, Urban Structure, and Regional Change* (New York: Praeger).

Richardson, H. W. (1972) Optimality in city size, systems of cities, and urban policy: a sceptic's view, *Urban Studies*, **9**, 29–48.

Richardson, H. W. (1973a) *Regional Growth Theory* (New York: John Wiley).

Richardson, H. W. (1973b) *The Economics of Urban Size* (Westmead: Saxon House).

Richardson, H. W. (1978) *Regional and Urban Economics* (Harmondsworth: Penguin).

Robert, G. (1975) L'Operation Fos, un test de l'aménagement capitaliste du territoire, *Urbanisme*, 63–76.

Rodney, W. (1972) *How Europe Underdeveloped Africa*. (Dar es Salaam: Tanzania Publishing House).

Rodwin, L. (1963) Choosing regions for development, in Friedrich, C. J. and S. E. Harris (eds.), *Public Policy: Yearbook of the Harvard University Graduate School of Public Administration*, Vol. 12 (Cambridge, Mass.: Harvard University Press).

Rostow, W. W. (1961) *The Stages of Economic Growth: A Non-Communist Manifesto* (Cambridge: Cambridge University Press).

Schopenhauer, A. (1966) *Le Monde Comme Volonté et Comme Représentation* (Paris: Presses Universitaires de France; (original 1819, Leipzig: Brockhaus).

Schumpeter, J. A. (1934) *The Theory of Economic Development* (Cambridge, Mass.: Harvard University Press; original 1912).

Seccchi, B. (1977). Central and peripheral regions in a process of economic development: the Italian case. *London Papers in Regional Science*, **7**, 36–51.

Seccchi, C. (1978) *Trade and Adjustment Problems Caused by the Enlargement to the Peripheral and Backwards Regions of the E.E.C.* Paper presented to the EADI General Conference, 'Europe's Role in World Development', September 1978, Milan.

Seers, D., B. Schaffer, and M. L. Kiljunen, (eds.) (1978) *Underdeveloped Europe: Studies in Core-Periphery Relations* (Hassocks, Sussex. Harvester Press).

Shapiro, D. (1979) *Industrial Relations in the Wilderness: Working for North Sea Oil*. Paper presented to the Urban Change and Conflict Conference, Nottingham University, 5–8 January 1979.

Siebert, H. (1969) *Regional Economic Growth: Theory and Policy* (Scranton: International Textbook, Co.).

Singer, D. (1970) *Prelude to Revolution: France in May 1968* (New York: Hill and Wang).

Smith, D. M. (1971) *Industrial Location: An Economic Geographical Analysis* (New York: John Wiley).

Soja, E. and C. Weaver (1976) Urbanization and underdevelopment in East Africa, in B. J. L. Berry (ed.), *Urbanization and Counter-Urbanization* (Beverly Hills: Sage Press).

Stöhr, W. B. and Tödtling, F. (1977) Spatial equity—some anti-theses to current regional development doctrine, *Papers of the Regional Science Association*, **38**, 33–53.

Stuckey, B. (1975) Spatial analysis and economic development, *Development and Change*, **6**, 89–101.

Sullam, C., M. Storper, D. Pittman, and A. Markusen, (1978) *Montana: A Territorial Planning Strategy* Working Paper 294, Institute of Urban and Regional Development, University of California, Berkeley.

Sunkel, O. (1969) Politique nationale de développement et dépendance externe, *Economie Appliquée*.

Sunkel, O. (1970) Desarrollo, subdesarrollo, dependencia, marginacion y desigualdades espaciales: hacia un enfoque totalizante, *Revista Latinoamericana de Estudios Urbanos y Regionales*, **1**, 13–51.

Sunkel, O. (1973) Transnational capitalism and national disintegration in Latin America, *Social and Economic Studies*, **22**, 132–76.

Sussman, C. (ed.) (1976) *Planning the Fourth Migration: The Neglected Vision of the Regional Planning Association of America* (Cambridge, Mass. : MIT Press).

Taaffe, J. Morrill, R. L. and Gould, P. R. (1963) Transport expansion in underdeveloped countries: a comparative analysis, *Geographical Review*, **53**, 502–29.

Tarrow, S. (1977) *Between Center and Periphery: Grassroots Politicians in Italy and France* (New Haven: Yale University Press).

Tarrow, S., P. J. Katzenstein, and L. Graziano (eds.) (1978) *Territorial Politics in Industrial Nations* (New York: Praeger).

Thompson, W. R. (1965) *A Preface to Urban Economics*. Published for *Resources for the Future*, (Baltimore: Johns Hopkins University Press).

Thompson, W. R. (1968) Internal and external factors in the development of urban economies, in H. S. Perloff and L. Wingo, Jr. (eds.), *Issues in Urban Economic*, Published for *Resources for the Future* (Baltimore: Johns Hopkins University Press).

Thorngren, B. (1970). How do contact systems affect regional development? *Environment and Planning*, **2**(4), p. 409–27.

Tornqvist, G. E. (1970) *Contact Systems and Regional Development*. Lund Studies in Geography, Ser. B, No. 35 (Lund: Gleerups).

United Nations Department of Economic Affairs (1951) *Measures of the Economic Development of Underdeveloped Countries* (New York: UNDEA).

Vernon, R. (1959) *The Changing Economic Function of the Central City*, New York: Area Development Committee of CED.

Vernon, R. (1966) International investment and international trade in the product cycle, *Quarterly Journal of Economics*, **80**, 190–207.

Weaver, C. (1978) *Planning and Willful Community Action: Some Epistemological Notes.* Paper presented to the Georgeville Seminar on Planning Theory, November 1978, University of Montreal. MS.

Weaver, C. (1979) *Rethinking the Regional Question: A Workable Response to Industrial Decline*. School of Architecture and Urban Planning, University of California, Los Angeles. MS.

Webber, M. J. (1972) *The Impact of Uncertainty on Location* (Cambridge, Mass.: MIT Press).

Weber, A. (1928) *Theory of the Location of Industries* (trans. by C. J. Friedrich from the 1st German edn., 1909) (Chicago: Chicago University Press).

Williams, P. (1977) The internal colony, *Planet Tregaron, Dyfed*, **37/38.**

Williamson, J. G. (1965) Regional inequality and the process of national development: a description of patterns, *Economic Development and Cultural Change*, **13**, 3–45.

Chapter 4

Basic-Needs Strategies: A Frustrated Response to Development from Below?

Eddy Lee

DEVELOPMENT STRATEGIES AND POVERTY

There is by now broad agreement that the economic performance of developing countries in the last 25 years has failed to live up to the initial expectations of development theory. It is true that, simply in terms of the growth of per capita GNP, the developing world taken as a whole has performed far better than expected.

> The GNP per capita of the developing countries as a group grew at an average rate of 3.4 per cent a year during 1950 to 1975. This was faster than either the developing countries or the developed nations had grown in any comparable period before 1950 and exceeded both official goals and private expectations (Morawetz, 1977, p. 12).

In spite of this, however, serious problems of poverty and underdevelopment remain. In part, this outcome can be attributed to the fact that growth has been unevenly distributed among developing countries:

> Thus, although it is true that per-capita income has roughly trebled for some 33 per cent of the people in the developing world during the past 25 years, it is also true that for another 40 per cent the increase in per-capita income has been only one or two dollars a year (Morawetz, 1977, p. 14).

If anything, however, these figures under-state the proportion of the population in developing countries that has experienced at best a slow advance in real income. Based as they are on country averages, the figures conceal important changes in the distribution of incomes within countries and the fact that the absolute incomes of a significant number of people may have actually fallen. Indeed, it has been argued that inequality and the incidence of absolute poverty

107

have increased, not only in countries with a poor growth performance but also in several fast-growing ones (ILO, 1977; Adelman and Morris, 1973).

Disenchantment with these accepted facts has focused on two interrelated issues: first, the nature of the world economic system, and the obstacles this system places in the path of those striving to conquer the problems of under-development and to narrow the economic distance between nations. The demands of developing countries for a New International Economic Order can certainly be regarded as one strand in the disenchantment with their experience over the past 25 years. The second issue is a concern with "alternative strategies" for national development to replace the discredited growth strategy.

Before we discuss these "alternative strategies" it would be useful to re-capitulate the main features of the discredited strategy, and analyse why it failed to live up to all that was expected of it. Although it would be inaccurate to suggest that development thinking and practice during the 1950s and 1960s was monolithic, it remains true that a dominant model can be identified and characterized. Such a model not only dominated development theory, but also constituted the basis for the policies of international development agencies as well as of the majority of countries in the Third World.

The main features of this model can be described in terms of targets set, the instruments used to attain these targets, and the outcomes predicted. The central target was the maximization of rate of growth of national income per capita—but it was not the only target. Rapid industrialization and the installation of the infrastructure for a modern economy were often the specific forms into which the growth target was translated. The drive to develop was expressed in terms of the need to overcome a shameful economic backwardness and to 'catch up' with former metropolitan powers and other developed nations. The accent was thus on modernization, on installing a modern economic structure, and diffusing modern attitudes and knowledge among the populace. These targets were regarded as being rapidly attainable and also mutually compatible; industry, especially heavy industry, was the dynamic and modern sector, and would thus simultaneously maximize both growth and modernization.

The instruments chosen to attain these targets sprang from the same ideo-logical perspective. Development was conceived to be an urgent and pressing task and processes therefore had to be telescoped. This implied the need for strong centralized control and direction over the key processes of capital accumulation and the allocation of investment. Thus in almost every developing country there is a Central Planning Agency and 5-year development plan. This spread of planning owed less to any ideological rejection of the market than to a feeling that in the interest of haste the market had to be supplemented by state intervention, especially in resource mobilization and investment. This planning machinery relied on some variant of a macro-economic growth model which emphasized the aggregate rate of investment and the capital–output ratio. The target of maximum growth could best be attained by increasing the level of investment and by reducing the capital–output ratio. The former involved resource mobilization, while the latter related to resource allocation and investment criteria. Development policies hence concentrated on manipulating fiscal and monetary instruments to increase the rate of savings, and guided by investment criteria which emphasized growth and reinvestment, on the diversion of resources to infrastructural and industrial investment. In addition commercial policies were geared to the fostering of import-substituting industrialization, and significant reliance was placed on attracting a fairly indiscriminate inflow of foreign aid, direct foreign investment, and foreign technology.

Clearly this package of instruments was geared overwhelmingly towards the maximization of growth, but it would be untrue to claim that the goal of attaining generalized economic progress did not enter into the consciousness of development planners. Rather, the growth strategy was based on a strong assumption that the fruits of economic progress would be automatically diffused throughout the entire economy. Not only was it believed that the problem of income distribution would take care of itself, but it was also held that the maximization of growth would in the long run be the most effective means of raising the incomes of the poor. The possibility of growing inequality in the short run was conceded, but even this was held to be desirable insofar as total savings were increased and hence a greater output would be available for later redistribution. Similarly a choice of capital-intensive technology was justified on the grounds of the greater reinvestible surplus associated with it. Furthermore, apart from these purely economic considerations, the growth and modernization strategy was also expected to pave the way for democratization of Third World societies and lead to greater popular participation (Amin, 1978).

In large measure, disenchantment with the growth strategy can be traced to the outcome of all these assumptions about the additional benefits that would automatically accompany a successful growth strategy. The 'spread' and 'trickle down' effects proved to be far less potent than originally anticipated. On the world scale, the hope that growth plus international trade would lead to narrowing of the economic distance between nations has not been fulfilled; the economic system has behaved quite differently from the smooth equilibrating adjustments predicted by the theory of international trade (Streeten, 1978). Similarly, within developing countries diffusion of the benefits of growth has been equally sluggish. Interregional inequality increased sharply in many countries. In some, such as the former Pakistan and Nigeria, this was a primary cause of political tension and discontent. In other multi-racial societies economic growth has not led to greater equality amongst racial groups, but has instead heightened communal tensions. Even more serious than this growth in interregional and interracial inequality was the accumulating evidence that inequality and absolute poverty may have actually increased in many countries. Of particular significance is the fact that the Kuznets hypothesis, which states that income inequality would increase in the initial stages of economic growth but would eventually narrow, no longer generates the same complacency about problems of income distribution and poverty that it used to. Not only is there a feeling that such a process may take an unacceptably long time, but there is also some question of the validity of the very hypothesis itself (Beckerman, 1978; Lee, 1977).

THE RE-EVALUATION OF DEVELOPMENT STRATEGIES

The disappointing outcome of growth strategy led to various attempts to find alternative strategies which could ensure growth while containing the worst manifestations of inequality and poverty. The best-known among these were the employment-oriented strategy of the ILO and the redistribution-with-growth strategy of the World Bank. Both these strategies started from recognition that it was necessary to broaden the objectives of development away from an excessive preoccupation with the rate of growth of GNP alone.

The employment-oriented strategy was based on the recognition that mounting unemployment and underemployment were the most serious manifestations of mass poverty, and that development strategies would need to adopt the generation of productive employment as a key target. This would imply that greater attention would have to be paid to the employment implications of the choice of product-mix and technology. Thus instead of allocating investment funds to projects that yielded the highest return irrespective of employment and income-distribution implications, it would be necessary to bias the choice in favour of projects which created the largest number of productive jobs. This was regarded as a necessary corrective against the prevalent preference for the most 'modern' projects using the latest and most capital-intensive technology. The main policy implications of this approach were that it would be necessary to shift the structure of production towards more labour-intensive goods, to choose appropriate technology and encourage small-scale and informal sector production which were seen as important sources of over-all employment generation. It was also argued that there was no necessary contradiction between the employment and the growth objective; a labour-intensive pattern of production would be more in line with the factor-endowments of poor countries, would save on scarce capital and foreign exchange, and, provided appropriate technologies were chosen, would enhance growth. Furthermore, increased labour absorption into productive activities would be a powerful means of raising the incomes of the poor and hence promoting greater equity (ILO, 1970).

The redistribution-with-growth strategy (Chenery et al., 1974) starts from the problem of inequality in income and asset distribution in developing countries. Since 'trickle down' or automatic redistributive effects have been found to be weak it was necessary for direct government intervention to achieve a more desirable outcome. After considering various alternative redistribution mechanisms it concludes that the best strategy would be to redistribute the incremental output generated through growth. Since it is acknowledged that a principal cause of poverty is that the 'poverty groups' lack productive assets, it is argued that redistribution should take the specific form of asset-generation aimed at designated target-groups of the poor. A prerequisite of a redistributive policy would thus be the identification of these target groups of the poor (landless agricultural labourers, small farmers, those in the urban informal sector, etc.) and then the designing of appropriate means for increasing access of these groups to productive assets. According to this approach therefore, the maximization of growth remained the valid paramount objective of development strategy, but it would need to be supplemented by direct intervention to redirect investment towards specific poverty groups.

It should be clear that neither of these approaches represents a fundamental break with pre-existing growth strategy. They both remained squarely within the paradigm of development 'from above', and in no way deviated from reliance on centralized macroeconomic planning and the manipulation of macro-economic aggregates. The employment approach argued for a modification of investment criteria to take into account considerations other than growth in output, while the redistribution-with-growth approach required that growth be supplemented by the use of some 'top-down' redistributive instruments.

More recently, however, the search for alternative development strategies has broken out of the confines of the 'development from above' paradigm and has addressed itself to a fundamental re-evaluation of the goals and processes of development. The Cocoyoc Declaration and the report of the Dag Hammarskjöld Foundation are the best-known statements of this 'alternative development' school of thought (Ghai, 1977). These, however, remain largely as rhetorical pronouncements, and do not yet have the status of a well-defined strategy of development which has been adopted by countries or advocated by international development agencies. Some of the ideas contained therein, however, have filtered through to the formulation of international development strategies. In particular the ideas about basic needs and mass participation have become key elements in the new development strategy advocated by the ILO at the 1976 World Employment Conference (ILO, 1976). Before analyzing the key elements of this strategy it will be useful to consider some aspects of the relationship between development and the satisfaction of basic needs.

DEVELOPMENT AND PERSISTENT DESTITUTION

Some Empirical Evidence

Part of the empirical evidence which forced a reappraisal of the growth strategy were statistics about the growing numbers of people in developing countries who suffered from malnutrition and debilitating disease; lived in grossly inadequate housing; and lacked access to essential services such as clean drinking water, sanitation, health care, and education. Although average indicators such as calorie intake, the percentage of children of school-going age at school, and average life-expectancy, tended to show some improvement, it was also clear that, because of population growth and the fact that the average figures may mask important distributive changes, the absolute numbers suffering from deprivation of their basic needs was growing (McNamara 1976).

Such a picture of destitution obviously offends humanitarian instincts and arouses an indignant desire that something ought to be done to eliminate this embarrassing adjunct to economic development. Sentiments such as these were probably an important inspiration for the advocacy of a basic-needs strategy of development which stresses satisfaction of basic needs of all, as a primary object-ive of development strategy. Stated in these broad terms, there is hardly any room for disagreement over the desirability of such a strategy; to dispute it would be to appear to deny the humanitarian sentiments upon which it is based. Beyond this stage, however, it is not easy to secure agreement over what such a strategy would imply in terms of specific economic policies.

One important reason why no immediate agreement is possible is that the nature of the relationship between satisfaction of basic needs and economic development is far from clear. For instance there appears to be no clear relation-ship between the satisfaction of basic needs and growth in per-capita income.

Only on five of the sixteen indicators—three on nutrition, infant mortality, and per-centage of dwellings with access to electricity—is there any significant relationship between improvement in the fulfilment of basic needs over time (1960–70) and growth in per-capita GNP. The conclusion is that—at least as the concepts are usually measured, at least for a period as short as ten years, and if the data are believed—GNP per capita and its growth rate do not seem to provide satisfactory proxies for fulfilment of basic needs and improvements in the same. (Morawetz, 1977, p. 58.)

On *a priori* grounds this result need not appear surprising or paradoxical. First, growth in income per capita is compatible with a worsening in income distribu-tion, i.e. a decline, stagnation or very slow growth in the incomes of the poor. In such cases, therefore, there would be no reason to expect any improvement in the fulfilment of basic needs. Secondly, an increase in the incomes of the poor is no guarantee of increased fulfilment of basic needs, since irrational consumer behaviour by way of spending on non-basic needs is possible. Moreover, some important basic needs such as preventive medical services, health care, and education are socially provided, and an increase in private incomes of the poor will not confer increased access to these needs if their availability is not in-creased. Finally it is possible that the relative price of basic commodities such as food may increase, or the composition of output may shift towards luxuries and away from necessities, and hence reduce their availability. An increase in the money incomes of the poor need not thus be automatically translatable into increased access to basic goods.

Another observation on the level of basic-needs fulfilment across countries could also be interpreted as confirmation of the lack of any necessary close relationship between basic needs and per-capita income. This relates to the fact that the level of basic-needs satisfaction in a few low-income countries or areas such as China, Sri Lanka, and Kerala are substantially higher than in countries with much higher levels of per-capita income (Streeten, 1977). These cases therfore establish that it is possible to meet basic needs at low levels of income, and that a careful study of such experiences can yield insights about appropirate policies to ensure the fulfilment of basic needs.

At a more general level, concern for the satisfaction of basic needs leads quite naturally to a basic question about development; 'Why is it that hundreds of millions are destitute in spite of unprecedented growth in the world economy?' The basic-needs concept offers an appealingly simple approach to defining an answer. Essentially this approach can be regarded as a 'transposition' of the definition of destitution. The destitute are those who do not have adequate access to basic human needs such as food, clothing, decent habitat, medical services, and education. Transposed, the question thus becomes 'why is it that the level of basic needs satisfaction for hundreds of millions falls short of requirements? Posed in this way, the focus tends to shift to questions of the definition of basic needs targets, the measurement of shortfalls from requirements and required increases in supplies of basic goods and services. It is important, however, that in this process we do not lose sight of the original question; an answer to it is a logical, necessary precondition for designing basic-needs-oriented development strategies. Otherwise we would be guilty of the fallacy of 'illegitimate isolation', of abstracting the problem from its social context and treating it unrealistically as a technocratic one of supply and delivery of basic goods and services. We shall therefore in the following concentrate on this basic question.

Why Basic Needs are not Met

We believe that an explanation of why basic needs are not met must be holistic, encompassing not only national economies but the international economic framework. The international economic system is characterized by great inequalities in global distribution of income, and consequently the demand of a few rich industrialized countries determines the pattern of production, trade, and the distribution of income. The impact of this on developing countries creates a situation of dependence where the pattern of production and trade is largely determined outside the Third World and beyond their control. The domestic pattern of production is distorted towards the production of minerals and cash crops, and away from the production of necessities or basic needs. Monoculture and other manifestations of over-specialization often condemn these countries to the vagaries of international market fluctuations and a consequent uncertainty about their exchange entitlement to necessities through international trade.

Coupled to this effect on the domestic structure of production is the fact that international inequality in the distribution of income is replicated within these peripheral economies. There is the same dominance of demand of the rich, and the basic needs of the poor come last. One of the saddest economic realities in the Third World today is that the market value on the life of a destitute is zero. A market system with a highly unequal distribution of income is clearly not a self-policing system in guaranteeing the right to life; even perfect markets do not necessarily create a 'minimum income level equilibrium'.

The effects of these two factors can be illustrated in several ways. Where 10 per cent of the population commands 50 per cent of the purchasing power, it is inevitable that the market will cater for the tastes of the rich, whether through production for the domestic market or through optimizing this consumption through world trade. Food provides the most important and visible example of these relationships.

Food is the most important of basic needs. Its importance arises not only from the fact of its obvious indispensability for survival, but also from its sheer weight in the consumption basket of the poor. Even for poor households who are on the margin of nutritional adequacy, food typically constitutes 60–70 per cent of total consumption.

We have already alluded to the distorting effect of the world market in terms of a shift away from food to cash-crop production. This affects not only the direct availability of food, but also more importantly increases the uncertainty of access of the poor in Third World countries to food. Once locked into the world market this access to food is subject to the endemic fluctuations in trade. Sharp short-term adverse changes in the terms of trade can mean hunger and even death for those on the margin of starvation. Moreover one additional facet of unequal exchange is that only the price of primary products fluctuates, not that of manufactures. While poor countries' exchange entitlements for food can fall perilously, this never happens in industrialized countries—which always have first claim in a system of rationing-by-income.

The specific form in which the external influence manifests itself can be either through direct capital penetration (e.g. multi-national agrobusiness), or indirectly through an influence in the choice of agricultural technology. Whatever the mechanism, the push of external influence is always towards the supplanting of a pattern of production with high use-value by one with high exchange-value. The nutritional pattern of output is invariably distorted. The Green Revolution has generated several examples of this process; while total output rises substantially, the impact of new varieties supplants production of traditional nutrition-rich crops like beans and pulses. Typically, large farmers who shift to new varieties and high-value crops outbid the poor farmers for land and other inputs, when it is the poor farmers who need the extra food most.

These effects are compounded by the existence of unequal agrarian structures. Under these conditions the burden of rent and usury ensures that even those of the poor who have access to productive assets can retain only a little of what they produce. As a consequence hunger sets in well before the average product is at the level of subsistence. For the large numbers of rural people without access to the means of producing their own food, the claim to food supplies is infinitely more tenuous. Growing landlessness means a decrease in the number of self-provisioning farmers, and reduced and more insecure access to food.

All these factors explain why the most important of basic needs of many are not met. There are in addition several other reinforcing mechanisms which ensure that not only food but also other basic needs are not met. Inequality on a world scale, and the consequent penetration of poor societies by the ideology and technology of the rich, results in the shaping of the tastes of the élite in poor countries. Imported ideology shapes the goals of poor societies, and through the actions of the élite it shapes the pattern of consumption and hence also production and technology. Luxury industrial goods, luxury western-style housing, medical services, etc., come to dominate the structure of economic activity. The penetration by multi-nationals of the production process and even the mass media, perpetuates and reinforces these tendencies. It results in situations such as the setting-up of Coca Cola factories before there are even adequate supplies of safe drinking water. The observant visitor to Third World countries can no doubt recount numerous examples of ludicrous distortions of local priorities as a result of such penetration.

Economic change in the context of such penetration often produces a variety of adverse effects on the poor. For example, the commercialization of agriculture could create cash needs to buy inessentials such as beer at the expense of subsistence requirements. Similarly the penchant for expensive goods which only the rich can consume, supplants traditionally produced goods to the detriment of the welfare of the poor. Furthermore expensive products entail expensive production costs: high salaries for managers and professionals and relatively high wages to a small labour aristocracy. Public expenditure would also tend to be geared towards the 'needs' of this small enclave in the form of motorways, international airports, and expensive technical education for a few. Through these mechanisms inequality becomes self-reinforcing, and the poor are the losers again, in traditional goods and jobs.

Other reinforcing 'circuits' exist. An important by-product of inequality is the rise of professionalization, often due to the impact of penetration of western educational standards. This creates an immense gulf between 'technocrats' and the poor, and the definition of needs in important areas like health and education is monopolized by this tiny coterie. 'Basic needs' become those which only highly trained professionals can satisfy. The effect of professional self-interest in standard-setting is also reinforced by the unequal pattern of demand. Provision is based on market valuation and is not need-determined. Thus research on diseases that affect the poor receive low priority, as do alternative systems of providing health care and education.

It seems clear that in view of the above we cannot speak of basic needs regardless of where a country is in relation to the world market, who defines goals for it, or the degree of inequality that prevails and what economic system it adopts. Analysis also reveals the main obstacles to the satisfaction of basic needs to be: (i) the international economic system, characterized by huge inequalities between rich and poor countries in the context of a market system; and (ii) replication of this inequality in the domestic structures of countries on the periphery.

KEY ELEMENTS IN A BASIC NEEDS STRATEGY

Basic Needs as Amendments to Conventional Planning

If one accepts that the main forces retarding the satisfaction of basic-needs are those described above, then this has clear implications in terms of what the key requirements of a basic-needs strategy are. To begin with, it would rule out any narrow interpretation that reduces a basic-needs strategy to a mere technocratic exercise involving the defining, quantifying, and delivering of basic needs. In this view, once shortfalls from basic needs are quantified, production targets for alleviating poverty are defined and hence the necessary supplies of basic goods and services can be generated. Something of this arid approach can be detected in those views which claim that it requires 'only x billion dollars' to satisfy basic needs within y years. Such an approach ignores how and in what context basic needs are to be produced and distributed. It forgets the danger of 'clientelism', of people being reduced to the status of mere consumers without consideration of important ethical questions such as how and by whom basic needs ought to be defined. Whether this is done from above or by the people themselves in decentralized, self-reliant units would affect not only what is set as a basic need, but also how people will obtain and use these goods and services.

A more elaborate variant of this approach would be to regard the basic-needs approach as a set of amendments to existing targets and instruments of development planning. Starting from the *status quo* in terms of economic structure, the distribution of assets and income and the institutional structure in which centralized planning operates, it is argued that appropriate modifications to planning criteria can lead to the full satisfaction of basic needs. These modifications centre around three elements. First, in terms of the planners' objective function, greater weight will need to be given to the satisfaction of basic needs. Thus in terms of project selection, commodities and services which contribute directly to the satisfaction of basic needs would be given priority. An extreme statement of this condition would be to assign a zero weighting to production of non-basic-needs items. Secondly, it implies the adoption of a

> high rate of time preference for the near future, reflecting the urgency of meeting basic needs soon, subject to maintaining achieved satisfactions of basic needs indefinitely. ... The operational implication of this is that measures to raise consumption of the poor now and in the near future, as long as they are conducted on a sustainable basis, will be acceptable even if they reduce capital formation for future consumption growth below what it would otherwise have been, but that this sacrifice is reduced by the bonus we derive from investing in future generations and reducing population growth. (Streeten, 1977, p. 3.)

Thirdly, it argues for greater selectivity and disaggregation in the choice of instruments for eradicating poverty; attention should be focused on getting specific goods and services to specific groups, instead of relying only on income growth.

These modifications, it is argued, will 'make it possible to satisfy the basic human needs of the whole population at levels of income per head substantially below those that would be required by a less discriminating policy of all round income growth' (Streeten, 1977, p. 4). First, resources would be saved on objectives with lower priority than basic-needs satisfaction. Secondly, the existence of complementarities between various basic needs will lower the total cost of implementing a basic-needs 'package' as compared to the alternative of attempting to supply basic needs in an unco-ordinated fashion. For instance, improved nutrition will contribute to greater labour productivity and better health, and hence will reduce the total resource cost of a basic needs programme. Thirdly, a shift in the composition of output towards more appropriate basic products is likely to raise employment, because such products will be produced more labour-intensively. This increase in employment in countries with underemployed labour, will be an important means of granting a claim to income and hence access to basic needs, as well as increasing production.

While these arguments are logically consistent and appealing, they do gloss over the institutional and political obstacles that stand in the way of the actual implementation of such a basic-needs plan. The point can be illustrated in several ways. It might be thought for instance, that satisfying basic needs consists of the relatively simple matter of implementing production targets for increasing the supply of basic goods and services. There are, however, serious obstacles to be overcome before this can be done. Given a scarcity of resources, the new production plan would involve a shift in the composition of output away from luxuries towards necessities. The existing structure of output, however, is not a mere accident of economic history. Where luxuries predominate they do so because they are the very *raison d'être* for the rich and are functional to the system. There would clearly be political problems involved in altering the composition of output, overcoming the vested interests of producers and importers of these commodities, diverting resources away from sophisticated medical and educational services towards systems geared to meeting mass needs, and overcoming the vested interests of professions committed to international standards of provision (Lee, 1977, pp. 69–70). It is also relevant that in a market system some basic-needs items such as education are 'positional goods' and that competition in terms of access to these goods will prevail.

Another way of presenting the obstacles to implementing a basic-needs plan without relying on fundamental structural changes would be in terms of the need to ensure consistency between production plans and effective demand. Even if it is assumed that the target bundle of basic-needs goods and services is produced, there would still remain the problem of ensuring the pattern of effective demand is altered, to ensure that the poor are actually able to purchase these. This immediately brings us back to the thorny political problem of how to effect an appropriate redistribution of incomes (Khan, 1977). The problem is further compounded if we consider that reliance on standard macroeconomic instruments for redistributing incomes may not be sufficient to ensure fulfilment of basic needs. For example, fiscal policies will have only a relatively limited impact in situations where income taxation captures only a small proportion of the economically active population. In addition there are a set of supply management problems which need to be tackled in a situation where a sharp redistribution of incomes is attempted within a short period. Raising the money incomes of the poor can lead to serious disequilibria in the markets for basic goods in situations where the supply elasticity for such goods is highly inelastic in the short run.

> The dynamics of the micro-economic adjustment processes are likely to be such as to cause a partial frustration of the attempted transition to a more egalitarian society and possibly even a macro-economic break-down. The disequilibrium situation will be manifest in three forms: an excess demand for wage-goods (or basic needs goods), an excess supply of goods consumed by upper income groups (or luxury goods) and, almost certainly, an excess demand for imports; that is, a redistribution of income in favour of the poor is likely to result in a pattern of demand which in the short run is inconsistent with the pattern of production and which, because of the magnitude of the problem, cannot be reconciled through international trade. In consequence... those low income persons who received the initial rise in money income will find that their real incomes will not have risen correspondingly because of the adverse change in relative prices. (Griffin and James, 1977, p. 3.)

Furthermore, the specific way in which income is redistributed can have differential effects on different subgroups of the poor. For example, high producer prices for food will benefit poor farmers but harm the urban poor, and measures to raise minimum wages will benefit the employed in the formal sector, but harm the unemployed and small employers. These standard macro-economic measures are thus blunt instruments for guaranteeing universal access by the poor to basic needs.

There are also other institutional constraints which are assumed away in the conception that it is possible to ensure the fulfilment of basic needs through a mere change in planning targets and criteria. The required basic-needs production may not be feasible because 'given the social institutions, some highly desirable technological alternatives would remain irrelevant from an operational standpoint' (Khan, 1977, p. 107). An example of this would be situations where 'the development of land through capital construction by otherwise unemployed labour is rendered impossible in a society with unequal distribution of land ownership' (Khan, 1977, p. 108). In addition, it is possible there may be institutional constraints to generating sufficient employment at 'high enough wages to ensure effective demand for the basic-need consumption bundle' (Khan, 1977, p. 105).

The Need for a Comprehensive Approach

It is clear from the preceding arguments that it is misleading to suggest that a basic-needs strategy consists of no more than a few amendments to planning criteria and that, once done, this would ensure fulfilment of basic needs within the existing economic structure. On the contrary, it appears that in many situations fulfilment of basic needs may be ensured only if there are changes in distribution of assets and the institutional structure.

Indeed, a primary advantage of the basic-needs strategy is that it leads to consideration of the total set of changes required to ensure sustained satisfaction of basic needs in a given situation. Although it is impossible to generalize, certain types of changes can be identified as being necessary parts of a basic-needs strategy in low-income countries characterized by the existence of mass poverty, substantial inequalities in the distribution of income and wealth, and a dependent relationship in the world economy.

In such a situation there would be several arguments for a redistribution of assets as part of a basic needs strategy. One is that this would be necessary in order to have a lasting redistribution of incomes. This arises because a transfer of income to the poor can be nullified by secondary effects if basic-needs goods are produced by factors of production owned by the rich (Griffin and James, pp. 18–19). Another reason why a redistribution of assets may be required is that it is a precondition for change in the organization of production and payments systems necessary to ensure the satisfaction of basic needs. Thus in a densely populated country with a high degree of landlessness and land concentration, a lasting solution to the problem of rural poverty will require a land reform that guarantees the rural poor access to land and livelihood. Furthermore, a redistribution of assets has also a pre-emptive function; it ensures against a sharp increase in inequality in the process of economic growth (Stewart and Streeten, 1976).

Changes will also be required in political, economic, and administrative structures. A central element is that there will need to be a shift towards a greater degree of decentralization in planning. Such a change would be towards a more self-reliant pattern of development which would not only allow a shift towards a needs-oriented society but, more importantly, reinforce the changes in that direction. Self-reliant communities within an egalitarian and decentralized system of planning would be conscious of their specific needs, both individual and communal, and able to respond to these needs by mobilizing local ingenuity, resources, and effort. Highly centralized delivery systems within unequal institutional structures are likely to fail because benefits are appropriated before they reach the poor, and often take the form of a distant planner/technocrat's conception of needs rather than that of the poor themselves. In contrast, self-reliance ensures that decisions on what to produce are the result of a direct decision-making process involving the poor, and not those of a 'handful of capitalist/entrepreneures marginally manipulated by planners' (Galtung, 1976). As a consequence, the structure of production and distribution would become more closely related to needs.

The above considerations in fact define an area where a basic needs strategy coincides with a 'bottom-up' strategy that stresses the importance of the spatial aspects of development. An emphasis on local self-reliance and the promotion of integrated economic curcuits within the developed regions is perfectly consistent with a basic needs approach. In particular, many basic services are territorially organised and 'can in fact be more efficiently provided at smaller scales'. (Stöhr, see Chapter 2).

There are other related reasons for greater decentralization. One is that

> it is beyond the capacity of the wisest central planning organisation to devise ingenious methods of mobilising unused resources at different local levels that differ widely with respect to local circumstances. Much more than in an advanced industrial society, planning for the removal of poverty in a poor society, requiring unorthodox methods, must be decentralised. (Khan, 1977, p. 110.)

Another reason relates to the fact that the majority of the poor in developing countries live in rural communities which are often isolated and lacking in adequate transportation links with major markets. For such communities a greater degree of self-sufficiency in food production than that dictated purely by calculations of comparative advantage would often be a better strategy for ensuring adequate levels of nutrition among the local population (UN, 1975).

It also bears repetition that administrative changes such as the institution of rationing and price control may be necessary during a transitional period to ensure equitable distribution of basic-needs goods (Griffin and James, 1977). In addition, a shift to a decentralized system of self-reliant communities will require changes in institutions for production and distribution at the local level, in order to ensure an appropriate balance between individual incentives and a guarantee of access to basic needs for all.

A system with highly unequal ownership of productive assets is unlikely to generate sufficient incentives for communal resource mobilization and production since benefits will be very unevenly distributed. In addition, a payments system based on wage labour in situations of highly unequal economic power and surplus labour will not guarantee adequate access to basic needs for all. Alternative systems would thus have to be devised which ensure distribution according to needs while at the same time maintaining individual incentives either through the use of 'non-material' incentives or by supplementing basic guarantees with a system of differential payments based on work effort (Stewart, 1978). The payments system in Chinese communes, for instance, involves basic guarantees based on needs, the use of non-material incentives and a system of differential rewards under the work-point system. Mechanisms would also need to be devised to deal with the problem of inequality between communities and regions in such a way that incentives are not impaired. A crucial issue in this connection would be to find an appropriate balance between the level of resource extraction from local communities, the need to preserve the collective incentives to such units, and the possible need for redistributive measures to correct the differences in resource-endowment between different communities.

A final element in a basic-needs strategy involves the re-evaluation of countries' economic relationships with the rest of the world. This does not mean a total de-linking, but rather greater selectivity in international economic relationships instead of regarding 'openness as a good thing *per se* and to let external links and foreign demand determine the direction and pace of the country's economic growth' (Diaz-Alejandro, 1978, p. 111). The pattern of trade will need to be adjusted to render it consistent with a new pattern of production under a basic-needs strategy. For instance, the import of luxuries needs to be curtailed, and in the interests of ensuring 'that basic needs of a country are not at the mercy of the caprice of international markets . . . a substantial margin of preference will be given for local production of these goods' (*ibid.*, p. 112). Similarly, 'direct and indirect controls will be substantial in the areas of foreign finance and direct foreign investment' (*ibid.*, p. 122). Such controls will be necessary to ensure that distortions in the domestic structure of production, distribution of incomes, and choice of technology caused by the activities of multi-nationals are removed. However, these are not arguments for a total rejection of comparative advantage. Rather it is argued that

> comparative advantage calculations should [not] be based on the assumption of the optimality of free trade. Such calculations should be based on the criterion of maximising the basic needs objectives from available resources. If specialisation and exchange leads to the attainment of a higher level of basic-need satisfaction from given resources then it is misplaced heroism to condemn the nation to autarchy. (Khan, 1977, p. 113.)

CONCLUSION

A central conclusion is that the basic-needs approach does not offer a simple short cut for solving the problem of mass poverty. Changes in planning targets and policies will need to be accompanied by a redistribution of assets and institutional change. These changes, however, can only come about with a restructuring of political power, and it is clear, even from the framework of basic needs analysis that we have used, that there are formidable obstacles to attaining this.

Apart from the rather obvious point that a redistribution of wealth and income will be resisted by strongly entrenched interests, it is also important to realize that seemingly innocuous prescriptions such as 'greater decentralization' and 'greater local self-reliance' conceal potential conflicts of interest. The old pattern of development based on 'top-down' centralized planning has created vested interests which would resist a shift towards greater devolution of power to regional and local units. Resistance would come from the centralized bureaucracy, from oligopolistic industrial interests created by import substitution, landlords, multi-nationals and the unionized labour aristocracy.

As an economic slogan 'basic needs' has its uses in showing the material deprivation of the poor more vividly than in aggregate measures of income inequality, and in focusing attention on the economic and political obstacles that stand in the way of ensuring the fulfilment of basic needs. As a political slogan, however, it is less appealing. To leaders of the Third World preoccupied with modernization and the extension of national power, 'basic needs' is unlikely to prove attractive as a rallying call to inspire national mobilization.

REFERENCES

Adelman, I. and C. Taft Morris (1973) *Economic Growth and Social Equity in Developing Countries* (Stanford: Stanford University Press).

Amin, S. (1978) "Development Theory and the Crisis in Development", in *Emerging Trends in Development Theory* pp. 13–18 (Report from a SAREC Workshop) SAREC Report R 3, 1978 (Stockholm; SAREC).

Beckerman, W. (1977) Some reflections on redistribution with growth, *World Development*, **5** (8), 665–76.

Chenery, H. H. S. Ahluwalia, C. L. G. Bell, J. H. Duloy, R. Jolly, (1974) *Redistribution with Growth* (London: Oxford University Press).

Diaz-Alejandro, C. F. (1978) Delinking north and south: unshackled or unhinged, in A. Fishlow *et al.* (eds.) *Rich and Poor Nations in the World Economy* (New York: McGraw-Hill).

Galtung, J. (1976) *Self-Reliance: Concepts, Practice and Rationale* (Mimeo, Geneva: UNCTAD).

Ghai, D. P. (1977) What is a basic needs approach to development all about; in D. P. Ghai *et al.* (eds.), *The Basic Needs Approach to Development*, pp. 1–18 (Geneva: ILO)

Griffin, K. and J. James (1977) *Supply Management Problems in the Context of a Basic Needs Strategy* (Mimeo, Oxford: Queen Elizabeth House).

McNamara, R. S. (1976) *Address to the Board of Governors* (Philippines: Manila) (The World Bank, Washington).

ILO (1970) *Towards Full Employment* (Geneva: ILO).

ILO (1976) *Employment, Growth and Basic Needs: A One-World Problem* (Geneva: ILO).

ILO (1977) *Poverty and Landlessness in Rural Asia* (Geneva: ILO).

Khan, A. R. (1977) Production planning for basic needs, in D. P. Ghai *et al.* (eds.), *The Basic Needs Approach to Development*, pp. 96–113 (Geneva: ILO).

Lee, E. (1977a) Development and Income Distribution: A Case Study of Sri Lanka and Malaysia, *World Development*, **5** (4), 279–89.

Lee, E. (1977b) Some normative aspects of a basic needs strategy, in D. P. Ghai *et al.* (eds.), *The Basic Needs Approach to Development*, pp. 60–71 (Geneva: ILO).

Morawetz, D. (1977) *Twenty-Five Years of Economic Development 1950 to 1975* (Washington: The World Bank).

Stewart, F. and P. Streeten (1976) New strategies for development: poverty, income distribution and growth, *Oxford Economic Papers*, **28** (3), 381–405.

Stewart, F. (1978) Inequality, technology and payments systems, *World Development*, **6** (3), 275–93.

Streeten, P. (1977) *The Distinctive Features of a Basic Needs Approach to Development* (Mimeo, Washington, DC: World Bank).

Streeten, P. (1978) Development ideas in historical perspective, *Internationales Asienforum*, **9** (1–2), 27–36.

UN (1975) *Poverty, Unemployment and Development Policy: A Case Study of Selected Issues with reference to Kenya* (New York: United Nations).

Development from Above or Below?
Edited by W. B. Stöhr and D. R. Fraser Taylor
© 1981 John Wiley & Sons Ltd.

Chapter 5

Growth Poles, Agropolitan Development, and Polarization Reversal: The Debate and Search for Alternatives*

FU-CHEN LO and KAMAL SALIH

INTRODUCTION

A debate has arisen over the approach in regional development appropriate to the needs and conditions of the underdeveloped countries in the Third World: It revolves around the role of growth poles in solving the problems of polarized development and its manifestations such as urban primacy, regional inequalities, and rural stagnation. In both its technical and ideological dimensions, the issue reflects the wider concern over the theory and practice of development itself, based on increasing recognition that the models and strategies of the early 1960s, given the crisis of world development in the 1970s, have not achieved the real goals of development, namely, the well-being of people and human development, particularly in the Third World. In seeking a new concept of development and alternative strategies for the 1980s, the universalism of earlier theories has now been abandoned for a plurality of approaches based on the needs, history, and prevailing conditions of the individual countries concerned.

Located in this proper and wider context, regional development—like economic growth, cultural development, ecological balance, and structural transformation—is an instrumentality, the necessary means to achieving the goals of human development.

The aim of this chapter, first within this *problematique* and second in an attempt to find alternatives in regional development in the Third World, is to critically examine the growth pole approach and its purported converse, agropolitan development, in the context of the continuing debate over the issue of

*This paper is a revised version of 'Regional disparity and city system: standing on one leg in Third World regional development', presented at World Regional Development and Planning Conference, University of Tsukuba, Japan, August 1978.

polarization reversal (UNCRD, 1976; Richardson, 1973, 1977). This issue is, to our minds, the essential core of the question posed in this volume of top-down versus bottom-up approaches in regional development, the former (Hansen, Chapter 1 in this volume) being associated with growth pole strategies and the latter (Stöhr, Chapter 2 in this volume) with approaches such as agro-politan development.

The thesis of our chapter is that existing solutions proposed for Third World countries are partial and exclusive, and that this is due to an improper articulation of the regional development *problematique*.

In the following sections, the polarization reversal issue is examined, followed by a critique both of the growth pole and agropolitan approaches. In the final sections, we outline a scheme which incorporates a macrospatial framework and a model of regional structure in which the issue of polarization reversal and the real *problematique* of regional development may be resolved. In conclusion, we outline a number of important considerations in seeking alternatives in regional development as an input towards a more appropriate theory and practice of development in general.

DEBATE ON POLARIZATION REVERSAL

The kernel of regional policy, which refers to choices regarding the nature, magnitudes, and form of allocation of investments among sectors and regions for different periods of a planning horizon, is founded on the thesis of divergence and convergence. The debate which ensued in the approach to Third World regional development counterposes on the one side proponents of market-equilibrating mechanisms to speed up the convergence of regional incomes without sacrificing aggregate national growth (Hansen, Chapter 1 of this volume), and on the other, proponents of a more radical intervention aimed at achieving greater equalization sooner, based on the view that market forces are disequilibrating and that regional convergence is not an automatic process (see Chapter 2, by Stöhr). This issue, especially the scale of intervention and the time-frame involved, is what Richardson (1973) has termed 'polarization reversal'.

The theoretical basis of the strategy which emerged is easily summarized. The process of spatial development in a national economy is said to consist of three components, namely (1) the onset of industrialization in a national economy is based upon economic expansion in one, two, or a few limited regions, leaving the rest of the economy relatively backward; (2) subsequent national economic development is associated at some stage with dispersion into other regions, a process which tends to integrate and unify the national economy; and (3) independent of the polarization and subsequent dispersion tendencies interregionally, growth within regions always tends to be spatially concentrated (Friedmann, 1976; Richardson, 1973).

As is by now well-known, two empirical studies by Williamson (1965) and El-Shakhs (1972), stood out, and lent some credibility to these classic works on polarized development. In 1965, the publication of Williamson's empirical test of the patterns of regional disparities as related to national economic growth supported the hypothesis of divergence and convergence introduced by Hirschman (1958) and others. About the same time El-Shakhs, in his work on the evolution of the city system, argued that urban primacy increases then declines in the process of development. This provided an empirical basis in the search for an optimal settlement policy for national development, and support for the role of growth centres and the city system in regional development. It is a sad commentary on the state of regional development theory that the link was never quite made between the city system and regional growth which would have served as a basis for a macro-location theory in regional development (Alonso, 1968, pp. 1–14; Stöhr, 1974).

The basis for regional development policy was thus theoretically and empirically established. It emphasized: initial concentration on fast-growing areas; out-migration from, together with ameliorative social programmes in, the lagging regions; letting market forces take their natural course for subsequent trickling down to the periphery; and eventual nationally integrated development when the economy achieves the state of full industrialization.

The record in many developing countries attempting to imitate and implement this strategy has, however, not been encouraging. Given the climate of changing concepts of development and the current challenge to established practice which characterize the development field in general there is, in fact, a certain crisis of confidence in the above strategy of accelerated regional development through the growth pole approach. For Asian countries, this is documented by UNCRD (Lo and Salih, 1978). In this connection, adoption of the growth pole approach by many Third World countries reflects two underlying forms of wishful thinking: first, that industrialization with modern technology can be decentralized to the benefit of rural areas; and, second, that national integration through the growth pole strategy can solve the problem of regional underdevelopment (*ibid.*, p. xiii). There is thus a great need to reassess the strategy, and to seek alternatives in regional development.

The broader developmental issue—namely the role of regional development strategy in achieving the goals of human development in which the question of development from above or below becomes critical—is still, however, unresolved. With the increasing concern about distributional issues in economic development which dominates development thinking today, this central issue has become even more relevant and can no longer be set aside shrouded in what can only be described as a *spatial mystification* of regional inequality. In order to examine it in more realistic terms, however, we need to consider its broader structural dimensions within the emerging new paradigms of development.

It can be shown that the processes behind polarization reversal are much more complex than outlined in earlier works. This complexity was obscured by the fetishism of space which, as alluded to earlier, is a common affliction of the regional planning profession. Consequently, research effort was diverted from examination of the crucial relationships between spatial and socioeconomic processes in development. In this instance, it is useful to see the connection by first dividing the phenomenon (of divergence and convergence) into its regional and personal (or household) income distribution components. It will be seen then that Williamson's (1965) findings regarding regional disparities, and El-Shakhs' (1972) data on city-size hierarchies parallel similar findings about the trajectory of personal income distribution with economic expansion of countries. In 1954, Simon Kuznets, on the basis of his studies of income and savings, enunciated what is now known as Kuznets' law, which states that: 'During the beginning of decades of a nation's entry into modern growth, the distribution of income tends to become unequal and only after some time will tendencies towards equality begin to dominate over forces tending to widen the distribution' (Kuznets, 1955, p. 62). Chenery extended this in his empirical studies of the patterns of transitional growth in developing countries, and is said to have provided some rational basis for the replicability, with local variations, of the western experience (Chenery, 1977 in Ohlin et al., 1977, pp. 457–89). The conclusion thus is obvious: to the extent that a universal pattern of development can be discerned, Third World countries should try to industrialize as rapidly as possible. Problems associated with inequality, in personal or regional income terms, are merely problems of transition. Thus all the measurable phenomena associated with polarization and polarization reversal—personal income distribution (Kuznets' law), regional income inequality (the Williamson hypothesis), and urban primacy (the El-Shakhs hypothesis)—are a function of maximal economic change, as indicated by the acceleration of income per capita through industrialization. Japan, for which data are available for all four indices over a long period, clearly demonstrates this experience, as shown in Figure 5.1. The turning points for the three measures of inequality occur roughly in the period early to mid-1960s.

Now, if all Third World countries are to follow this experience, two questions need to be asked, both important in context of the necessity to achieve polarization reversal and attain a more even pattern of development. First, what are the objective conditions for polarization reversal? Second, how soon can a turning point be reached? These questions have been asked before, but never satisfactorily answered. This has been due to the fact that to many authors, the reality of the trickle down process has been accepted almost as an article of faith. But the problem of persistent inequality is a serious one. The current situation, as can be seen in Table 5.1 and Figure 5.2, for the rest of Asia outside of Japan, a large part of the Third World, is potentially explosive and warrants immediate action. The current situation in these countries is different from the Japanese experience. Most underdeveloped countries are starting with both a lower per capita income and a higher degree of inequality. In fact, these simple indices of inequality conceal even deeper problems arising from the pursuit of rapid industrialization. Oshima argues:

In retrospect, this strategy of development together with the unprecedented growth of population appears to have produced a series of maladjustments, dislocations, and imbalances which make untenable the continuation of past policies. These are: subs-

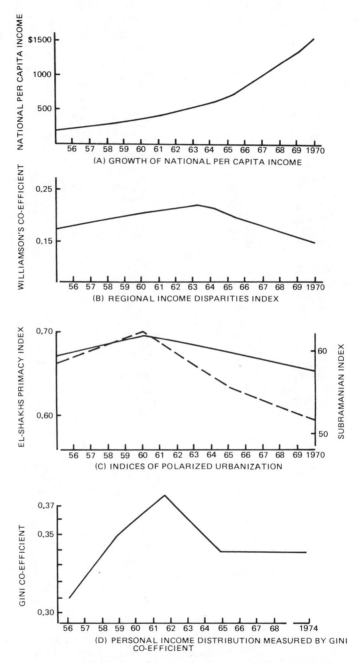

Figure 5.1 Trajectories of per capita income, regional disparity, urbanization, and income inequality for Japan, 1956–70 (Lo and Salih, 1978)

Table 5.1   *Gini coefficients of income inequality and per capita GNP for selected Asian countries*

	Year	Gini coefficient	Real per capita GNP ($US, base year 1970)	Growth rate per capita GNP, 1965–74
Southeast Asia				
Indonesia	1964/65	0.35	75	4.1
	1971	0.46	83	
Malaysia	1957/58	0.43	231	3.8
	1967/68	0.51	340	
	1970	0.51	380	
Philippines	1961	0.50	172	2.7
	1965	0.50	185	
	1971	0.49	216	
Thailand	1962/63	0.41	144	4.3
	1968/69	0.43	186	
	1971/73	0.50	212	
South Asia				
Bangladesh	1966	0.34	77	− 1.9
	1974	0.44	66	
India	1953–55	0.34	92	1.3
	1961/62	0.41	95	
	1968/69	0.43	102	
Pakistan	1966	0.36	96	2.5
	1968	0.34	100	
	1969	0.34	102	
	1970	0.33	100	
Sri Lanka	1968	0.38	99	2.0
	1973	0.35	118	
East Asia				
Japan	1956	0.32	504	7.0
	1959	0.36	704	
	1962	0.38	939	
	1968	0.34	1624	
	1974	0.34	2348	
Korea	1964	0.37	155	8.7
	1970	0.37	250	
	1975	0.41	373	
Taiwan	1964	0.33	288	6.9
	1971	0.30	460	
	1972	0.30	491	
	1975	0.30	600	

Sources:
 For Gini coefficients: S. Jain (1975) *Size Distribution of Income: A compilation of Data* (IBRD); H. T. Oshima and T. Mizoguchi, eds. (1977) *Income Distribution by Sectors and Over Time in East and South Asian Countries* (Tokyo: Council for Asian Manpower Studies).

 For GNP per capita: Asian Development Bank (1976) *Key Indicators*; UN, *1976 Statistical Yearbook*.

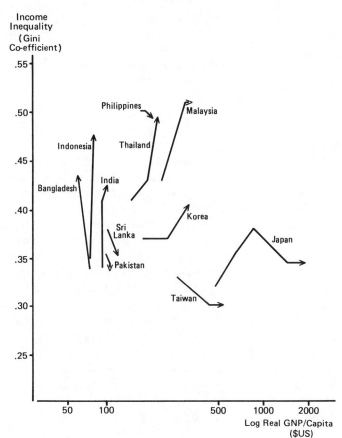

Figure 5.2 Trends in income inequality and per capita GNP in
selected Asian countries (for sources see Table 5.1)

tantial excess capacities in modern industries and insufficient food production; tendencies for imports to grow more rapidly than exports; unemployment in the urban areas,
underemployment in the rural areas, excessive rural-to-urban migration and the pro-
liferation of marginal jobs especially in the service sectors; regional disparities in the
availability of health, educational, cultural, welfare, and other social services. (Oshima,
1974, p. 8.)

Oshima posed two critical questions: How soon can Third World countries
achieve polarization reversal?; and has the increase in income so far achieved
trickled down to the lower income groups in the urban areas and diffused out to
the rural areas?

Taking all the relationships between personal and regional income into
account, while being aware of the possibility of ecological fallacies, it can be
argued that there are at least four objective conditions for polarization reversal:

(1) The attainment of economy-wide full employment;

(2) agglomeration diseconomies in the secondary sector in the core areas;
(3) achievement of interregional linkages and consequent spread effects on a mass scale;
(4) attainment of organizational complexity in corporate activities.

The issues subsumed under the latter two items are of course the basic foundation of much of the theoretical and empirical literature of modernization—diffusionist studies which advocate a top-down process within the framework of a system of growth centres. This has been widely dealt with in the literature with significant contributions being made by Pred (1973), Törnqvist (1973), Kuklinski (1975, 1978), Lasuén (1969), Berry (1972), Hansen (1972), Pedersen (1975), and Misra and Sundaram (1978), amongst others. But given the mounting evidence that spread effects from growth centres, even for 'natural' growth centres, have not contributed towards the reduction of regional disparities, this line of work cannot be overly promising, unless substantially modified in a manner suggested in this chapter. The first two conditions are perhaps the most important from the viewpoint of Third World countries and require further brief comment.

Regarding the first, the necessary conditions for the reversal of increasing regional disparities seem to reflect the maturation of the economy under the most favourable conditions and not merely a matter of spatial organization. The most fundamental of these appear to be the achievement of a turning point from a labour surplus situation to full employment (Oshima, 1977, p. 336), involving critical structural changes in the economy. These, however, cannot be analysed in isolation, viewing it from either the spatial or economic perspectives but in fact, for meaningful regional development policy, should be integrated into a common framework.

In this connection there is need to consider explicitly the question of employment possibilities in the non-agricultural sector and the employment absorption effect of the growth pole when discussing growth pole development policy (Lo and Salih, 1978). Because the rate of employment absorption in the modern industrial sector, which is occupied by the large-scale establishments, is always lower than that of the population increase, rural surplus labour, pushed out of the rural areas, flows to metropolitan or industrialized areas but cannot be absorbed by the modern industrial sector. As a result it forms a large social stratum of semi-unemployed (disguised employment) engaging in informal sector activities and leading to squatter formation. In short, the pattern of industrial development in Asian countries has engendered a pattern of spatial duality, reflecting specific population pressures and resource situations in each country and its stage of economic development.

The process of Asian urbanization thus resulted, through premature rural–urban migration and limited industrialization, in extreme duality in the primate city and continued underemployment in the rural areas. Given the high population growth rates anticipated in Asia which will have over half of the world's population by the end of the century—and that over half of this number, further-more, will be in cities of over a million—and given the long-run asymmetry of rural–urban relations under current structural conditions, the question of labour absorptive capacity, which currently concentrates only on a few cities, becomes a critical issue in any regional policy alternative.

The second condition relates to the debate on the possibilities and the appropriate strategy for polarization reversal centring on the problem of decentralization from primate cities. Originally focussing on optimum (or efficient) size of cities (see Alonso, 1968; Richardson, 1972, 1976; Mera, 1973; and Gilbert, 1976), the debate had been whether or not, and when, deglomeration economics will set in the primate city, which as always in the end becomes an ideological issue.

We have shown again elsewhere that this issue of optimum city size is in-correctly posed and, in fact, is irrelevant to the question of regional disparities (Lo and Salih, 1978). It was relevant only from the point of view of decentralized development, but not from the vantage point of other possible alternative strategies. By introducing the notion of comparative sectoral efficiency, rather than an aggregate measure of productivity of cities, it is possible to conclude that in considering the comparative advantages of agglomeration economies for industrial location in a system of cities, it is quite conceivable—even in the case in which the growth of a large metropolitan region exceeds the optimum size of the city in terms of maximum agglomeration economies for a certain type of industry—that decentralization of such industrial activities will not take place if the agglomeration economies of the smaller cities, presumably in the less-developed regions, still remain comparatively smaller.

Furthermore, even if a decentralization of industries is possible because of planned efficiency in an intermediate growth centre, the comparative efficiency of the tertiary vector between those two cities may still suggest a higher labour-absorptive capacity for the primate city. Thus labour released from rural areas under premature rural–urban migration may continue to locate in the major cities. Given the observed nature of major dualistic Asian cities, and the manner in which the growth centres in some countries are being planned, the above reasoning is sufficiently testable and will probably be borne out.

That this formulation is more realistic is supported by data which show the structural complexity of most Third World cities, as indicated by the size of the urban informal sector in selected Asian countries. Thus, no simple-minded implication regarding policy towards the big city, and therefore to the role of the city system from a top-down perspective, can be drawn in regard to the strategy of polarization reversal. On two counts, a simple rural-to-urban transformation inspired by the strategy of rapid industrialization fails to achieve its assumed result: first, such a shift consequent on urban investment does not lead to employment nor to reduction of regional income inequalities in the urban area. Second, nor, given the rates of population growth in rural areas, will such induced rural–urban shifts of population lead to reduction of regional disparities.

There is also another point which must be noted. In pursuit of a concentrated industrial location policy, and thus inducement of rural labour to migrate into the target urban area, we cannot commit the fallacy of assuming homogeneous labour, and that any influx of labour will be immediately absorbed. Todaro's work (Todaro, 1976), and the labour surplus situation which creates large informal sectors in the primate cities of Third World countries, suggest that the other condition of polarization reversal, namely that of full employment, will be far from being attained. There is therefore the possibility of a prolonged lag period before polarization processes are reversed, and concentration can continue in the primate cities in spite of agglomeration diseconomies for certain sectors.

In general, then, given the unlikelihood of the above four conditions being met, spontaneous polarization reversal in the developing countries is unlikely to occur in the foreseeable future. Furthermore, in many of the developing market economies of Asia, the enclave nature of industrial growth through the growth pole strategy of concentrated decentralization further inhibits trickling down effects. At the same time, the penetration of the upper-circuit sectors, through the internationalization of productive activity and the increasing commercialization of agriculture, into the extremities of the periphery, tends to accentuate leakage of surplus from rural areas. These are some of the questions relating to rural–urban relations and problems which must first be properly understood before the issue of polarization reversal as formulated can be resolved. What are their dimensions and form? What are the mechanisms involved, and how are these induced by the unequal relations of production and distribution?

When these issues are considered it becomes clear that, for regional theory to make a meaningful contribution to regional development in the Third World, there is a need to shift the bias from mere articulation of urban-industrial problems in regional policy toward the handling of issues of rural development. More particularly the theoretical framework must be able to incorporate both rural and urban processes in development—in other words, the whole complex issue of rural–urban relations and regional disparities within an expanded view of the concept of development itself. The task is compounded by the variety of differing historical experiences among Third World countries, by differences in current conditions, and by structural and other differences among these nations as a group and the advanced metropolitan countries.

GROWTH POLE STRATEGY OR AGROPOLITAN DEVELOPMENT

The traditional response to the problems of spatial polarization accompanying regional integration, which ironically was proposed in order to destroy the centre–periphery structure of most underdeveloped economies, has been the growth pole or growth centre approach now being characterized as a strategy of top-down decentralization. This approach, however, appears at best to be appropriate only at specific historical moments and may be extremely difficult to apply in the prevailing conditions of most Asian countries (Lo and Salih, 1978). Instead, a reversal of this top-down approach may be called for. The success of such an approach, however, is equally and critically related to the basis upon which regional linkages are encouraged. The analysis suggested throughout this chapter leads to the unavoidable conclusion that an approach to regional development from below requires some degree of regional closure (see Chapter 2, by Stöhr). Without the ability to internalize the benefits of their own resources, regional economies outside of the core areas will continue to be subject to overwhelming national and international forces leading to the capture of these resources outside of the region with little, if any, observable top-down spread effects.

> ...If only we could in some respects treat a region as though it were a country and in some others treat a country as though it were a region, we would indeed get the best of both worlds and be able to create situations particularly favorable to development. Their advantage consisted largely in their greater exposure to the trickling-down effects and in their ability to call for help from the larger unit to which they belong. Their disadvantage seemed to lie principally in their exposure to polarization effects, in their inability to develop production for exports along lines of comparative advantage, and in the absence of certain potentially development-promoting policy instruments that usually come with sovereignty. A nation attempting to develop its own backward regions should therefore provide certain equivalents of sovereignty' for these regions. (Hirschman, 1958, p. 199, see also Stöhr in Chapter 2, p. 46).

It is in this connection that, against the mainstream of ideas (see Chapter 1, by Hansen) we suggest that the growth pole strategy is partial, and is inappropriate for the development of underdeveloped, peripheral regions. For the strategy represents an approach to regional development from outside, when in fact the source should be internal, at least in the early stages until an overall initial evenness is achieved.

Without rehashing and belabouring the discussion on the critique of the growth pole concept, several points can be mentioned to underline the view that the concept faces severe problems in application. We have already mentioned that the appropriateness of the concept cannot be separated from the historical context within which regional and national development proceeds. In particular, both dependency and the nature of other non-spatial policies being pursued are crucial to the success of any decentralization policy.

Secondly, there is the problem of creating 'leading' industries in rural regions under conditions in which most sectors of the non-farm economy are dominated by firms in the metropolis and overseas which have advantages of scale, access to innovation (locally generated or controlled though patent rights attached to the borrowing of technology) and control of markets, and in addition are protected by the anti-rural biases of the state. This leads to a third point which stresses that under these conditions there is great difficulty in generating local linkages between industries in the growth centre and between industries in the centre and these hinterlands. Even industrial linkages to raw material extraction, where they exist, have limited—both directly and in terms of more general hinterland effects—possibilities for local complementarities which might enhance the growth of market relations in the peripheral economy.

Fourth, where development can be defined as technological progress and structural change, both among economic sectors and toward higher levels of urbanization, the principles of nodality and concentration enhancing the development process give expected advantage to the role of large cities. The absence of intermediate-size cities may severely constrain the decentralization of development activities (as partially explained by the comparative urban efficiency hypothesis earlier).

Fifth, the impact of linkage is expected to be uneven, both spatially and temporally. The key question is how long this expected 'short-run' asymmetry will last before the 'long-run' move towards equilibrium begins. Again, under current conditions of dependency and high regional leakage, no turning point can be expected within, perhaps, several decades, unless the basic unequal structural relations are transformed.

Sixth, the asymmetry suggested above is manifested in chronic core–periphery inequality, the solution to which involves strategies of decentralization. These, however, cannot easily be implemented because sources of growth are largely outside of both the national core region and the underdeveloped regional economy which tends to continually reproduce relations of exchange through external investment, aid, and trade as well as via the ownership of key technology, production assets, and other resources.

Seventh, the concept of growth poles is sensitive to, but is unable to generalize, the question of scale and its impact on development. The size of the centre, the possibility of a system of centres acting as a dispersed area for 'concentrated decentralization' and extending the effective scale of the impacted area of concentrated development are germane to major theoretical contributions to the growth pole concept, but are difficult to assess in terms of the relative importance of internal and external sources of development and consequent underdevelopment.

Finally, growth pole strategies depend on some degree of mobility of capital and labour. However, the consequences of different elasticities of mobility are important in assessing different regional strategies. The attraction of agglomeration economies causes leakage of capital and labour out of non-metropolitan regions as the private returns to both capital and labour continue to be higher in the core areas. An essential task of a growth pole policy is to be able to attract and absorb labour in the pole.

These eight points reflect different perspectives on a similar set of issues for regional development in Asian countries. The regional typology we have derived, and the partial analysis of its implications, leads to the conclusion that incentives imbedded in current spatial patterns will lead to increasing inequalities; will fail to absorb labour into increasingly remunerative employment; and may seriously distort the production–consumption relationships intrinsic to and inseparable from engagement in, or linkages with, the larger system of development. Economic efficiency itself cannot be separated from patterns of demand reflected in income distribution patterns and preferences of consumers. If regional production is increasingly separated from regional consumption, i.e. if demand for regional production is external to the region, the results can *inter alia* be a failure to produce even basic-needs requirements of local people. Thus closure, a recapturing and internalizing (rather than internationalizing) of development dividends by the region, appears quintessential if integration into the larger system is to lead to equal sharing of development. The ideological rejection of the growth pole–centre concept, especially in Latin America, is based on the assertion that integration into the world economy on an unequal, essentially exploitative basis, is untenable.

To this point, we have argued that linkage between regional economies at different levels of development, because of the particular relationships between them, tends to reproduce unequal or asymmetrical development due to associated leakage. What then are the contours of a regional development strategy capable of meeting the requirements for autonomous, or nearly autonomous, self-sustaining rural development of which the agropolitan approach is one possible suggestion (Friedmann and Douglass, 1978).

A necessary condition for the pursuit of rural development is, in keeping with the theme of this chapter, to reduce rural–urban distortions through the creation of rural–urban linkages on a symbiotic, equal basis at lower territorial scales of interaction. In this context, we have also argued that small towns provide an essential focus for reducing leakage from agrarian regions which arise from interregional linkage (Taylor, 1979). These rural–urban units may be called 'agropolitan districts'. Essential features of these units include: (1) a relatively small geographical scale; (2) a high degree of self-sufficiency and self-reliance in decision-making and planning, based on popular participation and co-operative action at local levels; (3) diversification of rural employment to include both agricultural and non-agricultural activities, emphasizing the growth of small-scale rural industrialization; (4) urban–rural industrial functions and their linkages to local resources and economic structures; and (5) utilization and evaluation of local resources and technologies. Friedmann and Douglass (1978) suggest a territorial unit encompassing a rural–urban population of from 50,000 to 150,000, depending upon local population densities, within a one-hour commuting range by generally available transportation. The central 'town' may initially have between 10,000 and 25,000 people in most countries. The urban population is, however, less important than the consideration of the district as a single unit with effective decision-making powers.

The setting of planning on an agropolitan basis, and the achievement of the objectives defined by the strategy itself, also aim at achieving the necessary conditions and benefits of regional closure: (1) the internalizing of multipliers and external effects through the emphasis on local linkages and complementarities between agriculture and industry so as to initiate broad increases in incomes; (2) aided by policies to equalize the ownership of productive assets, including land, capital and publicly created assets such as irrigation systems, the redistribution of income which would create demand for wage goods and thus equalize the benefits of development and promote the satisfaction of basic needs.

By way of expanding on the agropolitan approach, we can ask two questions, the answers to which provide a summary of our basic arguments: first, how can the objectives of internalizing the development process and increasing access to development resources and benefits be achieved in the face of polarization processes occurring on a national and international scale? How can spread effects be captured from local and extra-local activities to the benefit of local development?

These questions are raised to first address the barriers to creating a from-below development process. These include penetration by international and extraregional forces in to the regional economy which may destroy the competitive position of weaker local producers; unwillingness to decentralize decision-making powers on the part of many central governments implying the need for a redefinition of the role of the state in development; lack of local decision-making resources, and the necessary rural organizational and institutional development; lack of financial resources for undertaking local development (and, parallel to this, the question of central subsidy for local development) and putting this development on a self-reliant footing; unequal asset/income distribution and class (and ethnic) differentiation among the local population, leading to unequal access (social and political) to publicly created assets for local development and thwarting the call for popular engagement in decision-making.

In considering these difficulties, further analysis concerning which political and socioeconomic system, the ideology of the development process, is appropriate in order to ground planning in its real social structural context. Lastly, how can agropolitan plans be aggregated in a non-Tinbergen fashion to reflect both local decision-making and higher-level development objectives? Can such a process be initiated and maintained within the current institutional/bureaucratic framework, or is substantial restructuring required as a precondition?

The answers to these and other questions are difficult to give in the absence of experience with what we have called the agropolitan approach. Similar approaches have been and are being adopted throughout Asia; most, however, lack the key ingredients related to the internalizing of regional development which a policy of regional closure has called for. Some efforts have been made including Communes (China), Integrated Rural Development and Block Development (India), Kabupaten Programme (Indonesia), Saemaul Undong (Korea), FELDA/DARA (Malaysia), SADP (Nepal), Agrovilles (Pakistan), Village Production Committees (Sri Lanka), and New Village Development Programme (Thailand). Each of these efforts, however, requires careful analysis within its own national context before any final judgment can be made. It is argued, however, that emphasis on the role of urban functions and rural linkage to small towns may be an unrealizable potential if supporting policies to equalize social access to the production and consumption of society's goods, services, and welfare are not simultaneously implemented.

COPING WITH DEVELOPMENTAL COMPLEXITY: THE NEED FOR AN ANALYTICAL FRAMEWORK

Within the goal of such an expanded epistemology of regional development, it would appear that any formulation without reference to the larger complexity of real situations will be meaningless. Further, abstracting from such reality and conducting our analysis without critical evaluation of our assumptions and frame of thought is no more than an ideological act, even if we claim scientific objectivity in our approach.

For this purpose, we should be aware of the new thinking in, and major paradigms of, contemporary development analysis which provide the various possible bases for our approach. Over the past decade a major paradigm shift in development thinking appears to be occurring, which at least recognizes three different approaches; namely, the neoclassical, structuralist, and the neo-Marxist schools of thought. A similar rethinking in regional development has also begun, in line with the broader definitions of development goals and processes (UNCRD, 1977). It is rather clear now, given the experiences of many developing countries, and the structure of their domestic and international economic relations, that the problem is much more complex, and that there is argument for heterodoxy in approaches depending upon the historical and prevailing socioeconomic conditions of each country. It is thus necessary, in the context of the current efforts towards a redefinition of development and a search for development alternatives on a national and international scale, that current regional policy be re-evaluated and reformulated.

In rethinking regional development, a number of important dimensions must be taken into consideration and be consistently integrated into one framework which will allow us to analyse the various interactions between spatial categories and economic processes within concrete social formations which occur at local, regional, and national levels. We outline below some of the possible new dimensions in rethinking regional development policy.

A.   The first issue to note is that *regional policy cannot be separated from considerations of national development policy and alternatives.*

In this sense, regional policy cannot be merely reduced to spatial development strategies without considering the socioeconomic and political processes which influence them. Spatial rearrangements can be achieved, for example in decentralizing industry by creating a counterpole to the major urban concentration, but unless the structural linkages between the growth pole and its hinterland are also created—and this cannot be achieved just by building roads or new industrial estates—it is not likely that there will be major impacts on the rest of the periphery. However, to the extent that such a strategy depends upon an accelerated industrial development policy at the national scale, it appears that the growth pole strategy is only appropriate under certain national conditions.

We should be aware that, even in this instance, the contribution of the growth pole approach may be partial. A second prong appears to be necessary in order to ensure the proper balance between urban and rural development. This may be based on the agropolitan approach which takes certain spatial and political forms focused, among other things, on the provision of urban services and rural industrialization, geared to the mobilization of local resources and increasing agricultural productivity.

The choices open to different countries in formulating their regional policy clearly, therefore, depend on the objective conditions of their national economy. The question of national spatial integration or more regional self-reliance in spatial development policy (as in remote rural Asia) is a very complex one which, under different conditions, could lead to enclave development and economic leakage, while in others could lead to greater regional sharing of development. The objective of regional policy analysis should be to study these different conditions and thus reformulate particular alternatives appropriate to the conditions.

B.   A second issue is that *regional policy formulation depends on the particular structure of the regional economy in relation to the nation as a whole.*

Such a framework needed to analyse regional development problems should take cognizance that while there are broad national processes such as economic leakage, migration, and labour utilization, etc. which occur on a national scale, the formulation of regional policy must be particular to the condition of the regional target group. The policy for remote regions must obviously differ in this sense from a policy towards densely populated agricultural areas, and in turn must be differentiated from a policy towards areas within metropolitan-centred regions. However, such regional categories cannot be separately treated as well, but in fact they must be properly integrated within the broad national framework. Area-based policies should thus be considered within a typology of regional structures in order to consider properly the operation of regional labour markets, the penetration of external forces into the region, and the internal social structural changes which are possible. These are areas requiring considerable further study and utilization and integration of existing knowledge.

C.   A third issue in reformulating our thinking on regional policy is that *our understanding of regional development processes should be based on rural–urban relations.*

There is much urban bias in our thinking about regional development, and in the policies adopted by governments at the national level. Such policies may have had unintended anti-rural impacts (e.g. increased rural poverty, unemployment, and lack of basic needs). On the other hand, strategies focused on rural development *per se* may have ignored the implications of broader urban processes (marketing channels, off-farm employment, rural industrialization) and may have exacerbated rural contradictions, such as introduction of seed–fertilizer technology without appropriate changes in rural social and economic structure, e.g. land-tenure, and problems of access to inputs and subsidy.

A new regional or spatial development policy must thus consider these socioeconomic processes in rural–urban terms, and should seek to transform distortions which may have appeared through the specific bases of development policy at various levels, in order to foster mutually reinforcing rural–urban relations for local regional development.

D.   Corollary to the above, regional policy should also view *the role of the city system as a whole, and incorporate fully the different functions of the cities at different levels in the hierarchy*.

The current literature seems to be overly focused on the issue of the big city, perhaps in response to the controversy over growth pole decentralization. We have shown in this chapter that this is a non-issue, and that big cities have their role to play in absorbing labour in the tertiary sector, while the question of diseconomies of agglomeration is a matter of internal organization in both spatial and non-spatial terms. What is more critical to us in terms of regional policy is the role of the entire city system. In particular, because of the particular conditions of most Asian economies, an equal emphasis should be given to the role of small towns in regional development.

In this, the question of quasi-agglomeration economies of the resource frontier, or the more general notion of propulsive regions introduced by Higgins (1978) should be studied further for the implications it has on nodal response and the role of lower order centres, and thus the possibilities for regional development from below.

The urban bias in industrial and agricultural policies which have led to primacy development in most countries has also given rise to the destruction of village communities and the disarticulation of small towns in performing their development roles. Thus a second prong, and in some countries the priority prong, should be the reorganization and revitalization of the settlements at the lower levels of the urban hierarchy (Taylor, 1979). The small town policy in regional development must, however, be formulated in broader regional terms and be cognizant of rural–urban relations in the rest of the settlement hierarchy.

E.   It is in the above sense that the new regional policy should, for many developing countries, principally consider the *territorial organization and development of rural communities centred on enhancing their rural–urban linkages*. This is a necessary precept for the issue of development from below.

This implies a further articulation of the agropolitan approach in a multi-level framework of regional–national development (Friedmann and Douglass, 1978). While the development of such communities in a regional system must involve some assistance from central decision-making (political) units, the attention should be on greater internalizing of development processes, and enhanced self-reliance through utilization of local labour and raw-material resources, increasing the organizability and participation of the local population, and some reorganization of asset-ownership structures in order to achieve a better distribution of income and pattern of labour utilization.

The spatial implications of this would be how to territorially organize and promote the necessary development of rural industries, marketing networks, and better physical access to amenities. It is when sources of autonomous development in the local communities are identified and integrated into the whole regional development policy at the national level that a more even development pattern can be achieved from below in the long run.

F. The final dimension is *the question of focus of regional policy within the emerging new development paradigm, which must principally be concerned with the development of poor communities.*

In the context of the wider development processes, the new regional development policy should be concerned with what is happening to the poor. How have national development policies affected them, for instance through the impact on rural–urban wages, the prices of essential goods (inflation), technology and their productivity, the terms of trade between their products and imported goods consumed, their employment opportunities on and off-farm, and meeting of their basic needs? Development programmes must be specially discriminated to meet the needs of this target group. In spatial terms, regional policy must be geared to improving the lot of those left behind—namely the question of *in situ* poverty—through the territorial organization of their communities in concert with other economic and social policies.

The main purpose is thus to consider how regional policy can contribute to the goals of reducing poverty, unemployment, and inequality in development. In a broader sense, we need to define what regional development alternatives are open to each country within the context of regional and national conditions (Lo *et al.*, 1978).

The factual basis for much of the necessary reformulation in regional policy in developing countries is at present inadequate. Studies will have to be carried out on regional needs, and the impact of national development policies at the local–regional level. The study of rural–urban relations and their distortions appears to be the crux in our understanding of regional development and underdevelopment. With better factual data on patterns, processes, and needs, it should be possible to begin to formulate more effective regional policy in national development. For these, however, an analytical framework should be developed, capable of incorporating and elaborating on the dimensions above developed; this we outline below.

A MACROSPATIAL DEVELOPMENT FRAMEWORK FOR RURAL–URBAN RELATIONS

The requirements for such a framework are two-fold: it must first explain key dimensions of development but, secondly, must allow for sufficient disaggregation to account for the variety of prevailing national conditions, including resource endowments and population pressure on them, levels of adopted technology, socio-institutional relationships governing access to different productive activities, and stages or levels of development. It seeks to provide a more holistic view of national development processes, and following this, to consider each of the respective cells or dimensions of the model in greater depth. Of particular concern in turning towards a rural basis development are the socioeconomic relations within the agriculture sector and the role of rural towns and small urban places in directing a process of planning from below.

An understanding of the patterns, processes, and consequences of rural–urban distortions is not possible without recognition of the national and international setting in which development occurs. Figure 5.3 summarizes the basic characteristics of this setting. The essential components of the macro-spatial model are (1) external relations resulting in north–south dualism; (2) dualism between formal and informal economic activities, which is also related to (3) a dualism between urban and rural areas. Meiers (1977) has suggested a similar framework using the term 'triple dualism' and in a related study, Reynolds (1977, p. 378) develops a simple model describing the transformation from a colonial to an enclave economy which integrates the formal sector in concert with external relations as against an isolated indigenous economy.

Five essential components comprise the macro-spatial framework (Figure 5.3):
(1) a world market largely composed of developed countries buying primary products from the Third World countries and exporting manufactured goods, particularly modern technology embodying capital goods, to them;
(2) an urban formal sector dominated by enclave foreign and domestically financed modern manufacturing and business of the corporate type;
(3) an urban informal sector consisting of a wide range of traditional activities, small in scale and characterized by such occupations as hawkers, vendors, daily labourers, and services which are distinct from the enclave sector and its related professional and white-collar occupations;
(4) a rural export sector generated in many cases from the plantation economy developed during colonial rule together with post-independence natural resource exploitation such as mineral extraction, oil, timber, etc.;
(5) a rural peasant economy historically isolated from the national and world market and dominated by peasants and landlords engaged mostly in food-crop production.

The three dualistic relationships—north–south, urban–rural, and formal-informal—all related and integrated in a complex manner which differs from country to country depending on the prevailing conditions related to resource endowments, technology, demography, and development ideology—are the context in which economic growth and development proceed. Within this general set of dependent structures and relationships, however, there is need for further clarification before specific development paths can be outlined. One area of importance concerns the combination of natural resources (agricultural land, forest, and mineral resources) and the level of technology in use and adopted

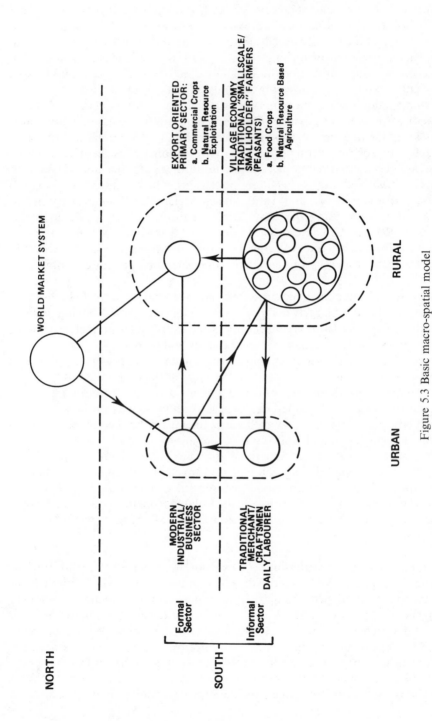

Figure 5.3 Basic macro-spatial model

on a massive scale. Nations with plentiful natural resources but with low technology levels, for example, will face a relatively different set of development issues than will nations with low resources and high technology. Migration in Indonesia, a high resource/low technology economy, may be characteristically distinct from that of South Korea, a low resource/high technology economy. Furthermore, these migration patterns may have different consequences for both growth and redistribution objectives of national policy. These issues, however, still revolve around conditions of dualism and the relationships within and between dualistic structures.

On the basis of this framework, then, we are able to discern four patterns (or models) of development for Asian countries. This could also without difficulty be used to classify other countries in Africa and Latin America: (1) countries which are rich in resources, relatively open domestic markets, and possess a relatively low level of technology (e.g. the ASEAN countries minus Singapore); (2) those with low per capita resource availability, large rural sector with low urbanization, and relatively low technology (e.g. the South Asian countries); (3) the centrally planned closed economy (the 'China' model); and (4) resource-poor countries with endogenized modern technology, and semi-open economy with protected domestic markets (e.g. Taiwan, South Korea and Japan) (Lo *et al.*, 1978).

The four models present the alternative development contexts within which approaches to a strategy of disengagement or integration may be examined. The different patterns of dependency with respect to the world economic system provide, more importantly, the basic structure within which processes of rural–urban distorations may be explained and regional policies toward their transformation analysed. But the impact of dependency relations and internal, unequal development cannot be separated when formulating such policies: they simultaneously provide the constraints and the possibilities for balanced rural–urban development in the rural regions.

The macrospatial model is useful for pointing to several propositions related to migration, rural–urban relations and internal development disparities. The more important of these are: (1) current international relations and national development patterns under strategies of accelerated industrialization in many cases increase rural–urban differentials; (2) these same relationships increase disparities among income groups; (3) as a result, urban and rural development become less articulate, and the capital city becomes the primary repository of vital economic and social functions; (4) although migration to the large urban centre is encouraged under this pattern of development, the results (largely due to formal–informal dualism and possible effects of development policies) may lead neither to an increase in rural productivity nor the absorption of labour into the urban economy.

TYPOLOGY OF REGIONAL STRUCTURES

The problem of evaluating the processes and structures of rural–urban relations and regional policy planning suggested in the macrospatial framework is that available models tend to treat external relations, dualistic structures, and spatial differentiation in growth and development as exogenous variables. In lieu of a general model to encompass these variables, the evaluation tends to be eclectic; policies are suggested which do not take into account the prevailing conditions in each Asian country. The necessary composite of approaches which would, for instance, allow for a more comprehensive treatment of labour mobility and national and regional development would include, as others have also suggested, a microeconomic analysis of the rural and urban labour markets, an analysis of rural–urban migration as the major linkage between rural and urban labour markets, and a macroeconomic analysis of employment as it is influenced by interactions in product and factor markets between dualistic sectors. Issues such as these, which are germane to analyses of rural–urban relations, imply the need for both an overview of the developing spatial system and a disaggregation of this system into relevant units for comparison and analysis.

While external dependency structures determine the patterns of rural–urban distortions on a macronational scale, rural–urban relations may vary in occurrence and degree on the basis of broad regional types within nations. In other words, in the later case the processes of rural–urban relations and their impacts are much more properly studied in the context of a typology of regions which spatially organize those relations. Thus, just as the policies for regional development will depend on the prevailing national conditions in relation to the world economic system, policies toward transformation of particular rural–urban distortions will also depend on the socioeconomic and political structures of a region in relation to other regions.

For these reasons, subnational regions must be distinguished by key dimensions capturing the dynamics of rural–urban relations. Four or possibly five regional types can be identified for this purpose:

(1) *metropolitan dominance*: the region comprising the primate city and the immediate surrounding countryside;

(2) *urban (metropolitan) shadow*: the region close to a major growth or dominant centre;

(3) *mixed rural–urban region*: symbiosis of a regional urban hierarchy within a relatively prosperous agricultural region—this category may be divided between regions having large urban centres (candidates for industrial decentralization) and those with large agriculturally based towns (which may act as centres to mobilize development from below);

(4) *rural dominance*: lagging region low in agricultural productivity and lacking in urban facilities and functions.

This typology is most fundamentally related to the concepts of accessibility and integration with the national centre. The metropolitan dominant region is in all cases the fastest-growing region of a nation; and along with the contiguous urban shadow forms the core region in the nation. In these regions, access to markets, resources, political power, and urban and rural functions are markedly superior to that of any other region in the nation. The urban shadow is the most likely candidate as beneficiary of any natural or programmed decentralization from the metropolitan region, including a highly advanced agriculture. But it is also likely to be the area in which industrial–agricultural land-use conflicts are most severe, as well as an area of chronic urban sprawl problems.

Between the rapidly advancing core and the relatively isolated periphery, regions with small and intermediate-sized cities and increasing agricultural growth may be found. As will be detailed later, the city size in these regions may critically affect the policy possibilities, especially those divided between a decentralization of industrial activity on the one hand and a town-centred rural resource-base development on the other.

In the rural-dominance regions, access to vital urban functions, markets, services, and institutions may be so low that a vast majority of the people have no means or incentives to increase local production to raise incomes and welfare in the region. These conditions may be further exacerbated by patterns of owner-ship of assets, monopsonistic markets, and institutional rigidities common in all regions but accentuated here by the very poverty of the people and the distance (social and physical) to sources of minimum support to advance local economies.

These four regional types—with some modification—serve to disaggregate the nation into regional economies reflecting structural relationships under current development patterns. To relate the regional types to labour markets and migration, the labour market may be described in each region by three characteristics: (1) industrial sector (manufacturing, service, and agriculture); (2) formal/informal dualism; and (3) location. The differential combination and transformation of the first two characteristics in space act both to explain migration patterns and to assess the effectiveness of existing labour markets in reducing internal disparities in the development process. The transformation itself is intricately associated with four dimensions, namely: (1) wage, income, and productivity differentials within and between regions and sectors; (ii) re-source and capital flows; (iii) production input and consumer markets; and (iv) social structures and institutions, including those affected by public policy. By explaining the influence of forces set in motion by, and in the context of, these factors, discussion can be moved toward an evaluation of policy alter-natives appropriate to the prevailing conditions and development opportunities of each region and nation. The four regional types from the basic structure of the national space-economy in which the urban hierarchy forms a basic skelton.

If the crux of regional policy is aimed at balanced urban–rural development in Asian and other Third World countries within the context of a new paradigm of development, then the framework to be developed should be able to focus on the extent of control and leverage we have over the interface between spatial structures and socioeconomic processes at different scales of interaction. At the level of national policy, the room for manoeuvre in transforming rural–

urban relations towards equitable development depends to a large extent on the different prevailing conditions in each country and on their relationships to the world economic system. It is clear that their different degrees of openness and dependency will define the options available to each country. These choices of policy will in turn have implications on internal dependency structures and the possibilities for reduction of rural–urban distortions arising from development process occurring at national and international levels.

At the regional level, on the other hand, rural–urban relations are conditioned by the basic socioeconomic structure of the region in question and its relationships with the national core areas. The pattern and degree of integration of various regional entities as territorial target groups within the national economy in the different cases will define, and in some cases constrain, or expand, the room for manoeuvre in regional development policy. In order to explain the patterns of urban–rural interaction and the level and nature of development in these different regional situations, the typology of regional structure needs to be made more explicit, refined and robust than the purely rural–urban distinctions found in the new development literature.

Together these two structural categories (the nation in the context of the world economic system and the region in the context of the national socioeconomic system) constitute the fundamental and integral spatial organization of rural–urban relations. This framework should allow us to examine the interface between development processes at the national–world level and the spatial structural interactions at the regional–national level, and thus better understand the processes of regional underdevelopment. This framework further represents the more important parameters within which policies for rural–urban transformation in regional development may be formulated.

REGIONAL LEAKAGE AND DEVELOPMENT FROM BELOW: TOWARDS A STRATEGY OF SELECTIVE REGIONAL CLOSURE

A major thesis of this chapter is that the dependency process of each Asian country has led to the fragmentation of their regional structures and thus to increasing rural–urban distortions. These distortions have been explained in the more 'classical' sense in terms of regional 'linkage' and 'leakage'. Friedmann and Weaver (1979), in discussing essentially the same phenomenon, have suggested that these distortions are a product of inherent conflict between territoriality and function with the current historical dominance by the latter under a world system managed by multi-national corporations. Regional economies are integrated or linked to a world economy on an unequal basis, leading to a polarization ('backwash') of development activities and leakage of vital regional resources out to the metropolis and abroad (Myrdal, 1957; Hirschman, 1958; Friedmann, 1966).

The regional policy question under the new development paradigm, which is aimed at reduction of disparities through regional self-reliant development, involves an understanding and solution of the leakage problem in a spatial context. The problem is essentially this: how can net rural–urban transfers of resources be channelled to favour the growth and development of rural areas? More narrowly, how can the agricultural surplus be retained in the rural areas, preventing it from flowing excessively to the urban areas to be reinvested for its own development? In this way, alternative employment opportunities can be created, enhancing local purchasing power, increasing the possibilities for rural industrialization and hence inducing the proper role for lower-order centres (see also Stöhr, Chapter 2).

In the question of regional policy, leakage implies two problem areas. First, the positive impacts of public expenditure programmes may accrue to groups outside the region, recalling here that regional policy is geared towards a territorially defined target population. Second, these impacts may benefit non-target groups within the target region, in a manner similar to the above discussion. Regional development programmes which are adopted and implemented without discriminating finely enough between target and non-target groups, and are unable to establish boundaries which exclude such non-target groups, are therefore bound to fail in their objectives if these two problems of external and internal leakage are not properly taken into account in the original allocation decisions.

It is thus crucial at this point to underline that in the problem of leakages the initial distribution of income and the institutional structure of the economy are critical elements determining the impacts of public expenditure (or pricing or any other strategy of public intervention) on target groups. In this respect, the effects of redistribution from growth, or an incremental approach, will be rather marginal. More radical programmes, or at least a radical mobilization of a fuller set of programmes which in the aggregate achieves such an effect, will be necessary to offset the in-built distorting market mechanisms for the distribution of surplus. What is required is an initial redistribution of income, or of assets (such as land) which determine income distribution. But the implications of such strategy go beyond the realm of economics into wider structural issues leading to solutions which society may or may not be able to absorb. Public expenditure in the less-developed areas serves as an instrument of regional transfer, which through hopefully localized multiplier effects would lead to a positive shift in the interregional income distribution and also intraregional inequalities. However, multiplier effects are not localized, and because of linkages with other regions, a net transfer of surplus value (through various mechanisms of the market, and the social institutions of exploitation) to the more-developed states (or regions) can occur.

The structures and mechanisms of leakage are complex (see Chapter 2). Leakage may occur through direct channels, such as the outflow of investment funds to the central region, or the repatriation of profits from enclave activities, or it may be 'hidden' as, for example, through internal terms of trade losses, through shifts in multiplier effects, through indirect taxes, and low compensatory public expenditure. But these processes vary in form and in effect on the local economy, depending on the agrarian structure and different degrees of regional integration and incorporation into the national economy, as well as by the physical and environmental conditions for production and surplus creation.

In searching for regional development alternatives under different conditions of production, surplus retention, and leakage, a concept of planning from below emerges to bring together discussion on, above all, ways of widening the basis of participation in development and to address the question of who is to benefit from it. In dealing with the problem of regional leakage and linkage, we are brought back to the suggestion by Hirschman (1957) as cited earlier, that the development of underdeveloped regions implies the need to protect it from the effects of regional polarization. In a general way we may describe this as a strategy of selective regional closure (see Chapter 2) whose specific forms will have to be defined with the experience and prevailing conditions of each country's situations.

It is not possible at this stage to elaborate on all aspects of such a policy of selective regional closure so as to achieve the required rural–urban transformation for regional development in one country.

We may, however, list the following elements:

(1) elements of temporary *economic protection* of regional–local industries, supported by a package of incentives (credit, marketing, subsidies, etc.) which enhances choice of appropriate indigenous technology and based on local resources;

(2) a unimodal strategy of agricultural development, associated with appropriate land reform, and appropriate agricultural incentives (price support, restructured land taxes, credit, distribution, etc.);

(3) creation of independent (non-governmental) peasant-based organizations which promote co-operative/collective participation in the management of development work and increased institutional access to income-generating assets;

(4) local provision of basic minimum needs (especially food and shelter protected from the necessary competition from outside) and urban public services (through enhanced urban accessibility);

(5) devolution of state-bureaucratic functions to increase participatory democracy and local-level decision-making regarding local–regional matters;

(6) evolving policy programmes that will disrupt continuous appropriation of capital by external forces and hence prevent excessive specialization of labour and therefore a fuller utilization of the total human resources within the region.

Implementation of programmes within a policy of selective regional closure recognizes that there should be an upper bound to the level of regional self-sufficiency. Within the context of local resource endowments, regional self-sufficiency should be pursued up to the level of provision of basic wage goods to meet at least minimum-need requirements, especially of food and shelter. In other words, a policy of selective regional closure must recognize that the region exists in a hierarchy of regions and supporting multi-level administrative apparatus. The design of this multi-level structure must allow for a proper allocation of discretionary power and co-ordination of activities of national authorities only to areas of jurisdiction which spill into other territorial authorities. In this connection, a system of fiscal and resource transfers may be developed in order to ensure that the elements of closure policy above are sustained.

It should be noted also that a multi-layer strategy of territorial closure within the international economic system would also be possible, in which national disengagement strategies, as well as systems of regional co-operation, will enhance even more the room for manoeuvre in subnational closure in order to promote regional development.

CONCLUSIONS: RESEARCH DIRECTIONS

We started this chapter by saying that its objective was to explore the critical dimensions of an essential problem of regional development, or the question of development from above or below, in order to redirect regional development theorizing towards relevant issues of development of peripheral nations in the Third World. In doing this, we emphasize that such a redirection involves recognition, as a basic element, of a new regional development paradigm and the inseparability of regional–spatial policy issues and development processes of the nation-state in a north–south world economic structure.

It was pointed out that, within the macrospatial framework we developed, the research for alternative regional policy is consonant with the particular policies of national disengagement from the world economic system in order to achieve self-reliance through following alternative independent paths of development. In this context we are suggesting that the room for manoeuvre in regional policy is constrained by the prevailing conditions and nature of integration of the national economy into the world economic system, where such incorporation has led to extreme unevenness in internal development, and in particular to distortions in rural–urban relations. However, the room for manoeuvre in transforming rural–urban relations in turn is constrained by the particular location of the region in question in relation to the typology of regional structures suggested.

Within the context of this framework, whose parameters must be empirically determined if it is to be of any use in fitting a particular regional experience, research should be directed to a number of areas which underlie the three key issues in urban–rural relations, namely, (1) the agricultural production system and its social relations; (2) the role of the small urban centres in articulating the process of primary accumulation in underdeveloped regions so as to enhance local production and distribution; and (3) structure of utilization and under-utilization of labour in fragmented labour markets. These areas of research include:

(a) the nature, extent, and consequence of internationalization (penetration) of productive activity on local enterprise within the particular settings of the regional typology;

(b) the impacts of relative public incentive structures (taxes, subsidies, tariffs, exchange rate policy, infrastructure development), and the role of state bureaucratic structures on different private and public allocation of resources (and pattern of distribution of surplus) between sectors and regions, and the consequences of these on personal income distribution.

(c) the role of social structure and change in determining the nature and extent of peasant participation in the development process and decision-making at the regional/local level.

The processes and structures which we are trying to identify are necessarily complex. However, in order to identify the real sources of autonomous development within regions, so that policy leverages may be properly located, it is important to separate causes from manifestations of problems. While the distinction is not entirely clear-cut, it is hoped that this chapter provides a conceptual framework for the organization of the diverse and complex processes which condition the pattern of uneven urban–rural development in the different countries of Asia in particular, and the Third World in general.

REFERENCES

Alonso, W. (1968) Urban and regional imbalances in national development, *Economic Development and Cultural Change*, **2**, 1–14.

Asian Development Bank (1976) *Key Indicators* (Manila: ABD).

Berry, B. J. L. (1972) Hierarchial diffusion: the basis for developmental filtering and spread in a system of growth centres in Hansen, N. M. (ed.) (1972) *Growth Centres in Regional Economic Development*, pp. 108–38 (New York: The Free Press).

Chenery, H. (1977) Transitional growth and world industrialization in Ohlin, B. *et al.* (eds.), *The International Allocation of Economic Activity*, pp. 457–89, (London: Macmillan).

El-Shakhs, S. (1972) City systems, primacy and development, *Journal of Developing Areas*, **7**, 11–36.

Friedmann, J. (1966) *Regional Development Policy: A Case Study of Venezuela* (Cambridge, Mass.: MIT Press).

Friedmann, J. and M. Douglass, (1978) Agropolitan development: towards a new strategy for regional planning in Asia, in Lo, F. and K. Salih (eds.), *Growth Pole Strategy and Regional Development Policy: Asian Experiences and Alternative Strategies* (Oxford: Pergaman Press).

Friedmann, J. and C. Weaver, (1979) *Territory and Function: The Evolution of Regional Planning* (London: Edward Arnold).

Gilbert, A. (1976) The arguments for very large cities reconsidered, *Urban Studies*, **13**, 27–34.

Hansen, N. M. (1972) (ed.) *Growth Centres in Regional Economic Development* (New York: The Free Press).

Higgins, B. (1978) Development poles: do they exist?, in Lo, F. and K. Salih (eds.), *Growth Pole Strategy and Regional Development Policy: Asian Experience and Alternative Approaches* (Oxford: Pergamon Press).

Hirschman, A. O. (1958) *The Strategy of Economic Development* (New Haven: Yale University Press).

Jain, S. (1975) *Size Distribution of Income: A Compilation of Data* (Washington: IBRD).

Kuklinski, A. (ed.) (1975) *Regional Development and Planning: International Perspectives* (Leiden: Sijhoff).

Kuklinski, A. (1978) *Regional Policies in Nigeria, India and Brazil* (The Hague: Mouton).

Kuznets, S. (1955) Economic growth and income inequality, *American Economic Review*, **XLV** (1), 1–28.

Lasuén, J. (1969) On growth poles, *Urban Studies*, **6** (2), 137–61.

Lo, F. and K. Salih (eds.) (1978) *Growth Pole Strategy and Regional Development Policy: Asian Experiences and Alternative Approaches* (Oxford, Pergamon Press).

Lo, F., K. Salih, and M. Douglass (1978) *Uneven Development, Rural Transformation, and Regional Alternatives in Asia*. UNCRD Working Paper 78–02, Nagoya, Japan.

Meiers, G. (1977) *Employment, Trade and Development* (Geneva: Institut Universitaire de des Hautes Etudes Internationales).

Mera, K. (1973) On the urban agglomeration and economic efficiency, *Economic Development and Cultural Change*, **21**, 309–24.

Misra, R. P. and Sundaram, K. V. (1978) Growth foci as instruments of modernization in India, in Kuklinski, A. (ed.), *Regional Policies in Nigeria, India and Brazil*, pp. 95–185. (The Hague: Mouton).

Myrdal, G. (1957) *Rich Lands and Poor: The Road to World Prosperity* (New York: Harper & Row).

Ohlin, B. P. O. Hesselborn, and P. M. Wijkman, (eds.) (1977) *The International Allocation of Economic Activity* (London: Macmillan).

Oshima, H. (1974) Perspectives in income distribution research, in *Income Distribution, Employment and Economic Development in Southeast and East Asia*, Papers and Proceedings of the Seminar sponsored by the Japan Economic Research Centre and CAMS, Tokyo.

Oshima, H. (1977) Comments on the CAMS–Hitotsubashi Income Distribution Seminar in Oshima, H. and T. Mizoguchi, (eds.), *Income Distribution by Sectors and Over Time in East and South Asian Countries* (Tokyo: Council for Asian Manpower Studies).

Oshima, H. and T. Mizoguchi, (eds.) (1977) *Income Distribution by Sectors and Over Time in East and South Asian Countries* (Tokyo: Council for Asian Manpower Studies)

Pedersen, P. O. (1975) *Urban-Regional Development in South America: A Process of Diffusion and Integration* (The Hague: Mouton).

Pred, A. (1973) The growth and development of systems of cities in advanced countries, in *Systems of Cities and Information Flows*, pp. 9–82. (Lund: Gleerup).

Reynolds, L. (1977) *Image and Reality in Economic Development* (New Haven: Yale University Press).

Richardson, H. N. (1972) Optimality in city size, systems of cities and urban policy: a sceptic's view, *Urban Studies*, **9**, 309–10.

Richardson, H. N. (1973) *Regional Growth Theory* (London: Macmillan).

Richardson, H. N. (1976) The argument for very large cities reconsidered: a comment, *Urban Studies*, **13**, 307–10.

Richardson, H. N. (1977) *City Size and National Spatial Strategies in Developing Countries*, World Bank Staff Working Paper No. 52 (April 1977).

Stöhr, W. (1974) *Interurban Systems and Regional Economic Development*, Resource Paper No. 26, Association of American Geographers (Washington, D. C.).

Taylor, D. R. F. (1979) *The Role and Functions of Lower Order Centres in Rural Development*, UNCRD Working Paper No. 79–07, Nagoya.

Todaro, M. P. (1976) *Internal Migration in Developing Countries* (Geneva: ILO).

Törnqvist, G. (1970) *Contact Systems and Regional Development* (Lund: Gleerup).

United Nations (1977) *1976 Statistical Yearbook* (New York: UN).

United Nations Centre for Regional Development (UNCRD) (1976) *Growth Pole Strategy and Regional Development Planning in Asia* (Nagoya: UNCRD).

United Nations Centre for Regional Development (UNCRD) (1977) *Ena Declaration on Regional Development* (Nagoya: UNCRD).

Williamson, J. G. (1965) Regional inequality and the process of national development, *Economic Development and Cultural Change*, **13** (4), 3–45.

PART II

Case Studies

a. Asia

Chapter 6

China: Rural Development—Alternating Combinations of Top-Down and Bottom-Up Strategies

CHUNG-TONG WU and DAVID F. IP

I. INTRODUCTION

It has been over 30 years since Mao Zedong declared 'the Chinese people have stood up'. Since then China has generally adopted development strategies which differ in approach and direction from those promoted in other developing nations. In contrast to the growing inequality and poverty rife in many developing countries, China seemed to have achieved, with little outside aid, the provision of basic needs for its vast population. Significantly, China made these impressive gains by emphasizing rural development as the foundation of its development programmes.

In the development literature, China's experience has sometimes been presented as if it had always single-mindedly pursued the strategies of rural development based on consensus administration, work-sharing and collective responsibility. Its policies in industrialization have largely been portrayed as one of emphasizing small-scale industries complementary to rural development. These development strategies, which in the context of this book might be called 'planning from below' (see Stöhr Chapter 2) were not the only type of strategies that China has followed. China has always maintained concurrently both policies of 'planning from above' and 'planning from below', and these have been emphasized differently at different times. There have also been variations in approaches to planning in different parts of the country. While it is tempting to characterize the concurrent approaches to planning as a 'struggle between two lines', it may be oversimplifying the situation. Elements of ideological struggle among different groups within the regime are certainly there, but the elements of the strategies cannot be so simply attributed. Nor is this purely a dichotomy of urban *vs.* rural development. Industrial as well as rural development policies are involved in both styles of planning. So it is not a matter of agricultural *vs.* industrial strategies. The two approaches represent differing treatment of: (a) the pursuit of equality among individuals and among rural and urban population; (b) the relationships between rural and industrial development; and (c) the administrative structure for development. Development strategies were of course formulated within a framework of socialism and made possible after a revolution where ownership of means of production had been collectivized. Thus discussions of the transferrability of the Chinese experience must take account of this framework.

'Planning from above' in China might best be exemplified by early efforts to collectivize agriculture and the imposition of collective farming on a large scale. Private plots as well as private markets in rural areas were discouraged. It culminated in the 'Great Leap Forward' with centralized planning of production targets, production schedules, and even methods. National and state control was paramount. Emphasis on heavy industries and state control of all enterprises marked early industrial strategies. During the Cultural Revolution and what is now known as the regime under the 'Gang of Four', egalitarianism was extolled through policies to retain large administrative units such as the commune or even the *xian* (county) as accounting units for rural areas and implementing equilization of individual incomes. At the same time, centrally directed industrial decentralization and state as well as provincial control of enterprises have been a continuing strategy.

'Planning from below' might best be exemplified by the better-known rural development strategies under a system of restructured communes with more responsibilities given to the brigade and in turn to the work teams. The accounting unit was decentralized to the brigade level. Private plots and markets are allowed. Decision about production methods, what to be produced, as well as the distribution of income and surpluses, are decentralized to the team level. Small-scale industries, managed along similar lines, are established to serve the needs of agriculture and the rural population.

The two planning approaches have tended to be applied at various times to industrial as well as rural development and could operate in different sectors at the same time (Reynolds, 1978). Along with the strategies of decentralized rural development and small-scale industries have been the policies of industrial decentralization of major enterprises, particularly those which are heavy and resource-related industries, very much centrally planned and administered. What occurs within China is therefore not so much a succession of approaches, but often a simultaneous implementation of 'from top down; from bottom up'.

Any discussion of Chinese development strategies need to be selective, and in this chapter the emphasis will be on the organizational aspects for development. The chapter attempts to identify the essential elements of China's 'planing from below' approach to rural development. Consequently, this chapter does not deal with such related topics as macroeconomic analysis of Chinese development nor the urban communes. The rest of the chapter examines the two planning approaches as practised in China, but with emphasis on the essential elements of 'planning from below'.

II. PLANNING FROM ABOVE

(a) Decentralizing industries

It is the contention of this chapter that, in China, strategies characteristic of both 'planning from above' and 'planning from below' have been practised during the 30 years since the establishment of the People's Republic. In China the best illustrations of 'planning from above' are the policies of industrial decentralization and collectivized agriculture. Both sets of policies represent a centralized approach to planning, implementation, and management.

Decentralization of industries is often central to many developing nations' development strategy; the strategy of growth centres to promote regional development being a recent example (Wu, 1979b). Though it is not clear whether 'growth centres' as such have been considered by the Chinese, the intentions of their programmes of industrial decentralization have been generally similar to that of most developing countries.

Industrial activities in pre-revolutionary China were concentrated in the coastal provinces, and the new regime made it a key policy to change this spatial concentration. In a 1956 address Mao declared that henceforth the majority of new industries should be located in the inland provinces to achieve ultimately an even distribution across the nation (Mao, 1977, p. 270). Pragmatic considerations, such as defence and inadequacy of internal transport networks, might have also influenced Mao's thinking. However the necessity of decentralizing industrial growth to the interior provinces did not mean straitjacketing existing industrial centres but using their resources and expertise to assist the lagging areas. The so-called basic industries, iron and steel for example, were still to receive priority treatment whenever conflict over supplies arose (Schram, 1974; Hua, 1978).

To understand the impacts of China's policies of decentralization, analyses have been made of estimates of provincial per capita gross value of industrial output (GVIO) data for selected years to approximate respectively: (a) the period immediately after PRC was established: (b) the period after the 'Great Leap Forward' but before the 'Cultural Revolution' and (c) the most recent year for which data are available (Wu, 1979a; Roll and Kung-Chia Yeh, 1975). For the purposes of the analysis, the three autonomous municipalities—Beijing, Shanghai, and Tianjin—which are directly administered by the central government, are referred to as provinces.

The result of the ranking exercise showed that the most notable changes in the ranking for the selected years were those of Nei Monggol and Gansu—both have experienced significant new resource development and are the locations of several new industrial cities (see Table 6.1). Much of the shifts in ranking occurred amongst the middle-ranked provinces but the pattern of change is inconsistent and does not seem to follow specific interventions such as the establishment of new industrial cities. The provinces of Xinjiang, Hunan, and Sichuan have all had new industrial cities established since 1949, but their relative positions have actually deteriorated.

Figure 6.1 The provinces of China. Note: Pinyin names according to the *Hanyu Pinyin Zhongguo Shouce* (Handbook of Chinese Place Names in Pinyin. Beijing: Ce Hui Chu Ban She, 1977)

The available estimates confirm that the coastal provinces, dominant in 1949, remain the central industrial core in spite of vigorous decentralization programmes.

The relative positions of the top-ranked provinces and those at the bottom have virtually remained static since 1952. Provinces with the lowest per capita GVIO have remained relatively the same since 1957, with minor shifts amongst the same few provinces. The very top four ranking provinces (Shanghai, Liaoning, Tianjin and Beijing) also remain precisely the same throughout the same period, though a small decline in the total share that the top four command is evident. The lion's share of the changes have been at the expense of Shanghai. A recent article in *Peking Review* (1978) reported that Shanghai supplied 41.9 per cent of the nation's total investments in capital construction between 1950 and 1976. It also reported that 'since liberation, Shanghai has sent hundreds of thousands of technicians and skilled workers to other parts of the country and helped train over 100,000 young workers for these places'.

Though the changes are modest they represent changes in the desired direction. Most importantly, higher concentrations have been prevented and this is a mark of success when compared to the generally worsened inequality in most

Province	Per capita GVIO as % of 1st rank	Province	Per capita GVIO as % of 1st rank	Province	Per capita GVIO as % of 1st rank	Province	Per capita GVIO as % of 1st rank
Shanghai	—	Shanghai	—	Shanghai	—	Shanghai	—
Tianjin	65.37	Tianjin	73.50	Tianjin	48.96	Tianjin	48.96
Beijing	24.53	Beijing	31.69	Beijing	33.06	Beijing	44.20
Liaoning	22.00	Liaoning	25.32	Liaoning	28.91	Liaoning	22.76
Heilongjiang	15.09	Heilongjiang	14.83	Nei Monggol	21.54	Nei Monggol	14.42
Jilin	9.18	Jilin	10.39	Heilongjiang	16.72	Heilongjiang	10.28
Shanxi	6.00	Shanxi	6.24	Jilin	12.53	Gansu	9.79
Guangdong	4.57	Guangdong	6.02	Guangdong	8.15	Hebei	9.16
Jiangsu	4.52	Jiangsu	5.52	Gansu	7.67	Jiangsu	8.43
Zhejiang	4.26	Zhejiang	5.10	Hebei	7.62	Jilin	8.11
Xinjiang	4.07	Xinjiang	4.99	Jiangsu	7.54	Guangdong	6.44
Hubei	3.37	Hubei	4.94	Guizhou	6.59	Qinghai	6.33
Fujian	3.30	Fujian	4.55	Zhejiang	6.15	Shandong	5.79
Nei Monggol	3.27	Nei Monggol	4.15	Shaanxi	5.82	Shanxi	5.44
Shandong	3.26	Shandong	4.08	Qinghai	5.69	Shaanxi	4.91
Shaanxi	3.00	Shaanxi	3.78	Hubei	5.49	Zhejiang	4.32
Hebei	2.42	Hebei	3.67	Shanxi	5.39	Guangxi	3.90
Sichuan	2.38	Sichuan	3.59	Jiangxi	5.00	Hubei	3.74
Jiangxi	2.30	Jiangxi	3.58	Shandong	4.97	Fujian	3.73
Qinghai	2.28	Qinghai	3.45	Fujian	4.89	Xinjiang	3.64
Gansu	2.19	Gansu	3.28	Sichuan	4.55	Jiangxi	3.64
Yunan	2.04	Yunan	3.13	Xingjiang	4.28	Hunan	3.34
Hunan	1.96	Hunan	2.73	Hunan	3.52	Henan	3.16
Anhui	1.85	Anhui	2.43	Anhui	3.42	Sichuan	3.13
Guangxi	1.70	Guizhou	2.18	Henan	3.32	Anhui	2.83
Guizhou	1.12	Guangxi	2.15	Yunnan	2.88	Guizhou	2.66
Henan	1.03	Henan	1.94	Guangxi	2.80	Yunnan	2.47
Ningxia	—	Ningxia	0.77	Ningxia	1.57	Ningxia	2.43
Xizang	—	Xizang	0.43	Xizang	0.68	Xizang	1.06

Note: The autonomous municipalities: Shanghai, Tianjin and Beijing are included as provinces to simplify the analysis.

Source: Wu (1979a)

Table 6.2   *Coastal provinces' share of national gross value of industrial output (GVIO) for selected years, in per cent*

Province	1949	1952	1957	1965	1974
Liaoning	8	14	14	13	12
Hebei	6	4	4	5	7
Shandong	6	6	5	4	6
Jiangsu	10	8	6	11	7
Zhejiang	4	3	3	3	2
Fujian	2	1	2	1	1
Guangdong	5	5	5	6	5
Beijing	1	2	3	4	5
Tianjin	5	6	6	5	5
Shanghai	21	19	16	17	15
TOTAL	68	68	54	69	65

Source: Wu (1979a), p. 10.

developing countries during the same period (Lardy, 1976). Whether these modest changes indicate the beginning of a trend towards more equal provincial or regional distribution of industrial output is problematic since size distribution of per capita GVIO over the period 1952 to 1974 still showed a very stable pattern. This is further confirmed by Lorenz curves constructed with the same data (Wu, 1979a, pp. 11–15). Due to the highly skewed distribution of GVIO among the provinces, the seemingly marked national changes are misleading as most of the changes actually occurred amongst the top-ranked provinces.

This analysis demonstrated that in spite of forceful implementation of policies aimed at a more even distribution of industrial activities across China, the dominance of the pre-revolution industrial centres remained largely unchallenged. The spatial pattern is almost unaltered. Continued expansion of the coastal industrial core was due to the necessity to expand heavy industries, most of them based in the existing core, in order to promote light and agro-based industries. Resource endowments also played a part. Much of the increased industrial output of interior provinces has been due to the development of

Table 6.3   *Percentage share of total gross value of industrial output (GVIO) for top four ranking provinces*

Provinces	1952	1957	1965	1974
Shanghai	19.3	16.3	17.1	15.4
Liaoning	14.1	14.4	13.5	12.0
Tianjin	5.5	5.6	4.9	5.1
Beijing	2.1	3.0	4.0	5.4
Total	41.0	39.3	39.5	37.9

Source: Wu (1979a), p. 11.

extractive industries, particularly petroleum and coal. Consequently, part of the shift is simply due to location of natural resources and are not necessarily reflective of policy successes.

The stability of the spatial pattern of industrial output in China confirms the more recent thinking among development theorists that redistribution of urban-based industrial activities is unlikely to bring significant changes in spatial inequality (Friedmann and Douglass, 1978; Lo and Salih, 1978; Appalraju and Safier, 1975). Lasuén (1969) has argued that no significant changes in the spatial system could be expected at different stages of the development process. He hypothesized that growth would simply distribute over the existing urban system, leading to a stable pattern of growth. This seems to have been confirmed by the available information on China.

Centrally directed industrial decentralization as the major component of national development is characteristic of the 'top down' type of development strategy. It has been criticized for its 'urban bias', that is, for focusing investments in urban centres and in activities, such as industries, which inevitably favour urban locations (Lipton, 1976). Regional development strategy based on decentralization of industries assumes that growth would decentralize from the selected points and development of the surrounding areas would follow. It further assumes that surrounding areas of such centres would be benefited by greater degree of connectivity with the centres (Richardson, 1978; Rondinelli and Ruddle, 1978; Rodwin, 1978). China's experience of decentralizing major industrial activities reflects a similar locational approach though it is not clear whether her planners subscribe to the assumptions which underlie such policies (Paine, 1976). From what we know about urban growth in China, it is clear that one of the impacts of the policies of industrial decentralization has been large population increases in provincial and prefectural centres.

Between 1953 and 1970 population in five of the 25 provincial capitals tripled while that in another five doubled (Chang, 1976; Chen, 1973). Those in the north-west and south-west regions grew more rapidly. Of those which tripled their population, two, Lanzhou and Xining, are capitals of northwest provinces.

Capital investments in new industrial complexes were deliberately located in capital cities. For example, Chang reported that six of the nine largest cotton textile factories in the provinces of Shaanxi are located in the capital city of Xi'an. Two out of four of the factories producing electric generators in Jiangxi are in Nanchang, the capital, and the transformer factories in Anhui are in Hefei. The significance of new development policies is particularly evident at the municipal levels. Nine of the prefectural level municipalities, and nine others at the *xian* level, are entirely new cities with mineral extraction industries as the dominant activities. Prefectural cities are generally the sites of medium- and large-scale industries (Chang, 1976, p. 140). A number of these cities experienced rates of industrial growth well above the provincial averages (Lardy, 1978, p. 160).

Rapid urbanization is not exclusively the result of a 'top down' type of development strategy. The rapid population growth associated with industrial decentralization merely underlines the fact that economic growth may not necessarily spread to other parts of the region since urban concentration generally results from such industrial development. Again, the Chinese experience in this respect is similar to that of many developing countries.

(b) Collective Farming

'Planning from above' has not been limited to industrial planning. During the first decade of China's independence, the overall economic planning model was that of the 'big push', and policies were aimed at the maximum mobilization of resources, particularly to support rapid industrialization (US Congress, 1975; Oksenberg, 1973).

Soon after the revolution, agricultural development policy consisted chiefly of land-reform measures which redistributed surplus land from the rich peasants, but it was not a complete reform aimed at equality of ownership (Wong, 1973; NFRB, Feb. 20, 1953). Rich and middle-peasants continued to flourish, and polarization amongst the rural population continued. Schurmann (1968, p. 437) observed that 'land reform did not lead to an economic revolution, for production patterns in the village did not change fundamentally'.

Persistent poverty in the rural areas prompted the leadership to progressively experiment with collectivization of agriculture. These were taken in three major steps (Schurmann, 1968, p. 491). In the first phase, Mutual Aid Teams (MAT) of up to ten households were established and they pool their resources, draft animals and farm implements for joint agricultural activities, especially during peak seasons. The means of production, land and implements, etc., remained privately owned and each member was free to dispose of his own produce.

From 1952 to 1954, the next step was taken and this came about in two phases. The initial phase involved the formation of the elementary Agricultural Producer's Cooperatives (APC), averaging about 20 households in size (Model Regulations, 1976). Land and other means of production were pooled for collective use, though land was still considered to be privately owned, and the members of the cooperative received dividends on the basis of their share in the collective. This is the so-called 'central management but private ownership' phase.

The advanced phase placed all land and means of production in collective ownership and the APC became much larger units, some three times the initial stage and about ten times the size of MATs. With the exception of small plots of household land, and poultry as well as small tools, all other means of production became collectively owned and payment on land dividends ceased (*RMRB*, 15 June 1957).

In spite of problems and debates, this stage was implemented with great speed (*NFRB*, 2 November 1955; *NFRB*, 5 June 1955; *RMRB*, December 1955). Between 1955 and 1956, 20,000 advanced stage APC's were formed in the province of Guangdong alone (*NFRB*, 8 October 1956). These large units of cooperatives were thought to be better able to cope with peak season demands in labour, better able to support required draught animals, better able to provide larger farm implements such as water pumps and mills and to provide more efficient use of land by pooling fragmented land holdings; better able to marshal savings and provide better social security (Ip, 1979, pp. 125–7). Reports of increases in yield, income, and area under cultivation were encouraging (*NFRB*, 10 July 1957 and 15 September 1956).

The early successes with the agricultural producers' cooperatives encouraged the leadership, and in 1958 they announced the next step—the formation of people's communes by reorganizing the collectives. Communes had an average of 5,000 households, and were aimed at self-sufficiency and combining political, social, economic, and cultural functions. It included experiments with mass mess facilities and 'all-people ownership' of means of production as means to attain egalitarianism (Walker, 1965). In Guangdong alone, more than 20,000 cooperatives were merged into 1104 communes in one year (*Ren-Min Kong-She* . . ., 1961).

The communes were often larger than most market town (sometimes even doubling their size). The communes were organized into large production brigades of about 250 households each, and these were further subdivided into smaller production brigades (later called teams) of about 40 households. Thus all of the units were much larger than a village neighbourhood, and villagers were often grouped together irrespective of previous social ties or traditional animosities. The system of organization vested management of all resources and decisions for production at the commune level. Communes cadres were often ignorant of farming requirements and unable to make efficient use of farm labour. Often extra off-farm work, such as tending to small iron refineries, was required at the peak of seasons when labour demand on the fields was the highest (Hong-qi, 1959; Zhang, 1959). Production levels for communes were also fixed by plan directives which merely reflect central planning goals rather than being related to local production capabilities.

Central management of economic planning was based on the idea that industrial development should be the basis of national development, and that further collectivization would be the best way of mobilizing resources for this purpose. The need for investments in rural areas was ignored, with the result that during the years 1953–57 only 6.2 per cent of the state budget was invested in the agricultural sector (Wheelwright and McFarlane, 1970).

Inflexible central management and planning, combined with ignorance of local conditions; the lack of incentives; and three years of exceptionally bad weather—floods and droughts—brought chaos to agricultural production and suffering to the countryside. Falling production, food shortage, and growing urban unemployment and unrest due to large rural-to-urban migration forced the leadership to reconsider its industrialization-first policy (Erisman, 1975; *RMRB*, 15 March 1954; *NFRB*, 30 December 1955; *NFRB*, 18 February 1957 and 16 March 1957).

The Great Leap Forward was a milestone, for in the subsequent years rural development strategies were completely revised and national development strategy became focused more on rural development as the first priority and a different style of planning became more prevalent.

III. PLANNING FROM BELOW

(a) Rural Development

Collective farming, large communes vested with major decision-making powers, and the pursuit of egalitarianism marked the early Chinese approach to rural development. The severe problems created by this approach led to changes which shared decision-making on resource use and distribution at the lower level of rural organization—the work team. Recognition of local customs, social organizations, and traditional ties marked the new approach to organizing for rural development. These essential elements are hallmarks of 'planning from below' in which much of the control of development is vested in local units but within an overall structure which is related to the national system.

The Chinese approach illustrates the need for a clear and understandable national framework, but one which is flexible enough to allow adaptations, adjustments, and experimentation sympathetic to local conditions. While the general approach outlined in the early 1960s remained intact throughout the 1970s, there are important variations, such as the experiments pioneered at the Dazhai commune, which became well-known during the Cultural Revolution years. It serves to illustrate that the Chinese approach has its regional and local variations as well as different emphases at different times, and should caution against over-generalization on a single Chinese model (Green, 1978; Reynolds, 1978). The national framework was so structured that these variations and experimentations are possible, but the major elements of 'planning from below' remained constant.

This approach to rural development was not strictly limited to the agricultural sector, but involved the development of small-scale rural industries complementary to agriculture and needs of the rural population (Sigurdson, 1977). However, it is the organization for rural development which has made the most far-reaching impacts, and this section examines this in detail.

(i) Reorganizing the Communes

Just as they learned from the mistakes of the early land-reform programmes, the leadership of the young PRC learned from their experience with collective farming and commune organization. Beginning at the end of 1960, a series of reviews resulted in several documents which culminated in fundamental changes in the communes. The first of these was the November 1960 Central Committee 'Urgent Directive on Rural Work', commonly known as the 'Twelve Articles' (*Nong-cun* . . . , 1965). This banned egalitarianism in distribution and sanctioned the contract between the brigade and the teams by the 'four fixes and three guarantees system', under which the brigades are allowed to 'own' land, manpower, draught animals, and farm implements and to 'fix' these for the production teams to use. In essence, the commune could no longer withhold any of the means of production. The brigades were to entrust decision on output quotas, cost, and use of manpower to the production teams. The teams became the key decision-making units. Private plots for household sideline production and free markets were restored.

During 1961 and 1962, further directives, commonly known as the 'Twelve Articles' banned coercive measures, allowed suspension of mess hall, stipulated that the brigade could allot private plots, and prohibited commune authorities from interfering in family sideline occupations. However, it was not until 1965 that the final form of the new communes was consolidated.

The commune's scale, ownership, structure, and functions were substantially altered. The most noticeable change was in its size. Averaging 1,600 households, each commune approximated the size of a marketing community or a town surrounded by about 20 villages and conforming with precollectivization ecological and social areas (Stavis, 1974, pp. 29–42; Parish, 1975a; Crook, 1975). Brigade boundaries were also redrawn so as to include natural features historically used by brigade members, and to coincide with their customary trading, mutual aid, and intermarriage ties (Parish, 1975b). Team boundaries were redrawn so that member households were on good terms with each other. Crook estimated in 1975 that there were, in all, 750,000 brigades and five million teams in China while an average commune has some 3,140 households and about 13,800 people divided into 15 production brigades and 100 production teams (Crook, 1975).

A three-level ownership was established. The commune owned industrial enterprises, motor vehicles, fertilizer mills, farm-tool repair shops, and other large-scale or more highly specialized installations that were beyond the strength of the brigades and teams. The brigade, on the other hand, owned some farm tools and facilities—larger means of production too expensive for the average team to buy (such as tractors or larger irrigation and drainage equipment) and certain service, industrial, and specialized food-production facilities. The production team became the basic unit for agricultural production. It owned land, draught animals, and basic tools. Each level of the commune thus became a corporate body in itself by owning some property. Most significantly, without approval of the *xian* (county), the commune or brigade could not freely appropriate properties of the team.

(ii) Administrative Structure

The revised administrative structure of a commune was to consist of a management committee (called the Revolutionary Committee after the Cultural Revolution) and an elected supervisory committee on each level of the unit (except at the production team level) (see Figure 6.2). The most important function of the revolutionary committee was to implement party–state policies and to

Rural Commune Organization[+]

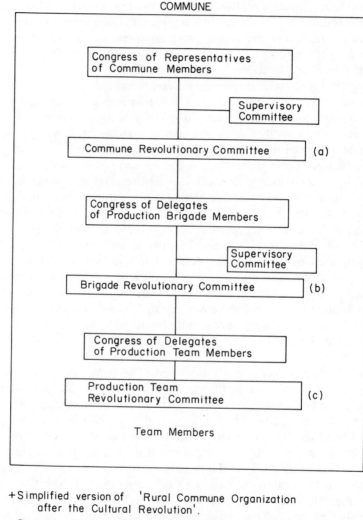

COMMUNE

Congress of Representatives
of Commune Members

Supervisory
Committee

Commune Revolutionary Committee　　(a)

Congress of Delegates
of Production Brigade Members

Supervisory
Committee

Brigade Revolutionary Committee　　(b)

Congress of Delegates
of Production Team Members

Production Team
Revolutionary Committee　　(c)

Team Members

+Simplified version of 'Rural Commune Organization
after the Cultural Revolution'.

Source: Zhong-gong Nian-bao (Yearbook on Chinese
Communism) 1969 (Taipei: Institute for the
Study of Chinese Communist Problems).

Figure 6.2 Rural commune organization in China

assure that food and industrial crops could be procured for rural and industrial
needs. This function was fulfilled through the formation of a national economic
plan which called for the production of certain crops.

As the plan was passed down from the *xian* to the team, production quotas
were based on the units' demonstrated productive capacity rather than, as

during the Great Leap Forward, on national needs or ideal standards of local productivity. Units were left to build slowly on the basis of their combined subsistence–commercial agricultural economy and the handicraft economy. The state also guaranteed a reduction in levied taxes for any year in which natural disasters reduced yields markedly below normal (about 30 per cent below) to assure a minimal level of unit income (Li, 1975). It also promised to maintain the same absolute burden of taxation for ten years and to pass on real increases in productivity to the unit itself. Thus each unit can look forward to benefit from the surpluses generated by its own work.

The commune or the Revolutionary Committee also coordinated the brigades and teams by providing and generating new inputs for agriculture. While the commune provided instructions on more advanced agricultural techniques and other extension services, and gave advice on organizing large joint projects— such as irrigation construction—the brigade developed plans for small and medium-scale projects such as the construction of reservoirs and canals, and organized labour and materials for such projects.

Civil administration was also the responsibility of the commune, which is staffed by full-time state functionaries. Population registration and marriage licences were handled by the office of the commune. By keeping household registration and issuing travel permits associated with food coupons, the commune was able to control the migration of people.

The brigade then was charged with the task of supplementing the commune on the one hand, and coordinating the teams on the other. Its management committee, which was similar to—but smaller than—the commune's, was responsible for assuring that the teams met the state Plan. For this purpose, it carried out administrative coordination by sponsoring joint projects for irrigation, forestation, and small industry; producing both consumer goods and services such as brick-making, flour mills, tool maintenance, and food-processing on a self-sufficient basis.

The brigade's cadres were only part-time functionaries drawing half of their salary from administrative work and the other half from productive work, both of which were estimated in workpoints. The brigades received financial subsidies from the teams, approximately 1 per cent of each team's total income.

In the restructured commune system the work team became not only the basic decision-making unit, but also the basic accounting unit, with the result that team members could easily identify their income with their own efforts. This was a reversal of the earlier arrangements in which the much larger commune was the basic unit of accounting.

Dazhai, the model commune, while noted for its self-reliance efforts to expand arable land and controlled irrigation (Maxwell, 1975) also became innovative in its insistence on using the brigade as the basic account unit (Tsou et al., 1979a, pp. 9–18). It also initiated discussions of using the commune as the accounting unit in its drive for egalitarianism, much against national and even local trends. During the Cultural Revolution and the years till the fall of the 'Gang of Four', Dazhai was hailed as the model to be emulated. What it demonstrated was that within China variations in commune organization and management exist, and the 'models' need to be examined with care.

(iii) Production Team Management and Organization

Changes at the production team level were particularly crucial. Production teams enjoyed autonomy in production. The team's elected management committee consisted of a chief, one or more deputy(ies), a work-point recorder, an accountant, a cashier, a warehouse custodian, and cadres responsible for sideline production, women, and production—none of them were exempt from labour work. The committee, in consultation with the peasants, had the autonomy to make decisions on their production plan: what crop to grow, what special techniques and imported producer goods were to be used, what work-point system was to be paid for each piece of labour, and the like—to meet the production target determined by the county government. Coordination with brigades and commune plans for joint projects, investments, and other endeavours were permitted as long as the team met state targets, produced feasible projects, and did not damage natural resources.

The labour force of the team became self-organized. All peasants, including women and youngsters, became team members (Barnett, 1974). Work was allocated to the team members within a framework of collective planning and supervision; self-responsibility replaced the previous collective responsibility method. The labour force of the team was divided into work groups of about eight to ten households. However, the work group must refer to the production team for any decision it may want to make. A recent discussion of the production team's responsibility confirmed that the key role of the production team has been renewed since the fall of the 'Gang of Four'.

> The production team must grasp five powers: the power to control the means of production; the power to decide production plans and important measures; the power of unified assignment of manpower; the power of unified handling of assets; and the power of unified distribution of products and cash. (Guangdong, 1979, p. 43.)

Groups could be formed on the basis of personal ties, residential contiguity, or even kinship relations. Again, the size and organization of the work groups were recently reaffirmed by discussions on the drive for agriculture modernization (Hong-qi Commentator, 1979; Guangdong 1979; Xia-men, 1978; Zhou, 1979; Zhongyang, 1979).

Two methods of work assignment were in use. The first is known as 'bao-gong' (fixed work) in which a work group is given the responsibility of specific fields for at least an entire season, and except during the peak periods of transplanting and harvest they are responsible for all the work on the assigned fields. Each member household, therefore, has a considerable stake in ensuring good performance, as it will affect the value of the work-points that they earn and future reassignment to better fields or work (Hsu, 1974, pp. 434–35). The second method of assigning work was 'pai-gong' (assigned work) in which smaller tasks which could be performed by one or few individuals may be assigned on a permanent basis. Assignments were based on consultations with experienced peasants to ensure that the most qualified person would be assigned to the most suitable job.

The collective income, however, was distributed in a thoroughly collective manner. First of all, labour was evaluated by the work-point system. There were various work-point systems (Hsu, 1974, pp. 436–39). Before 1962, the time-rate system awarded fixed daily work-points of about ten to the members of the teams and brigades for their number of days worked in a month. This system was found wanting since it failed to take into account the individual's productivity.

Another method was adopted later—the labour 'base-point' in which the team members were classified into three labour grades: first, second, and third, each of which have different daily work to take into account the difference in labour productivity. However, it proved difficult to classify each peasant into one of the three grades, since performance varied with specific conditions and types of jobs.

The 'labour-norm' system was also practised. This system rated every possible piece of work in light of its nature, skill required, and hardship involved. It required ingenuity and experience for the team to rationally evaluate each task. The system accounted for each peasant's performance, thereby providing work incentives for one's efforts and talents. This particular system of assigning work-points according to the difficulties of the task, and how well the task has been preformed, lost favour during the Cultural Revolution and subsequent years under the 'Gang of Four', but it is now being re-emphasized as the one system that should be applied. Modifications are made to take account of those who are weak and elderly and those who may have other difficulties. The emphasis is on 'an-lao fen-pei' (distribution according to labour) and is seen as the best means of pushing ahead modernization in agriculture (Zhongyang, 1979, pp. 13–16).

The Dazhai Brigade devised a unique system combining model work-point, self-report and public evaluation. Here, the brigade's work-point recorder registered each member's performance and the number of days he worked every month. At the end of the month, the brigade selected a model peasant who had shown not only the best work performance but also the best political attitude, and decided upon how many work-points he deserved for a day. Using this model peasant, each member discussed each other member's claim and finally decided allocations case by case. Under the movement of 'In Agriculture, Learn from Dazhai', this system was promoted.

Whatever system the team adopted, the team recorder would note each member's work-points in his books and make them public each month. The points were also recorded in the 'work-point handbook' which belonged to the members. At the end of the year, after computing the gross team product, the team would first set aside grain for agricultural tax, the quantity required for sale to the state, for production expenses, for reserve grain for the needy, for public accumulation fund for capital projects and for welfare fund for emergencies (Lippit, 1977).

It was after setting aside all these that the team distributed the remaining grain and cash to its members. The income of each member was determined by the total sum of work-points earned by working members, and the value of each work-point in turn was determined by dividing the team's entire distributable income by the sum of work-points.

The most important reorientation of the commune system after 1962 was the permitting of private plots: peasants could hence derive income from the private family sector by their involvement in some 'sideline' occupation. Each production team could allot 5–7 per cent of the total cultivated area to families for private gardens, referred to as zi-liu-di (private plots). It was generally allocated to a family on a fixed per capita basis which would not be affected by the subsequent changes in family size through birth, death, or marriage. Only if the work would not disrupt the daily collective routine work hours was the family free to grow whatever they wanted on their plots. They could also pool their resources and operate collectively. The produce they derive from these plots was either sold to the state purchasing agency or in the free market (chen-xu).

The peasants are thus provided with legitimate outlets for their private interests, satisfying the peasant's demand for diverse goods and cash. The system also supplemented deficiencies in the collective and shifted the burden of certain crucial production tasks, especially the raising of non-cereal subsistence foods for the rural population and the supply of animal product which the collective could not yet handle effectively.

Opportunities for commune families to supplement their own needs, as well as selling the surplus from their private plots in the free markets, have been severely criticized during the 'Gang of Four' years, but have since been reaffirmed as desirable policies (Zhongyang, 1979, pp. 47–52). However, the free markets are different from the periodic markets which William Skinner (1964) had described under Republican China. A number of measures have been instituted to ensure that the free markets do not become venues for private capitalism. Only individual commune members and some urban residents are allowed to sell things or attend the markets. Items for sale must be produced by the individual seller, and these items must be ones which are not included on the list of planned commodities and not prohibited for sale. Price ranges are also regulated to prevent excessive profit taking (Zhongyang, 1979, pp. 53–58).

(iv) The Supporting Structure

The commune's organization was supported by a series of policies aimed at ensuring a stable national framework for rural development to proceed. It involved a new system of distribution and purchasing; free markets, new credit facilities, changes in education, and allocation of manpower. Along with the state's assurances on lowering tax and favourable prices for produce these provided a necessary and stable environment for rural development.

Distributing facilities were organized into a nationwide network under the All China Supply and Marketing Cooperatives in order to facilitate the state's 'unified procurement' (tong-gou) system (*FEER*, 1967). Under this system, various Supply and Marketing Cooperatives (SMC) were set up and extended to commune offices, while branches were also established in brigades and market places. Agents with the SMC were sent to the teams. Under this 'unified procurement' system, the SMC's concluded procurement contracts with the teams, on behalf of the state. The peasants also participated in these cooperatives by buying a certain number of shares. These cooperatives were the only agencies supplying them with daily necessities, purchasing produce not banned from the free markets, and fertilizers. These cooperatives thus served as a central exchange for goods between urban and rural areas by bringing goods to the villages and supplying produce to the cities.

Free markets, which were traditionally known as *chen-xu* and were held periodically as centres of rural life, were strictly prohibited during the early days of communization. Under the Sixty Articles of 1962, these were also brought back and tolerated for meeting local needs and increasing opportunities for individual exchange (Hsu, 1974).

Credit facilities were also extended down to the brigade level. The People's Bank and the Agricultural Bank were operating in the brigades to provide not only routine financial transfers, but also to grant long-term (extending over two years) and short-term (due within a year) loans with either no or relatively low interest. Credit cooperatives, operating mostly on the brigade level, were set up to promote mutual self-help among the peasants and there were no national agencies administering these cooperatives. These cooperatives handled about 60 per cent of the peasants' deposits and loans (Ahn, 1975).

Several educational reforms were practiced until the early 1970s. Primary schools were taken over by brigades and run on a self-sustaining basis. Middle schools, previously the responsibility of the county, were now entirely run by communes. The school years for primary universal education had been shortened to five years and those for middle schools to two years. (In some cases, middle and high schools had been combined with the four-year secondary schools.) The curriculum was revised to include more production knowledge to serve the peasants' practical needs (Stavis, 1974). Moreover, applicants from rural areas were given priority in university education, although in recent times consideration was also given to the applicant's entrance examination results (Teiwes, 1974; Prybyla, 1975; Cassella, 1975; *Peking Review*, 1977). But in either case, graduates from high schools as well as college were made to work in the rural areas—the former being referred to as *xia-xiang* (going down to the countryside) while the latter was a policy of *she-lai she-chu* (from the commune, back to the commune).

The cadres in the communes were the key to successful policy implementation. Their role, thus, was to implement directives from above but at the same time to reflect local conditions from below. To ensure that they performed this role well, these cadres were required under the Sixty Articles to participate in labour:

the commune cadres had to perform labour production 60 days yearly, the brigade cadres 120 days, and the team cadres were not exempted from labour at all. Higher level cadres were also sent down to the basic level (*xia-fang*) so that they could take part in labour and investigate the local situation. They were also required to "squat at selected points" (*dun-dian*) for extended periods of time so that they could understand the local condition thoroughly (*GMRB*, 1974).

(b) Towards Rural Equity

The organizational aspects for rural development have been given in detail to emphasize the institutional requirements for implementing policies aimed at 'development from below'. Institutional changes need to produce visible results, to achieve acceptance and continued support. The remarkable advances in production and more equal income distribution in China is analysed and described in a large body of available literature. Interested readers are referred, for example, to the works by Aziz (1978), American Rural Industries Delegation (1977), Blecher (1976), Gurley (1975), Stavis (1974).

That China has made great strides does not mean an egalitarian rural society has been achieved. On the contrary, the leadership is quite conscious of the remaining inequality brought about by the life-cycle of each family, the vagaries of natural endowment of each commune, and the differences between communes which are brought about by a combination of population and natural conditions.

The main source of inequality between communes, and between brigades and teams within communes, is the difference in their respective available means of production. Short of ownership by the whole people, equality among various collective units and the constituent households is impossible to achieve, since the natural endowments of soil fertility, available arable land, irrigation, man—land ratio, accumulated tools and machines—as well as quality of the labour force—are very different among the various units. To abolish this source of inequality, ownership of land and resources must be pooled and the rental income distributed equitably. In other words, in order to achieve equality, the ownership of land and other assets must be transferred from lower levels of collectives to higher levels. A shift towards the higher collectives as accounting units has occurred in parts of China: teams in some areas have ceased to be the basic accounting units; this function has moved one tier up to the brigade. This is especially the key to the evolving model of Dazhai production brigade— but increasingly under criticism in recent years. Most recent policy statements again reaffirm the work team as the basic accounting unit (Hong-qi, 1979; Xia-men, 1978). At the same time, the rates of accumulation out of income generated at the brigade and commune levels have generally been higher than that at the team level (Khan, 1977). These measures would certainly help to reduce the sources of inequality between units within a commune, although they would have no effect on differences between communes. Nevertheless, they represent the direction and the course of transition of ownership from the level of the communes to that of the whole people.

Other policies are aimed at reduction of intra- and inter-commune inequalities. Before the ownership system is finally transformed and transferred, the emphasis is on self-reliance. The state refrained from using its power of taxation to redistribute income from the wealthy to the poor. It did not restructure the agricultural tax, so richer areas have to share their wealth with other areas, but insisted the poorer communes or teams should rely on themselves instead of relying on outside resources. The emphasis is thus not on supplying capital for relief, but on providing administrative and material assistance to those who are 'lagging behind'. Administrative assistance usually takes the form of sending cadres of the commune party committee from the advanced commune to work in poorer communes and teams in order to improve their managerial efficiency — to retrieve manpower and implements employed in sideline production elsewhere, to teach them how to get the best results in their production, management, and organization of their resources. All these aim at passing on experience in 'grasping revolution' to build up self-reliance rather than at supplying short-term social welfare.

Where communes are lagging behind because of shortage of capital construction funds, interest-free loans are available. With such loans, the commune is obliged to appropriate the funds for developing an infrastructure for agricultural development, such as the purchasing of farm implements (Wertheim, 1974; Shenk-Sandbergen, 1973). The experience of Lu-yu production team in Guangdong best exemplified the effects of such assistance:

> The Lu-yu production team ... is situated in a mountain ravine with poor and thin soil. In the initial period after the establishment of the people's commune, this production team was comparatively backward economically. The commune revolutionary committee and the brigade revolutionary committee, to help this team to develop production, mobilized its members to launch the movement of 'in agriculture, learn from Dazhai', make progress on their efforts, fight against nature, reclaim the mountain gully, transfer 20 mu of muddy fields and expand the acreage of paddy-rice by 50 mu. This team had not much labour power. To make good use of the farming season, every year the brigade sent its tractors into the gully to plow land and harrow fields for the team members, and help them to complete the job within the season. In this way, this production team had pushed up its output of grain continuously year by year. Last year its per-mu yield leapt from some 200 catties in 1964 to 1,300 catties. Next the brigade again helped this production team to buy water wheel machines, rice-grinding machines, and crushing machines, mechanize the processing of farm and sideline products, and raise the productivity of labour. To change quickly the appearance of poverty of Lu-yu production team, the brigade again led this team to develop forestry production by cultivating some 500 mu of forests and running 200 mu of fruit gardens. In the last year, the team received a total income of more than 35,000 yuan from farm and sideline undertakings, 28,000 yuan more than that of 1964, each household of commune members increased its cash income by four times (China News Agency, 1973). [1 mu = 0.0666 hectares; 1 catty = 0.5 kilogram; 1 yuan = US$0.5.]

Rural industrialization represents another effective measure for the reduction of rural inequalities among production teams. This is possible because of the methods of allocating surplus and distributing income in the commune- and brigade-run industrial enterprises, and because of higher factor income in industry than in agriculture. The commune and brigade retain only part of the surplus and distribute the rest among the production teams according to the proportion of their contribution to the total labour inputs of the enterprise. Instead of direct income from the factory, the workers receive work-points from their own production team and a small amount of subsidy from the factory for food and travelling from home to the factory. Their wages are paid directly from the factory to their production team as a source of collective income. Thus income disparity between the factory workers and the farmers of the same production team is reduced, and all team members also benefit from progress of rural industry (Ng, 1976; Xia-men, 1978). Given this method of income distribution of commune-run industries, the commune may recruit labour for industry mainly from poorer production teams for reducing intra-commune inequalities. This may lead to a decline of collective farm income, but it will be more than offset by an increase in industrial income. Indeed, the relative importance of man–land ratio as a source of rural inequalities has declined, and inequalities can be more effectively reduced in communes with a number of industrial enterprises.

The major cause of overconsumption in households, and the problem of individual inequalities, is essentially a function of the stage of the life-cycle, that is, at any one time some families have more favourable hand – mouth ratios than others. And since most families do experience the various stages of the life-cycle, the effects on individual households of inequality resulting from the hand–mouth ratio will eventually even out over time as most households' ratios change over the years and their incomes move up and down the income distribution as a consequence (Blecher, 1976). It is, therefore, more important to see how these inequalities are moderated than to exaggerate the conceptual importance of inequalities resulting from these differences in hand–mouth ratio. Directives were issued to rural cadres to find ways to limit the proportion of households which overconsumed, especially whose overconsumption was not due to the ratio of family members in the labour force to total family size but rather a result of negligence of work (Zhong-fa, 1971). In some areas, the poorest families could also apply for small amounts of relief grain or cash, and would be eligible for certain benefits such as special assistance for private sideline production (Zhong, 1974). And as Whyte and Parish both pointed out, the grain-over-consumption system is in itself a sort of welfare mechanism which allows poor families to consume beyond their earnings in any particular years (Whyte, 1975, p. 689).

There are certainly inegalitarian aspects in rural China under the commune system, but the system has certain institutional constraints on the polarization of inequality. The collective ownership system implies that a rich collective unit is unlikely to exploit the poor one through the transfer of land ownership or the employment of cheap labour. At the same time, the rural communes also provide many more items of consumption collectively, as well as the distribution of private consumption income. Within a commune access to health, education and other services is probably the ideal suggested by Khan (1977) and indeed the distribution of such services seems to be designed to improve the distribution of consumption within a commune in that the fees for medical and educational services are charged on the basis of ability to pay. The quality of such services provided by the communes also appears to differ less between communes than distributed income. Moreover, irrespective of membership of specific communes, all have access to services provided by the provincial and central government.

A number of regulations and mechanisms were instituted further to minimize inequality between teams, brigades, and communes. These include prohibitions on owning land, draught animals and major tools of production, and restrictions on marketing of grains and other produce. The institution of payment systems in which differentials are set by consensus of the team members ensures that members must be able to justify to one another decisions in both economic and ethical terms if they were to continue to work and live together harmoniously. Furthermore, new mechanisms such as apportioning a greater share of collective income for public accumulation and a smaller share for direct distribution to individuals for consumption, can be found to minimize the effects that higher collective income may have on increasing inequality (Dazhai, 1974).

More recent policy decisions abandoning egalitarianism and allowing individual teams to become prosperous first (mao-jian) rely on these same mechanisms to ensure that such a policy would not lead to polarization but instead would have the beneficial result of providing more incentives for individuals to contribute to the best of their ability (Jiang, 1979; Li, 1979; Zuo, 1979).

IV. CONCLUSIONS

During the last three decades, China has practised both 'planning from above' and 'planning from below' simultaneously, though the emphasis might be on one or the other at different times and involves both agriculture and industrial policies. The two planning approaches do not necessarily reflect conflict between those emphasizing agricultural and those emphasizing industrial policies of development; nor do they simply reflect an urban or rural bias. The present drive for 'Four Modernizations' and the associated new initiatives deserve separate detailed examination. However, based on the history of the evolvement of development strategies in China, one might characterize the more recent policies as part of a continuing dialectic process of redefining development goals as well as the means to achieve them.

Past attempts at 'planning from above', especially the policies of collective agriculture and industrial decentralization, have not generated the desired results. The Chinese rural development policies have been more successful and are of greater interest to those concerned with 'planning from below'. The essential features of the Chinese approach to 'planning from below' for rural development might be summarized in the following six elements.

Firstly, the strategies focused on the provision of basic needs to ensure that the population has productive work, is supplied with essential commodities, and provided with medical and health care. Additional policies ensure that education is under the control of local communities, and that those who have been sent away for further education or special training would be reassigned to where they originated.

The second element is a well-defined national framework for local control of decisions regarding use of resources, planning for production, infrastructure improvements, work values, work distribution, and the deployment of surpluses. Within this administrative framework these decisions were devolved to the production units, the work teams, and the brigades. Consequently, there is direct control by those who are involved in production, are accountable for quality of their work, and who generate the economic surplus. Local control was implemented under a system of collective ownership of the means of production, but given enough flexibility to allow for individual incentive and enterprise.

A third element involves the set of national policies which consistently support the rural sector through reduction of agriculture tax, guarantee of low taxation, price support ensuring favourable terms of trade for the agricultural sector, as well as provisions for purchase price bonuses for production above planned quotas. These ensure sufficient incentives to increase production and generate surpluses which can be used for bolstering commune reserve funds, for distribution among members of team or brigade, for reinvestments, or which can be retained for other projects.

A fourth element is the promotion of small-scale industries which are complementary to the rural sector. These industries are generally under the administration of the brigade, the commune or the *xian*. They may be engaged in the production of inputs, such as nitrogenous fertilizers, for improvement of agricultural production, in the repair of agricultural machinery, or in processing of agricultural products. Others manufacture consumer products required by the rural population, or for 'export' to other areas. These industrial activities provide off-farm employment opportunities and have a levelling effect on urban and rural wage differentials.

A fifth element is local control of surpluses generated by agriculture and small-scale industries under the management of the brigades and communes. Surpluses from the industrial enterprises are generally ploughed back for reinvestment. According to Griffin and Ghose (1979) one of the major causes of continuing rural underdevelopment in many developing countries is not the lack of productivity nor the lack of generated surpluses but the lack of local control over the surpluses. Local control of production, the incentives for increased rural production, and the development of rural small-scale industries in China are all structured to provide the generation of rural surpluses as well as their retention in the same local unit for the use and benefit of the local population.

A sixth element is the emphasis on the use of local resources, manpower, and knowledge; that is, self-reliance. Financial aid is only available to the most abjectly poor regions, but tax reliefs are available whenever natural disaster strikes. Other than these needy cases, the accent is on self-reliance and on making the best of available local resources, as well as mutual learning among communes and brigades. Even research on agriculture is decentralized to maximize adaptation to local conditions, and supported by a national network to facilitate rapid dissemination of knowledge (Jiang, 1965; Morehouse, 1976).

These six integrally related and complementary elements provide the basis for a new set of interdependent urban and rural relations. The Chinese experience suggests that a 'rural-bias' and a class-oriented strategy is necessary to address the existing inequalities and to promote interdependent urban–rural relations. The Chinese strategy does not rely primarily on the efficient allocation of resources or distribution of activities; instead, it is based on the masses—what might be called 'planning from below'. The rural development programmes implemented through the commune organization in China are integrated ones, viewed as essential for the development of the total society and not as an isolated sectoral, short-term selective development aiming at increasing the performance and complexities of a particular sector.

The Chinese experience also gives support to the concept of 'selective regional closure' (Stöhr and Tödtling, 1977; Lo et al., 1978; see also Chapter 2 and 5 of this volume), as a strategy to promote regional development. A key element of this concept is regional control over the penetration of its economy by outside (national and international) elements as well as control over economic surpluses generated within the region. None of the proposals so far have given specific attention to the context in which 'selective regional closure' might be implemented: partly because it would be difficult to specify this in a general but meaningful way. The Chinese experience with rural development, as summarized in the six elements, supports the basic ideas of regional closure as a means of ensuring rural development. Two features of the Chinese approach, self-reliance and its national support system, are particularly relevant for any proposals for 'selective regional closure'.

In China, regional self-reliance under collective ownership and decision-making allowed for regional variations in style of management, in selecting programmes which are suitable to regional level of development and available resources. Devolution of power—of decision-making—to levels of organization as close as possible to the individual, is essential. Self-reliant development did not mean the closing-off of options but the creation of opportunities, such as agriculture-oriented small-scale industries, which are suitable to the local context.

Self-reliant rural development was possible only with the implementation of carefully designed national support systems, including favourable pricing of agricultural commodities, tax reductions, favourable terms of trade, as well as a unified purchasing and distribution system. Price support, tax reductions, and favourable terms of trade make it possible to increase rural income and consequently to generate and accumulate surpluses. Unified purchasing and distribution regulates the interactions between the region and the national, as well as international, systems.

The Chinese experience suggests that selective regional closure strategies would have to include elements of these policies and institutional arrangements. The Chinese development design reflects a firm commitment to provide an institutional framework in which peasants and the masses can have sustained participation. It recognizes that only when the peasant masses have seized power are they motivated to articulate their aspirations fully and to develop their political perspectives, their consciousness, and to become a force in rural development as well as development as a whole.

REFERENCES

Abbreviations used in Text:
FEER   Far Eastern Economic Review
GMRB   Guang-Ming Ri-bao
NFRB   Nan-fang Ri-bao
RMRB   Ren-min Ri-bao

Ahn, Byung-Joon (1975) The political economy of people's communes in China: changes and continuities. *Journal of Asian Studies*, 34 (3) (May), 631–58.
American Rural Industries Delegation (1977) *Rural Small-Scale Industry in the People's Republic of China* (Berkeley: University of California Press).
Appalraju, J. and M. Safier (1975) Growth centre strategies in less-developed countries, in A. Gilbert (ed.), *Development Planning and Spatial Structure*, pp. 143–68 (London: John Wiley).
Aziz, S. (1978) *Rural Development: Learning from China* (London: MacMillan).
Barnett, A. D. (1974) *Uncertain Passage: China's Transition to the Post-Mao Era* (Washington, DC: Brookings Institute).
Blecher, M. (1976) Income distribution in small rural communities, *The China Quarterly*, 68 (December), 797–816.
Cassella, A. (1974) Recent development in China's university recruitment system, *The China Quarterly*, 58 (June), 297–301.
Chang, Sen-dou (1976) The changing system of Chinese cities, *Annals of the Association of American Geographers*, 66, 398–415.
Chen, Cheng-Siang (1973) Population growth and urbanization in China, 1953–1970, *Geographical Review*, 63, 55–72.
China News Agency (1973) *News Release*, No. 6849, 16 July.
Crook, F. (1975) The commune system in the People's Republic of China, in US Congress, Joint Economic Committee *China: A Reassessment of the Economy* (Washington, DC: Government Printing Office), pp. 366–410.
Dazhai, Hong-qi (The Red Flag of Dazhai) (1974, Peking: Ren-min Chu-ban-she).
Erisman, A. (1975) China: agriculture in the 1970's. In US Congress, Joint Economic Committee, *China: A Reassessment of the Economy* (Washington, DC: Government Printing Office), pp. 328–29.
Far Eastern Economic Review (1967) 11 May, pp. 30–32.
Friedmann, J. and M. Douglass (1978) Agropolitan Development: Towards a New Strategy for Regional Planning in Asia, in F. Lo and K. Salih (eds.), *Growth Pole Strategy and Regional Development Policy: Asian Experiences and Alternative Approaches* (Oxford: Pergamon).

Green, R. H. (1978) Transferability, exoticism and other forms of dynamic revisionism, *World Development*, 6 (5), 709–13.

Griffin, K. and A. K. Ghose (1979) Growth and impoverishment in the rural areas of Asia, *World Development*, 7(4/5) (April/May), 361–83.

Guangdong (1979) Provincial CCP Committee's General Committee, Investigation and Research Office (1979), View on output responsibility system in agriculture, *Hong-qi*, 4 (3 April), 40–43. Translation from Joint Publications Research Service, 073650, pp. 66–72, June 1979.

Guangdong-Sheng tu-di gai-ge wei-yuan-hui diao-cha-zu (1953) Tu-gai-hou nong-cun sheng-chan-li di fa-zhan ching-huang (Development of rural productivity after land reform), *Nan-fang Ri-Bao*, 20 February.

Guang-ming Ri-bao (1974). 7 December.

Gurley, J. G. (1975) Rural development in China 1949–72, and the lessons to be learned from it, *World Development*, 3 (7/8) (July-August), 455–71.

Hong-qi (1959) No. 1 (1 January), p. 14.

Hong-qi Commentator (1979) It is imperative to respect a production team's decision-making power, *Hong-qi*, No. 2 (2 February), pp. 16–18. Translated in Joint Publications Research Service, 073304, pp. 25–30, 25 April 1979.

Hsu, Tak-ming (1974) *Wen-ge Hou Di Zhong-Kong Jing-ji* (The Chinese Communist Economy After the Cultural Revolution) (Hong Kong: Union Research Institute).

Hua, Kuo-feng (1978) Unite and strive to build a modern powerful socialist country! (Report to the First Session of the Fifth National People's Congress, 26 February 1978). English version provided by the People's Republic of China Mission to the United Nations. Reprinted in *Population Development Review*, 4 (1978), 167–79.

Ip, D. F. (1979) *The Design of Rural Development: Experiences from South China 1949–1976* Unpublished PhD. Dissertation, Department of Anthropology and Sociology, University of British Columbia, Canada.

Jiang, Yi-xhen (1965) Ji-yi kai-zhan yi yang-ban-tian wei zhong-xin di nong-ye ke-xue shi-yan yun-dong duo-kuai hao-sheng di fa-zhan nong-ye ke-xue wei nong-ye sheng-chen fu-wu (Energetically unfold the experimental movement in agricultural sciences, develop agricultural sciences with greater, faster, better and more economic results in service for agricultural production), *Zhong-guo Nong-ye Ke-xue*, No. 4.

Jiang, Yue-shao (1979) Rang yibufen sheyuan xian fu qilai, (Let some commune members become prosperous first), *Hong-qi*, No. 4 (April), 44–45.

Johnson, G. (1973) Rural economic development and social change in China (Mimeo). A Report submitted to the International Development Research Center, Ottawa, Canada.

Khan, A. R. (1977) The distribution of income in rural China. In International Labor Office, *Poverty and Landlessness in Rural Asia* (Geneva: ILO), pp. 253–80.

Lardy, N. R. (1976) Economic planning and income distribution in China, *Current Scene*, 14, 1–12.

Lardy, N. R. (1978) *Economic Growth and Distribution in China* (Cambridge: Cambridge University Press).

Lasuén, J. R. (1969) On growth poles, *Urban Studies*, 6 (2), 137–61.

Li, Cha (1975) *Zhong-Kong Shui-shou Ji-du* (Taxation system of Communist China) (Hong Kong: Union Research Institute).

Li, Qïan He (1979) Yi-bu-fen xian fu-yu he gong-tong fu-yu, (Collective prosperity and prosperity for a few), *Re-min Ri-bao* (15 April).

Lippit, V. (1977) The commune in Chinese development, *Modern China*, 3 (2) (April), 229–55.

Lipton, M. (1976) *Why Poor People Stay Poor* (London: Temple Smith).

Lo, F. and K. Salih (eds.) (1978) *Growth Pole Strategy and Regional Development Policy: Asian Experiences and Alternative Approaches* (Oxford: Pergamon Press).

Lo, F., K. Salih and M. Douglass (1978) *Uneven Development, Rural Transformation, and Regional Alternatives in Asia*. UNCRD Working Paper 78–02, Nagoya, Japan.

Mao, Zedong (1977) *Collected Works* Vol. 5 (Beijing: Remin Chu Ban She).

Maxwell, N. (1975) Learning from Taichai, *World Development*, 3(7/8) (July/August), 473–95.

Model Regulations for an Advanced Agricultural Producers' Cooperative (1976) (Peking: Foreign Language Press).

Morehouse, W. (1976) (Notes on Hua-tung commune), *The China Quarterly*, 67 (September), 583–96.

Nan-fang Ri-bao, 20 Feb., 1953.

Nan-fang Ri-bao, 5 June, 1955.

Nan-fang Ri-bao, 2 Nov., 1955.

Nan-fang Ri-bao, 30 Dec., 1955.

Nan-fang Ri-bao, 15 Sept., 1956

Nan-fang Ri-bao, 8 Oct., 1956.

Nan-fang Ri-bao, 18 Feb., 1957.

Nan-fang Ri-bao, 16 March, 1957.

Nan-fang Ri-bao, 10 July, 1957.

Ng, G. B. (1976) *Rural Inequalities and the Commune System in China* (Geneva: International Labor Office) World Employment Program Research Working Paper, WEP 10–6/WP-11, October.

Nong-cun Ren-min Gong-she Tiao-li Cao-an (1962) (Draft regulations concerning rural people's communes). (Taipei: The National Security Bureau, Nationalist Chinese Government) 1965, reproduced.

NFRB (1955) Nong-ye he-zuo-she zen-yang zhu-yong geng-niu di ban-fa (Ways of renting draught animals in agricultural cooperatives), *Nan-fang Ri-bao*, 5 June, 1955.

Oksenberg, M. (ed.) (1973) *China's Developmental Experience* (New York: Praeger).

Paine, S. (1976) Balanced development: Maoist conception and Chinese practice, *World Development*, 4 (4) (April), 277–304.

Parish, W. L. Jr. (1975a) The commune as a social system: Kwangtung Province 1970–1974. Paper Presented at the Association for Asian Studies Annual Meetings at San Francisco, March.

Parish, W. L. Jr. (1975b) 'Socialism and the Chinese peasant family, *Journal of Asian Studies*, 34 (3) (May), 613–30.

Peking Review (1977) The new educational enrollment system, 11 November, pp. 16–18.

Peking Review (1978) Shanghai: a coastal industrial base, 4 (27 January), 11–13.

Ping-yi ru-she sheng-chan zi-liao di zheng-ce jie-shuo (1955), (Explanatory notes on the policy of evaluating production resources turning into commune property), *Nan-fang Ri-bao*, 2 November.

Prybyla, J. S. (1974) Notes on Chinese higher education 1974, *The China Quarterly*, 58 (June), 271–96.

Ren-min Kong-she You-yue-xing Wen-da (1961) (Questions and answers on the superiority of the people's commune). Gunagzhou: Ren-min Chu-ban-she.

Ren-min Ri-bao, 15 March, 1954.

Ren-min Ri-bao, 15 June, 1957.

Reynolds, B. L. (1978) Two models of agricultural development: a context for current Chinese policy, *China Quarterly* (December), No. 72, 842–72.

Richardson, H. W. (1978) Growth centers, rural development and national urban policy: a defence, *International Regional Science Review*, 3 (2), 133–52.

Rodwin, L. (1978) Regional planning in less developed countries: a retrospective view of the literature and experience, *International Regional Science Review*, 3 (2), 113–31.

Roll, C. R., Jr. and Kung-Chia Yeh (1975) Balance in coastal and inland industrial development. In US Congress, Joint Economic Committee, *China: A Reassessment of the Economy* (Washington, DC: US Government Printer), pp. 81–93.

Rondinelli, D. A. and K. Ruddle (1978) *Urban Functions in Rural Development—An*

Analysis of Integrated Spatial Development Policy (Washington, DC: Office of Urban Development, Technical Assistance Bureau, Agency for International Development, US Department of State).

Schram, S. (1974) *Mao Tse-Tung Unrehearsed* (Harmondsworth: Penguin).

Schurmann, F. (1968) *Ideology and Organization in Communist China* (2nd edn.) (Berkeley: University of California Press).

Shenk-Sandbergen, L. Ch. (1973) How the Chinese people remove polarity within their countryside *Eastern Horizon*, 12(3).

Shen-zhong chu-li geng-niu nong-ju ru-she wen-ti, (Carefully handle the problem of turning draft animals and farm implements into commune properties), *Ren-min Ri-bao*, 6 December 1955.

Sigurdson, J. (1977) *Rural Industrialization in China* (Cambridge, Mass.: Harvard University Press).

Skinner, G. W. (1964), Marketing and social structures in rural China, *Journal of Asian Studies*, 24 (1) Nov., 3–43; 24 (2), Feb. 1965, 195–228; and 24 (3), May 1965, 363–99.

Stavis, B. (1974) *People's Communes and Rural Development in China* (Ithaca, New York: Cornell University Center for International Studies, Rural Development Committee).

Stöhr, W. B. and Tödtling, F. (1977) Spatial equity—some antitheses to current regional development doctrine, *Papers of the Regional Science Association*, **38**, 33–53.

Teiwes, F. C. (1974) Before and after the cultural revolution, *The China Quarterly*, 58 (April/June), 332–48.

Tsou, Tang, M. Blecher and M. Meisner (1979) Organization growth and equality in Xiyang County—a survey of fourteen brigades in seven communes,

Part 1, *Modern China*, 5 (1) (January), 3–39.

Part 2, *Modern China*, 5 (2) (April), 139–85.

US Congress, Joint Economic Committee (1975) *China: A Reassessment of the Economy* (Washington, DC: US Government Printer).

Walker, J. (1965) *Planning Chinese Agriculture* (Chicago: Aldine).

Wei, Sheng-feng (1957) Zhong-shan xian nong-min sheng-huo zhuang-kuang diao-cha (An investigation on the living conditions of peasants in Zhong-shan County) *Nan-fang Ri-bao*, July 10.

Werthheim, W. F. (1974) Polarity and equality in the Chinese people's communes, in *Revue des Pays de l'est* (Editions de l'Universite de Bruxelles).

Wheelwright, E. L. and McFarlane, B. (1970) *The Chinese Road to Socialism* (New York: Monthly Review Press).

Whyte, M. K. (1975) Inequality and stratification in China, *The China Quarterly*, 64 (December), 684–711.

Wong, J. (1973) *Land Reform in the People's Republic of China*. (New York: Praeger).

Wu, Chung-Tong (1979a) Development strategies and spatial inequality in the People's Republic of China. WP: 79-06, United Nations Center for Regional Development, Nagoya, Japan.

Wu, Chung-Tong (1979b) Economics or politics—the new regional planning. Paper prepared for the Waigani Seminar, Port Moresby, Papua New Guinea, 24–28 September.

Xia-men Da-xue Jing-ji Xi Diao-cha Zu (1978) Fa-hui re-min gong-she 'san-ji suo-you, dui wei ji-chu' zhi-duo de you-yue-xing—Fujian sheng Long-hai-xian Jiao-mei gong-she de diao-cha (Make known the superior quality of the "three tier ownership, team as the foundation" system in people's communes—an investigation of Jiao-mei commune, Long-hai county, Fujian province) *Jing-ji Yan-jiu*, 10 (October), 54–57.

Xin-hui xian nong-cun jing-ji zhuang-kuang diao-cha (1956) (An investigation on the agricultural economy in Xin-hu county), *Nan-fang Ri-bao*, 15 September.

Zen-yang tuan-ji nong-ye-she li di shang-zhong-nong (1957) (How to unite the upper and middle peasants in agricultural cooperative), *Ren-min Ri-bao*, 15 June.

Zhang, Fo-jian (1959) Tui-an-pai lu-dong-li di ji-dian yi-jian (A few suggestions for managing manpower), *Ji-hua Yu Tong-ji* (Planning and Statistics), pp. 328–29.

Zhong, Dian-ho (1974) *Tan-tan Nong-cun Ren-min Gong-she Fen-pei Zheng-ce* (Let's talk about the policy on distribution in rural people's communes) (Guangdong: Guang-dong Ren-min Chu-ban-she).

Zhong-fa (1971) Document of the CCP Central Committee No. 82, (Reproduced and translated in Zhong-gong Yen-jiu, *Studies on Chinese Communism*), September.

Zhong-yang ren-min guang-bo dian-ti nong-cun zu (1979) (Central people's broadcasting station village section) *Nong-cun Jing-ji Zheng Ce Jiang Hua* (Talks on village economic policy) (Beijing: Ren-min Chu-ban-she).

Zhuo, Yue Li (1979) Luo-shi nong-cun jing-ji zheng-ce de ji-ge wen-ti (Several problems concerning implementation of rural economic policies), *Hong-qi*, 4 (April), 34–39.

Zuo, Mou (1979) Lun mao-jian (Treatise on being ahead) *Jing-ji Yan-jiu*, 7 (July), 14–17.

Development from Above or Below?
Edited by W. B. Stöhr and D. R. Fraser Taylor
© 1981 John Wiley & Sons Ltd.

Chapter 7

Thailand: Territorial Dissolution and Alternative Regional Development for the Central Plains

MIKE DOUGLASS*

Regional development planning has been called upon to assist a strategy of 'development from below' (see Chapter 2). In Third World countries, the formulation of a regional development approach appropriate to such a strategy encounters both practical and normative problems. The practical concerns arise from dissatisfaction with policies of 'top-down', concentrated decentralization (the growth pole approach), tied to the development strategy of urban-based industrialization. Failure of spread effects to reach the rural areas where the vast majority of the people live is indicative of the urban bias which influences development planning (Lipton, 1976). This bias, engineered by the industrialization strategy, has resulted in the stagnation of rural economies, the exhaustion of land and natural resources, and continued food deficits in spite of regional pockets of 'green revolution' success. With rural demand for urban-manufactured goods chronically depressed, import-substitution industrialization has reached a saturation point domestically as well; but continued and increasing protection of markets in the north against non-primary products from countries in the south has made clear the political and economic reality: industrialization in most Third World countries will have to continue to search for domestic rural markets; imports and balance of payments will depend in most cases not on changes from an import-substitution to an export-substitution strategy (Paauw and Fei, 1973) but rather upon accelerating agricultural production and rural development. The practical problem is how to put agriculture and rural development on a broad, participatory basis so that

* Grateful acknowledgement is given to the United Nations Centre for Regional Development and to the National Economic and Social Development Board of the Government of Thailand for providing the opportunity to conduct research in Thailand. The responsibility for the contents of this chapter remains solely with the author.

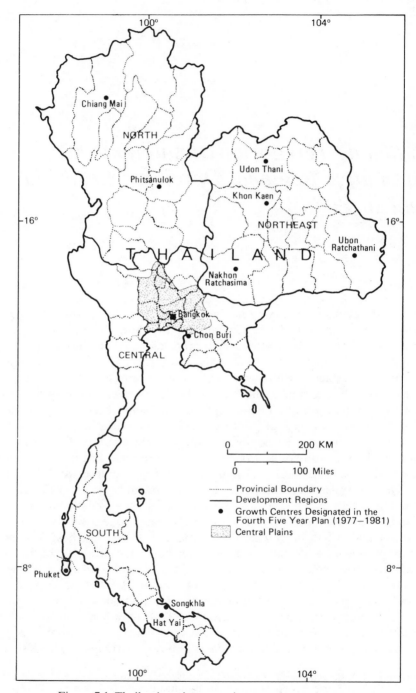

Figure 7.1 Thailand: regions, provinces, and growth centres

it can generate a more inward-looking growth of mutually supportive rural and urban economies.

The normative issues have been raised because of the failure of this same industrialization strategy to induce a growth process capable of redistributing the dividends of development on a path towards greater equality. This failure has led to a reassessment of the meaning of development itself (Seers, 1969). In raising these equity issues (which were thought to be either beyond the scope of development planning when defined as economic growth planning; or were understood to merit consideration only after increases in value-added could be guaranteed), discussion has quickly moved beyond that of limited modification of a growth model (Chenery *et al.*, 1974) to more far-reaching considerations of 'another development' (*Development Dialogue*, 1975), basic needs (ILO, 1976), and self-reliant development. In what can only be described as an explosion of research chronicling the growing proportions of people existing below a basic-needs and subsistence level and without access to remunerative employment (Onchan and Paulino, 1977; ILO, 1977), the question in search of response has now become: Who should be the protogonist in Third World development, and who should benefit? Growth has become a necessary by-product of participatory development in this reappraisal, rather than the first and only consideration (Higgins and Abonyi, 1978).

In responding to these practical and normative questions, spatial and regional development planners and theorists have begun to make thorough reappraisals of their own. Some have suggested the formation of small politically empowered urban-agrarian districts to solve the impending crisis of development and the designing of development processes which respond to local needs (Friedmann and Douglass, 1975; Friedmann and Weaver, 1979). Others have returned to the debate initiated by Myrdal (1957) and Hirschman (1958) and elaborated on a core–periphery model of development (Friedmann, 1966) concerning necessary conditions for polarization reversal. Their policy conclusions have pointed towards selective regional closure (Stöhr and Tödtling, 1977) to curtail the leakage of surplus and capital from rural to metropolitan regions and to protect 'infant' agriculture (see Chapter 5). Still others have kept alive Johnson's (1970) plea for investment in small urban places to act as foci for the spatial organization of village and rural development (Taylor, 1978). These and other proposals—such as those for integrated rural development—present a similar logic aimed at bringing a variety of development functions down to local levels to co-ordinate resource uses, to retain capital for local investment, to protect specific sectors from external control and competition, and to make population and distribution accountable to rural and urban areas and people who live in them.

Although these responses have been designed to solve the basic problem of how to devolve planning authority and economic power into the area of local development, they have not fully confronted the social dimension of the change such regional development policies and regional integration will bring. Neglect of this aspect runs a two-fold risk: that devolution of power will merely enhance the position of local elites, and that as a result, regional development policies will continue to be irrelevant to meet required development goals of Third World countries. By the end of the 1970s the ample evidence shows that attempts to organize the peasantry in order to implement a strategy of development 'from below' simply failed to make headway, without any thorough-going social and political reform accompanying this increased access to economic functions. In lieu of such reform, conditions of these people's lives have generally deteriorated (ILO, 1977).

In drawing upon a case study of regional development in Thailand, the purpose of this chapter is to incorporate the analysis of the processes creating social inequalities and the renewed concern over distribution of political and economic power on a sub-national, territorial basis and also the building-up of local development functions based on spatial organization principles. First discussing the social dimension, the argument will proceed by drawing upon a case study of the Central Plains of Thailand, and propose that the incorporation of local economies into national and international trade patterns has resulted in the loss of communal control over use of development resources.

It will be shown that this same historical process has resulted in the formation of two circuits of access to social, political, and economic power, the continued expansion of which has made clear the limitations of current development patterns and the inequalities they produce in distribution. This will be followed by a review of the development policies adopted in Thailand which have a bearing on increasing access of the peasantry to resources and institutional support for production. A summary will inquire into the logic of a regional solution to the problem of social inequality, suggesting a reformulation of both the regional concept and the metaphors being used in contemporary proposals for regional development.

THE DISSOLUTION OF COMMUNAL LIFE: THE CASE OF THE CENTRAL PLAINS

Prior to 1850 the majority of Thai people lived in closed, self-reliant villages. Interaction outside the village was limited to rudimentary regional networks of exchange. Within these self-reliant villages, communal life was maintained by right-of-access to goods and services necessary to sustain life and livelihood, by mutual assistance through labour exchange for production, and through other systems of reciprocity. The fate of the household could not be significantly different from the fate of the village itself. To accrue wealth and not share with others—especially in times of need—was immoral (Scott, 1976). With the inclusion of economic exchange in an all-encompassing communal organization of production and distribution, a variety of religious, social, and political sanctions were available to make economic relations accountable to specific household needs.

Changes following the signing of the Bowring Treaty in 1855, which opened

Thailand to international trade, were spectacular. Their greatest impact was on the Central Plains, the rice bowl of Thailand lying adjacent to the Bangkok metropolis and the rural area most completely incorporated into the Bangkok and international economy. In 1850 only 5 per cent of the rice grown in Thailand was exported. By 1930, 50 per cent was being exported and production had increased 26-fold (Ingram, 1971). While the north and north-east regions remained relatively isolated from the change-over from production for home and communal use and self-reliance (Potter, 1976; Moerman, 1968) to production for exchange on international markets and for comparative advantage, production in the Central Plains quickly became specialized. Rural non-farm industry, especially the making of textile products in village households, declined rapidly (Bell, 1970).

In spite of increasing separation of rural and urban life and the inequalities arising from crop specialization, however, the natural bounty of the Central Plains continued to mask the simultaneously induced process of rural social fragmentation until well into the last half of this century, a full hundred years after the process had been set in motion in Thailand. By then rural stagnation had begun to cast doubts on the continued pursuit of the national development strategy. Also by this time the accumulation of land and wealth in the Central Plains, and to a lesser degree in the outlying hinterlands, had begun to take the place of merit-making as an indicator of social status, reducing form of primary a redistribution at the village level (Piker, 1972). Land became an exchange commodity, marking a significant departure from the people's usufruct rights enjoyed in the nineteenth century. Tenancy and landlessness began to appear, and access to money and credit became essential for even subsistence-level production. Cooperative production and labour exchanges were replaced by hired labour and wage employment. Under such conditions, only household and patron–client relations bound together the 'loosely structured' social system (Hanks, 1975; Janlekha, 1955); other sources of reciprocal arrangements became 'virtually meaningless' (Piker, 1968, p. 138).

These social changes set in motion in the 1850s were paralleled by equally far-reaching transformations of political institutions at the end of the nineteenth-century. In 1892 King Chulalongkorn implemented a series of radical administrative reforms which completely absorbed local governments into the bureaus of the Bangkok government under a newly created and powerful Ministry of the Interior. This created a new élite of government bureaucrats who, along with the military and prosperous Chinese businessmen, would dominate national political power and development planning efforts to the present day. Under the reforms all provincial governments were brought under direct control of the centre, the number of local officials was kept at a minimum, while the central government's presence was extended into the peripheries of the nation and also reached below provincial level to the more than 500 districts created for administrative purposes. All rewards and punishments, including arbitrary transfer of officials, were delivered from Bangkok. Villages—even those which still maintained communal production and distribution systems—were left without any retaliating powers.

With the rapid centralization of power and the dissolution of communal relations, local systems of governance atrophid (Neher, 1975; Turton, 1976). The result is an extremely low level of local organization for development. (Jacobs, 1971). In a comparative study of local organizations in 18 countries, Thailand was found to be amongst the least organized, ranking seventeenth in terms of organizational linkages and characterized by strong downward control, few upward channels, and almost no village linkages (Uphoff and Esman, 1974). It was rated last in terms of electoral participation, control of bureaucracy, scale of rural development and influence in policies of resource allocation for rural development. Areas such as the Central Plains, completely absorbed in the ever-growing sphere of the central government's domain, also experienced the greatest weakening and dislocation of political and social institutions:

'In the central plain region...there is very little correspondence between the villages' political and social systems, on the one hand, and their administrative system on the other hand. The integrated, self-governing village has largely disappeared. (Neher, 1975, p. 235.)

CIRCUITS OF ACCESS TO PRODUCTION AND DISTRIBUTION

This incorporation of village production into the national and international political economy, and the ensuing dissolution of communal reciprocity systems, accountability, and moral claims on distribution were the causes of the formation of new linkages, which can be described in a simplified way as two circuits of access to Thailand's production and distribution systems. The circuit imagery implies that increased spatial integration of local economies has been essentially bimodal—with each circuit having different production opportunities, institutional arrangements, and labour markets—and resulting in growing inequality in welfare. It does not imply a unity of classes within each circuit and indeed such unity is lacking, in the lower circuit especially; this lack is part of the explanation for its chronic persistence. Nor does it imply isolation of the two circuits from each other, which is commonly assumed in many dualistic development models (McGee, 1978); rather, the two circuits are well-integrated as is seen by the linkages created through credit markets and assest transfers, seasonal and permanent employment through migration, product markets and consumption patterns. The division between the two circuits is manifested by the general increase in capital, income, and welfare in the upper circuit and the continuing subsistence level or even declining welfare levels in the lower circuit.

The upper circuit is composed of farmers with land resources capable of producing marketed surpluses on a scale allowing for capital accumulation; landlords (in the Central Plains one-quarter or more of landlords may be absent from the districts where they hold land: MAC, 1974); and in urban areas, white-collar workers, skilled labourers, and sales people. Many of those in the sales category comprise the so-called informal sector (ILO, 1972). In turning away from village and traditional forms of social accounting, those in the upper circuit in rural areas, in spite of urban biases exacerbated by public policy, have gained by ties forged through regional integration. For the lower circuit, however—the peasantry tilling small areas of land, usually in tenancy, the landless, and the urban unskilled labourers—this integrative process has meant a general loss of social, political, and economic prestige and power. Loss of access to social and political institutions has been discussed above. Loss of economic power has been manifested in at least five ways: decline in access to productive assets and a rise in landlessness; continued low occupational mobility of lower-circuit rural–urban migrants; stagnation of wages available in the lower circuit; increasing inequalities in the income distribution between the two circuits; and continued low levels of access to basic needs.

The distribution of land, the most important productive asset and the ultimate source of rural inequality, has been characterized in recent years in the Central Plains by an acceleration of transfers from lower to upper circuit. Data covering the 1963–74 period indicate not only that tenancy is concentrated in the small-farm sector (of less than 2.5 hectares) but that this has increased, at a time when the proportion of small farms has declined (Douglass, 1978). Paradoxically, at the same time when tenancy is the highest in the nation, farms in the Central Plains have an average size larger than any other region, including the low-land productivity areas of the north-east (Ahmad, 1976). Ingram (1971) has shown that, until the early 1960s, rapid expansion of transportation and the resulting rural–rural migration to frontier areas actually witnessed a decrease in the percentage of farms in tenancy in the nation as a whole. The near-exhaustion of these land frontiers (IBRD, 1978), especially in the Central Plains, has meant an acceleration of land transfers from the small farm sector to large farms and to rural and urban landlords (Amyot,1976).

The relationship between land holdings and access to institutions supporting production is obvious. Where a unimodal or small farm development strategy (Johnson and Kilby, 1975) could make these small farms economically viable (as the limited 'green revolution' in the Central Plains has shown), in practice they become uneconomical through lack of access to institutional credit, fragmentation of holdings, the loss of water resources to larger farms through irrigation projects, high tenancy rates, and a squeeze on agriculture which deflates the price of rice to one-half and raises the price of increasingly essential inputs such as fertilizer to twice that on the international market. In such situations land may be sold or lost through credit dealings, creating—for Thailand a recent phenomenon—landlessness (Kitahara, 1977), which however already accounts for 10–25 per cent of farm households in the *changwat* (provinces) of the Central Plains (MAC, 1974). Children of small tenant farmers and the landless, of course, inherit no land at all, and swell the ranks of an apparently unlimited supply of surplus labour which, with the loss of land frontiers, forges linkages with the metropolitan lower-circuit labour markets of general labourers and service workers. Net five-year migration between the Central Plains and other regions doubled between 1960 and 1970. During the latter five years, however, out-migration to other regions remained the same, while migration to the Bangkok metropolis increased four-fold (Douglass, 1978). By the early 1970s out-migration from the Central Plains became essentially a counterbalance to natural population increase in the richest agricultural area of the country (Pakkasem *et al.*, 1978). This demonstrates not only the low labour-absorptive capacity of the rural transformation, but also the stagnation of rural central places now growing at rates lower than natural population increase, and which are net contributors to the Bangkok-bound migration steams, rather than sources of off-farm employment for increasing rural populations.

The two-circuit character of regional integration has become more clearly defined by the expanding migration streams. Except for small farmers working in seasonal wage employment, and the landless forming rural squatter areas in forest preserves, along and on rivers and near upland plantations, most rural–rural migration—including land acquisition—is by large farmers who have the wherewithal to purchase and operate upland cane- and corn-producing plantations and farms (Pakkasem *et al.*, 1978). Those in the lower circuit struggle to maintain meagre assets, resist resettlement schemes which risk their hold on these assets and local patron–client relationships, and increasingly send family members to the metropolis where they search for and often find seasonal if not permanent wage employment in the city. Remittances from lower-circuit urban workers back to rural relatives form key linkages between the metropolitan and Central Plains economies and may, in fact, be the major source of transfers to rural areas, especially to the lower circuit (NSO, 1977).

As the simplified two-circuit metaphor suggests, rural–urban labour market linkages are highly structured. With the exception of farmers, 90 per cent of all occupations taken up in the metropolis are similar to those held before migration in the rural areas and small towns themselves (Douglass, 1978). In 1976, 80 per cent of those classified as farmers in rural areas became either labourers in manufacturing activities (54 per cent) or service workers; another 7 per cent found wage employment on farms in the metropolitan area. Contrary to conventional suppositions these migrants did not become petty entrepreneurs (hawkers and vendors) in the city (NSO, 1977a). As other studies on Asian countries have shown (McGee and Yeung, 1977), capital requirements, organizational and territorial agreements, and ethnic, regional, and other entry requirements make such sales occupations a minor source for absorption of rural migrants. Four-fifths or more of rural-to-Bangkok migrants entered into wage employment. With the simultaneous increase in numbers and loss of social, economic, and political power in rural areas, these migrants present urban employers with an unlimited supply of labour. The result is that there is little upward pressure on wages (Sarkar, 1974). Real wages for lower-circuit rural and urban occupations have shown no increase over the past decade, while those in the upper circuit have enjoyed increases (IBRD, 1978). Resulting income disparities lie behind of aggregate measures of inequality which are among the highest in Asia (Lo *et al.*, 1978) and which are increasing (Wattanavitukul, 1977).

Low mobility from the lower to the upper circuit through migration is further demonstrated by studies showing that even after 15 years in the metropolis, migrants from rural areas with low education levels fail to advance beyond general labour employment (Tirasawat, 1977). Studies of the slums of Bangkok also confirm that in spite of a wide range of activities and 'specializations' of virtually everyone in the household, combined incomes remain at or below official nutrition-related poverty standards (Thammasat, 1971; NSO, 1977b). This coincides with various estimates of basic needs and poverty which indicate that for those in the lower circuit, the level of welfare for many has fallen and the proportion of people below this level has increased (Ministry of Health, 1977; Krongkaew and Choonsiri, 1975). Estimates for the Central Plains show that as much as 40 per cent of the population may be living below subsistence level (ILO/ARTEP, 1977), a percentage equal to estimates of the size of the lower circuit in the rural area of this region (Douglass, 1978).

Finally, Figure 7.2 shows that within the two-circuit dichotomy between linked rural and urban labour markets there is also a friction of distance, which essentially bottles up much of the poverty in rural areas. In each type of location—the metropolis, urban places, and rural areas of the Central Region—

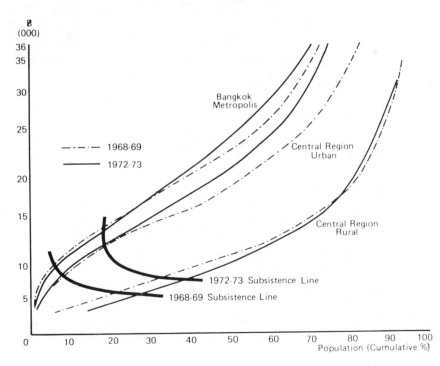

Figure 7.2 Cumulative distribution of household annual income, Bangkok metropolis and central region, 1968–69 and 1972–73

crossing of the lines on the graph for the years 1968–69 and 1972–73 indicates that those with higher incomes have in general benefited during the five-year period, whereas those with lower incomes have experienced deteriorating economic welfare. Using a poverty line set at a level at which debt would be incurred to obtain basic food requirements for both time-periods, the figure shows that poverty increases as the line extends from the metropolitan core to its rural periphery. The situation appears to have worsened in recent years, with rapidly increasing inflation outpacing increases in income. Chronic debt problems can

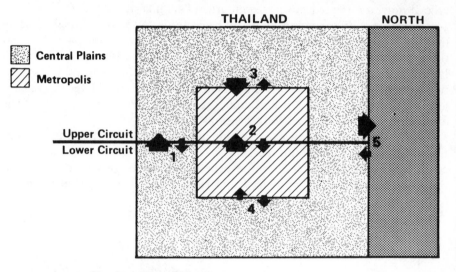

Lower Circuit — Upper Circuit

1 asset transfers, crop sales and labour from the lower circuit in exchange for credit, sales receipts, wages and patronage

2 casual and permanent unskilled labour supplied from the lower circuit in exchange for wages and shelter

Periphery — Core

3 capital and investment transfers, crop sales, allegiance from the rural upper circuit in exchange for subsidies on credit and inputs, rural development projects, sales receipts and legitimization

4 support of seasonal and permanent migration from the rural lower circuit in exchange for remittances from the urban lower circuit

Thailand — North

5 primary product sales, investment returns from Thailand in exchange for sales receipts, investments, development advice and assistance, military support

Figure 7.3 Production and exchange relations in Thailand: circuits, regions, and the international political economy

only increase, as evidence shows average rural incomes in the Central Region to be below average expenditures by 1976, (NSO, 1978).

Figure 7.3 summarizes the propositions contained within the argument that the dissolution of communal relations has been accompanied by a two-circuit formation, and increasingly unequal production and exchange relations at various scales. With the exception of exchanges between the rural and metropolitan lower circuits, in every instance these relations are marked by larger flows —'leakages' in the more benign terminology—from a lower or peripheral status to a higher one in a national and international core. Couched within these unequal relations is a transformation of patterns of access to the means of producing and receiving even the most basic commodities. While upper-circuit and core regions accumulate capital, economic, and political power, the lower circuit emerges as a pool of unlimited labour supply, earning subsistence-level wages and dependent upon a meagre and declining hold on assets for home production, and upon the seemingly capricious production requirements and the good graces of the upper tiers of society, from village level to global scale.

NATIONAL AND REGIONAL DEVELOPMENT PLANNING IN THAILAND

Development planning in Thailand has played a decisive role in the dissolution of communal capabilities for local development, and the associated formation of circuits of unequal access to production sources. The purpose as well as processes of planning and development have, however, come under greater scrutiny in recent years. The result has been proposals for more decentralized development, greater political authority at local levels, and enhancement of cooperative development efforts. Rather than viewing these proposals as current reactions to perceived crises, they may best be seen as a contemporary collage of minor traditions and themes of the Thai political economy, which have yet to receive any major commitment from those in government who manage the use or power.

In Thailand as in many underdeveloped countries, development planning officially began in the 1950s and was associated with approaches advocated by such international agencies as the World Bank whose report was instrumental in setting up the National Economic Development Board (NEDB)—later called the National Economic and Social Development Board (NESDB)—in 1959 to replace the Thai Technical and Economic Commission and the National Economic Council which were both established in 1950 to help plan, administer, and implement projects through financing largely made available through foreign aid. The first national development plan appeared in 1961 and covered a five-year period during which the strategy of import-substitution—urban-based industrialization (Lewis, 1954; Fei and Ranis, 1964)— was assembled from both traditional policies (the transfer of rural surpluses for metropolitan development) and new policies (promoting and protecting import-substituting infant industries). This basic strategy was carried into the Second Five-Year

Plan (1966–71) (Marzouk, 1972) which increased the role of the Board of Investment created during the first plan. By the end of the 1960s, full duty-exemption on capital goods and raw material imports was given for protected industry, which almost without exception chose to locate in the metropolitan area.

Domestic investment resources for the industrialization strategy were made possible, especially in the early years of the strategy's implementation, by restructuring the age-old systems of taxing agricultural production through creation of a complex revenue collection system summarized as the 'rice premium'. The effect of the rice premium has been and still is to put the tax burden squarely on the shoulders of the rice farmers, as costs are pushed back from exporters to middlemen and down to the farm gate (Ingram, 1971), keeping domestic rice prices substantially below international prices, and farm-gate receipts at less than 40 per cent of international prices.

The rice premium was rationalized not only as a convenient form of revenue collection in a country where direct income taxes are difficult to assess and collect; it has also been promoted as a means of suppressing urban wages— i.e. wages of the lower circuit—by keeping urban food prices low. To ensure stability of both wages for, and the political life of, lower-circuit wage-earners, and to engineer a high rate of profits and savings through a low-wage bill, labour unions were banned at about the same time as the setting up of national planning machinery. While peasant organization disintegrated in the villages, its urban counterpart was prohibited.

Throughout the time of the first two development plans there was no policy to deal with those inequalities which resulted from the planned development path, nor was there any explicit regional planning or development policy. Protection of urban industry, the 'rice premium', and the relative neglect of rural areas formed the *de facto* social and spatial development policy. Spatial consequences of this strategy were as dramatic as the social consequences. The heavy dependence on imported raw materials and capital goods coming in through the metropolitan port, incentives for locating industry in the metropolis (which included both the agglomeration economies of the capital itself and the increasing political and economic power of this centre), and the continuing low demand level for central-place and manufacturing functions in the rural areas, saw 86 per cent of all new, promoted industry locating in the metropolitan urban area between the years 1960 and 1973 (IDE, 1976). Two-thirds of all urban growth between 1960 and 1970 was in the metropolis, and the size of the city doubled from two to four million which increased its share of total urban population from 52 per cent in 1960 to 63 per cent in 1973. In 1960 Bangkok was already 26 times the size of Chiangmai, the nation's second-largest city. By 1976 it was 54 times its size, exhibiting the highest level of primacy in Asia and becoming one of the 15 fastest-growing cities in the world (Todaro, 1976).

This rapid growth, and the metropolis's resulting inability to manage traffic congestion, allocation of services, and sanitation in a city where 30 per cent or more of the households live in squatter settlements and slums (UN, 1975), became a primary rationale for 'top-down' spatial planning in the 1970s (NESDB,

1977). The management crisis of the Bangkok metropolis was, however, only one of several problems and events receiving public attention which found expression in development planning in the 1970s. Withdrawal of the American armed forces brought the loss of an important artifice which had filled trade gaps and stimulated urban development in Thailand's poorest region, the north-east, where the US bases were located. Internationally the economic order was in disarray, with resource nationalism and protectionism working to increase import over export prices and to block efforts to diversify Thai exports, especially of processed and manufactured goods. Terms of trade, using 1970 as a base of 100, fell to 80 by 1977 (IBRD, 1978). Finally for the first time in modern Thai history, the government was forced out of power by non-military, non-bureaucratic forces through student demonstrations in 1973. Although still largely contained within the powerful elite groups which have dominated Thai politics since the Revolution of 1932 created the constitutional monarchy, these events have brought about a high level of government instability. While recognizing the dramatic domestic and international changes which have taken place, this instability has generated conflicting ideas regarding the form of response to be undertaken, especially the course regional development should follow.

Recognition of the changing national and international political and economic environment has been obvious in the development plans of the 1970s. Under the Third Five-Year Plan (1972–76), for example, social justice was stressed, which for the first time marked the inclusion of objectives other than economic growth in development planning. This was carried over to the Fourth Five-Year Plan, which for the first time adopted an explicit spatial development strategy.

In principle, the central concept of the spatial approach to development planning adopted under the Fourth Five-Year Plan is straightforward application of the gorwth-centre strategy of concentrated decentralization (UNCRD, 1976; Salih et al., 1978; NESDB, 1977). Nine regional centres have been selected, with four objectives in view: curtailment of the growth of Bangkok; decentralization of administrative power; an accommodation of population growth; and creation of non-metropolitan centres of change (NESDB, 1977). With the exception of three centres, all are under 100,000 in population (1977) but are to be upgraded to intermediate-size cities of between 100,000 and 300,000 through proposed land-use controls in Bangkok, and tax and other incentives to locate in these municipalities.

Although it is premature to comment on the success of this spatial strategy, a number of observations raise serious questions concerning its ability to meet stated regional development objectives. First, the policies are of the same type as adopted elsewhere which have been ineffective in creating necessary incentives for private industry to decentralize (UNCRD, 1976). Secondly, the ability of eight or nine relatively small cities and towns to serve as centres of change for a national population currently greater than 45 million, and likely to double by the end of the century, must be seriously questioned, as each of these small cities would be called upon to provide services and act as industrial and administrative centres for population of up to double the current population of Bangkok. These centres and their hinterlands cannot be expected to be any less costly to expand and provide services for than the metropolis. If it is to act in aiding rural development, a more rural-oriented infrastructure—such as would be required to maintain agricultural research stations and agro-based industries—could be as costly as those structures required for industrial estates and import-substitution industries (Myint, 1975).

Third, and most importantly in regard to local planning capabilities, recent research has argued that the central obstacle in carrying out the intentions of the growth-centre idea is the low level of authority and competence of local administration in planning and implementing appropriate policies (Hennings and Kammeier, 1978). With heavy concentration of administrative and political power in the metropolis, the study concluded that the low self-governing capabilities of local government would 'render decentralized development very difficult' (Hennings and Kameier, 1978, p. iii). The ability of Bangkok to guide such development is also in doubt, as local needs are unlikely to be well understood, coordination is lacking amongst the various ministries and line agencies, and financial resources are also extremely limited (Bangkok Bank, 1978).

The strategy is typical in essence of the urban (metropolitan) bias in spatial planning, where events in rural areas are ignored in discussions on policy formation and implementation. The primary agenda seems to be to secure political control over rural peripheries and decentralize Bangkok. Plans for the centres themselves are urban-oriented in that they are primarily concerned with urban infrastructural problems such as traffic congestion, rather than with rural development functions to be provided by the towns. In what manner the centres are expected to link rural and urban development has not been made clear; how these centres are to promote the development of rural regions has not yet been articulated.

All these observations lead to the conclusion that the newly adopted spatial development policy must be seen as at best a partial approach. In its present form, it is more one of a series of town-planning schemes, rather than a strategy for regional development. Other policy changes not strictly spatial in stated form require some comment, therefore, in order to evaluate the prospects of linking them together to form a decentralized regional development strategy capable of generating a process of planning and development 'from below'. Policies to be discussed are those relevant to pursuing social and institutional reform and creating systems of access to political and economic functions which could reverse the process of social fragmentation, loss of political voice, and continued low level of welfare of peasants subsisting in what has been here termed the lower circuit of underdevelopment.

The government has responded to the problem of social fragmentation primarily through attempts to build farmers' cooperatives. The stated objective of these efforts has been to make credit available to the peasantry. It is significant that most problems related to indebtedness have been clustered in the Central Plains Region (Thisyamodal *et al.*, 1972; *Business in Thailand*, 1974; AID, 1973) where capitalist modes of production are strongest and traditional forms of labour exchange and communal sharing of agricultural tools and other inputs are weakest. For large farmers, however, indebtedness is generally not a serious problem, as they have incurred debt to expand commercial production, rather than simply to maintain subsistence production levels. Moreover, it is to these farmers that virtually all the institutional loans at low interest rates have been made available. For a majority of the peasantry and the landless, credit is still to be obtained only from so-called informal sources—relatives, neighbours, landlords, local merchants and traders—at rates multiple of those charged by government institutions. Only 10 per cent of all agricultural credit is from formal money markets; commercial banks in fact provide less than 1 per cent of their total deposits for agricultural credit even though 65 per cent of all bank offices are located outside of Bangkok (*Business in Thailand*, 1974). Under such conditions where banks are acting primarily as sources of leakage of regional resources, rather than as sources for rural development, where tenancy and seasonality of production make credit needs chronic, and where indebtedness has grown rapidly, the setting-up of cooperative organizations to create credit markets in the lower circuits appear absolutely essential.

Although government efforts to build such cooperatives began in the early years of this century and have been a major component of both the Third and Fourth Plans, by 1976 only 24 per cent of farm families were members of cooperatives. Many rural people are in fact not qualified to join because of membership requirements based on the criterion of land ownership. Landless and marginal farmers are left outside the officially sanctioned production organization and credit markets. In the early 1970s, for example, less than 20 per cent of loans from the Bank for Agriculture and Agricultural Co-operatives (BAAC) went to households with incomes under regional average (AID, 1973).

Rural inequalities ultimately rest upon access to land, as indicated by data concerning membership in cooperatives, and access to credit and the systems which support production. In 1975 for the first time in Thai history a land reform act was adopted, signifying official recognition of problems of increasing tenancy, indebtedness and, according to some sources, the rise of organized protest amongst peasants, especially in the Central Plains region (Ahmad, 1976; Senarak, 1974). To date, however, commitment seems low to any rapid implementation of land reform.

In its present form the act is also irrelevant to the growing number of landless, and does not address root causes of underdevelopment of the peasantry centring on land relations:

The historical process of land relations [in Thailand] has been as follows: the rise of export-oriented commercial production of rice which has converted the traditional pattern of subsistence economy into one which is considerably monetarized, with Thai peasants gradually beginning to depend on cash and credit. Such trends have resulted in farm indebtedness and particularly after the Great Depression, the disintegration of the peasantry has been considerably intensified. The number of tenant-farmers and the area of rented land has increased in core agricultural regions and rural wage earners have emerged with the increase in population pressure. (Ahmad, 1976, p. 92.)

The above efforts and others to devolve social, political, and economic power into rural regions to reach the peasantry in the lower circuit, can be characterized by lack of any sustained commitment by the government, but they may also be viewed as minor themes which, because they are responding to manifestations (especially urban ones) rather than the underlying forces of rural underdevelopment, fail to form a countervailing strategy capable of accelerating development of rural regions. Included among these schemes are recent proposals to give governors of the 71 *changwat* greater discretion over development expenditures, and the revival of the *tambon* (sub-district) development scheme. Although indicating a perceived need to devolve development planning authority, the low priority given these efforts is made clear in the case of the *changwat*-level proposal by the low level of per capita funds, which averaged $0.32 in 1970 and were clearly skewed towards Bangkok (USOM, 1970), and the fact that governors will continue to be appointed by the Ministry of the Interior rather than elected by their constituencies. The *tambon* scheme in its original form was to include the revitalization of *tambon* councils, first established in 1914 as the final administrative link between Bangkok and the rural areas of the nation, by allowing for their development into politically responsible development units. Proposals for change in this direction have remained the same since the early years after establishment of the constitutional monarchy in 1932 and the aborted reforms attempted by the Pridi-designed Municipalities Act of 1933 (Wilson, 1962). The continuing ambivalence over any radical decentralization is evidenced by the current form of this plan as a pilot 'new village' programme for the insurgency-troubled north-east, rather than as a national village development effort.

To make note of the repeated occurrence of proposals to implement policies of devolution along with this ambivalence is to stress how efforts to orchestrate incremental changes in the periphery from the centre dominate over efforts to broaden the participatory base of growth and development processes. The assembling of these minor efforts within a regional development framework has as a result never been at issue, because of this overriding orientation towards organizing systems of control rather than systems of participatory development; urban development rather than rural development; industrial growth over agricultural growth; and core expansion rather than periphery development. Thus increasing social and spatial inequalities have been generated, which have begun to weigh heavily on the prospects for prosperity in the nation as a whole and especially for those in the lower circuit.

RESTATEMENT: REGIONAL DEVELOPMENT FOR RURAL DEVELOPMENT

One of the most innovative features of the Fourth Five-Year Plan was the introduction of policies to decentralize development away from the metropolis and towards rural regions of the nation, where 85 per cent of the population resides. The spatial development component of this strategy, however, indicates a continuing 'top-down', urban-oriented development which places any efforts to 'effectively decentralize administrative power' (NESDB, 1977) into extremely moderate programmes to increase fiscal responsibilities of appointed prefectural governors, town-planning efforts for eight or nine growth centres, and a spatially restricted experimental *tambon* development scheme. Complementing these efforts are both new and existing ones to provide rural credit through co-operative societies, to institute land reform, and, in selected areas, to carry out expensive 'green revolution projects of land consolidation and water control. A review of these policies indicates a low commitment to rural development in general.

A possible framework, however, for a participatory spatial and regional rural development exists, and could be constructed through an assembly of these policies. The machinery for regional development for rural regions has been set in place for almost a century. The existence of a hierarchy of administrative units, especially at the *changwat* and *amphoe* (district) levels, could if activated as politically responsible and responsive units, begin to serve to integrate 'top-down' and 'from below' initiatives, by giving greater voice and bargaining power to rural districts. *Amphoe*, for example, would on average organize the rural population into units of 100,000 or 200,000 within easy commuting distance to centres of political and other activities; they would form the basic units of an 'agropolitan' development strategy aimed at organizing development on a participatory basis (Friedmann and Douglass, 1975). In addition, there exist more than 800 lower-order centres with populations between 2000 and 40,000 which could serve to transform the urban growth-centre strategy into a spatial strategy to create intraregional development networks linking rural to urban development. They could, by generating non-farm employment opportunities, increase incentives for local investment and production which might retain those rural surpluses now being transferred to the metropolis and abroad.

Regional development, especially in current attempts to make it relevant to rural development, must, however, go beyond the mere articulation of a spatial development strategy. Contemporary urban-oriented concern for spatial planning involving transportation networks, urban functions, the location of industry, industrial linkages, and innovation diffusion, are of limited value where the urban systems themselves are only rudimentarily developed (i.e. where a trickling down of development impulses is unlikely to occur, and where primary bottlenecks to development are political and institutional and must be confronted in their rural setting as well as in the metropolis).

Regional development especially needs to aid in creating systems and access networks for the peasantry. It should make accountable to them the social, political, and economic forces controlling the productive resources and their use, and the distribution which results from those controlling forces. The major task thus may be that of turning minor efforts towards cooperative development, land reform, *tambon* development, political reform, programmes for development of rural regions, and the devolution of administrative power, into a major theme—in which regional development means the creation of a more inward-looking, internally sustained development.

Nowhere is this task more urgent than in the Central Plains where most of these inward-looking, self-reliant development capabilities have eroded. This area, long considered the wealthiest region of the nation, is also the region where local development functions are most limited (Ingle, 1974). It has even been ignored in the new growth-centre approach; yet urban centres have been stagnating here, growing more slowly than the population of the region itself. Local government is virtually under the complete control of Bangkok, and 84 per cent of *changwat* financing is channelled through Bangkok with most of this allocated for education and with only 9 per cent of budget going to rural development (Bangkok Bank, 1978). Few indigenous organizations remain in this region which would be capable of creating a base for social, political, or economic power for the peasantry. At the same time the traditional relief routes from the 'irreducible claims' of an expanding system dominated by utilitarian relations based on exchange values, (Scott, 1976; Migdal, 1974) have all but vanished: where protest once was made simply by migrating to virgin land (Ingram, 1971) it is now reduced to a struggle to obtain rural wage work or employment in the metropolis in a lower circuit of underdevelopment. There is no automatic solution and no turning-point in sight.

The argument for policies to help solve the problems of poverty, inequality, under-employment, basic needs, and the necessity for more self-reliant development, has until recently—and especially in Thailand's national development plans—been seen primarily in normative terms, as an issue of social justice. For it to be raised to a major theme, with sufficient political commitment for success, it will also have to be seen as a compelling and practical alternative to the elitist, urban-based strategy of accelerated industrialization. In more abstract terms it will have to join its logic to the forces making contemporary history.

The compelling argument for a rural-based participatory form of development in practical as well as normative terms has been put forth at the outset of this chapter. Rural stagnation can no longer be tolerated for purposes of national development, and the limits of support for urban-based development are in view. Between 1961 and 1972 agricultural production advanced at only 0.5 per cent per capita. Paddy yields increased by only 0.23 per cent, the lowest in Asia (*Asiaweek*, 5 May, 1978). Natural population growth and the resulting rise in domestic demand for rice have severely reduced production available for export, traditionally one-third or more of the nation's total exports. More pessimistic estimates are that domestic demand will equal domestic production within the coming decade. In the early 1970s a ban was temporarily placed on rice exports during poor seasons, an action taken for the first time since the signing of the Bowring Treaty more than a century ago. 'Top-down' planning and 'green revolution' projects have failed to reverse this process significantly, as they are dependent upon foreign management, and virtually without active participation by rural people themselves (*Investor*, 1975)—especially in such regions as the Central Plains where individualism has replaced communal production efforts (Ingle, 1974). These plans and projects are expensive, extremely limited in scope, and too slow in implementation to keep pace with the growth of social and economic inequalities.

If regional integration has resulted in a dissolution of local planning capabilities, and if centralized planning systems cannot cope with (and may even abet) the formation of social inequalities and an expanding circuit of poverty-sharing, the question for consideration turns the mirror around, and asks how regional development can be expected to coordinate the necessary themes and efforts to put agricultural and rural development on a broad participatory basis. A partial answer lies in a reconsideration of the concept of a 'region' and the framework it provides for policy formulation. Within this framework such proposals as those of regional closure as well as the pitting of a 'top-down' against a 'from below' strategy can be discussed. Finally, without an abstract solution, how can such a regional framework be filled out in real terms?

The regional concept has not only been overwhelmed by an addiction of spatial planners to the elixirs of urbanization, which makes such a concept thereby virtually irrelevant to countries where four-fifths of the population still reside in rural areas; it has also been obscured by related traditions adopted over the past several decades, defining regions as evanescent enities with misty boundaries identified through descriptive statistical analysis which searches for homogeneity and nodality. These traditions also grumble over interference of political and administrative boundaries which cut through statistical regularities; even though working within politically designed units has been welcomed by some with a sigh of relief by some who consider that arbitrary bases of selection through quantitative techniques is confusing the purpose of a regional development planning concept (Richardson, 1978). The obscured commonality of these traditions closely involves political considerations, and this becomes obvious from such concepts as those describing regions as areas within a national economy with greater internal than external unity, coherence, and, presumably, purpose (Boudeville, 1966); those which claim regions to be areas sufficiently comprehensive in structure to be capable of functioning independently (Richardson, 1978, after Czamanski, 1973); or concepts suggesting that regions should be defined in such a way that they can internally comprehend and work towards resolution of planning issues (Keeble, 1969); or stressing a region as a unit capable of managing complex decisions through a common approach and an awareness of its problems (Klaassen, 1965); and, reaching back towards earlier traditions of cultural geography, those which view one purpose of sub-national planning to be a taking-into-account of 'regional consciousness' (Glasson, 1974) and a shared territorially-bound destiny (Friedmann and Weaver, 1979).

Territorial unity, the capacity for independent functioning and internal resolution of planning issues—especially those related to social inequalities and the competing objectives of a variety of interest groups—the pursuit of a common approach and an awareness of these issues which takes into account the existence or building-up of a shared territorial consciousness, all require regional-level structures for political discourse and accountability to be counterminous with the level of economic and social complexity intrinsic to the regional concept. The (re)insertion of the requirement for political competence in turn calls for a setting of regions into real space, a taking-on of political, social, and economic life within jurisdictional constraints identified by the division and limits of power meeting at commonly accepted boundaries. The usual caveats would still apply under this concept: not all decisions nor all functions would be within the sole domain of regional authorities; the ideal of combining physical networks and nodal structures within these units (Friedmann and Douglass, 1975) would still be sought, and may be more feasible in Third World countries where decentralized planning is embryonic (Richardson, 1978). Regions, however, must be viewed as entities endowed with political and administrative capacities to bring public discource within reach of rural people. As mentioned, in Thailand the existence of *changwat* and *amphoe*-level administrative units, most of which contain at least one of the municipalities of the nation, would amply serve to bring this capacity for participation down to local levels.

It is through the limitation of regions as political units, however, that current interest in regional closure and 'planning from below' can best be assessed. Both may ultimately be misleading concepts. Regional closure, unless it is simply a matter of reforming pricing policies in the capital city by reversing terms of trade in favour of 'infant' agriculture (as Japan, and more recently Korea, have done), rests on effective regional power able to curtail leakages— most of which go beyond terms of trade into external control of assets and technologies, central identification of development projects and value added trails relaying multipliers to the capital city and abroad. In this regard, however, a more appropriate solution to the issue may not so much centre on the forming of regional trade and tariff barriers by selectively closing off the region. But in empowering regions with the means of controlling institutional arrangements and ownership patterns internally, and bargaining externally for allocation of national resources on a par with metropolitan and other regions. The policy of regional closure begs the question of the role of the state in a political economy, by remaining silent on the process of class formation and access to political functions in the region. A lowering of the scale of resolution of territorial power does not carry with it the automatic resolving of social inequalities; rather it may serve to lower the scale of that elite monopoly over the use of power evidenced at national level by the persistence in Asia and elsewhere, of those entrenched non-elected regimes which came into being because of a similar call for territorial disengagement. It may even be suggested that the failure of national closure, either (1) to be realized under most of these governments because of the interest of those in the upper circuit in increasing international linkages; or (2) where adopted (by choice or by default), to produce any internal resolution to the problem of unequal development, without equally profound shifts in increasing internal systems of political and economic access—has, ironically, been most responsible for raising the issue of regional closure, as an escape from internal colonization.

The essential lesson to be drawn is that in gaining the regional 'tokens of sovereignty' (Hirschman, 1958), the regional purpose is in turn to endow its citizenry with an equitable share in it. In situations where high tenancy, indebtedness, growing landlessness, institutional barriers prohibiting access to development functions to those in the lower circuit all exist, the mere shutting-off of regions is unlikely to make territorial solutions coincide with social solutions to the normative and practical problems which currently dominate policy planning. And without a solution which can raise rural incomes above subsistence levels, the awaited demand for urban functions, regional manufactures, and the resulting rise in off-farm employment to eliminate 'surplus' labour—in order to engineer a polarization reversal and provide the mechanism which could implement spatial policies aimed at stimulating growth of lower-order centres would simply not be forthcoming.

The above observations are further elaborated by questioning the usefulness of pitting 'top-down' against 'from-below' development strategies by dichotomizing them as polar opposites. Inquiries into the mechanism for a successful local development plan have insisted, for example, that participatory development depends very much on a high frequency of development impulses both 'top-down' and 'from below'. Local self-reliance in isolation provides little leverage for development (Uphoff and Esman, 1974). Furthermore, regions as subnational systems will require hierarchical arrangements of development functions which are extraregional as well as contained within the regional domain: to provide adjudication of interregional disputes, juridical and monitoring systems to ensure the attainment of national welfare minima, to pursue projects of national interest, and create a complexity of information systems (many of which may be redundant) needed to support development as a learning process and to provide linkages between and among both territorial and non-territorial interest groups in order to integrate regional development networks into the broader national space. It is perhaps for these same considerations that advocates of regional closure (see Chapter 2 and 5) have carefully avoided pushing their proposals to a position of regional autarchy (absolute sovereignty), or even autarky (complete self-sufficiency).

The commonality, and the point to be taken from the ideology of regional closure and planning 'from below', is the need for a devolution of political and economic power itself within a territorial framework, a task that includes simultaneous pursuit of social and institutional reform and a more equal integration, rather than separation, of local development systems.

In finer detail, the logic of aligning regional development with rural development calls for consideration towards the formalization of political institutions and local assemblies (Friedmann, 1978), rural public works programmes which give ownership rights on newly created assets to the rural labourers hired to build them (Griffin, 1978), land reform to at least provide security of tenure if not a more radical transfer of ownership rights to tenants, reform of credit institutions to include tenants and the landless, changes in pricing policies to support rather than penalize the agriculture sector, creation of a system of widely dispressed agriculture experimental stations (Ruttan, 1974), the improvement of village-to-market town linkages and policies to encourage growth of lower-order centres (Johnson, 1970); and the strengthening of peasant cooperative to create systems for risk-sharing, capital formation, and a political voice in resource allocation and production.

Arrayed in this fashion, the list leads to the question not of how to devise policies, but rather where to begin. The 'where' suggested here is at the regional level, with a purpose to reverse the processes of polarization, social fragment-ation, and underdevelopment of the peasantry. It is by giving this regional focus to rural development that the policies become practical, and normative issues become resolvable through active participation of the people whom development planning has always, in word if not in deed, been designed to aid.

To the extent that the issues are social and political rather than strictly econo-mic and spatial, the ultimate inquiry is into the linkages between logical impera-tives and the uses of political power. In Thailand it is only in recent years that the imperatives for an inward-looking, rural-based approach to development have become apparent. It is perhaps no accident that it has only been in recent years, too, that non-traditional political forces have breached the citadel of those forces that have controlled the exercise of power for the last century or more. Whether or not existing or alternative sources of political and social organ-ization will be capable of giving a sustained commitment to what would be a radical departure from the past, cannot be answered in the abstract. Rural impoverishment has had a poor record for bringing about such a commitment in other Asian countries, and urban manifestations of the impoverishment of the peasantry have tended to raise concern over the management of metropolitan growth, rather than over the more fundamental rural locus of underdevelop-ment.

It is suggested that without attention to the regional basis for rural develop-ment, continued spatial integration will simultaneously bring about an ex-panding lower circuit of unorganized household labour, whose main form of participation is in subsistence-level production and wage labour largely contained in rural areas, but increasingly directed and linked to the metro-politan economy. The case of the Central Plains has been taken in order to explore this basic dimension in the process of spatial integration—a dimension often neglected when regional analysis is directed towards 'lagging' or 'poor' regions where harshness of physical environment often disguises the harshness of transformation of social environment. It has served to display the contradic-tions in the coexistence of rising regional prosperity and rising inequalities. It also points up the limitations reached in the pursuit of previous national develop-ment policies—limitations resulting from insufficient attention paid to rural development in general and to the century-long process of disenfranchisement of the people in rural regions from access to a social accounting in the systems of production in particular. Policies contemplated under the Fourth Five-Year

Plan have given notice of the need to redirect new as well as often-voiced proposals to resolve these contradictions. It may be too soon either to applaud or reject them, but this redirection in itself is significant—if not for level of commitment or perception of purpose, then at least because it gives sustenance to minor themes searching for ascendancy.

REFERENCES

Ahmad, Z. (1976) *Rural Employment and Land Reform Policy* (Geneva: ILO).

AID (1973) *Spring Review of Small Farmer Credit (Thailand)* (Thailand: Bangkok).

Amyot, J. (1976) *Village Ayutthaya; Social and Economic Conditions of a Rural Population in Central Thailand* (Chulalongkorn University, Social Research Institute), (Bangkok).

Bangkok Bank (1978) The budget—business as usual, *Bangkok Bank Monthly Review*, 19 (10), 445–59, (Bangkok).

Bell, P. F. (1970) *The Historical Determinant of Underdevelopment in Thailand* (New Haven: Yale Economic Growth Center).

Business in Thailand (1974) Rural credit: a continuing lesson in massive exploitation, July, (Bangkok).

Boudeville, J. R. (1966) *Problems of Regional Economic Planning* (Edinburgh: Edinburgh University Press).

Chenery, H. B. *et al.* (1974) *Redistribution with Growth*, IBRD.

Czamanski, S. (1973) *Regional and Interregional Social Accounting*, (Lexington Books), (Mass.)

Development Dialogue (1975) What now; the 1975 Dag Hammarskjöld Report, 1 (2), (U.N., New York.)

Douglass, M. (1978) *Migration Labour Absorption and Income Distribution in the Bangkok Metropolitan Subregion of the Central Plains* (Nagoya, UNCRD, mimeo).

Fei, J. and G. Ranis (1964) *Development of the Labor Surplus Economy: Survey and Critique*, (Irwin, New York).

Friedmann, J. (1966) *Regional Development Policy: A Case Study of Venezuela* (Cambridge, Mass.: MIT Press).

Friedmann, J. and M. Douglass (1975) Agropolitan development: toward a new strategy for regional planning in Asia. Paper presented at the UNCRD Seminar of Industrialization Strategies and the Growth Pole Approach to Regional Planning and Development: The Asian Experience (Nagoya, Japan), 4–13 November.

Friedmann, J. (1978) The active community; toward a political–territorial framework for rural development in Asia. Paper presented at the UNCRD Seminar on Rural Urban Transformation and Regional Development Planning (Nagoya, Japan), 31 October–10 November.

Friedmann, J. and C. Weaver (1979) *Territory and Function: The Evolution of Regional Planning* (London: Edward Arnold).

Glasson, J. (1974) *An Introduction to Regional Planning* (London: Hutchinson).

Griffin, K. (1978) Growth and impoverishment in the rural areas of Asia. Paper presented at the UNCRD Seminar on Rural–Urban Transformation and Regional Development Planning (Nagoya, Japan), 31 October–10 November.

Hanks, L. M. (1975) 'The Thai social order as entourage and circle', in G. W. Skinner and A. T. Kirsh, eds., *Change and Persistence in Thai Society: Essays in Honor of Lauriston Sharp* (Cornell University Press).

Hennings, G. and H. Kammeier (1978) *The Development of Secondary Urban Centers in Thailand* (Bangkok: German Agency for Technical Cooperation).

Higgins, B. and G. Abonyi (1978) 'Basic needs' and the 'unified approach' : making them operational, *UNCRD Working Paper* (Nagoya).

Hirschman, A. O. (1958) *The Strategy of Economic Development* (Yale University Press).

IBRD, 1978, Thailand: *Toward a Development Strategy of Full Participation* (Washington, DC).

Institute of Developing Economies (IDE) (1976) *Performance and Perspectives of the Thai Economy* (Tokyo).

ILO (1972) *Employment, Incomes and Equality: A Strategy for Increasing Productive Employment in Kenya* (Geneva).

ILO (1976) *Growth, Employment and Basic Needs: A One-World Problem* (Geneva).

ILO (1977) *Poverty and Landlessness in Rural Asia* (Geneva).

ILO/ARTEP (1977) *An Illustrative Exercise in Identification and Qualification of Basic Needs for Thailand* (Bangkok).

Ingle, M. (1974) *Local Governance and Rural Development in Thailand* (Cornell University, Rural Development Committee).

Ingram, James C. (1971) *Economic Change in Thailand, 1850–1970.* (Stanford: Stanford University Press).

Investor (1975) The tambon scheme; involving the villagers, 7 (6) (June), 21–33. (Bangkok).

Jacobs, N. (1971) *Modernization without Development; Thailand as an Asian Case Study*, (New York: Praeger).

Janlekha, K. O. (1955) *A Study of the Economics of a Rice Growing Village in Central Thailand* (Bangkok: Ministry of Agriculture).

Johnson, E. A. J. (1970) *The Organization of Space in Developing Countries*, (Cambridge, Mass. : Harvard University Press).

Johnston, B. F. and P. Kilby (1975) *Agriculture and Structural Transformation; Economic Strategies in Late Developing Countries*, (Oxford University Press).

Keeble, L. (1969) *Principles and Practice of Town and Country Planning*, (Estates Gazette, London).

Kitahara, A. (1977) *The Forms of Rural Labor and Their Role in the Rural Economy of Thailand* (Tokyo: IDE).

Klaassen, L. (1965) *Area Social and Economic Development* (OECD, Paris).

Krongkaew, M. and C. Choonsiri (1975) Determinants of the poverty band in Thailand, *Warasrn Thammasat*, 5–1 (June–September), 48–69. (Bangkok).

Lewis, W. A. (1954) Economic development with unlimited supplies of labour, *The Manchester School*, 22.

Lipton, M. (1976) *Why Poor People Stay Poor: A Study of Urban Bias in World Development* (London: Temple Smith).

Lo, F., K. Salih and M. Douglass (1978) Uneven development, rural–urban transformation, and regional development alternatives. Paper presented at the UNCRD Seminar on Rural–Urban Transformation and Regional Development Planning (Nagoya, Japan), 31 October–10 November.

Marzouk, G. A. (1972) *Economic Development and Policies; Case Study of Thailand* (Rotterdam University Press).

McGee, T. G. (1978) Doubts about dualism, *UNCRD Working Paper* (Nagoya).

McGee, T. G. and Y. M. Yeung (1977) *Hawkers in Southeastern Asian Cities* (IDRC, Toronto).

Migdal, J. (1974) *Peasants, Politics and Revolution* (Princeton).

Ministry of Agriculture and Cooperatives (MAC) (1974) *Agricultural Land Tenure 1974* (Bangkok).

Ministry of Health, Government of Thailand (1977) *National Policy on Food and Nutrition* (Bangkok).

Moerman, M. (1968) *Agricultural Change and Peasant Choice in a Thai Village* (University of California Press).

Myint, H. (1975) Agriculture and economic development in the open economy. In L. G. Reynolds (ed.), *Agriculture in Development Theory* (New Haven: Yale University Press, Berkeley).

Myrdal, G. (1957) *Economic Theory and Underdeveloped Regions* (London: Duckworth).

Neher, C. D. (1975) Thailand. In R. N. Kearney (ed.), *Politics and Modernization in South and Southeast Asia* (John Wiley).

NESDB (1977) *The Fourth National Economic and Social Development Plan* (1977–1981), Bangkok.

National Statistics Office (NSO) (1977a) *The Survey of Migration in Bangkok Metropolis, 1976* (Bangkok).

National Statistics Office (NSO) (1977b). *The Utilization of Labor in Thailand, 1975* (Bangkok).

Onchan, T. and L. Paulino (1977) *Rural Poverty, Income Distribution and Employment in Developing Asian Countries: Review of Past Decade* (Bangkok: Kasetsart University).

Paauw, D. S. and J. C. Fei (1973) *The Transition in Open Dualistic Economies: Theory and Southeast Asian Experience* (New Haven: Yale University Press).

Pakkasem, P. S. Photivihok, U. Khaothien, U. Kerdpibule, and M. Douglass (1978) *Impacts of Green Revolution and Urban-Industrial Growth in the Bangkok Metropolitan Dominance Subregion* (Bangkok: NESDB).

Piker, S. (1968) Sources of stability in rural Thai society, *Journal of Asian Studies*, 24 (7), 777–90.

Piker, S. (1972) Post-peasant village in Central Plain Thai society. Paper presented at the 1972 meeting of the Association of Asian Studies (New York).

Potter, J. M. (1976) *Thai Peasant Social Structure* (Chicago: University of Chicago Press).

Richardson, H. W. (1978) *Regional and Urban Economics* (Hamondsworth: Penguin).

Rosen, George (1975) *Peasant Society in a Changing Economy* (University of Illinois Press).

Ruttan, V. (1974) Rural development programs: a skeptical perspective. Paper presented at the Colloquium on New Concepts and Technologies in Third World Urbanization, (Los Angeles; UCCA).

Salih, K. and F. Lo (1979) Polarization reversal and a policy of regional closure. In W. Stöhr and D. R. F. Taylor (eds.), *Development from Above or Below? The Dialectics of Regional Planning in Developing Countries* John Wiley.

Salih, K. P. Pakkasem, E. Prantilla, S. Soegijcko, 1978, "Decentralization Policy, Growth Pole Approach, and Resource Frontier Development: A Synthesis Response in Four Southeast Asian Countries," in F. Lo and K. Salih, eds., *Growth Pole Strategy and Regional Development Policy*, Pergamon.

Sarkar, N. K. (1974) *Industrial Structure of Greater Bangkok* (Bangkok: UNADI).

Scott, J. C. (1976) *The Moral Economy of the Peasant; Rebellion and Subsistence in Southeast Asia* (New Haven: Yale University Press).

Seers, D. S. (1969) The meaning of development, *International Development Review*, 11 (4), New York.

Senarak, B. (1974) *Land Alienation of the Farmers: A Case Study in Amphoe Bangmun Nak Changwad Phichit Prior to 1974* (Bangkok: Faculty of Economics, Thammasat University).

Stöhr, W. and F. Tödtling (1977) Spatial equity—some antitheses to current development strategies, *Papers of the Regional Science Association*, 38.

Taylor, D. R. F. (1978) The role and function of lower order centers in rural development. Paper presented at the UNCRD Seminar on Rural–Urban Transformation and Regional Development Planning (Nagoya, Japan), 31 October–10 November.

Thammasat University (1971) *Klong Toey, A Social Work Survey of a Squatter Slum* (Bangkok: Faculty of Social Administration).

Thisyamodal, P. *et al.* (1972) *Agricultural Credit in North Thailand* (Bangkok).

Tirasawat, P. (1977) *Economic and Housing Adjustments of Migrants in Greater Bangkok* (Bangkok: Institution of Population Studies, Chulalongkorn University).

Todaro, M. (1976) *Internal Migration in Developing Countries* (Geneva: ILO).

Turton, A. (1976) Northern Thai peasant society: twentieth century transformations in political and jural structures, *Journal of Peasant Studies*, 3 (3), (April), 267–98.

United Nations (UN) (1975) *World Housing Survey.*

UNCRD (1976) *Growth Pole Strategy and Regional Development Planning in Asia.* Proceedings and papers from the Seminar on Industrialization Strategies and the Growth Pole Approach to Regional Planning and Development, November 1975.

Uphoff, N. and M. J. Esman (1974) *Local Organization for Rural Development in Asia* (Cornell University: Rural Development Committee).

USOM (1970) *An Analysis of Local Government Revenue and Taxation Data for 41 Changwats* (Bangkok).

Wattanavitukul, S. (1977) *Income Distribution of Thailand* (Thammasat University: Faculty of Economics).

Wilson, D. A. (1962) *Politics in Thailand* (Cornell) University Press, Ithaca, New York.

Chapter 8

Papua New Guinea: Decentralization and Development from the Middle

DIANA CONYERS

Since achieving political independence in 1975, Papua New Guinea has embarked upon an ambitious programme of political and socioeconomic reform, designed to redirect the focus of development away from the national level towards the majority of people in the country's rural areas. This programme is, however, rather different from the approach to spatial planning and development adopted in most countries.

The most obvious difference is that the main emphasis is on the reorganization of the political and administrative structure within which planning occurs— rather than the use of specific spatial planning tools, such as, for example, growth centre strategies, the relocation of industry or integrated rural development projects. This does not mean that such tools are not considered important. It merely means that the main effort at present is concentrated on the development of a political and administrative structure which will allow maximum participation in development and provide a viable framework for the introduction of more specific planning tools.

The importance of establishing this basic organizational structure has often been underestimated in the past. This has resulted in the failure of many regional planning projects because of the lack of support provided by the political and administrative system in the region and the lack of involvement and commitment from the people themselves.

However, more recently, the importance of organizational structures has begun to be recognized. This is reflected in the growth of the science of 'development administration' (Schaffer, 1969; Robinson, 1971; Hyden *et al.*, 1970); and in some of the more recent literature on development planning in general, and regional planning in particular. For example, Friedmann and Douglass, in their attempt to define a new strategy for regional planning in Asia, include as one of the eight components of their strategy the need:

To devise a system of governance and planning that is ecologically specific and gives substantial control over development priorities and programme implementation to district populations..., to enable them to take advantage of ecological opportunities where they exist..., to harness the richly personal, embodied learning of local inhabitants to the more formal, abstract knowledge of specialists..., and to encourage a growing sense of identification of local people with the enlarged communal space of the agropolis. (Friedmann and Douglass, 1978, p. 189.)

This is, in effect, what Papua New Guinea is attempting to do in its own particular fashion.

The other significant feature of Paupa New Guinea's approach is that it does not focus on either the national level or the village (or 'grass roots') level, but on the intermediate administrative level of the province. It might thus be called 'development from the middle'—rather than development from above or below.

The Papua New Guinea approach involves the creation of a new tier of government, known as provincial government, and the decentralization of a large number of government powers and functions to the new provincial governments. These powers include the responsibility for planning and implementing provincial development programmes, limited only by very broad policy and financial constraints set by the national government.

This chapter describes the reasons for introducing the decentralization programme and the form which it takes, and then discusses the impact which it may be expected to have on spatial patterns of development in the country. Before considering the programme itself, however, the chapter looks briefly at the Papua New Guinea environment and at past and present spatial development patterns and policies. For further basic information on Papua New Guinea, readers are referred to Nelson (1972) and Ward and Lea (1970).

THE PAPUA NEW GUINEA ENVIRONMENT

Papua New Guinea, with a land area of just under 500,000 square kilometres and a population of about three million, is a small country by world standards and one might wonder why there is any need for decentralization. However, it is considerably larger than most of its South Pacific neighbours (with the exception of Australia and New Zealand) and it is also very fragmented, both physically and culturally. Both these factors have contributed to the move towards decentralization.

The country consists of the eastern half of the large island of New Guinea (the western part being the Indonesian province of Irian Jaya) and a large number of much smaller islands scattered over several hundred kilometres to the north and east (see Figure 8.1). Most of the country is very mountainous, with many peaks between 4000 and 5000 metres high. The large number of islands and the mountainous topography create a very fragmented country, in which transport and communication are major problems. There are very few roads and most communication is by air, sea—or foot.

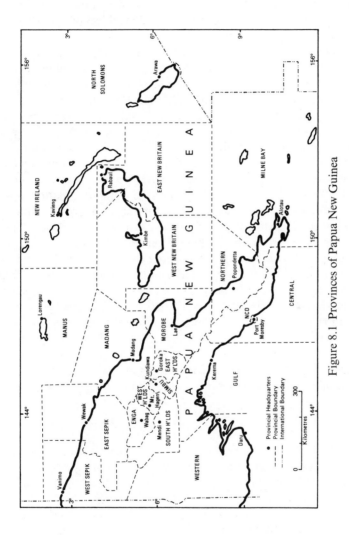

Figure 8.1 Provinces of Papua New Guinea

The physical fragmentation has to a large extent been responsible for the cultural diversity and fragmentation which also exist. The three million people are divided into 700 distinct language groups and there are marked physical and cultural differences in the people, from one area to another. Although communication is now improving and there is considerable interaction between groups, especially in urban areas, regional and local feelings are still very strong and play a major role in social and political life.

The urban population is relatively small, accounting for less than 15 per cent of the total; but as in most developing countries it is growing rapidly, probably at a rate approaching 5 per cent per annum, as compared to a growth rate of 2.7 per cent for the population as a whole. (Accurate population statistics are difficult to obtain in Papua New Guinea because the last census was in 1971 and it was only a sample census. The figures used here are based on projections, which are notably unreliable. There have been a few recent surveys and censuses of some urban areas but not enough to make over-all generalizations about the urban population. For census purposes an urban area is a non-village area with more than 500 people.) The distribution of the urban population is more even than in many countries, mainly because communication problems have lessened the tendency to migrate to a single urban centre. Nevertheless, the capital, Port Moresby, accounts for more than 25 per cent of the total urban population and its growth rate is more rapid than those of other centres. A survey carried out in 1977 indicated that the city had a population of about 107,000 and an annual growth rate of 5.8 per cent. There are seven other major urban centres with populations between 10,000 and 45,000, and many smaller centres.

The country is divided into 19 major administrative units, formerly known as districts but renamed provinces at independence (Figure 8.1). The provinces vary considerably in population, from less than 30,000 (Manus) to nearly 300,000 (Morobe). Port Moresby City and its immediate environs constitute a separate administrative area known as the National Capital District.

COLONIAL POLICY

Colonial administration in Papua New Guinea lasted nearly 100 years. Until the Second World War the country was administered in two parts: New Guinea in the north and Papua in the south. New Guinea was originally administered by Germany but became a United Nations Trust Territory, administered by Australia, after the First World War. Papua was originally occupied by the British but became an Australian colony in 1905. Although Australia was responsible for the administration of both Papua and New Guinea from the end of the First World War, they were administered separately until after the Second War and even then their official status was different, Papua being an Australian colony and New Guinea a Trust Territory of the United Nations. This historical difference has not been forgotten and, together with differences in people, resources, and patterns of development, still causes tensions between the two areas. The division between Papua and New Guinea is thus another factor adding to the fragmented nature of the country and periodically the national government is pressurized by a Papuan separatist movement, known as Papua Besena.

The Australian colonial administration had two main effects on spatial patterns of development in the country. One effect was the imposition of a highly centralized political and administrative structure onto a society which had known nothing more than clan or village government in the past. Two overseas consultants who advised on the preparation for provincial government in 1974, declared that:

> in our experience of political systems in Asia, Africa and the Caribbean, we have not come across an administrative system so highly centralized and dominated by its bureaucracy. (Tordoff and Watts, 1974, p. 2/2.)

The Australians did introduce a system of local government councils which still exists in most parts of the country but the councils were not given the powers or resources to become effective governing bodies. Moreover, since they bore little or no relation to traditional systems of government, they were not strongly supported by the people.

The situation was worsened by the proximity of Australia, which resulted in a much higher proportion of decision-making powers being retained in the metropolitan country than was usual in colonial situations (Ballard, 1973). The official ties with Australia were broken at independence, but the centralization of decision-making within the country remained and the recent decentralization programme is in part a reaction against this. It is also an attempt to return to a system of government which is administratively more effective and more in keeping with the country's traditions.

The other major impact of the Australian administration was the development of regional inequalities in levels of economic development, social services, and administration. These inequalities arose as a result of two factors: variations in economic potential and variations in the rate at which the Administration opened up (or 'contacted') different parts of the country.

Australia, like most colonial powers, concentrated on the development of those areas which would yield the greatest return. In the case of Papua New Guinea these areas were mainly the island provinces (North Solomons, East and West New Britain, New Ireland, and Manus), where copra and cocoa plantations flourished, and the areas around the major ports (particularly Moresby, Lae and Rabaul). Much later, attention was also given to the Highlands provinces, in part because of the potential for coffee production. However, since most parts of the Highlands were not 'discovered' by the Australians until the 1930s and were not opened up until the 1950s, the Highlands provinces are still struggling to 'catch up', especially in the provision of basic social services such as health and education. In the provinces along the north and south coasts of the mainland, the situation is the reverse of that in the Highlands. These areas have benefited to some extent from many years of 'contact' but have suffered because of their lesser potential for cash cropping or other forms of economic development.

Table 8.1 *Socioeconomic indicators by province*[1]

Region[2]	Province	Population (1971 Census)	Index of health status[3]	Index of health services[4]	Index of educa-tion[5]	Index of land trans-port[6]	Index of small-holder cash income[7]	Exten-sion-staff per 1000 popula-tion[8]	Govern-ment expend-iture (Kina per capita)[9]
South Coast	Western	82,930	48.7	57.3	22.9	5.6	4.5	10.4	41.2
	Gulf	65,240	43.0	52.7	30.2	4.2	8.7	10.5	52.9
	Central	122,111	61.9	74.6[10]	53.7	20.2	3.9	18.5	52.7
	Milne Bay	124,300	51.9	66.5	42.0	2.9	13.0	11.2	38.4
	Northern	77,430	61.9	53.6	38.9	12.3	15.7	12.2	54.6
High-lands	Southern Highlands	210,430	2.1	14.5	6.5	18.6	10.6	6.3	40.5
	Enga	162,683	23.6	30.5	6.7	14.6	18.2	5.1	34.4
	Western Highlands	230,069	13.1	36.0	8.1	28.4	41.7	8.2	41.0
	Chimbu	166,420	8.2	21.3	8.7	24.1	42.4	7.6	36.6
	Eastern Highlands	269,170	19.2	38.3	6.6	31.2	31.0	6.6	32.6
North Coast	Morobe	284,280	34.6	53.6	35.1	20.6	18.2	10.8	42.1
	Madang	191,570	36.7	53.0	30.4	11.8	16.5	10.1	42.6
	East Sepik	198,370	16.6	20.0	23.9	17.5	17.4	9.3	40.4
	West Sepik	101,760	1.0	11.1	20.0	5.7	10.1	7.8	37.9
Islands	Manus	30,160	60.1	70.5	55.7	15.1	12.8	14.7	67.0
	New Ireland	67,260	91.3	91.0	65.0	100.0	29.9	13.1	63.5
	East New Britain	118,276	100.0	98.5	83.7	30.5	60.7	18.9	64.7
	West New Britain	80,180	62.1	72.5	41.1	16.8	35.4	12.3	70.1
	North Solomons	111,580	75.8	83.4	49.0	55.3	70.2	14.5	70.9

Source: Data prepared by the National Planning Office, PNG; collated by Rosemary Lynch.

Notes:

[1] The data in this table are provisional and subject to considerable error. However, they are sufficiently accurate to indicate the relative status of each province. Most of the data were compiled in 1977 and, unless otherwise stated, are based on the most up-to-date figures available at that time.

[2] Regions are not official administrative units, but they provide a useful basis for comparative analysis.

[3] The index of health status is based on the average of the following indicators, each of which has been weighted from 0 to 100: rural life expectancy; rural child mortality; child malnutrition (based on weight measurements).

[4] The index of health services is based on the average of the following indicators: health extension officers per 1000 population; population per aid post; travelling time to aid post.

[5] The index of education is based on the average of the following indicators: adult literacy; percentage of 7-year-olds in Grade 1; percentage of 13-year-olds in Grade 7.

[6] The index of land transport is based on the capital value of roads and bridges per square kilometre and per capita.

[7] The index of smallholder cash income is based on per capita income from ten cash crops and from livestock.

[8] Extension staff include staff involved in health, education, agriculture, general administration, and works.

[9] Expenditure includes expenditure on health, education, agriculture, general administration, and capital works.

[10] The index of health services for Central Province underestimates the level of services available because many people use health services in the National Capital District.

Table 8.1 shows the variations between provinces in a number of indices of socioeconomic development as a result of the colonial pattern of development. The inequalities between provinces are not as great as in many countries but they do exist, and they are enough to create regional tensions. Although the figures in Table 8.1 give a reasonable indication of the variations between provinces, it should be noted that there are also marked variations in levels of socioeconomic development within many provinces. An analysis of inequalities within provinces has been made by Wilson (1975).

Colonial policy has had a more obvious impact on regional inequalities than on inequalities between classes; but its impact on class structure cannot be ignored (Weeks, 1977). Compared with many developing countries the traditional social structure in Papua New Guinea was relatively egalitarian. Although every community had (and still has) its leaders—referred to as 'big men' in the *lingua franca* of Melanesian pidgin—leadership was not hereditary in most areas and there was no rigid caste or tribal system. Furthermore, the distribution of land was relatively egalitarian and there were no landlords or landless peasants.

Although there are some signs that a rural elite of small-scale cash crop farmers and businessmen is beginning to emerge (Donaldson and Good, 1977), colonialism has not brought about radical changes in this structure in the rural areas. Moreover, because there is adequate cultivable land in most parts of the country and climatic conditions are favourable, the extreme poverty found in the less advantaged sections of rural communities in many African and Asian countries does not exist in Papua New Guinea. Although there is some malnutrition among young children due to the lack of a balanced diet, starvation or food shortage occur very rarely and most families in the rural areas are able to meet basic needs with relatively little output of energy.

However, colonialism has resulted in the emergence of an urban elite, dominated by national politicians and senior public servants, many of whom also have business interests. The members of this elite group have been strongly influenced by their colonial predecessors and, as in many other countries (Fanon, 1967; Dumont, 1969), there is thus a tendency for the white colonial elite to be merely replaced by a black indigenous one.

POST-INDEPENDENCE POLICY

The first Papua New Guinea government was formed in 1972 and during its period of office the transition to self-government in 1973 and full political independence in September 1975 took place smoothly. During this period there was also a noticeable reorientation in national development policy, which has continued in the post-independence period.

The first noticeable change in policy was the announcement in December 1972 of a set of basic national goals, known as the Eight Aims. These Aims emphasize Papua New Guinean control of the economy, equality, decentralization, self-reliance, the use of appropriate technology, participation of women, and greater involvement of government in the economy.

Soon after the announcement of the Eight Aims, a parliamentary committee was established to prepare a constitution to come into effect at independence. The committee, known as the Constitutional Planning Committee, put a great deal of thought and effort into its work and paid particular attention to the basic philosophy of development which it felt should underlie the structure of government. One outcome of this was the incorporation in the constitution of a set of National Goals and Directive Principles. These National Goals include the principles inherent in the Eight Aims but adopt a broader approach to development, emphasizing human and social goals as much as economic objectives. The five basic goals in the constitution can be summarized as: integral human development; equality and participation; national sovereignty and self-reliance; environmental conservation; and the use of traditional Papua New Guinean ways.

Another outcome of the work of the Constitutional Planning Committee was the introduction of a leadership code, imposing restrictions on the behaviour of government ministers, senior public servants and certain other specified leaders. Three years later, in 1978, the Prime Minister, Michael Somare, considered the introduction of legislation to strengthen the code in order to prevent such leaders (or their immediate families) from owning any business interests. However, this move was postponed due to lack of political support at the time.

These moves may suggest that Papua New Guinea has a clearly defined development policy not unlike those adopted by countries which are aiming for a socialistic approach to development, with modifications to suit their particular customs and traditions. In practice, however, this is not really the case in Papua New Guinea; the approach to development is not nearly as clear-cut as the Eight Aims and National Goals imply.

In the first place, although several of the stated development goals suggest a socialist approach to development, there is also a strong and open commitment to the continuation of a basically capitalist socioeconomic system. The Prime Minister himself, in explaining his reasons for wanting to strengthen the leadership code in 1978, emphasized that it was designed to prevent corruption and conflicting interests—not as an attempt to introduce socialism (Somare, 1978).

Secondly, the government's success in implementing the Eight Aims and National Goals has so far been somewhat disappointing. For example, one of the most important goals is economic self-reliance. However, during the period 1972–73 to 1975–76, the ratio of imports to gross domestic product actually increased from 0.39 to 0.48. Similarly, in 1976–77 Australian aid still account-ed for 40 per cent of national revenue, the same proportion as indicated by statistics for 1972–73 (Berry, 1977); and in 1979 it still accounted for 33 per cent. It must be acknowledged that the 1972–73 figures underestimate total Australian contributions, since a number of 'hidden' costs were met by Australia at that time; however, this is at least partially offset by the considerable increase in financial assistance (both grants and loans) from overseas sources other than Australia in recent years.

The government in 1978 introduced a fairly sophisticated system of 'rolling' economic planning, known as the National Public Expenditure Plan, in the hope of controlling and redirecting government expenditure more effectively. It remains to be seen what effect this will have. However, there is some danger that, as in many other developing countries where a sophisticated planning system has been introduced, it will tend to increase the demand for skilled man-power to plan and monitor projects and the main benefit will accrue to those organizations and provinces which have the greatest capacity to plan and justify projects (Berry and Jackson, 1978).

Another example of lack of progress is in the area of social stratification. Unfortunately there are no quantitative data to indicate precise changes in the class structure. However, qualitative evidence suggests that, in spite of the leadership code and talk of equality and equal opportunity, there has been a steady increase in the size and well being of the urban elite which began to emerge during the colonial period. One reason for this is the regular (although not unduly large) increases in urban wages and salaries—currently guaranteed at 3 per cent per annum, combined with the government's somewhat controver-sial 'hard currency' policy, which has resulted in a series of currency revalua-tions which primarily benefit the urban consumers, at the expense of primary producers in the rural areas.

The rather slow rate of progress in implementing the stated national goals is due both to the practical problems of undertaking any substantial socioecono-mic reform and to the lack of adequate political support for the basic con-cepts underlying the goals. This is a situation which is, of course, not unique to Papua New Guinea by any means. Many countries (both developing and developed) have a similarly admirable set of aims and ideals but have great difficulty implementing them.

However, Paupa New Guinea has made very real progress in the implement-ation of one of its national goals—the goal of decentralization. In this respect, Papua New Guinea's experience is perhaps unique and of particular relevance to any discussion of spatial patterns of development.

THE CONCEPT OF DECENTRALIZATION

The concept of decentralization through the establishment of provincial govern-ments was developed by the Constitutional Planning Committee between 1972 and 1974, although there had been some vague discussion of decentralization prior to the formation of the Committee (May, 1975; Conyers, 1976). The Committee emphasized that:

> Power must be returned to the people. Decisions should be made by the people to whom the issues at stake are meaningful, easily understood and relevant. The existing system of government should therefore be restructured and power should be decentralized, so that the energies and aspirations of our people can play their full part in promoting our country's development. (Papua New Guinea, 1974, p. 10/1.)

It recommended that the best way of doing this was through the establishment of governments at provincial level.

The recommendations of the Constitutional Planning Committee on decentralization aroused a somewhat mixed reaction from members of Parliament, when its report was discussed at a series of Parliamentary meetings. There was a prolonged debate on the advantages and disadvantages of provincial government and it was eventually decided that the initial version of the constitution, which came into effect at independence in September 1975, should not include any provisions for its establishment. However, a year later the constitution was amended to provide for the establishment of provincial government. This was followed by the introduction of a special 'organic' law (a term used in Papua New Guinea to mean a law which supplements the constitution and, like the constitution, requires the approval of two-thirds of the members of Parliament to introduce or amend), which gave very substantial powers to the new tier of government (Conyers, 1976).

The prolonged debate over the introduction of provincial government reflects two factors. Firstly it reflects the natural reluctance of national politicians and senior public servants to relinquish a significant part of their newly won powers and privileges to another level of government. Secondly, it reflects the fact that there are two fundamental objectives behind the moves for decentralization in Papua New Guinea and, although these aims are closely related, they are also potentially conflicting, particularly in terms of their effect on spatial patterns of socioeconomic development.

The main objective of decentralization, and the one expounded by the Constitutional Planning Committee, is to increase participation in government and, through the strengthening and coordination of political and administrative structures at provincial level, to increase the capacity to plan and implement development programmes in each province. The proponents of provincial government believe that because of the physical and cultural fragmentation of the country the only way to achieve national unity is to allow each area a considerable amount of autonomy in making decisions about its own development.

There is also a strong feeling in Papua New Guinea that participation in government is an important end in itself. This may be attributed partly to traditional systems of decision-making, which involved people at village or clan level. However, it may also be due to the relative 'subsistence affluence' of

Papua New Guinea society, which allows more time and resources to be devoted to politics and participation in decision-making for its own sake (Fisk, 1962). Activities which in many countries would be classed as 'leisure' or part-time activities are valued highly in Papua New Guinea and involvement in local level politics is one of the most important of these. In fact, to many Papua New Guineans it would be regarded more as full-time work than as a part-time or leisure activity.

Equal distribution of social and economic development is not explicitly stated as an aim of decentralization. However, it is one of the main national aims and many supporters of provincial government believe that it should follow as a result of the increased capacity to plan and implement development programmes within each province.

The other, secondary, objective of decentralization is to ensure that individual areas or groups of people are not dominated by others and receive their fair share of the country's resources. This objective is seldom expressed openly but it lies behind some of the moves, particularly by the more advantaged provinces, to obtain provincial government and it is reflected in the financial arrangements prescribed in the Organic Law on Provincial Government.

The province which was most concerned to obtain provincial government was North Solomons (previously known as Bougainville). It is an island province in the extreme south-east of the country, near the Solomon Islands, a former British colony which became independent in July 1978. The support for provincial government in the North Solomons can be attributed to a genuine feeling of separation from the rest of the country, based on the province's geographical location and the distinctive physical characteristics of its people, and to a belief that the province was not getting its fair share of resources. North Solomons provides a very large part of the country's internal revenue, mainly from Bougainville Copper Limited, a subsidiary of the Rio Tinto–Zinc Corporation which operates a very large open-cast mine on the island, but also from the production of cocoa and copra. The Bougainvilleans felt that they had to suffer all the negative effects of the mine (such as loss of land, environmental pollution and social disruption) but received very little benefits. Thus the demand for provincial government in North Solomons was a demand for both greater autonomy in decision-making and more financial resources (Mamak and Bedford, 1974; Conyers, 1975a).

North Solomons (then known as Bougainville) played an important part in the debate over the introduction of provincial government. Before the Constitutional Planning Committee had made any final recommendations on decentralization, the Bougainvilleans persuaded the national government to allow them to establish their own interim provincial government. However the national government was not prepared to decentralize any significant powers to the Bougainville government until decisions had been made about the form which decentralization would take in the country as a whole. The Bougainvilleans then became frustrated and, shortly before Papua New Guinea's independence, attempted to secede from the rest of the country, declaring themselves to be the independent Republic of the North Solomons. This resulted in a movement against provincial government at the national level and, consequently, the omission of provisions for provincial government from the initial version of the constitution.

However, six months after independence, there was a reconciliation between the national government and the North Solomons, the latter agreeing to remain part of Papua New Guinea provided provisions were made for the introduction of provincial government on terms acceptable to it. The constitution was thus amended and the Organic Law drafted, taking into account the conditions demanded by North Solomons.

The political climate is now very different from what it was at the time of the drafting of the Organic Law. There is now much more political support for decentralization at the national level and correspondingly less pressure from provinces like North Solomons. Consequently, the main objective of decentralization is now uppermost in the minds of most people and steps are being taken to counteract the inequalities built into the original provisions for provincial government. However, there will continue to be a potential for conflict between the two basic objectives of decentralization, and the impact of this on spatial patterns of development will be examined in more detail later in this chapter.

THE DECENTRALIZATION PROGRAMME

Technically Papua New Guinea's system of government might be described as a 'quasi-federal' system; the provincial governments have many powers similar to those of state governments in a federal system but national legislation can ultimately override provincial legislation and the creation and suspension of provincial governments is a national government power. In practical terms, however, most Papua New Guineans do not look upon their government in terms of a rigid legal division of powers, as is normal in a federal or quasi-federal system. They look at it more as a flexible sharing of the powers and responsibilities of government, in which problems and disputes are solved wherever possible through discussion and compromise rather than through legal channels. This reflects the Papua New Guinean approach to politics in general. This approach, which is often called the 'Melanesian way', is described by the Prime Minister, Michael Somare, in his autobiography (Somare, 1975).

The 19 provincial governments are full governments in the sense that they are elected bodies and they have the power to pass legislation and make other policy decisions regarding a wide range of activities in the province. These activities include agricultural extension, small-scale industrial and commercial development, formal education up to Grade 10 (subject to national policy on curriculum), non-formal education, rural health services, community development, and provincial roads and other capital assets of provincial rather than national importance. They are also responsible for the development of local or community-level government within their own province. Some of these powers are automatically provincial government powers under the Organic Law on

Provincial Government, while others have been delegated to them by the national government.

The decentralization of political and legal powers has been accompanied by a major administrative and financial decentralization. Because of the small size of the provinces, they have not been allowed to establish their own public services. However, the national public service has been decentralized through the establishment of a separate department of the public service for each province, in addition to the normal national departments. A provincial department includes all those public servants in the province who are involved entirely or mainly in the execution of provincial government powers and functions. It is headed by a senior public servant of departmental head status (known as an Administrative Secretary) and it is politically responsible to the executive of the provincial government, not to a minister of the national government. However, its members are subject to all the normal conditions of the national public service and can transfer from one province to another just as from one national department to another.

This administrative structure is designed to meet two objectives: to ensure that public servants are responsive to local political needs, and thus hopefully avoid the need for provincial public services, and to improve administrative coordination and management at provincial level. The structure itself is probably unique to Papua New Guinea. The only other country known by the author to have introduced a similar administrative system—but within a centralized political structure—is Tanzania (Conyers, 1974 and Lundqvist, Chapter 13 in this volume) although the Solomon Islands is planning a somewhat similar decentralization. It is not entirely coincidental that in both Tanzania and Papua New Guinea the administrative decentralization resulted at least in part from the advice of an international consultant firm, McKinsey and Company. However, there was in fact very little communication between the two McKinsey projects, or between the governments of the two countries. It will be interesting to see what effect administrative decentralization has in these countries, particularly in terms of improving coordination and encouraging integrated planning at provincial or regional level. The Papua New Guinea system may prove particularly interesting because it is combined with a somewhat unusual form of political decentralization. The possible impact of the administrative decentralization in Papua New Guinea is considered in the next section.

The financial decentralization is also of major importance for the planning and management of provincial development. Provincial governments have relatively few independent revenue sources. They are allowed to levy a few local taxes (including a retail sales tax, land tax and head tax) and the proceeds of certain national government taxes, including royalty payments made by companies developing natural resources in a province, are refunded to provincial governments. However, with the exception of a province like the North Solomons, which receives a sizeable income from mining royalties, provinces are unlikely to generate much revenue from these sources and the national government has retained control over major forms of taxation, such as income tax.

For most provinces, by far the most important sources of revenue are unconditional grants from the national government. In 1978 grants allocated (or, in the case of provinces which had not yet assumed full financial powers, eventually to be allocated) to provincial governments on an unconditional basis accounted for about 20 per cent of the total budget. With these grants the provinces are expected to fund all provincial activities carried out by the provincial departments and the construction and maintenance of capital assets classified as a provincial responsibility. This gives the provinces a great deal of autonomy in managing their own affairs and encourages a system of financial planning and management which is almost impossible to achieve under conventional departmental funding arrangements.

The main problem inherent in the financial arrangements is that the size of the grant given to each province tends to favour the more developed provinces. As a result of the conditions demanded by the North Solomons province, the national government is required by law to give each province no less than the real value of money spent on provincial functions prior to the establishment of provincial government, provided that there is not a decrease in the real value of national government revenue. In addition, provinces receive certain extra grants based on their contribution to national government revenue. Both these provisions, and also the limited taxation powers available to provinces, tend to favour the more advantaged provinces and perpetuate existing inequalities.

However, the national government may also give additional grants (either conditional or unconditional) to provinces if it wishes to do so. The present policy of the national government is that these extra grants should be used to reduce inequalities between provinces and to support projects which are in line with national goals. The national government is also providing special assistance in planning and financial management to the less-advantaged provinces. In the short run, this sort of assistance is probably more important than additional grants because in many of the poorer provinces the problem is not only the lack of financial resources but also the incapacity to utilize the existing financial resources effectively. The possible counteracting effect of assistance from the national government to the poorer provinces is discussed further in the next section.

THE IMPACT OF DECENTRALIZATION

Unfortunately it is impossible to assess the impact of Papua New Guinea's decentralization programme on spatial patterns of development at this stage, because the programme is only in its infancy. However, it is possible to make some predictions based on the objectives and form of the decentralization, and past and present experiences.

This section of the chapter examines the possible impact of the decentralization on the spatial patterns of two aspects of development: participation in government and socioeconomic development. In each case, the impact of regional differences, rural–urban differences and class differences is considered. This is because the stated aims of development in Papua New Guinea, and in most current international development programmes, include the reduction of inequalities between regions and between rural and urban areas, and the provision of more assistance to the poorer sections of the population. Since there has as yet been very little attempt to evaluate the impact of the decentralization programme, the following comments are based mainly on the author's personal observations and the unpublished views of others familiar with the decentralization.

Participation in Government

More equal participation in government and decision-making is the main aim of decentralization in Papua New Guinea. As explained earlier this is regarded not only as a means of achieving social and economic development at provincial level but also as a very important end in itself.

It is obvious that the introduction of provincial government has increased participation in government and provided a much greater opportunity for people to participate in decision-making in their own areas. A very crude measure of this is that there are now about 400 provincial politicians in addition to the 107 members of the national parliament. This in turn provides a much greater opportunity for people either to become politicians themselves or to bring grievances to their local politician.

What impact does this have on differences between areas and groups of people? The most obvious impact is that decision-making is no longer centred in the capital city of Port Moresby. Each province now has its own decision-making body, and it is encouraging to notice in many provinces a marked increase in activity and enthusiasm among both politicians and public servants at provincial level. Equally important is the fact that the provincial governing bodies have equal powers and opportunities to make decisions about their own areas, whereas in the national government there is a tendency for decision-making to be dominated by a few politicians, often from the more affluent provinces.

However, it is more difficult to tell what impact this will have on the vast majority of the people who will never become either provincial or national politicians. There is of course a danger that decentralization will merely mean the development of 19 small, self-centred bureaucracies at provincial level and a new elite of provincial politicians. It is too early to predict whether or not this will be so. A lot will depend on the background and attitudes of the provincial politicians and the efforts they make to extend the influence of their government below the provincial level, through local-level government, improved extension services, and other forms of communication. The Constitutional Planning Committee in its recommendations on provincial government emphasized that:

> Effective government at both national and district [= provincial] levels requires that there be strong links with the villages. The government and development of our country depend upon the village people. We attach the very greatest importance to government at the village level. (Papua New Guinea, 1974, p. 10/3.)

Initial experiences do indicate that provincial politicians are more concerned than the national government to develop an effective system of local-level government and to improve other links with the villages, partly because they are nearer to their people and so more conscious of their needs—and more subject to pressure from them. However, it is too early to make any definitive judgments at this stage.

Socio-economic development

It is more difficult to assess the likely impact of the decentralization programme on patterns of socioeconomic development, partly because it is regarded as a secondary rather than a primary aim of the decentralization and partly because effects cannot be expected immediately.

The impact on regional patterns of development will depend mainly on the outcome of the inherent conflict between the national aim of equalization and the provisions which tend to favour the more-developed provinces. The main factors likely to perpetuate or increase existing inequalities are the financial arrangements prescribed in the provincial government legislation, which guarantee minimum grants based on previous levels of expenditure and contributions to national revenue, and the tendency for the more affluent provinces to have a greater capacity to plan and implement development pro- grammes and (within the limited powers available) to generate their own internal revenue. The latter factor is particularly important at present because of a serious shortage of skilled manpower in the country.

On the other hand, it can be argued that the inequalities between provinces have become much more obvious as a result of provincial government, and this has resulted in increasing pressure on the national government to improve the status of the less-developed provinces. In the past, very little government expenditure was broken down by province and so no one knew the extent of the inequalities. Politicians in particular were unaware of the real differences because the budget which was published and debated in Parliament was sub- divided by department rather than by province. In contrast, since decentraliza- tion the annual budget indicates clearly the size of the grants to each province. Provincial governments also provide a voice for the poorer provinces to express their demands and grievances, whereas in the National Parliament their re- presentatives are often dominated by the more vocal and influential members from the more advantaged areas.

There are also indications that decentralization is increasing the planning and management capacity of the less-developed provinces, by improving administrative coordination and in particular by attracting skilled manpower from the provinces to return and work in their home areas. This is happening in all provinces, but the effects are particularly marked in the poorer provinces where administration has been very weak and skilled manpower extremely difficult to obtain.

Finally, the national government is at present committed to trying to reduce existing inequalities by the redistribution of additional grants and by providing technical assistance to the poorer provinces. Both the Department of Decentralization, which is responsible for the coordination and monitoring of the decentralization programme, and the National Planning Office are very much aware of the importance of such action, and a preliminary study has suggested that the existing inequalities could be reduced very significantly in a short period of time if every effort was devoted towards it (Garnaut, 1978). A great deal will depend on the national government's ability to withstand pressures from the wealthier provinces and its capacity to provide the financial and technical assistance required. This in turn will depend on the country's overall financial situation (which is itself dependent on such factors as external markets and Australian aid policy) and on the priority which is given to provinces in the allocation of financial and manpower resources. There is considerable potential for the reduction of financial and manpower commitments at the national level, thereby releasing more resources for assisting the provinces; but there has as yet been little attempt to realize this potential.

In addition to the factors already considered, which are clearly working either for or against equalization, there are other factors which may work either way. The most important of these, and one of the most unpredictable, is differences in natural resource endowment. In some of the more-advantaged provinces a favourable natural resource base will encourage further development and generate additional revenue. However, some of the less-developed provinces also have a great potential for natural resource development. For example, the potential for agricultural development is good in the Highlands provinces, while the country's next major mineral developments are likely to be in two of the least-developed provinces, Western and West Sepik. Many forms of natural resource exploration are only just beginning in Papua New Guinea and new discoveries and developments could drastically change the current patterns of resource utilization and, therefore, the distribution of benefits (and problems) resulting from such utilization. Moreover, although such large-scale developments are often dependent on capital provided by private, foreign-owned companies, the national government now has considerable control over the location and conditions of operation of these companies, so it can influence the pattern of resource development and, therefore, the benefits accruing to different areas.

The situation is thus very complex, and decentralization is only one of many factors likely to influence regional variations in socioeconomic development in the future. However, if the national government has both the political will-power and the financial and technical capacity to maintain its fundamental policy of equalization, and thus provide the assistance needed by poorer provinces, inequalities may gradually be reduced.

The impact of decentralization on rural–urban differences will depend to a large extent on the priorities set by the provincial governments. However, the legal provisions for provincial government will at least ensure that a reasonable proportion of government funds is spent outside Port Moresby, either in the smaller urban centres or actually in the rural areas. This in itself is important, given the disparities between Port Moresby and other urban centres; moreover, in many cases expenditure in the smaller urban centres is of benefit to the surrounding rural areas, while expenditure in Port Moresby is much less likely to have any positive effect on the rural parts of the country.

Within the provinces it is likely—although by no means inevitable—that the provincial governments will give more, rather than less, attention to rural development. When sitting in an air-conditioned office in Port Moresby it is very easy to forget that there are such things as villages; but in a provincial centre that is much more difficult to do. Provincial politicians tend to maintain closer links with their electorates than their national counterparts, and consequently to be more responsive to their needs. Similarly, most public servants at provincial level spend much of their time in the rural parts of the province, while village people often visit the provincial headquarters for advice and assistance.

Furthermore, the administrative and financial decentralization which has accompanied the establishment of provincial governments should provide a basis for better planning and implementation of rural development programmes. Many rural development programmes in Papua New Guinea and elsewhere have failed because of the problems created by a centralized administrative system, in which activities are compartmentalized along departmental lines and most decisions are made by the headquarters of each department or agency (Conyers, 1975b). The two main problems are lack of coordination between the various departments or agencies concerned, and inadequate contact or communication between the project area and the administrative headquarters where decisions are made. These problems affect any form of development programme but their efforts can be particularly serious in the case of rural development. Most rural development projects involve several different departments or agencies, and it is essential that they work very closely together. Similarly, such projects also require an intimate knowledge of local conditions and attitudes and direct involvement of local officials and the people themselves. It is very difficult to implement this sort of programme through a centralized, compartmentalized bureaucracy.

Many countries have responded to these problems either by demarcating special planning or development areas, in which a special administrative structure may sometimes be established, or by creating a special department or agency (such as a department of community development), which cuts across traditional departmental lines. (This concept of spatial planning areas was first used in industrial countries—e.g. the United Kingdom, United States, Italy—and was later extended to developing countries. For a general review of this approach to planning, see the various studies of regional planning in different parts of the world produced by the United Nations Research Institute for Social Development, and Johnson, 1970. The community development department established in India in the 1950s is a good example of the latter approach to development planning).

Papua New Guinea, however, has reorganized the whole system of government in order to establish in each area a political and administrative structure which has the power and resources to plan and implement its own rural development programmes within broad national guidelines. In terms of rural development planning this is perhaps the most significant aspect of Papua New Guinea's decentralization, and the one that is most likely to be of interest to other countries.

Unfortunately, it is too soon yet to evaluate whether the new structure will have a marked effect on the quality of rural development. Much will depend on the quality of political leadership and administrative support in each province. Those provinces with good leadership and capable administrative staff have already shown a noticeable improvement, especially in the planning and co-ordination of development programmes. The less well-endowed provinces, on the other hand, are still trying to come to grips with their new political and administrative arrangements and have as yet had little or no impact on social and economic development. However, whatever the difference in capacity between provinces, there is at least an administrative structure in each province which should encourage rather than hamper attempts to improve coordination and planning.

The impact of the decentralization on class differences is even more difficult to assess at this stage. Again it will depend very much on the attitudes of provincial politicians. On the one hand, the provincial governments should be more aware of the needs of the poorer sections of the community and more responsive to their needs than the national government. On the other hand, members of the provincial governments will, like all politicians, be tempted to use their power and influence to improve their own social and economic well-being, rather than the welfare of the less-privileged majority. This is already indicated by the interest which some provincial politicians are showing in high salaries, expensive cars, and overseas trips.

However, the initial experience with provincial government suggests that although provincial politicians are by no means 'perfect', their motivations and actions are at least no worse than those of national politicians. In fact, in 1978 provincial leaders voluntarily agreed to accept the restrictions placed upon them by the proposed strengthening of the leadership code, while national politicians appeared likely to oppose them. However, once again it is too early to judge at this stage. Moreover, it is important to remember that although inequalities between classes appear to be increasing, at least in the urban areas, they are still small compared with those in most developing countries, and the extreme poverty found in many other countries does not exist in Papua New Guinea and is probably unlikely to exist in the foreseeable future.

CONCLUSION: A BLOODLESS REVOLUTION

Papua New Guinea's decentralization programme does not guarantee radical changes in the spatial patterns of socioeconomic development in the country. Its impact on these patterns will depend to a great extent on the attitudes and priorities of individual provincial governments, and on the political will-power and technical and administrative capacity of the national government to pursue its goal of equalization. However it does have a very significant impact on development patterns in two other ways.

First, it guarantees much wider participation in decision-making and ensures that every part of the country has a great deal of autonomy in determining its own development priorities and programmes. Furthermore, if provincial governments succeed in decentralizing their decision-making powers below the provincial level, it could result in participation in government right down to the community or village level. In evaluating the importance of this achievement, it must be remembered that this is the main aim of the decentralization programme and that in Papua New Guinea, participation in politics and decision-making is valued particularly highly.

Secondly, the decentralization programme has created a political and administrative structure at provincial level which could, if used effectively, provide an excellent base for the planning and implementation of integrated rural development programmes, thereby affecting indirectly the patterns of socioeconomic development. This structure has important implications for rural development planning in other countries because it represents an alternative to the conventional centralized, departmentalized administrative structure which has created so many obstacles to rural development in the past.

Papua New Guinea's decentralization programme has been implemented very quietly and peacefully. It has received very little attention overseas— except for the attempted secession by the North Solomons Provincial Government in 1975—and even within Papua New Guinea it has not been greatly dramatized. However the Minister for Decentralization, who was also one of the main architects of the system, has described it as a 'bloodless revolution'— and in many respects he is right. It is certainly having a revolutionary effect on the structure of government in Papua New Guinea, and if it proves to be reasonably successful, it could have very significant implications for the planning and administration of development in the Third World.

REFERENCES

Ballard, J. A. (1973) The relevance of Australian aid: some lessons from Papua New Guinea. 45th Congress of the Australian and New Zealand Society for the Advancement of Science, Perth.

Berry, R. (1977) Some observations on the political economy of development in Papua New Guinea (Port Moresby). *Yagl-Ambu*, 4, 147–61.

Berry, R. and R. Jackson (1978) Inter-provincial inequalities: the State and decentralization in Papua New Guinea (Port Moresby: Waigani Seminar on Decentralisation).

Conyers, D. (1974) Organization for development: the Tanzanian experience, *Journal of Administration Overseas*, 13, 438–48.

Conyers, D. (1975a) *The introduction of provincial government in Papua New Guinea: Lessons from Bougainville.* Port Moresby, New Guinea Research Unit, Discussion Paper No. 1.

Conyers, D. (1975b) *Planning for District Development in Papua New Guinea*, Port Moresby. New Guinea Research Unit, Discussion Paper No. 3.

Conyers, D. (1976) The Provincial Government Debate, Port Moresby, *Institute of Applied Social and Economic Research, Monograph 2*.

Donaldson, M. and K. Good (1977) *Class and Political Organization in Papua New Guinea*, University of Papua New Guinea working paper.

Dumont, R. (1969) *False Start in Africa*, Chapter 6 (New York: Praeger; revised edition).

Fanon, F. (1967) *Black Skins, White Masks* (New York: Grove Press).

Fisk, E. K. (1962) Planning in a primitive economy: special problems of Papua New Guinea, *Economic Record*, 38, 462–78.

Friedmann, J. and M. Douglass (1978) Agropolitan development: towards a new strategy for regional planning in Asia, in Lo, F. C. and K. Salih, *Growth Pole Strategy and Regional Development Policy*, pp. 163–92 (Oxford: Pergamon Press).

Garnaut, R. (1978) Report to the Minister for Decentralisation on Provincial Government (Port Moresby), unpublished.

Hyden, G. *et al.* (1970) *Development Administration: the Kenyan Experience* (Oxford: Oxford University Press).

Johnson, E. A. J. (1970) *The Organization of Space in Developing Countries* (New Haven: Harvard University Press).

Mamak, A. and R. Bedford (1974) *Bougainvillean Nationalism: Aspects of Unity and Discord* (New Zealand: Christchurch) Bougainville Special Publications No. 1.

May, R. J. (1975) The micronationalists *New Guinea*, Vol. 10.

Nelson, H. (1972) *Black Unity or Black Chaos* (Hamondsworth: Penguin Books).

Papua New Guinea (1974) Final Report of the Constitutional Planning Committee (Port Moresby).

Robinson, R. (1971) Difficulties in public administration. In Robinson, R. (ed.), *Developing the Third World* (Cambridge: Cambridge University Press).

Schaffer, B. C. (1969) Deadlock in development administration. In Leys, C. (ed.), *Politics and Change in Developing Countries* (Cambridge: Cambridge University Press).

Somare, M. (1975) *Sana* (Port Moresby: Niugini Press).

Somare, M. (1978) Speech at the opening of the Morobe Provincial Government, unpublished.

Tordoff, W. and R. L. Watts (1974) *Report on Central–Provincial Government Relations* Government of Papua New Guinea, mimeogr, (Port Moresby).

Ward, R. G. and D. A. M. Lea (1970) *An Atlas of Papua New Guinea* (Port Moresby: University of Papua New Guinea and Collins-Longman).

Weeks, S. G. (1977) *The Social Background of Tertiary Students in Papua New Guinea* (University of Papua New Guinea), Educational Research Unit Report No. 22.

Wilson, R. K. (1975) Socio-economic indicators applied to sub-districts of Papua New Guinea (Port Moresby), *Yagl-Ambu*, 2, 71–87.

Chapter 9

Nepal: The Crisis of Regional Planning in a Double Dependent Periphery

PIERS BLAIKIE
(with J. Cameron and D. Seddon)

INTRODUCTION

The objective of this chapter is to describe the status of regional planning and its practice in Nepal. Regional planning is not very well developed in Nepal, and such regional policies as exist seem to have a negligible effect upon the problems of the vast majority of the Nepalese people. This situation prompts an explanation, which requires an analysis of the Nepalese state and its internal relations, especially its class structure and the state apparatus. External relations between class interests in Nepal and elsewhere, especially India, are significant. It is suggested that the people of Nepal face a deepening crisis; increasing food shortages, environmental stress, and an undermining of the integrity of the Nepalese state itself, are already present and these conditions are accelerating. The relevance of regional planning to the averting or even the palliation of this crisis must be seriously questioned.

A brief theoretical discussion of the nature of space in social organization, and its 'causal' role in social change, is required to begin with. Previous analysis of the Nepalese political economy facilitates an examination of the nature of Nepalese space, an aspect of social organization which is both the manifestation of, and to a lesser degree a causal factor in, the process of social change. The chapter ends with a review and prognosis, and concludes that regional planning in Nepal is by itself an inadequate instrument on theoretical grounds for the necessary structural changes to avert the impending crisis, because space in the final instance is both determined by, and given its rationale by, the political economy. Thus spatial policies are ineffective radical policies. Secondly, regional planning is inadequate on practical grounds, given the specific nature of the Nepalese state in particular as well as of the State of Nepal in general, which is seen as not actually embodying the collective will, nor being able to transform the economy in order to aid the mass of the people.

REGIONAL PLANNING IN NEPAL

A review of regional planning activity usually focuses on the ideology, policy formulation, implementation, and consequences of the regional allocation of scarce resources. However, in the case of Nepal there is little evidence of policies or their implementation, although a number of policy documents exist which state the need for regional planning and outline guidelines for the spatial allocation of resources in the most general terms. The prime objectives of regional development in Nepal are seen to be (a) to secure a greater degree of economic and political integration; (b) to ensure optimal use of all available resources; and (c) to improve income distribution—to which objectives few would take exception.

Hitherto, much of the implementation of regional development policies has been restricted to road building (National Planning Commission 1974). Most road construction projects are foreign, and such projects (financed largely by India and China, both for strategic reasons), can be easily carried out since their construction need not call into question the implementation capacities of the Nepalese government. Also, road construction in a country where travel is extremely time consuming assists in the central control of outlying and hitherto inaccessible areas (or, in other analytical terms, the possibility of coercive action on the part of the ruling class). Thus road construction represents a convergence of the aims of the aid-donor and controlling government interests, quite apart from its role in any development strategy.

Since Harka Gurung's influential paper for the National Planning Commission in 1969, regional development strategy has been closely linked to road construction. The concept of the 'growth axis' was developed from this time and has been an important element in regional planning thinking ever since. These axes are

'a series of north—south growth axes or development corridors, linking diverse regions. The juxtaposition of a wide range of resources (the *terai*, hills and Himalayas) within a common development corridor will permit economic viability and generate greater inter-regional circulation of goods, services and people.... Each growth axis will either have a road or a road is presently being constructed or planned in the area. These roads will link a series of growth centres where development efforts will be concentrated in order to achieve full economies of scale and encourage agglomeration economies'. (Gurung, 1969: 12–13).

Figure 9.1 shows the location of these growth axes in Nepal, which all cross the boundaries of the three geographical regions of the country—the mountains, the hills and the *terai* (the plains area). Indeed the vast majority of flows of goods, money, and people is in a north—south direction, and there is little in the way of hill-to-hill or *terai–terai* trade. In effect Nepal is made up of a number of separated regional economies, each one dependent upon its railhead connection with India. As subsequent explanation will demonstrate, a positive historic balance of trade which Nepal enjoyed with India until the beginning of this century, has changed into a large deficit. Furthermore, the hills have become

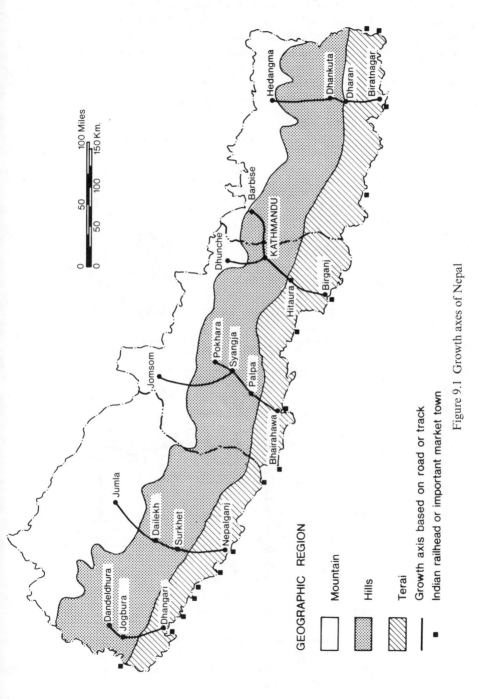

Figure 9.1 Growth axes of Nepal

grain deficient, and are obliged to export labour to the *terai* and to India. The mountains are sparsely populated, and trading has long been carried out to supplement income from pastoralism and agriculture. Thus in each of the regional economies, predominantly north–south trade flows have persisted. Road construction has altered the location of flows and break-of-bulk points, rather than their substance (see Figure 9.2). In fact, as a recent report has shown (Blaikie *et al.*, 1976), road provision in west-central Nepal has accelerated in

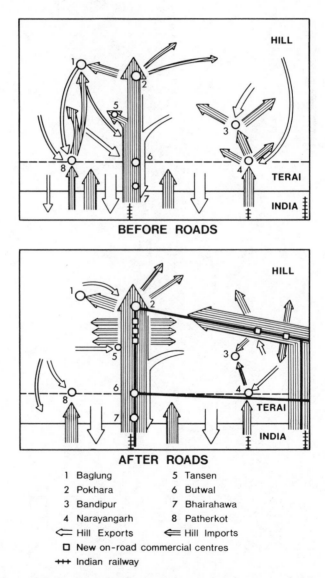

1	Baglung	5	Tansen
2	Pokhara	6	Butwal
3	Bandipur	7	Bhairahawa
4	Narayangarh	8	Patherkot

⇐ Hill Exports   ⇐ Hill Imports
☐ New on-road commercial centres
+++ Indian railway

Figure 9.2 An example of a dependent regional economy in Nepal

some instances the long-established decline of the hill economy (particularly in the case of artisans, craftsmen, and occupational castes). A number of separated dependent economies in effect exist, with their Indian connection of controlling importance (Ojha and Weiss, 1972). Some exceptional flows linking different regional economies also exist, but many of these represent government efforts to distribute food grains to particularly heavily deficient hill areas (e.g. the flow of millet from Chitwan to Pokhara, see the east–west flow along the road in Figure 9.2), or the construction of buildings associated with the enormous increase in the bureaucracy at the zonal and district level.

Until 1974, and then only in outline, regional planning was in no way connected with agricultural planning and rural development (although the main scope for regional planning must lie in the sector of economic activity in which 93 per cent of the people are employed). The 1974 National Planning Commission document for the first time linked the Small Area Development Programmes (SADP) with regional planning, through a consideration of sites for the SADP. This programme had six major objectives, namely (i) intensive development of arable horticultural and livestock farming, (ii) improved transport facilities, (iii) afforestation and soil conservation, (iv) minor irrigation projects and water management, (v) improved social services, and (vi) development of cottage industries and other non-agricultural activities (NPC, 1974, p. 3). The location of these schemes was to be in the existing 'growth areas', one in each ecological zone of each growth axis, so that no area of the country should be neglected.

The area, intensity, and location of such schemes (as well as the subsequent integrated rural development projects for which funds are being sought from donations of international aid) have mutually contradictory aims. The problems of poverty, deprivation, ecological decline, lack of physical infra-structure and personnel within a political economy tend not to be conducive to a purposeful solution, and these problems are all so pervasive (as the rest of this chapter will attempt to show), that large and effectively deployed resources are required in order for there to be any noticeable and lasting impact. Hence, 'SADP should be large enough to make a noticeable impact, but small enough to be manageable' (NPC, 1974, p. 2); and the Nepalese government is reportedly exerting pressure to expand the areal extent of those schemes financed through foreign assistance. The experience of donors in the field of rural development however, is that considerable time and a very small 'target population' are needed in order to sustain any improvements in productive capacity. The Kathmandu valley continues to absorb an enormous share of foreign aid in terms of location of projects, with indirect spin-offs from the location of increasing numbers of foreign agencies in the capital, such as investments in transportation, medical care, and hotels. Much remains to be done to provide an adequate data base for allocative decisions, although several publications have pointed out the need for a system of data collection, retrieval, and analysis. The result has been to locate rural development projects in each of the administrative regions representing a rough-and-ready equity consideration. Currently, Integrated Rural Development Projects, largely financed by foreign aid, are being planned, one

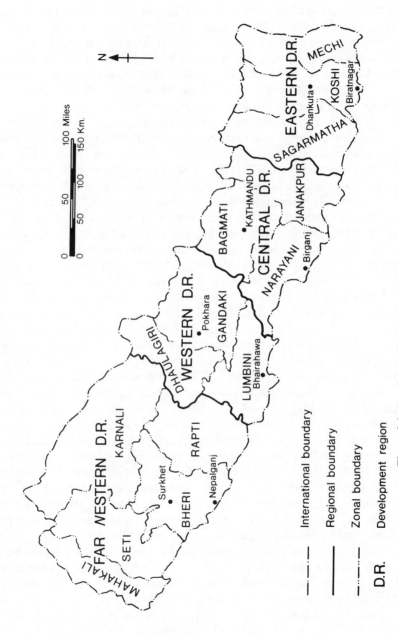

Figure 9.3 Rural development regions and zones of Nepal

for each of the four development regions. However, the location of one such project within the far western region (the Rapti Zone) has been explicitly discussed (e.g. APROSC, 1977, p. 15), and the criteria used are: extreme poverty of the whole area, further recent decline in economic opportunities due to prohibition of the hashish trade, and the existence of a road project in the area. Targets identified by this extremely innovative project are not so much people in specific regions, but people as defined by their relations of production and caste (emphasizing social relations rather than spatial considerations—a significant point to which this chapter returns later.)

The administrative machinery of regional planning is, on paper, quite comprehensive. The development region, the largest administrative unit in the country (Figure 9.3) only exists for planning purposes. Each region has a headquarters and a planning office. However, in spite of the rhetoric of decentralized decision-making and budgetary control, these offices as yet only have a coordinating role. Most important decisions are still taken by the Ministries of Agriculture and Fisheries, and Home and Panchayat Affairs in Kathmandu. Each development region is made up of zones (*Anchal*) which are day-to-day administrative zones, and below this level, districts (*Zilla*) each with its district headquarters and staff (see Figure 9.3).

The inability of the regional planning administration to intervene is all the more regrettable, given the outstanding regional inequalities in Nepal, principally between the hills and the *terai* but also between different development regions. Table 9.1 shows the much greater population pressure on land in the hills than in the *terai* in terms of the amount of *khet* (irrigated land capable of growing paddy, and also wheat in winter) per holding. The preponderance of extremely small holdings of irrigated land in the hills, and the existence of relatively large holdings in the *terai* (particularly the western *terai*) is striking. As early as 1962 the estimated size of holding in the hills (including both irrigated and non-irrigated land) was only a quarter of that in the *terai* (Central Bureau of Statistics Sample Census of Agriculture 1962). Productivity and income are also much higher in the *terai* (see Tables 9.2 and 9.3) although Table 9.3 indicates very considerable variation between districts, particularly in the hills. Part of this variation can be explained by the persistent anomaly of the Kathmandu Valley

Table 9.1 *Percentage distribution of khet (irrigated) land classified by size of holding (hectares) and region, 1963–69*

	Less than 0.5 hectares	0.5 –1.0	1.0 –2.0	2.0 –4.0	Larger than 4.0	No. of observations
Hills	83	13	3	0.2	0.2	422
Eastern Terai	13	17	29	23	18	400
Western Terai	1	6	9	18	66	435
Central and inner Terai	4	7	18	39	32	419

Source: Central Bureau of Statistics, Kathmandu (1970).

Table 9.2   *Indicators of productivity and income, 1970–71*

Region	Agriculture GDP 1970–71	Percentage GDP contributed by agriculture	Per capita GDP (Rs)	Agricultural GDP per economically active population agriculture (Rs)	Per capita GDP ($)	Agricultural GDP per economically active population agriculture
Mountain	460,255	85.20	426	735	40.57	70.00
Hill	2,169,111	67.77	539	867	51.33	82.00
Terai	3,410,634	63.91	1,288	2,380	116.95	226.67
Totals	6,040,000	66.54	785	1,319	74.76	125.62

Source: Central Bureau of Statistics.

Table 9.3   *Distribution of districts by GDP per capita as a percentage of GDP per capita for all Nepal*

Region	Below 40%	40–60%	60–100%	Above 100%	Total
Mountain	2	9	6	—	17
Hill	4	24	6	4	38
Terai	—	—	1	19	20
Totals	6	33	13	23	75

Source: A New Dimension in Nepal's Development, ed. by Prachandra Pradhan (1973, p. 4).

Table 9.4   *Selected development indicators*

Region	Roads (all weather) (mileage)	Airports (all weather) (No.)	Power (kW)	Irrigated land (hectares)	No. of colleges	Hospitals No.	Hospitals No. of Beds
1. Eastern Mountains	Nil	Nil	Nil	Nil	Nil	1	11
2. Eastern Hills	Nil	Nil	240	8,579	6	3	40
3. Eastern Plain	124	1	1,895	54,812	7	7	185
4. Central Mountains	38	Nil	10,050	1,160	Nil	2	40
5. Central Hills	249	1	32,140	17,282	20 + 1	11	1,085
(Kathmandu Valley)	(114)	(1)	(4,270)	(11,646)	(20 + 1) University	(7)	(1,000)
6. Central Plain	190	1	1,100	43,722	4	7	195
7. Western Mountains	Nil	Nil	Nil	Nil	Nil	Nil	Nil
8. Western Hills	156	1	1,180	6,216	6	9	205
9. Western Plain	55	1	830	19,252	1	6	125
10. Farm Western Mountains	Nil	Nil	Nil	575	Nil	1	15
11. Far Western Hills	Nil	Nil	Nil	10,793	1	2	30
12. Far Western Plain	26	Nil	50	18,935	3	4	95
Totals	838	5	47,485	181,266	49	53	2,024

Source: A New Dimension in Nepal's Development, ed. by Prachandra Pradhan (1973, p. 4).

(Row 5 'Central Hills' in Table 9.4) which absorbs the dominant share of invest-ments of local and foreign origin in both industrial and agricultural production as well as in social services. The origin and continuation of this economic and political dominance of the Kathmandu Valley is explained in the following section.

In summary, government intervention is not particularly effective, and the implementation of regional planning as a means of channelling this is still in its infancy. The most important development efforts in agricultural and rural development have not been closely linked to either regional planning policy or to implementation by regional planning authorities, in spite of marked regional inequalities and the pervasive failure to produce enough. An explanation, in terms of the Nepalese state itself, follows—which in turn will suggest the type of regional planning appropriate for the future.

THE HISTORICAL DEVELOPMENT OF THE NEPALESE STATE, AND ELEMENTS OF CRISIS

Linked to the inability of the state apparatus to intervene effectively is the impending survival crisis in Nepal, already manifested in chronic and in-creasingly severe food shortages; and it is in the context of crisis that the need for regional planning must be examined. An analysis of the Nepalese state also allows an examination of the nature of spatial organisation in Nepal and the construction of a theoretical base on which any future regional planning must rest.

The Development and Preservation of the Tributary State

The current situation of the Nepalese state can only be adequately explained in an historical context, but only a very brief historical sketch can be provided here (this has been dealt with in more detail elsewhere, e.g. Regmi, 1971; Blaikie, Cameron and Seddon, 1977; 1980, forthcoming). The major thesis of this section is that Nepal today still has many characteristics of a tributary state (the term is explained below), albeit in a depleted and dependent form, and that as such the state apparatus is unable to intervene effectively in the production process within its own borders to reverse the increasing population pressure, dwindling resource base, and declining output-per-unit of land and labour, nor to break out of the dependency relationship it is in with regard to India.

The area that is now Nepal was divided between numerous petty states during the seventeenth and eighteenth centuries, and ruled mainly by immigrants of the Thakuri caste from India claiming descent from the ancient Rajput families who ruled Rajasthan before the Muslim invasions in the tenth and eleventh centuries. The nobility were predominantly high caste Hindus of Indo-Aryan extraction (the castes of Brahmin and Chhetri today), while the subject classes, who were taxed and forced to serve as soldiers in order to maintain the state and the ruling classes, included the majority of the indigenous Mongoloid population (Magar, Gurung, Rai, Limbu, etc.), whose tribal structures had been only partly destroyed by establishment of these Hindu kingdoms. The immigrants brought with them a new technology, both of production and warfare. The prevailing systems of agriculture prior to their arrival had consisted of pastoralism and shifting cultivation using hand implements only, while the new arrivals from India brought the plough and the techniques of wet rice production—which brought about both alterations in production relations, and thus class structure, as well as a regional differentiation between hillside (dry land agriculture and pastoralism) and valley bottom (rice culture). The hill people had also previously relied in war upon easily fabricated bows and arrows, while the newcomers brought swords and other metal weapons which required special skills to manufacture and repair. Hence this period of Nepalese history saw gradual class formation, and with it some small-scale regional differentiations.

These changes in the political economy, together with others such as the introduction of maize and the potato, generated a decisive change in the political structure of the area, setting the stage for the speedy conquest and unification of the country by Prithvi Narayan Shah, the founder of the present royal dynasty in the eighteenth century. Originally he ruled the petty kingdom of Gorkha in what is now west-central Nepal, and managed to extend control from the Tibetan border to the *terai* (which although malaria infested, was still a source of taxable agricultural production). Also in 1744 he was able to annex the small kingdom of Nuwakot through which an important part of Tibeto-Indian trade passed, and on which the three Kathmandu valley kingdoms largely relied. These latter petty kingdoms had their capitals in the fertile Kathmandu valley, where, as nowhere else, flourished a genuine urban economy and society which produced commodities such as cloth and metalware, and was the base for major trading operations between Tibet and India. As soon as this latter source of revenue was tapped, the Gorkha king was able to purchase firearms and capture the valley, and within twenty years, the rest of what is now Nepal. By 1789 he had pushed both south into the fertile *terai*, and as far west as Kashmir. Prithvi Narayan Shah transferred his capital to Kathmandu, setting in motion a type of pre-capitalist centre–periphery relationship between the valley (the focus of the ruling class where appropriated surpluses were lavishly expended on palaces, gardens, and conspicuous consumption, the whole fuelled by what could be termed 'the royal multiplier'), and the rest of Nepal, where subject classes were obliged, often under conditions of slavery, to apply their labour to the land.

The state was maintained largely by the appropriation of surpluses in the form of taxes from direct producers, the peasantry, and often from local lords who were granted the right to collect taxes and to appropriate a considerable portion for themselves before passing them on as tribute to the centre. Those granted rights of taxation were generally obliged to keep local law and order, maintain trails and ferries (without which military and political control from the centre would have been impossible), and to raise levies of troops when required. The outlines of the tributary state had thus been drawn, and have been preserved until the present day.

British Imperialism and the Nepalese Ruling Class

It was inevitable that Nepalese expansion would eventually confront the equally expansionary British East India Company, and the ensuing war of 1814–16 over control of parts of the *terai* and the Tibeto-Indian trade, led to a costly British military victory and the important Treaty of Sugouli in 1816, which was not abrogated until 1923. Under this treaty Nepal ceded some important territory, but thereafter its territorial integrity was assured by Britain; the East India Company had the right to recruit soldiers to police its growing empire to the south, and certain markets were opened up to British enterprise in Nepal. For nearly a century then Nepal was a kind of political dependency upon imperial Britain, an arrangement which had benefits both for imperial interests (a self-manning buffer state against possible hostile powers to the north, a regular supply of soldiers, raw materials and primary products from Nepal) as well as for the ruling class of Nepal who were guaranteed virtual autonomy from outside pressure for change.

It is helpful to consider Nepal in this period as a 'semi-colony'—a term Lenin used to describe non-capitalist societies which preserved juridical independence, but were dependent economically upon imperial powers (China, Japan, Turkey, Ethiopia, Thailand, and Persia are other examples). The concept of semi-colony is useful for understanding contemporary Nepal, and some of the dilemmas facing those who make policy decisions there, for two reasons. First, it emphasizes that under-development is not a consequence of isolation from the world, but comes from incorporation within it. For the semi-colonial state this incorporation involves many of the disadvantages of colonialism with none of its advantages (such as those deriving from investment in productive capitalism). Secondly it illustrates the difficulty or impossibility of the ruling class under those conditions breaking away from their self-rewarding dependence on the imperialist power; and also a chauvinistic ideology which incorporates a romantic sense of history where the aristocracy plays the leading heroic roles, quite detached from considerations of economic development. Thus the preservation of an antique, semi-feudal state was assured, in which a landowning aristocracy centred in Kathmandu was able to keep out progressive capitalism (it knew only to well the historically progressive changes which industrial capitalism would entail), and could maintain its role in the extraction of surpluses without any changes in productive capacity, under the watchful surveillance of a British Resident in Kathmandu.

Population Growth

Emergence of the present crisis was intimately related to the subsequent development of the 'Indian connection', and to critical changes in the productive capacity of labour within Nepal. Given the static nature of the agricultural technology and the virtual exclusion of industrial capital, and the state's sole interest (the maintaining of law and order, and the extraction of revenue), it is not surprising that population growth began to reduce outputs-per-unit of scarce arable land and of labour, and hence the taxability of the peasantry itself. Private property rights had progressed slowly, and were only officially recognized in 1923 (thus producing a land-owing aristocracy in the strictest sense of the term); and slavery was abolished in the same year, reflecting the decline in the need to keep scarce labour from escaping back into the forest (due to population growth), rather than any new liberal sentiment. The slow increase in population in the twentieth century, from $5\frac{1}{2}$ million in 1911 to $12\frac{1}{2}$ million in 1974, was offset by the private expansion of cultivable land up the hillsides onto the steeper slopes, where marginal returns started to decrease rapidly. This crisis of a vanishing surplus was temporarily averted for the ruling aristocracy by the timely arrival of a demand for timber for railway sleepers with expansion of the railway network in north India after the turn of the century. The forests from the *terai* ('sal' or *shorea robusta*) met this demand, as witness the fact that the majority of palaces in the Kathmandu valley date from this time.

Some Reponse to the Crisis

For the peasantry such an alleviation was impossible, however. The manifestations of population pressure are difficult to date (that is, a population growth wherein development of the forces of production is only feeble), but such manifestations probably began to appear in some localities between the two world wars. As this brief historical perspective has already shown, new agricultural technology in terms of fertilizers, new seed, new crops, and mechanization was practically unknown here until the 'fifties, and even today only about 8 per cent of cultivators in West Central Nepal use, even in tiny quantities, any one of the innovations mentioned above (Blaikie *et al.*, 1976). Therefore any increase in total food production which an increasing population demands, must be met by more intensive use of existing arable land, or an extension of arable land into the forest. Without additional sources of nutrients, such as chemical fertilizer, the arable land must start to lose fertility, and if transference of fertility is insufficient from forest to arable land (with fodder and bedding for stall-fed stock, whose manure is applied to the fields), the only way to maintain—let alone increase—food supply would be a further extension of cultivation into the forest. This situation rapidly becomes a vicious circle, since the balance between a self-regenerating forest (to provide nutrients) and arable land becomes even more seriously disturbed. Soil erosion, the drying-up of perennial water sources, and wholesale geomorphologic alterations further exacerbate the situation. Various authors have recently commented upon this (Enke, 1971; Eckholm, 1976; Rieger, 1977); for example:

'Visual evidence of already existing population includes deforestation, erosion and silting. Within the last decade, wooded hill-tops have been cut down or severely depleted, terraces have been extended to the tops of hills, and cattle have had to graze further away. In some areas, hill-top terraces have leached out, have been abandoned, and have started to collapse on terraces below. Villagers often have to go much further to cut fodder for animals. The complex interaction of wood for fuel, cattle for manure and draught, and manured terraces for rice etc., is becoming increasingly vulnerable to overcrowding of the hill areas. If conditions worsen, areas now cultivated... will have to be abandoned'. (Enke, 1971, p. 20).

Agricultural holdings have been diminished, and are very small, even by Asian standards, as Table 9.1 shows. A recent World Bank report observes that 'population density per sq km of arable land is probably as high as 1100, a concentration similar to that found in certain Asiatic deltas, but where in contrast, the soil is more fertile and the climate allows two or three crops a year' (IBRD, 1973, p. 4).

The over-all trend of the economy has been extrapolated by the FAO, and Table 9.5 succinctly outlines the increasing deficits of foodstuffs.

There have been four major kinds of response by the peasantry to this deteriorating situation. These responses tend to be individual rather than collective, and characteristic of a peasantry rather than of a rural proletariat. They are circumscribed by the development of historical processes already outlined, and the sum of their individual actions is collectively self-defeating, in the medium- and long-term. Only collective action to change the structural determinants of this crisis will have some chance of success. The responses of the peasantry are: (1) increased use of the public economy (the forest) to supplement insufficient privately-owned land; (2) intensification of current arable land use, replacing non-cereal (and also non-food) crops by basic cereals (causing an increased dependency upon imported cloth, tobacco, cooking oil, etc.); (3) increased capacity in petty commodity production; and (4) emigration to the *terai* and India.

Table 9.5 *Extrapolation of past trends up to 1990 in respect of current indicators*

	1970	1980	1990
Population growth rate (per cent per annum)		2.3	2.3
GDP growth rate (per cent per annum)		2.2	2.2
GDP per caput		0.1	0.1
Total cereal production (1000 metr. t)	3793	4190	4628
Total cereal domestic demand (1000 metr. t)	3427	4283	5415
Cereal balance surplus (+) (1000 mt) deficit (−) (1000 mt)	+ 366	− 93	− 787
Per caput supply (kg per caput)	355	296	258

Source: FAO, *Perspective Study of Agricultural Development for Nepal*, 1974 p. 8.

None of these responses seek to change the structural determinants of under-development. Thus any policy, regional or otherwise, which seeks to improve or reform piecemeal the conditions under which these responses are made, will at best alleviate only temporarily an already critical situation. Temporary allevia-tion is not to be despised, but merely recognized for what it is.

Dependency Perpetuated—the 'Indian Connection'

The 'Indian connection' has been significant in the development of under-development in Nepal. It remains to examine briefly the present form which this takes and its formation after India regained her independence in 1947. The new class configuration of an independent India had profound implications for Nepal, still broadly recognizable today. The Indian state, which represented a variety of class interests in regard to Nepal, wished primarily to guard her northern border against China and a liberated Tibet. This in turn led to various attempts to coerce the ruling élite in Nepal into accepting a direct military presence, and also into bringing about its downfall when it was per-ceived to be the source of political unrest. The last thing India wanted was a popular revolt against a landed aristocracy outside Indian political influence. Thus close links with the emergent Nepali Congress party were forged, and in the ensuing civil war in Nepal during 1951 India gave strong logistic support to the rebels. India also continues the recruitment of Gurkha soldiers—not only because of their famed fighting qualities, but also as a strategic and military measure. Remittances and pensions from army services can further 'prop up' the ailing hill economy, and at least postpone a break-down in a strategically very sensitive area.

Other interests in India are more regional in character, and these involve: (a) maintenance of access to land in the Nepalese *terai*, now cleared of malaria since the early 'sixties and available for large-scale forest clearing and settle-ment; (b) access for Indian labourers to seasonal employment by Nepalese land-owners in the *terai*, which provides a valuable additional income for many landless or near-landless labourers in Uttar Pradesh, Bihar, and West Bengal; (c) access for the Indian industrial and commercial bourgeoisie to Nepalese markets, which although absorbing only 1 per cent of total Indian exports by value (and 5 per cent in such sectors as cotton fabrics, transport equipment, and pharmaceutical products), are still appreciable for north Indian capitalist concerns; (d) access for Indian land-owners to water for irrigation, of consider-able importance for the agricultural development of northern India, but which is more easily controlled in Nepalese territory than in Indian.

It is not within the scope of this chapter to provide an analysis of how these interests are articulated (by such agencies as local politicians in state legislatures of north Indian states, national negotiators in Indo-Nepalese trade and transit agreements, etc.), but all of these have deep implications for the future of Nepal, and also for the scope that regional planning may have for effective action.

The flat land of the Nepalese *terai* represents the last frontier of Nepal, and a safety valve for refuges from the deteriorating economy of the hills. It is usually highly fertile, and had hitherto been cultivated in cleared patches by indigenous people called Tharu (who have a partial resistance to malaria), both as peasants and as labourers where large land grants had been made to absentee landlords (the present representatives and inheritors of the semi-feudal productive relations of the formerly tributary state). With the population density of Indian districts of the *terai* being three or four times higher than the Nepalese (Indian and Nepalese Censuses, 1971), with an 'open' frontier, and a population on either side of this frontier similar in language, caste, racial, and other characteristics, it is not surprising that representatives of all agrarian classes from India settle the land, the 'Indian' landowners by means of outright purchase (and the purchase of obligatory Nepalese citizenship, a requirement for legal ownership of land in Nepal), and the small Indian peasant or labourer by squatting illegally in the forest, like his Nepalese counterpart from the hills. This process leads to very rapid settlement of the *terai*, and considerable conflict between the interests of the Nepalese state (wishing to preserve the forest and realize part of the timber revenues accruing to forest contractors and land-owners) and those of the poorer settlers of Indian and Nepalese nationality, who are burnt out of their temporary homes and harassed by police and forest guards.

It is in the interest of large employers of agricultural labour in the *terai* to keep the frontier sufficiently open to allow Indian labour to migrate in order to fill the seasonal labour demand for paddy transplanting and harvesting (and increasingly, wheat harvesting—where winter wheat has been adopted). The wage rate of 4.8 kg of paddy per day has remained unchanged for at least four generations, but the profitability of cereal production has markedly increased with the advent of new inputs such as high-yielding new varieties, pesticides, and chemical fertilizers.

This issue of markets too results in the maintenance of a 'leaky' frontier, where class interests on either side serve both to preserve the frontier and to penetrate it. A long history of penetration of the Nepalese markets by mass manufactures from India is very clear. The substitution of mass-manufactured products for locally-produced goods (made either by skilled artisans in small Nepalese towns, by members of occupational castes such as leather-workers, blacksmiths and tailors, or by members of peasant households themselves) is widespread and includes such items as kitchenware, shoes, cotton thread and cloth, dye, fuel for lighting (kerosene, replacing clarified butter), soap, agricultural tools, paper, and general purpose containers for storage and carriage. The process came about through a number of influences, namely, 'internal' pressures of population, already mentioned; a stagnant technology and increasing necessity for foodgrain production; by an increased monetization of the economy following out-migration and a regular flow of returning migrants carrying cash or gold to the hills; by conscious and aggressive policies of Indian capitalists to increase sales; by the complicity of a Nepalese merchant class who stand to gain enormously by handling importation of these goods; and by a bureaucracy which reaps rewards by its 'policing' of the frontier itself. (On both

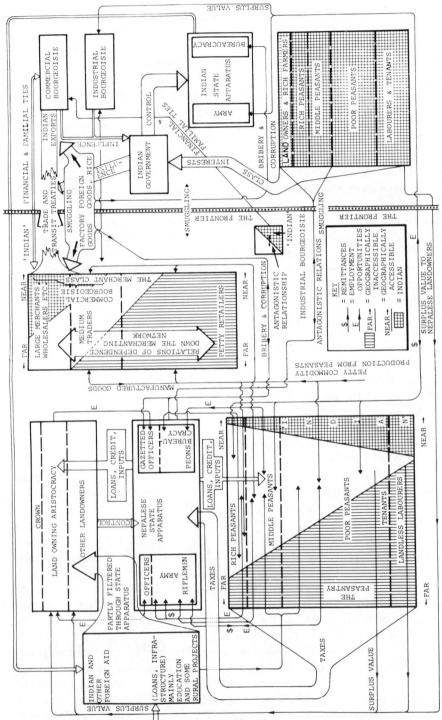

Figure 9.4 Nepal's leaky frontier and the class interests which maintain it (Blaikie and Seddon, 1978)

sides of the border some towns, such as Bhairahawa, Birganj, Forbesganj, and Biratnagar, over their importance to smuggling alone). The implications of the 'leaky frontier' for regional planning are taken up later, but it is sufficient here to emphasize the 'international' component (which must be firmly linked to internal processes in any analysis) in the problems currently facing the Nepalese people. Figure 9.4 (from Blaikie and Seddon, 1978) summarizes in cartographic form the role of the leaky frontier in Indo-Nepali relations.

Is there Relevance in Regional Planning under Crisis Conditions?

The previous two sections seem to give a gloomy prognosis for Nepal. However, the preceding analysis must not lead to passive pessimism, as so much writing on 'dependency' seems to do. The situation is undoubtedly grave, but possible solutions must consider an alliance of forces which have some chance of putting into practice suggestions based on analysis. There are contradictions within the Nepalese state, and policies should seek to make use of these contradictions. While the prognosis may be serious, at least the method employed here is optimistic: to recognize the brighter side of the assertion, that the state cannot be considered a monolithic structure faithfully reflecting only the interests of a ruling class, and that there are cleavages here which offer some hope for action. And it is not enough to look for the 'historically-grown social structures' (Stöhr, Chapter 2) in a static sense. Instead an advantageous alliance must be sought in the continuous transformation of these structures, to bring about those social and economic changes which are so necessary. Hence regional planning as a policy instrument cannot be ruled to be irrelevant alongside the other policies on the grounds of hopelessness.

In such a serious situation, regional planning may seem not only a luxury but worse, a pretence and prevarication, as a means of expanding the bureaucracy to provide employment for the otherwise unemployed (and politically problematic) graduate student population; as a rhetorical device to attract foreign aid (by demonstrating that implementing machinery actually exists and that government is actively thinking about regional inequalities and other allied conventional wisdoms presently espoused by international aid agencies); and perhaps as a means to channel attention away from exploitation and oppression directly deriving from the class struggle, to something less embarrassing and immediate, namely their spatial structures. While analysis here suggests there is some validity in according mixed objectives to the regional planning activity, its main justification must lie in the necessity for an informed spatial allocation of resources according to both means and objectives, which could avert the Nepalese crisis. Clearly the present form of regional planning does little to solve present problems, but there are several good reasons why regional planning must be a part of any effective programme of action. It is the contention of this chapter however, that regional planning can never be radical in the important sense of being a 'leading sector' in social change. The logical form of spatial pattern, actual spatial configurations, spatial policy, all follow (though they do not faithfully reflect) the political economy. Regional planning in Nepal therefore must follow in the wake of, and be an instrument amongst others in, the necessary transformation of Nepalese society. It must cease to be 'reified', and to exist as a separate bureaucratic activity outside economic, social, and political realities. This view is founded both on analysis of the Nepalese state outlined in the previous section, and on a particular view of the relationship between space and the political economy which follows.

SPACE AND THE POLITICAL ECONOMY

The Nature of Space

Too often in the literature of regional science and human geography, the laws of motion of the social organization which both defines space ideologically and creates spatial forms and relations, are thought to fall outside the area for study. Thus many studies tend to concentrate upon a particular manifestation (space) of a particular social organization (capitalism) (Blaikie, 1978, p. 268). Or as Santos put it:

'One should be concerned with world space as a whole and not just aristocratic space where the only flows are those of giant firms and persons of leisure' (Santos, 1974, p. 2).

Much of the underlying theory and assumptions of regional planning in non-socialist countries rests upon the logic and laws of motion of one particular type of economic system or, more accurately, the capitalist mode of production. What is usually taken as a universal is in fact specific to that particular mode of production. In all societies, space is what the political economy makes it: if the political economy comprises—and it almost always does even in developed countries—more than one mode of production, then it will be possible to identify more than one kind of space. In order to appreciate fully the nature of space in any particular society or in any relationship between two or more societies (e.g. to grasp the nature of centre–periphery relations), it is important to recognize that different spatial logics may coexist and overlap; different spatial logics will imply different spatial structures, corresponding to the co-existence and articulation of different and distinctive modes of production.

Spatial structures derive first from the mode of production's base, in a material, infra-structural sense. Industrial conglomerations, roads, railways, retail activities, and rapidly growing cities are some obvious examples of material structures whose locational rationale may be found within the broad logic of capitalist production methods, and the particular shape of material production forces and class relations that develops under industrial capitalism. It is crucial to recognize however that the location of such material structures may not result from the logic of the capitalist mode of production alone, but rather from the complex interplay of more than one mode of production. The alignment of a highway for example, in a country like Nepal, may reflect a complex logic that derives not merely from economic considerations as defined by conventional economics or economic geography, but also from other considerations related to the coexistence of non-capitalist production relations within the Nepalese political economy.

This may not always be apparent, for space tends to be ordered in the interests of a dominant class, and to reflect the pre-eminence of one particular mode of production in a given political economy. This may be true both of predominantly capitalist societies (as David Harvey has ably demonstrated in his *Social Justice and the City*, 1973), and of predominantly non-capitalist societies such as Nepal. Space thus becomes both a manifestation of a mode of production, and a crystallization of it in concrete form. The crystallization of the laws of motion in regard to a certain manner or way of production in its spatial structures however, is not a simple one-way process, because the spatial logic so created and the material structure thus generated, themselves affect development of the manner of production (e.g. distance expressed as a cost which affects the location of economic activities).

Nepalese Space

Indo-Nepalese relations can be roughly characterized in terms of a pre-capitalist but capitalist-dominated social structure or society—Nepal—connected in complex fashion with another spatially and politically distinct social structure in which capitalism predominates—India. At this basic level of analysis there are two identifiable sets of spatial logic and spatial structure: the political economy of Nepal (characterized by predominance of non-capitalist production relations), and the political economy of India (characterized by predominance of a capitalist mode of production, itself connected or articulated in complex fashion with the more developed economies of Western Europe and the USA, amongst others). India is not suggested to be capitalist in any simple or total fashion, it is rather that those interests which are most clearly and powerfully expressed in identifying the connection between the two social formations and political economies, can be characterized as capitalist. The explanation has certain similarities with what Lo and Salih (Chapter 5) describe as a 'low per-capita resource semi-open economy'. In Nepal however, the extent of under-development and the impoverishment of the people have reduced her 'import-substituting industries with modern technology' (the typical category into which most South Asian countries fall) almost to vanishing point. As well the importance of petty commodity production alongside the production of large landholders in contributing to a 'limited commercial crop export potential' is a significant variation in the case of Nepal.

The Spatial Logic of Non-Capitalism in Nepal

As described, Nepal has existed as a tributary state for at least two hundred years. Here as in other tributary states (e.g. Ethiopia, pre-colonial Morocco), a physically centralized aristocracy appropriates surpluses from the peasantry while directly controlling their means of production, including land, by means of intermediaries, usually members of the nobility, who extract these surpluses through coercion, and pass a portion on as tribute to the 'centre'. The vast poportion of surplus in non-capitalist societies comes from agricultural labour, and its appropriation means there must be control over the area or physical space. The surplus is often in the form of food grains, and is collected centrally by a spatially concentrated section of the ruling class, for reasons outlined below. Hence fundamental contradictions arise from the need to extend territorial control and extend the economy on the one hand, and increasing problems of political and military control at the margin which result from such an extension, on the other.

The peasantry in Nepal therefore found themselves part of a political economic 'map' drawn by the ruling class. In such circumstances the only physical access problems the peasant faced were (1) 'good' access, from the ruling class's point of view, of revenue officials and recruiters of corvée labour and slaves; and (2) proximity to the small range of resources needed to sustain life (water, pasture, wood, and arable land), all of which until relatively recently, were available everywhere. However the imposition of taxes, and the peasantry's inability to support themselves in the face of growing population and a stagnant technology, together pushed them into the production of small amounts of goods to sell in the market, to meet cash needs. As soon as exchange values started to subvert the natural economy, new problems of physical access arose for the peasantry (chiefly access to markets), as did the associated problem of choice of product appropriate both to the peasant's physical access position, and to his control over other resources, such as labour and land. This redefinition of spatial structure by the extension of exchange values into the peasantry is discussed at length in an earlier portion of this chapter.

The tendency for the ruling class to be spatially concentrated requires explanation. First, political control over dissenting factions encourages spatial concentration of the powerful sections of the ruling class. Alliance-making and breaking, court conspiracies, marriage brokerage, and occasional recourse to assassination require constant presence on the part of the protagonists of leading factions. Any history of palace politics, whether in the Mughal Empire, Ethiopia, Morocco, Afghanistan, Thailand or Nepal is one of intrigue and murder (activities significantly less commonplace amongst a spatially dispersed population), and the removal of a particular faction from the scene often precedes downfall. Secondly, some locations offer a higher potential for extraction of surpluses than others (by virtue of richer soils, or superior climatic conditions for example), which can be compounded when a site tends to attract the rudiments of state apparatus: army barracks, the mint, perhaps becoming a centre of religious activity, as Marx observes in his study of *Pre-Capitalist Economic Formations*. Once there is a reason for concentration therefore, the acquisition of surpluses in this kind of society tends to take the form of palaces and lavish luxuries; and since these things tend to tie their owners to a particular location, and also to promote concentration of builders, stonemasons, jewellers and other artisans and craftsmen, the 'royal multiplier' in the case of many tributary states encourages a concentration of the aristocracy, simply so that they too can avail themselves of the advantages of agglomeration in the practice of lavish living. There is too the added advantage of an audience for display of status-conferring luxuries—usually imported, such as motor vehicles, foreign clothes, hunting gear, as well as such characteristic luxuries as gold, silk, and slaves.

Articulation of Modes of Production in Nepal, and Associated Spatial Structures

There is little need here to outline the spatial logic of capitalism, since most of the regional science and geographical literature, based as it is on neo-classical economics, is familiar (too familiar, our argument maintains, see also Friedmann and Douglass, 1978; Friedmann and Weaver, 1979, for similar views); and since it is extended as the only logic, and remarks pre-capitalist structures in its own image (just as in wider context the capitalist mode modifies, dissolves and destroys the pre-capitalist modes). Thus this section can take the logic of capitalist space for granted, and move on to an analysis of the articulation of pre-capitalist and capitalist modes in Nepal.

Nepal's existence as a politically independent state with its necessary spatial expression on the frontier, has hindered articulation of non-capitalist Nepalese production with the dominant production mode of India. If Nepal were not a separate state, its population would have been subjected to the same forces as parts of hill India (e.g. Assam to the east, or Almora, Nainital, Garhwal or the Beas Basin to the west). In one important sense the filtering of the effects of capitalism on the part of the Nepalese ruling class has reduced any benefits of capitalism which might have accrued to a peasantry long imprisoned by its economic and political structure as a tributary state. In the Indian hills, where for the peasants simple transport costs rather than the problems of a frontier and an independent state hinder communication, capital has been invested in agriculture in the form of for example, tea estates, and more recently apple orchards, often owned by big corporations. The products of these enterprises are then transported by refrigerated lorries to the major metropolises of India. In comparison to Nepal, a non-capitalist and under-developed state (under-developed in part because of metropolitan capitalism itself), India has provided its own peripheral areas with hospitals, schools, and the rudiments of a delivery system for high-yield varieties of seeds, fertilizers, and agricultural extension. These areas however are also under-developed, in the sense that they are reduced to providing raw materials, labour, and a small (though not inconsiderable) market for manufactured goods exported from core regions. The forests of the foothills had been carefully preserved by the British administration through a series of Forest Acts dating from the turn of the century, which had the effect of forcing up to a third of the peasantry to migrate. These forests are now being felled by large timber contractors, and thus one of the last resources of the hills is being used up. These areas, like Nepal, (where a very poor peasantry can find little non-agricultural employment because of under-development of industry and the spatial flow of capital towards investment in industrial regions of northern India) are heavily dependent on the export of labour.

Nepal has all the disadvantages of spatial peripherality but none of the advantages—at least none that the peasantry can enjoy. The ruling Ranas benefited from without from the under-development of Nepal—selling the forests of the *terai* for example, to British firms extending the rail network through the northern United Provinces and Bengal. Merchants benefited from monopoly control of the import trade in Indian-manufactured goods. The peasantry bore the brunt of the disadvantages—a stagnant productive base, next to no services being provided by the state, little employment either inside or outside an unproductive subsistence agricultural base, and the consequent need for out-migration.

While the frontier, as a spatial phenomenon, appears dependent for its creation and maintenance upon the existence of two nations, the frontier's specific form as well as its location are controlled by class interests on both sides. Frontiers can have a 'lag' effect on the territorial balance of power of rival ruling classes. In the early nineteenth century the boundaries of Nepal expanded and contracted across the *terai* during the various campaigns with semi-autonomous Indian states and the East India Company, all adversaries aiming to maximize the populations from which revenues could be collected. Once the costs involved in fighting and defence against possible attack had become prohibitive compared to probable income for both sides, a treaty was signed and the frontier stabilized. Under present conditions of change, this frontier plays a very important role of itself in setting a framework for the shifting patterns of class interests.

The frontier is now a necessary device to protect the interests of various powerful classes in Nepal. It prevents Indian merchants from taking over Indian—Nepalese trade as a whole from Nepalese merchants. Although there are contradictions between Nepalese and Indian merchants within Nepal itself, the frontier protects the ultimately fragile strength of the latter. It also filters Indian immigrants in search of land, allowing those who can afford to 'purchase' Nepalese citizenship to pass—certain departments within the Nepalese bureaucracy too, have reason to support operation of a 'leaky' frontier! In this sense there is an antagonistic relation between the Indian landless and the Nepalese state apparatus which manipulates the frontier to its own advantage. It is also in the interests of Nepal's landed aristocracy, who still benefit from the rewards of high office, to preserve the forest (by means of state intervention) for controlled felling rather than the illegal burning by settlers (*sukhumbasi*) which seriously threatens an important source of revenue. Lastly, the leaky frontier also serves the interests of petty traders who smuggle goods both ways (foreign goods and rice into India, or machinery into Nepal) by using the price differential resulting from customs restrictions and tariffs at the frontier and the various other price regulations, and by taking advantage of its physical openness (there is no barbed wire, merely a line of pillars at wide intervals), to smuggle goods through. The bureaucracy, especially customs and excise, and the police too appear to derive lucrative incomes from 'allowing' this illegal traffic to pass. Evidence from a report (Blaikie *et al.*, 1976) shows a healthy balance-of-trade surplus with India in terms of local currency transactions at banks near the border, while the current official figures for Nepal show a large deficit. With other data from production and consumption surveys and surveys of traffic, it becomes clear that smuggling takes place on a tremendous scale (smuggling is not a deviant activity carried on by a few criminals, but rather the coexistence of a frontier and smuggling are complementary and serve a number of powerful class interests). Here the increase in 'distance' caused by existence of a national frontier is clearly related to specific class interests. On the other hand, the fact that the frontier has been incapable as a tariff barrier of protecting indigenous Nepalese industry demonstrates the very considerable influence the Indian industrial bourgeoisie has over their government's policy, as reflected in the Trade and Transit treaties India has contracted with Nepal over the last twenty years. The few manufactured exports which had value added in Nepal but which competed with Indian goods, were banned; and low tariffs for Indian exports were won by the Indian side—often with a show of crude strategic strength (for example delaying the importation of fuel, which brought road freight to a halt in a matter of days, during the protracted negotiations of the 1971 Trade and Transit Treaty). In this way the industrial faction of the Nepalese bourgeoisie is seriously under-developed, and any surpluses accumulated from commerce have not gone to develop local capitalist manufacturing or even into agriculture, but into importing the very manufactured goods which directly contravene any alternative use of those surpluses. In fact, a considerable portion of the industrial bourgeoisie in neighbouring areas of India have strong financial and family ties across the border, and undoubtedly are involved in a variety of legal, quasi-legal and illegal transfers of profits to India.

Surpluses generated within Nepal are strongly attracted into business and retailing, and some of the most prosperous merchants in Nepal have amassed considerable personal fortunes out of the across-the-border trade. The trading network itself re-defines space in terms of centre–periphery relations. The largest operators, many of them Indian, have the best access to industrial output at the factory (often in the form of franchises), and the best access to high-level bureaucracy both Indian and Nepalese to help evade regulations, or at least to ease the passage of consignments and avoid otherwise indefinite delays. These large-scale merchants in turn, often operating 'up-country', supply the larger Nepalese shopkeepers who travel to border towns to purchase consignments for shops in the hill towns (consignments of over Rs 1 million are not uncommon for a single shop, twice or three times a year). Credit, 'sale or return', and other arrangements frequently bind these travelling shopkeepers to wholesalers in the border towns. These up-country traders (usually established at roadside locations to take advantage of bulk handling by lorry and concentrations of potential customers) in turn distribute to the petty retailers in their locality and to remoter areas in the hills. Geographical accessibility to sources of supply (particularly the frontier) and also to roadside locations at lower levels of the hierarchy, is of fundamental importance in determining a particular merchant's ability to accumulate, and aids the tremendous concentration of profits in the hands of those few whose class position gives them simple spatial access and the ability to minimize the 'distance-creating' effect of the frontier by their ready access to the bureaucracy manning the frontier, and to the industrial bourgeoisie of India.

REGIONAL PLANNING AS A COURSE OF ACTION

A Review of the Nepalese Situation

The previous sections have described the situation in which the people of Nepal find themselves, the main elements of which are:

(1) A crisis of stagnant production and growing population, which is yielding deteriorating living standards to all, and to many great difficulty in avoiding starvation without out-migration. The *terai*, at present the safety valve for surplus population, is being rapidly filled up by in-migration from the hills and from India and by natural population increase. The peasantry is dying on its feet, rather than being rapidly proletarianized by market relations and an emerging capitalist class in the countryside—as is the case in some developing countries.

(2) There are very marked spatial inequalities in Nepal, both between the hill areas, the mountains and *terai*, and between hill and valley bottom although both these pairs of regions have strong links between them—in the first instance a flow of cash and people from the food-deficient hills and mountains to the *terai*, and in the second case in the hills themselves, cash flows from the hillside dry-farming area to the rice-growing valley floors. There are also especially poverty-stricken areas in both hills and mountains which suffer from severe environmental deterioration.

(3) The vast bulk of the population is engaged in strategies for survival, though these strategies usually do not involve any collective action which might change the nature of production or of the state in general. Their strategies are individual responses aimed at acquiring or preserving their status as individual peasants. Because of this there has been little pressure from rural people for stronger political representation. The State in the eyes of the peasantry and landless labourers, is usually repressive, and upholds creditors' claims by jailing debtors, and is instrumental in driving *sukhumbasi* (illegal squatters) off unregistered land. The present system of '*panchayat* democracy' restricts the bulk of the people from making their own decisions through political processes.

As has been suggested, spatial structures (including planned spatial allocation of resources) follow—although they do not faithfully reflect—the political economy. Thus, spatial (or in planning parlance 'regional') planning cannot be radical planning, in spite of spatial manifestations of the political economy which have been described (e.g. hill–*terai* and hillside–valley bottom spatial inequalities, Kathmandu–rest-of-Nepal inequalities, and Indo-Nepalese relationship inequalities and the associated nature of the frontier). Regional planning within the present system can only therefore aid in a minor way, allocating resources within existing production and property relations. And thus, it is maintained, it will not avert the coming crisis of survival. On the other hand, regional planning is always necessary, since in any social formation spatial relations exist. Regional planning therefore must be part of a package of more radical transformations, in order to aid Nepalese society.

Such a package must face the realities of the present class structure in Nepal. 'Planning from below' is otherwise no more than a Utopian dream, a pleasing but empty device to be used cosmetically by academic and government institutions. Alliances must be looked for and used to promote real change; and these alliances will also shift, as class contradictions alter.

New Directions

One of the most fundamental changes needed to raise production as well as to provide employment, food and shelter, would be to permit increasing participation in production decisions to those who produce the wealth of the country, the peasants and the landless or virtually landless labourers (see Stöhr, Chapter 2). It might seem as though the peasantry already make their own production decisions, and participation is only denied to those who work for others for a wage. In reality, however,

'the process of development, in general. . . (has) bypassed the people—the small farmers and agricultural labourers. Thus the need for a change in the objectives and approaches towards agricultural development is imperative' (Dhital and Yadhav, 1976, p. 4–5).

and also:

'Till now the main lacuna in our objectives and strategy of agricultural development is an omission of the equity factor which must be deliberately imbedded itself rather than wait for the 'invisible hand' to take care of it' (Yadhav, 1976: 14).

Thus control over decision-making in the production process is considered here not only to be a fundamental basic need in its own right, but also as a means of increasing production, since the many needed changes in the production process will require wide participation in their inception and implementation. For example, a rational programme of re-forestation and rehabilitation of grazing land is at present impossible, since it would deprive the landless and the small peasant of a vital part of his livelihood. Thus re-forestation programmes and the guarding of existing forests under present circumstances become merely a policing exercise, discriminating against the already very poor. A fundamental question therefore arises regarding private property rights and the current form of access to means of production. Nor is it enough to implement an effective land reform programme. There is not enough permanently utilizable land anyhow to perpetuate an independent peasantry, and the collectively self-defeating survival strategies previously described would continue to the point of common destruction of the means of livelihood for all.

Thus a change in property rights must be considered. Already *panchayat* land ownership has been publicly suggested by an eminent Nepalese scholar, M.C. Regmi, and recent programmes (notably the Small Farmers Development Programme, see FAO Bulletin 122, 1978; and APROSC's feasibility report on the Rapti Integrated Rural Development Project) as well as the outline of the Sixth Five-Year Plan (NPC, 1978) explore communal enterprises and the pooling of land resources amongst small groups of farmers and labourers.

Identification of a target group for production development by its social relations (where caste status is used by landowners and employers as a means of discrimination to reinforce those relations), has already been emphasized by one major rural development project (in Rapti Zone, APROSC, 1977). In this project however, regional variations in extent of exploitation, oppression and discrimination have also been mentioned. The hills exhibit a lesser degree of social situations where such discrimination occurs than does the *terai*, but none-theless most hill farmers suffer generally similar problems of under-production, under-employment and deteriorating environmental and social conditions.

Such a regional policy, where problematic relations in the production structure can be identified, acknowledging intra-regional inequalities, is certainly a welcome innovation, and one where for once the conventional wisdom of international aid agencies may have a progressive role to play. Although the present author has no illusions about the role of aid, in certain situations such as in Nepal (90 per cent of whose budget is financed by such aid), outside donors could lend support to the more aware and progressive technocrats in the Nepalese government, its intelligentsia and commentators, who have hitherto been systematically excluded from major decision-making in government.

Political representation of the majority of the Nepalese people, small peasants and labourers, is also essential to the restructuring of centre-periphery relations in the different regions of Nepal, as well as the restructuring of her external relations with other countries, particularly India, and international institutions (such as FAO and UNDP). Only in this way can the gross inequalities between the Kathmandu valley and the rest of Nepal—maintained in part by the locational filtering of aid programmes—be reduced. The location of the present capital is only important as a locale for the garnering of surpluses by various factions of the ruling class. A change in composition of that ruling class together with a fundamental shift in production relations (and distribution of the surpluses they generate) will automatically alter the 'locational' imbalance. A call for acceleration in the process of turning over decision-making from large to small territorial units (Stöhr and Tödtling, 1978; p. 112), or the institution of communal decision-making (Stöhr, Chapter 1) 'to harness the richly personal embodied learning of local inhabitants to the more formal, abstract knowledge of specialists' (Friedmann and Douglass, 1978, p. 170) is warmly welcomed. But the means by which the necessary shift of power from one class to another or others could take place, has to be painstakingly worked out.

Indo-Nepalese relations as well can only be re-negotiated to the advantage of the majority of the Nepalese people by new prevailing class interests, which must replace those already described. The frontier will then automatically change its form and role. Thus it is not merely suggested here that the frontier should be better guarded and policed since this will not be done until new initiatives are taken. Here, the favoured position of Nepal with regard to foreign aid, and the extent to which Nepal can extract aid from India and China by maintaining a political 'balancing act', could be used to offset some problems arising from initial withdrawal from dependency upon Indian capitalism. These problems include interruption in the flow of household goods, chemical fertilizers and machinery from India, and a rise in cost-of-living for already poor rural households. A temporary alleviation of such symptoms could take the form of subsidised alternative supplies of these goods, as well as the initiation of protection for individual Nepalese concerns, which has indeed taken place, financed by international aid. In effect, the frontier must be used to protect Nepalese industry which will always need it (the frontier could be used effectively by new class interests, replacing those of merchant capital and a bureaucracy who would wish it to continue to 'leak') in order to hold its own against competition from the larger and more efficient Indian manufacturing enterprises.

Regional planning in Nepal today is therefore merely one element in the general rhetoric of decentralization, social justice and equity. But it has had a role to play, in spreading what resources are available into different parts of the country more equably than if regional planning had never been heard of. Such resources, however, are inadequate and cannot, under existing production structures and given the low level of popular participation, even alleviate the extreme poverty of the majority of the people. There are policies with spatial dimensions which can be pursued, but the fundamental root of the problem which these policies must tackle does not lie in the domain of space, but in economic relations. By transforming these, space can likewise be transformed, and it must then be planned in its new image.

REFERENCES

APROSC (1977) *Prefeasibility Study for Integrated Rural Development Project (Rapti Zone)*, 190 pp.

Bhasin, K. and B. Malik (1978) Small Farmers of Nepal Show the Way, *FAO Bulletin 122* (Rome: FAO).

Blaikie, P. M. (1978) The Theory of the Spatial Diffusion of Innovations: a Spacious Cul-de-sac, *Progress in Human Geography* 2, (2), 268–295.

Blaikie, P. M., J. Cameron, D. Feldman, A. Fournier, and D. Seddon (1976) *The Effects of Roads in West Central Nepal*, Report to Ministry of Overseas Development (U.K.).

Blaikie, P. M., J. Cameron, and J. D. Seddon with R. Fleming (1977) *Centre, Periphery and Access in West-Central Nepal: Approaches to Social and Spatial Relations of Inequality* (UK, Unpublished report to Social Sciences Research Council).

Blaikie, P. M., Cameron, J., and J. D. Seddon (1979) *Nepal in Crisis: Growth and Stagnation at the Periphery* (Oxford University Press).

Blaikie, P. M. and J. D. Seddon (1978) A Map of the Political Economy of Nepal, *Area* **10** (1), 30–31.

Central Bureau of Statistics—see census of Nepal 1952–54; 1961; 1971.

Dhital, B. P. and R. P. Yadhav (1976) Agricultural Modernisation in Nepal. Seminar on *Science and Technology for the Development of Nepal*, September 5–9, 10 pp.

Eckholm, E. P. (1976) Losing Ground: Environmental Stress and World Food Prospects (New York: W. W. Norton), 74–83.

Enke, S. (1971) Projected Costs and Benefits of Population Control in *Seminar on Population and Development* (July 28–30, 1971), Centre for Economic Development and Administration, Nepal.

F.A.O. (Food and Agriculture Organisation of the UN) (1974) *Perspective Study of Agricultural Development for Nepal:* Statistical annex (66 tables).

Friedmann, J. and Weaver, C. (1979) *Territory and Function: the Evolution of Regional Planning* (London: Edward Arnold).

Friedmann, J. and M. Douglass (1978) Growth Pole Strategy and Regional Development Planning in Asia in Lo, F. C. and K. Salih (eds.), *Growth Pole Strategy and Regional Development Policy* (Oxford: Pergamon Press), 163–192.

Gurung, H. (1969) *Regional Development Planning for Nepal*, National Planning Commission, 24 pp.

Harvey, D. (1973) *Social Justice and the City* (London: Edward Arnold).

IBRD (1973) *Economic Situation and Prospects of Nepal*, World Bank Report No. 125-NEP, Asia Region, South Asia Department.

Lo, F. C. and K. Salih (1977) *Rural-Urban Interactions and Regional Development Planning: a Progress Report of ADIPA Collaborative Research.* Paper prepared for the 3rd. Biennial Meeting, Association of Development Research and Training Institutes of Asia and the Pacific, the Fort Aguada Beach Resort (India: Goa), Sept. 19–24, 1977.

Ministry of Food and Agriculture, Kathmandu (1972), *Agricultural Statistics of Nepal*, 166 pp.

National Planning Commission (1974) *Regional Development Policies: Report of the Task Force* (Kathmandu). 54 pp.

National Planning Commission (1978) *The Sixth Five Year Plan (draft)* (Kathmandu).

Ojha, D. P. and D. Weiss (1972) *Regional Analysis of the Kosi Zone, Eastern Nepal* Vols. I (171 pp.) and II (Documentation 112 pp.), (Kathmandu: German Development Institute and Centre for Economic Development and Administration).

Okada, F. E. (1970) *Preliminary Report on Regional Development Areas in Nepal* (Kathmandu: National Planning Commission) 101 pp. plus appendices.

Rana, S. J. B. R. (1974) *Regional Planning in Agricultural Development*, Seminar on Agricultural Planning and Project Analysis for the Least Developed Countries, FAO-NMG (Nepal). November 17–December 6, 9 pp. and comments.

Regmi, M. C. (1971) *A Study of Nepali Economic History 1768–1845* (New Delhi: Manjusri Publishing House).

Reiger, H. C., F. Bieri, H. Eggers, W. Goldschact, and J. Steiger (1976) *Himalayan Ecosystems Research Mission: Nepal Report* (Heidelberg: South Asia Institute, University of Heidelberg).

Santos, M. (1974) Geography, Marxism and Underdevelopment, *Antipode*, **6**, 1–9.

Stöhr, W. and F. Tödtling (1978) An Evaluation of Regional Policies: Experiences in Market and Mixed Economies, in *Human Settlement Systems: International Perspectives on Structure, Change and Public Policy*, ed. Hansen, N. M., (Cambridge, Mass. : Ballinger Publishing Co.)

Development from Above or Below?
Edited by W. B. Stöhr and D. R. Fraser Taylor
© 1981 John Wiley & Sons Ltd.

Chapter 10

India: Blending Central and Grass-Roots Planning

R. P. MISRA and V. K. NATRAJ

INTRODUCTION

India has a record of a quarter-century in the field of planning. The mainspring for the Indian attempt at planning is not very different from those which have served as the launching-pad for planning, or at least, development-oriented activities, in other developing countries. The main forces which propel ex-colonial countries to adopt some form of planning are factors such as the existence of mass poverty, glaring inequalities in income and wealth, low rates of growth, and excessive dependence on the primary sector, together with lop-sided industrial development. While these characteristics are generally common to most developing countries, there are nonetheless differences both in socio-economic parameters and, more especially, in the kind of planning effort attempted. These differences relate to and derive from the specific historical antecedents of the countries in question, and the composition of the ruling class which came to power after gaining independence from colonial rule.

India perhaps had an advantage, in that several political parties took an active interest in socioeconomic development of the country even before in-dependence. Mention should be made of the efforts of the National Planning Committee of the Congress Party (Shah, 1947); the People's Plan (prepared by the Indian Federation of Labour, then headed by M. N. Roy) (Bannerjee *et al.*, 1944); the Bombay Plan (drafted by a group of industrialists) (Thakurdas *et al.*, 1944); the Gandhian Plan of S. N. Agarwala (Agarwala, 1944); and the efforts of Sri M. Visvesvaraya, former Dewan and engineer–statesman of the progressive princely state of Mysore (Visvesvaraya, 1936).

The chief merit of all these plans lay in the fact that they made planning respectable in the eyes of the people, although they all basically suffered from a series of unrealistic assumptions—a natural consequence of drawing up plans from outside the closed system of colonial administration. However they had the salutary effect of making planning an accepted word. The first independent

259

Figure 10.1 The States of India

national government therefore was not faced with the task of converting members of its party to the idea of planning. This implied, not surprisingly, that planning was armed with political teeth—an important factor given its lack of success in countries where planning has been largely a technocratic affair without political support. However, this consensus or apparent consensus on planning has been a mixed blessing as we shall see.

IDEOLOGY OF PLANNING

Planning without ideological underpinnings tends to lose direction. Its objectives and goals get blurred, its contents disorganized. In appraising the ideological foundations of any planning effort it should first be recognized that planning itself is an ideology. However the term often means different things to different people. Agreement on the need to plan does not necessarily imply agreement on goals. Even vociferous agreement on goals often does not mean agreement on the *modus operandi* of achieving those goals. The Indian experience has quite vividly illustrated this dichotomy, with the inevitable consequence of goals being imprecisely defined—or when defined precisely, being covered with caveats of various kinds, and in the end not being properly implemented.

The major points in the ideological underpinnings of Indian planning are as follows: first, since the first five-year plan there has been great stress laid on unity of purpose. This can be taken to mean the need for a nationally agreed consensus on what to plan for. It also expresses a commitment to a democratic frame of operation. Secondly, there has always been insistence on public participation. That the people must be involved in the planning process has been elevated to an article of faith. And finally, the avowed and often reiterated goal of Indian planning has been reduction of inequality of income, wealth, and socioeconomic power. The latter, which may be described as the ultimate goal, or 'goal of goals' of Indian planning, is formally enunciated in the plans, and also derives its sanctity from a resolution passed at the Congress session in 1955 at Avadi which declared that the aim of planning should be the establishment of a socialistic pattern of society.

These three may rightly be regarded as the fundamental tenets of Indian planning. Almost all other objectives added on with each five-year plan are in essence further explications of these tenets. Moreover, these basic postulates themselves have historical antecedents.

The emphasis on unity of purpose and consensus derives from the simple fact that the National Planning Committee of the Congress Party, which had the longest and most systematic record for analysing development issues prior to independence, had problems to contend with. In their thinking the Congress Party was not homogeneous on the orientation of development to be pursued when India became independent. There were important differences on the basic question of industrial development. The Gandhian group voted for small and cottage-type industries and essentially stood for a village-centred approach to development. Nehru, on the other hand, looked to industrialization planned from above as the key to the problem. The other major differences related to what may be called the distributive question. Nehru, supported by the socialist group, argued for greater state control; while another group was lukewarm about such a policy. Conflicts were resolved by a compromise solution which in an omnibus fashion attempted to integrate all these disparate views. Nehru confessed that as the party was then constituted, it was worth retaining the support of large sections therein, even at the cost of toning down radical components of the strategy. The consensus approach sanctified in pre-independence days was carried over to the planning exercise in independent India.

The next component, namely public participation, derived from two sources. Firstly it was an axiomatic corollary of the commitment to a democratic frame of reference. Secondly there is hardly any doubt that this along with decentralized planning, was a direct derivative of the Gandhian influence. However we should note that what plans up to now have attempted to do is to borrow the idea of decentralization as a technique, rather than as a philosophy—which it is in the Gandhian scheme. Much of the confusion prevalent over the mechanics of decentralization can be traced to this. The plans glorify the idea of local-level planning but offer few if any prescriptions which could make such an exercise practicable.

The third component, namely the socialist pattern, also derives from several sources. There is first the Congress Party which despite internecine disputes was always attracted by the notion of distributive justice, at least in theory. Gandhi, whatever one may say about his practical approach to the problem, had considerable empathy with the poor. The Congress Party also had quite a strong socialist lobby before independence. Furthermore Nehru, the main architect of planning in the party, was influenced to a considerable degree by the example of the USSR, although he had some grave reservations about the Soviet experiment. Some consensus, even though uneasy, thus emerged or was made to emerge on the need for a more egalitarian society. The methods by which to achieve this laudable aim were, however, rarely discussed to their logical conclusion.

After independence when planning was formally launched, this question had to be faced in order to make concrete policy decisions. The shaky foundations of the consensus then became obvious. Faced with the task of devising policy instruments to translate the aims of an egalitarian society into reality, the policy-makers vacillated. Given the institutional frame of an essentially capitalist economy with a semi-feudal social structure it became difficult to arrive at a meaningful compromise between growth which appeared to demand investment incentives, and equity which demanded institutional reforms which would affect the power base of the system itself. The second and third plan documents in particular are replete with examples of what A.H. Hanson calls 'the schizophrenia of Indian planning' (Hanson, 1966). Formal adoption of the socialistic pattern as the goal of planning did little to solve the problem.

The fourth and fifth plans have spoken of the need to satisfy basic needs. These in reality ought to have been the first charge on a free India's first budget and indeed on the first five-year plan. Yet since policy-makers were found wanting in their commitment to institutional reform, the provision of basic needs continues to remain peripheral, rather than central to the planning theme, recent re-emphasis on them notwithstanding. The argument against basic-needs-focused planning is that it may retard progress. Those who advocate that growth (efficiency) should precede distribution justice (equity), do not appreciate that growth is not an abstract concept. Growth consists of a package of goods and services. In other words we should speak of commodity-specific growth. Therefore it is incorrect to argue that 'growth' whatever its commodity composition is a good thing. A particular package of commodities may help fulfil basic needs while another may not (Kurien, 1977).

The first five-year 'rolling' plan (1978–83) launched in 1978 carried this dilemma further. It assumes that poverty can be banished from the country in a decade without far-reaching institutional reforms. Mere increased allocation for sectors like agriculture, social service, etc., the planners think, would strengthen the hands of the poor, exploitative semi-feudal institutional framework notwithstanding.

Theoretical Bases

The theoretical foundation of our plans has also suffered from the fuzziness surrounding the ideological basis of our development policies. The first plan was largely a collection of existing projects, and did not have any theory or model to guide it. It nevertheless achieved significant results in agriculture and hence in the growth of the national income. Inspired by this success, the second plan was a bigger and bolder venture. At this time the Soviet model of industrial development, and also the socialistic pattern of development, were specifically built into the five-year plan. However the dilemma alluded to above was not resolved. The second plan could have achieved its targets only if required social structure changes in consonance with the socialistic pattern of social order were effected. Only then would it have been possible to channel the available resources in full in the direction the plan envisaged. The third plan, beginning in a more sombre atmosphere, argued for more attention to agriculture without neglecting industrial advance. The fourth plan, while not giving up the fundamental objectives described earlier, came out more vehemently in favour of growth plus social justice. The fifth plan, drafted ambitiously, has had to contend with political upheavals and has remained a holding operation. The sixth plan, otherwise known as the five-year (1978–83) 'rolling' plan, still in process of discussion and debate, reverts once again to agricultural fundamentalism. The new government has now discontinued the quarter-century-old tradition of having five-year plans. Instead it has opted for a rolling plan effective from 1978, the last year of the fifth five-year plan. In the debate on the goals of planned development we now find much greater emphasis being placed on the urban–rural contradiction. It would appear that those who insist on a resolution of this contradiction feel that this supervenes the poor–rich contradiction which is surely more fundamental (Hanumantha Rao, 1978).

REGIONAL DIMENSIONS

Reduction of interpersonal inequality, as we have seen, has been an expressed aim of Indian five-year plans. Policy statements to this effect have also been articulated, although the political will to take hard decisions has always been weak. The same cannot be said of regional balance. As an issue, balanced regional development extended into the sphere of actual planning much later than the over-all inequality issue, and has not been debated with the same heat and intensity. All the country was prepared to do was to accept river basins as areal units for development of water and power resources, which were essential for rapid industrialization. The Damodar Valley is an example of this recognition.

Balanced regional development as an openly articulated plan objective came in 1961 with the third five-year plan. This is surprising for two reasons: in the first place when the country became independent, regional differences were quite evident. These were the result of differences in natural endowment, locational advantages for trade and industry, and the fact that colonial powers had done little to redress these imbalances. Moreover large parts of the country were ruled by princes, not many of whom were interested in development *per se*, while several of them ruled over very small geographical areas which were hardly viable units for development. In this sense,

> at independence, to a substantial degree, the foundations for future differences in rates and patterns of regional growth had already been laid. In the nature of things, attempts to accelerate development would be expected to bring larger opportunities for growth to areas and groups which had already made an appreciable start. (Singh, 1974, p. 48.)

The second point is that soon after independence a clamour arose for dividing states on a linguistic basis. It is reasonable to suggest that this demand arose not only for cultural (language) reasons but also because minor language groups in large states felt they were not receiving their due share of gains from development. This sense of deprivation was a major contributory cause for the agitation for separate linguistic states. The first major thrust in this direction came from the Telugu-speaking area of the then composite Madras State. This agitation culminated in the formation of Andhra in 1953 which comprised twelve districts of Madras State.

In spite of this, both the first and second five-year plans were silent on the issue of regional development. It may of course be argued that the plans kept intentionally silent, in order not to increase existing tensions. The history of such tensions in India could be interpreted in two ways. It could be said that had the idea of linguistic states not been accepted, then border disputes among states would not have arisen. Equally forcefully it may be contended that the course of wisdom consists in recognizing sociocultural differences where they exist and affording equal opportunity to all to share in the gains of development. People's China, for example, has deliberately followed a policy of recognizing the problems of the non-Han minorities and giving them special treatment. There is necessarily no contradiction between reducing regional imbalances and reducing interpersonal inequalities. In fact the two ought to go together. It is also possible that a high incidence of absolute poverty may be found in regions of extreme backwardness. However the two problems, although related, are not the same. A government that attempts to correct regional imbalances may also in the process correct extreme inequalities in income distribution, but this is by no means a certainty. A reduction in areal imbalance does not necessarily indicate reduction in interpersonal inequality. There is no guarantee that the benefits of development (or more likely those of growth) will be uniformly spread in the backward area especially chosen for development. Distribution of gains from development is likely to be much the same as in the rest of the economy in the absence of positive measures geared towards income distribution.

The majority of less-developed countries show a marked presence of interpersonal inequalities. Many of them profess that a major aim of their planning is to reduce these inequalities and strive for more equity. Even the most sincere of such protestations generally adds a caveat by drawing attention to the conflict between equity and efficiency, and in essence opts for some kind of compromise. A good example of this is the Indian second five-year plan. Interpersonal inequalities clearly are difficult to resolve and often depend for their success on instruments of policy like taxation, control on private investment, and consumption, etc. All these are politically sensitive issues and turn largely on the value-judgments which a society has made for itself.

At the same time no underdeveloped country today is in a position not to benefit from using the rhetoric of equity. This concept has an instrumental value in terms of political mobilization. It is therefore at least plausible that societies which do not fully subscribe to equity in income distribution, or who feel it difficult to use the policy instruments required for equity, emphasize regional balance as a goal. As mentioned earlier, the Indian five-year plans have not been too explicit on the issue of regional imbalances. Even though Indian plans have shown a high degree of sophistication in terms of internal consistency, model-building, etc., and have also after a fashion discussed the equity issue, they have not openly addressed themselves to the regional question, and have yet to devise—they are now in the process of doing so—a frame for multi-level planning. Yet there are countries such as the Philippines where organization for regional development has been greatly refined. For instance there are 12 administrative regions for planning in the Philippines, each encompassing several provinces. The National Economic Development Agency's (NEDA) regional office is entrusted with preparation and implementation control for all plans and projects in the respective regions (A detailed survey of regional planning in several Asian countries is given in Gruchman, 1978). An almost similar situation obtains in Indonesia. Thailand too has quite an elaborate structure for regional planning. All these countries have a more-or-less authoritarian political framework in which the reduction of interpersonal inequalities has not been a prime consideration. Indeed the extent to which such a political framework can permit such a consideration is itself a debatable issue.

The clamour for linguistic states in India which reached its zenith in the 1950s could be regarded as an expression of, or a reaction to, regional disparities. The demand for a separate Andhra State to be carved out of the old Madras Presidency originated from the feeling of neglect which the Andhra (Telugu-speaking) region had over a period of years. In fact a scheme for the development of Rayalaseema—the most backward area in the Presidency, a Telugu-speaking area as well—was proposed before Andhra State became a reality. The same was true of the demand for Karnataka, the Kannada-speaking parts of Bombay, Hyderabad, and Madras States feeling that their economic interests were always subordinated to those of the majority-language groups. Paradoxically, adding weight to this interpretation, some 20 years after formation of linguistic states, contradictions have appeared amongst the different regions of these states. Perhaps the best example is the state of Andhra Pradesh which is clearly divisible into three regions: Coastal Andhra, Rayalaseema and Telangana. The demand for special protection from the Telangana area led to serious strife in the late 1960s and early 1970s. It may therefore be inferred that when real or even imagined regional disparities exist even the unifying force of language tends to lose some of its strength.

Notwithstanding this the fact remains that the issue of regional disparities was not brought up for quite some time in the five-year plans. Also, given the varying historical antecedents of different regions, the plans naturally had regionally differentiated impacts. This is why Tarlok Singh suggests that the commencement of planning actually led to further regional inequalities (Tarlok Singh, 1974). What he means of course is that the existing equilibrium was disturbed by the entry on the scene of planning which had different consequences for different regions. Tarlok Singh also argued that the demarcation of states on a linguistic basis also contributed to regional imbalances owing to, among other things, differences in quality of administration. It is interesting to recall that the States Reorganization Commission did not proceed only on the basis of language in suggesting formation of new states. Economic viability was to the Commission an important criterion and in the case of Telangana was used to recommend that this area not become part of the proposed Andhra Pradesh. This recommendation was not acceptable to the Government and the new state of Andhra Pradesh (including Telangana) was formed in 1956.

The first real reference in terms of planning to the need for balanced regional development is found in the Industrial Policy Resolution of 1956. This resolution referred to the need for 'securing a balanced and co-ordinated development of the industrial and agricultural economy in each region' so that the entire country could 'attain a higher standard of living'. Even this should properly be recognized for what it is—merely a hopeful suggestion appearing as a 'Shadow' in some sectoral policies, to use a picturesque expression employed by K. V. Sundaram (Sundaram, 1977).

Apart from this, there were other sporadic attempts at grappling with the regional issue. Some efforts were made to identify backward areas; plans were drawn up for developing inter-state water resources, and power and transport plans were formulated with regional perspectives, etc. Many of these got bogged down primarily due to the low level of development and rate of growth, and were further hindered by endless inter-state disputes on use of inter-state resources, principally water. To this day the major river systems, which are all inter-state in character, are subjects of disputes. Interestingly there is legislation covering all these points, but for one reason or another practical results are not impressive.

The third five-year plan was the first to discuss specifically the question of regional balance. While it emphasized the need for balanced regional development, not much was offered by way of policy instruments to achieve this balance. However, the fourth plan period and subsequent exercises have seen some advance in this direction. It is at least possible that planners were made anxious by widening interregional disparities. The decade between 1960 and 1970 witnessed, for example, an increase in differences in per-capita net domestic product from 1.9 : 1, to 2.6 : 1. Furthermore planners began to appreciate the fact that in most cases backward areas are largely inhabited by severely underprivileged peoples such as tribal groups. All these factors brought into sharp focus the immediate need to develop special programmes for backward areas. In spite of this, the central mode of planning has remained sectoral rather than regional.

Table 10.1 presents interesting data on differentials in growth rates and productivity over regions (Bhalla, 1979). These results are shown in map form in Figure 10.2.

It is interesting to note that regions with high productivity and high growth rates are confined to a small area in the north-western part of the country—entire Punjab, eastern Haryana and Western U.P. In these areas, a fast increase in irrigation coupled with the intensified application of the new technology and increase in cropping intensity led to growth rates of output higher than 5 per cent compound per annum. These areas account for about a quarter of the increase in agricultural production. In more widespread areas, particularly in Rajasthan, Madhya Pradesh and parts of U.P., Andhra Pradesh and Karnataka, growth in agricultural production was evident at a much lower pace than in the first set of areas, the growth rates ranging from 2.5 to 3.5 per cent compound per annum. On the other hand, regions characterised by low productivity and low growth rates cover an extensive area of the country in the states of Karnataka, Andhra Pradesh, Maharashtra, Rajasthan and Tamil Nadu. 37 of the low level productivity districts have registered negative growth rates, and twelve of the low level productivity districts have shown a slow rate of growth. In all, these 75 districts perhaps constitute the 'regions of sub-marginal subsistence'. (Sundaram, n.d., p. 5.)

Recognition of this problem implied that a suitable policy for allocation of resources amongst states should be devised. In a large federal country this is no easy task since criteria for allocation of resources have to be informed both by equity and efficiency. Present criteria for interregional allocation of central assistance consist of a weighing for population, per capita income, tax effort, special problems of the state, and commitments with respect to major continuing irrigation and power projects. This theory has built-in limitations. The basic difficulty consists in 'the absence of a method by which central assistance can be linked with the levels of development and sectoral requirements' (Sundaram, n.d., p. 6).

There is a more serious issue involved which relates to the whole gamut of centre–state relations. The question is whether backward areas can be identified at all and therefore assisted in a special way, from the remote heights of the national level. Should this not rather be the task of state governments? If so it

Table 10.1 *Agricultural growth rates and productivity by region*

GROWTH RATE OF AGRICULTURAL OUTPUT

		High	Medium	Low
PRODUCTIVITY OF LAND	High	Punjab (entire state) Eastern Haryana Western U.P. (Assam Hills)	Kerala—Northern Tamil Nadu—Coastal northern W. Bengal—Himalayan —Central plains —Western plains Karnataka—Inland southern	Kerala—Southern Tamil Nadu—Coastal southern —Inland W. Bengal—Eastern plains Andhra Pradesh—Coastal Maharashtra–Coastal Karnataka–Coastal and ghats
PRODUCTIVITY OF LAND	Medium	Karnataka—Inland eastern U.P.—Himalayan	Assam—Plains Bihar—Southern Western Haryana M.P.—Eastern U.P.—Central	Orissa (entire state) Andhra Pradesh—Inland —Southern Bihar—Northern central U.P.—Eastern —Southern Gujarat—Eastern
PRODUCTIVITY OF LAND	Low	Rajasthan—South-eastern Gujarat—Kitch	M.P.—Inland Eastern —Inland Western —Western —Northern Rajasthan—North —Eastern —Southern	Karnataka—Inland northern Andhra Pradesh—Inland northern Maharashtra—Inland western —Inland northern —Inland central —Inland eastern —Eastern Rajasthan—Western

Source: Bhalla, 1979.
Notes:
 Productivity: High = Rs. 1300 or more per hectare
   Medium = Rs. 700–1300 per hectare
   Low = Rs. 700 or less per hectare.
 Growth Rate: High = 4.5 per cent or more
   Medium = 1.5–4.5 per cent
   Low = 0–1.5 per cent.
 Abbreviations: U.P. = Uttar Pradesh; M.P. = Madhya Pradesh

would mean a change in the financial equation between the centre and the states. This view must be juxtaposed with the fact that on the whole what we find in India is an interstitial pattern of backwardness, not a clear division of advanced and backward areas—although there are whole states which are backward compared to the national average.

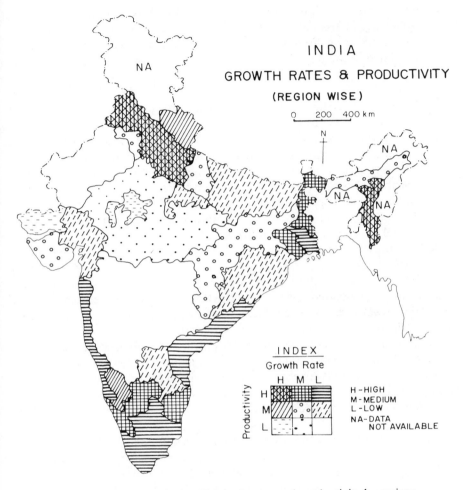

Figure 10.2 India: agricultural growth rates and productivity by regions

The existing framework has not explicitly provided for a multi-level planning frame. As mentioned, some efforts are now being made in this direction. The debate between 'development from above' and 'development from below' has also gone on in India. The second plan devoted a whole chapter to 'district development administration', and grass-roots planning has been considered almost sacrosanct in Indian thinking. However, the political–administrative implications of such a strategy have not been fully appreciated. In the Indian context it is first necessary to work out a system of relations between the centre and the states, in which planning functions of the two are clearly delineated and these functions matched with the required powers. This essential first step is still fraught with conflict, especially in the sharing of financial powers. Generally few states have been willing to share power meaningfully with the districts, etc.

The following inferences emerge from our analysis. Regional development with a view to reducing inter- and intra-state imbalances have only recently been explicitly stated in the plans.

To recapitulate, the main elements in regional strategy are:

(a) Attempts to locate large-scale industrial projects in industrially lagging areas through central sector investment.

(b) Allocation of central assistance to states on the Gadgil formula with weighting for population, per-capita income below the national average, tax effort in relation to per-capita income, outlays for selected continuing irrigation and power projects costing over Rs. 20 crores each, and existence of specified special problems in the state. The weights are 60 per cent/10 per cent/10 per cent/10 per cent/10 per cent, respectively.

(c) Special allocation of funds to backward areas.

(d) Measures to encourage private investment to move into backward areas.

Evidence indicates that the location of large industrial projects in backward areas has not generated all-round growth—much less development—and propulsion towards development has been centripetal (periphery to core) rather than centrifugal (spread effects from core to periphery).

All these instruments of regional development currently in vogue aim essentially at reducing interregional disparities at the national level. In other words planners and policy-makers obtain a synoptic vision of regional differences, and attempt to achieve some balance through redistributive policies worked out at the top and conceived in macro-fashion. Here we have another illustration of the 'top-down' paradigm.

The fundamental issue seems to be: can we still persist with the sectoral approach, and expect regional imbalances to be reduced merely through the medium of the measures referred to above which are not yet central to the planning process? In one sense the minimum-needs programme and the regional question are similar, yet the main centre of the plan is quite different. Plans are addressed to an increase in GNP, with emphasis on sectoral targets. The intention of sharing in gains of development has not been shown to be of primordial concern in the plans, but if such intentions are to be fulfilled they must become the central focus around which the entire plan is formulated. (Hanumantha Rao, 1978). Sectoral planning is an essential component of planning in any country; but if other considerations are regarded as important then an areal strategy must be devised within which sectoral balance is important, but not the sole factor. Regional balance depends upon the ability of planners, and more importantly policy-makers, to devise a frame of operations within which the different components of an integrated plan can be handled with precision at different inter-areal levels. This brings us to the question of multi-level planning in India.

MULTI-LEVEL PLANNING

Regional development has in general been sought to be achieved through special plans or programmes for problem areas. The Drought-Prone Areas Programme, the Small and Marginal Farmers' Development Programmes, the Hill-Area Development Programmes, and the Tribal Area Development Programmes all fall into this category.

The ushering-in of these programmes also led to creation of a multi-level planning frame. From the beginning Indian planning has stressed the need for decentralization. This is a remnant (along with popular participation) of the Gandhian influence—so far more honoured in preaching than in observance. Only in the last few years has a multi-level planning frame emerged. Any real development requires attention to detail, i.e. to micro-indicators. Specific locational decisions are called for. Coordination, both vertical and horizontal, is essential, implying creation of new agencies and mechanisms of coordination.

Planning Regions

In the past many attempts were made to divide India into planning regions, and these attempts are well documented (Misra *et al.*, 1974), but none could be adopted as the basis for regional planning for three main reasons: first, these regions were based on criteria sound from scientific and theoretical viewpoints, but placing too great a demand on the colonial administrative system to make any headway; secondly, the whole question of regional planning was looked at from a macroregional angle having no relevance to grass-roots development; and finally, such a macroregional framework could not find favour with states carved out in 1956 on a linguistic basis and which for all practical purposes constituted macroregions of the country.

In due course political and administrative compulsions made it necessary to forget the scientific basis for regionalization, and adopt existing administrative areas for regional planning. The states were accepted as planning units from 1950 onwards. Each state prepared its own second five-year plan, and this tradition continues today. Central assistance is given to states as planning units.

In the early stages of Indian planning an attempt was also made to encourage inter-state planning through zonal councils. These attempts did not go beyond the idea stage. Only in the states of Assam, Mizoram, Meghalaya, Nagaland, and Manipur did inter-state planning make some headway, and the zonal councils are dead.

Another experiment in inter-state regional planning was conducted by the Town and Country Planning Organization in the resource-rich region of Bihar, West Bengal, Orissa, Madhya Pradesh, and Uttar Pradesh. A plan was prepared for the so-called South-East Resource Region under the leadership of the Joint Planning Board. The plan has yet to be finalized—let alone implemented—despite the fact that work on it started in 1968.

Realizing that state-level planning is no different from national planning in matters of approach and methodology, a beginning was made in district planning in the third five-year plan period (1961–66). A three-tier multi-level planning system was devised, with its main objectives being:

(1) removal of inter-district and intra-district imbalances in development; and

(2) providing opportunity for each district to attain full development in potential, available manpower and other resources.

At this stage it is worthwhile to look back to the early 1950s when community development (C.D.) programmes were launched all over the country. One fundamental idea behind these programmes was to decentralize planning and bring it to the doorstep of the common man. The C.D. block was conceived as the lowest administrative unit for execution of rural development projects. Each block consisted of about 100 villages. Choice of villages and headquarters of blocks were dictated by local pressures and decided at state level. In the course of time the community development programme, however, degenerated into a bureaucratic exercise aimed at nothing but the routine work of agricultural development. The utter failure of the C.D. block as a planning unit left the impression that a block is too small a unit for planning, and what the country needed was district-level planning, especially because the necessary administrative infrastructure for implementing the plan existed at this level alone.

After experimenting with district planning for the last decade or so thinking has now come full circle. The block has been reintroduced as the basic unit of rural area planning in 1978. During the following five years 2000 blocks were to be covered by scientific planning aimed at full employment. All blocks are to be covered within the next ten years, and the 1978–79 target was 300 blocks.

Block-level planning is not an exercise in disaggregating the targets fixed at state and district levels, but is meant to revive and recapture the concept of planning 'from below'. In this framework it has vast potential, provided it does not go the way of the old discredited community development programme.

Thus a four-tier multi-level planning system has now evolved: (1) nation, (2) state, (3) district/metropolitan/urban area, and (4) block. In theory at least the first tier should look at planning from the top down, while the last views it from the bottom up. State and district levels provide the organic links between nation and block. There are states like Karnataka where the block is co-terminous with the Tauk (the administrative unit next to the district).

Planning Functions

At the apex, the National Planning Commission is the chief planning body. Headed by the Prime Minister as chairman, with a well-known economist as Deputy Chairman, the commission formulates and coordinates development activities of the central ministries and states. It prepares three types of plans: a perspective plan for 15–25 years, a five-year 'rolling plan', and annual plans. The perspective plan is a set of projections and directions for each sector of the economy. The five-year plans are medium-term attempts to lead the economy in the direction set by the perspective plan. Each five-year plan has to be appraised mid-way to enable the Commission to readjust allocations to meet challenges of new and emerging situations. Annual plans are essentially devices to allocate funds for projects proposed by central ministries and states to implement programmes of the five-year plan.

An important role the Commission plays is in promoting scientific planning at state and other territorial levels.

States, however, are expected to carry out far more detailed planning and project formulation. Furthermore they are expected to relate the plan more intimately to the sociocultural ethos of the people, and to lay greater emphasis on spatial and equity aspects of planning, as they are nearer the people and have more dependable data on interpersonal and interregional disparities in income, wealth and quality of life. States are also supposed to initiate grass-roots planning to revamp development administration and carry out required institutional and structural reforms in a developing society.

In the Indian federal set-up, most structural reforms fall in the state sector. The centre can only offer incentives and guidelines. Even where the central government has specific roles to play, implementation is by-and-large left to state governments.

Districts are not statutory or constitutional units in the Indian federal system, they are administrative units and can be created or abolished by the state at will. The reasons for choosing the district as the third-level planning unit were threefold. First, the district is the only level below the state where adequate administrative and technical expertise is available; secondly, *Zila Parishads* (district-level bodies for planning and development activities, elected by the people) had adequate popular leadership to carry out development tasks; and third, it was the only grass-roots territorial unit where adequate finances and information necessary for planning were readily available. Popularly elected *Zila Parishads* do not exist in all states, however; in many like Karnataka there is only the district development council, mainly composed of *ex officio* and nominated members with little power, less prestige and almost no expert staff for development work.

When the Janata Party assumed power in 1976 there was a further decentralization of planning processes in India and blocks (which in some states correspond to *taluks*, which rank below districts as administrative and revenue units, and in others are parts of a district originally grouped together for extension work) became the lowest-level units of planning.

One major change in the newly initiated block-level planning scheme is the recognition that planning has to be a 'bottom up' process, and blocks should be the basic grass-roots planning unit. Thus a block has to be the basis for district planning, and should aim at:

(1) optimum realization of growth potential of the area;

(2) ensuring that larger than proportionate gains in development accrue to weaker sections of the population;

(3) fulfilment of the minimum needs programme—health and medical facilities, drinking water, housing, education, and the supply of essential commodities through a public distribution system;

(4) reorienting existing institutions or organizations to protect the interests of the poor;

(5) promotion of a progressively more egalitarian structure of ownership of assets;

(6) augmenting the duration and productivity of employment of the poor and under-employed in their existing occupations, *inter alia* through up-grading of technology, imparting of skills and the setting-up of non-exploitative institutions of credit, marketing and extension; and

(7) alleviating residual unemployment through employment on public works (Planning Commission, 1978, pp. 1–2).

Block level planning is to be viewed not as an isolated exercise but as a link in a hierarchy of levels from a cluster of villages below the block level to the district, regional and state level. Its relevance as a unit of planning is based on the following reasons:

The block is distinguished by a certain community of interests. It is sufficiently small in terms of area and population to enable intimate contact and understanding between the planners, those responsible for implementation of the plan and the people. It provides an observation platform in close proximity of the beneficiary group and thus helps to:

   (i)   understand more clearly the felt needs of the people and factors inhibiting the uplift of the weaker sections;

  (ii)   ascertain area (block) specific physical and human resource potential;

 (iii)   identify constraints inhibiting socio-economic and technological growth; and

 (iv)   expand the area of people's participation in the preparation and implementation of plans. (Planning Commission, 1978, pp. 4–5.)

An illustration of activities which can be planned and executed at the block level without sacrificing planning efficiency is given below:

    (i)   agriculture and allied activities;

   (ii)   minor irrigation;

  (iii)   soil conservation and water management;

  (iv)   animal husbandry;

   (v)   fisheries;

  (vi)   forestry;

 (vii)   processing of agricultural produce;

(viii)   organizing input supply, credit, and management;

  (ix)   cottage and small industries;

   (x)   local infrastructure;

  (xi)   social services, such as drinking water supply, health and nutrition, education, sanitation, housing, local transport, and welfare programmes;

 (xii)   training of local youths and unskilled population. (Planning Commission, 1978, pp. 5–6.)

Each block plan is supposed to take note of development activities being planned and executed at supra-block level. It is expected to get plans executed through popular organisations like village *panchayats* (*panchayat* is a local popularly-elected assembly for a village or group of villages, depending on population size). *Panchayat Raj* institutions have to be restructured for this purpose (Mehta, 1978).

> The logic of increasing locational specificity and growing technological complexity of development projects would necessitate the district being considered the first point of decentralisation with the state governments devolving most of their district level developmental functions and the corresponding financial and administrative resources upon the *Zila Parishads*.

> The spatial focus of most of the development prospects in the coming decades would be multi-village rather than mono-village and it would be difficult to provide the requisite technical inputs for each and every village. The Indian villages were increasingly developing extra-local linkages with other villages as well as small and medium towns and these inter-linkages would necessitate the establishment of democratic institutions below the district and the block to handle the emerging patterns of functional relationships in such fields as input supplies, product marketing and credit flows. The spatial dispersal and functional decentralisation of the local level rural development would make the establishment of *Mandal* (group village) *Panchayats* a functional necessity. (Mehta, 1978.)

Thus the Ashok Mehta Committee in essence recommends a fifth tier in multi-level planning schemes.

DEVELOPMENT FROM BELOW

In the foregoing critical review of the planning process in India it is apparent that planning and its various manifestations have reached a stage where old premises and models are being challenged. The accent is shifting from sectoral growth to a judicious blending of growth and distribution; from centralized planning to multi-level, if not decentralized; from expenditure-oriented to physical-achievement-oriented; from a bureaucratic and mechanistic approach to popular participation; and from economic development to human development. No major departure from the past is expected in the short run; nor is there any distinct possibility of a new planning culture which would steer clear of fixed notions and models. What is apparent is a rethinking, a questioning, a desire and willingness to change and experiment with something unconventional, and a willingness even to revert to those approaches and models which are unpopular but which were born out of the national experience and ethos.

Each five-year plan has generated new thinking and opened new vistas for re-articulating the planning process. From the high-pitched centralised planning of the 1950s we have now come to local-level planning. In one sense we have started integrating planning 'from above' with planning 'from below' although we are still unsure of how to go about it. Experiences of the last three decades have made India realize that Western models of growth do not adequately answer her development problems, for the obvious reason that the experience of the nineteenth century cannot be transplanted into the twentieth century, nor can 'Western' Europe's experiences be transplanted to 'Eastern' India.

In its drawn-out search for an indigenous model of development, India looks back now once again to Gandhi, the man who led the country to freedom and who offered a new model of development for humanity as a whole. It is partly to recapture the Gandhian image of development that India appears to have stumbled onto the idea of 'bottom-up' planning, with blocks of about 100 villages as the basic unit of planning.

The current debate on planning, its political overtones notwithstanding, centres partly on the conflict between the bottom-up approach of Gandhi and the centre-down approach of Nehru. It is rightly argued that centre-down organization never allowed planning to reach the grass-roots level. Instead it strengthened semi-feudal class relations, placed the commanding heights of the economy in fewer hands, further strengthened structural rigidities—spatial, sectoral, and others—and kept millions of people out of the modern production system—and made the rich richer, the poor poorer.

This is not to say that centralized planning achieved nothing. It put the national economy on the rails in macro-terms: agricultural production increased by 250 per cent and industrial production by about 500 per cent. Improvements in the tertiary subsectors are equally significant. But all these achievements fade into insignificance or even irrelevance when one observes the almost equal increase in the poverty of the masses. India now looks to Gandhi not so much to reverse this trend, as to find a model which can help eliminate poverty without adversely affecting the growth of the economy.

Block-level planning as presently conceived is rooted, as the Janata leaders say, in the Gandhian approach to development. For Gandhi the village was the basic unit of development. He wanted each village to be self-sufficient in many of its needs, including food, clothing, and energy—not that he did not envisage any inter-settlement linkages, not even that he wanted to abolish urban centres; what Gandhi wanted was a new society consisting of small communities organically linked with each other but undivorced from nature and work. His whole concept of village development is inextricably linked with the national economic and social policy and the grand design of a new humanity that he thought the world needed 'for eternity'.

My idea of village *Swaraj* [independence] is that it is a complete republic, independent of its neighbours for its vital wants, and yet interdependent for many things in which dependence is a necessity. Thus every village's first concern will be to grow its own food crops and cotton for its cloth. It should have a reserve for its cattle, recreation and playground for adults and children. Then, if there is more land available it will grow useful money crops, thus excluding Ganja, Tobacco, Opium and the like. The village will maintain a village theatre, school and public hall. It will have its own water-works ensuring clean water supply. This can be done through the controlled wells and tanks. Education will be compulsory up to the final basic course. As far as possible every activity will be conducted on a co-operative basis. . . . The government of the village will be conducted by the *Panchayat* of five persons annually elected by the adult villagers male and female, possessing minimum prescribed qualifications. I have not examined here the question of relations with the neighbouring villages and the centre, if any. My purpose is to present an outline of village government. (Gandhi, 1942, p. 238).

The Gandhian concept of decentralization is clearly far more comprehensive than mere devolution of authority from the centre. Gandhi wanted local communities to take full responsibility for development, using local and regional resources to the fullest extent and providing opportunities for each individual to develop his personality to the fullest.

Unfortunately Gandhi has been thoroughly misunderstood, in India especially and one might even say, intentionally misinterpreted by vested interests including his own disciples. His 'package' was torn to pieces and discredited as 'romantic', irrelevant to the world of the twentieth century, and impracticable. Gandhi once said:

> According to me the economic constitution of India and for that matter that of the whole world should be such that no one under it should suffer for want of food or clothing. In other words, everyone should be able to get sufficient work to enable him to make two ends meet. And this ideal can be realised only if the means of production of the elementary necessities remain in control of the masses. (Planning Commission, 1973, p. 6.)

Gandhi was not against industrialization *per se*. However he wanted the size of industries to be within manageable limits in human scale.

> As a moderately intelligent man, I know that men cannot live without industry. Therefore, I cannot be opposed to industrialization. But I have a great concern in introducing machine industry. The machine produces much too fast and brings with it a sort of economic system that I cannot grasp. I do not want to accept something when I see that its evil effects outweigh whatever good it brings with it. I want the dumb millions in our land to be healthy and happy. I want them to grow spiritually. As yet, for this purpose we do not need the machine. There are too many idle hands. But as we grow in understanding, if we feel the need for machines we certainly have them once we have shaped our lives on *Ahimsa* [loosely translated, 'complete non-violence'] we shall know how to control the machine. (Gandhi, 1952, p. 348.)

Gandhi was not against industries: he was against industrialism.

> I have no quarrel with steamship and telegram... [but] we must not suffer exploitation for the sake of steamship and telegraph. They are in no way indispensable for the permanent welfare of the human race. Now that we know the use of steam and electricity, we should be able to use them on due occasion and often we have learnt to avoid industrialism. Our concern is therefore to destroy industrialism at any cost. The present distress is unsufferable. Pauperism must go. But industrialism is no remedy. (Gandhi, 1926, p. 348.)

Nor was Gandhi against mass production.

> Mass production, certainly, but not based on force.... If you multiply individual production to millions of times, would it not give you mass production on a tremendous scale?... When production and consumption thus become localized, the temptation to speed up production indefinitely and at any price disappears.... Mass production, then, at least where the vital necessities are concerned, will disappear (Gandhi, 1934, p. 301.)

The supreme consideration [according to Gandhi] is man. The machine should not tend to make atrophied the limbs of man. . . . Take the case of Singer sewing machine. It is one of the few useful things ever invented, there is a romance about the device itself. Singer saw his wife labouring over the tedious process of sewing and scanning with her own hand, and simply out of his love for her devised the sewing machine in order to save her from unnecessary labour. (1924, p. 378.)

As we read through his other writings, we realize that the Gandhian model is an integrated whole and includes socioeconomic reforms of a far-reaching nature. Gandhi pleaded for equality of wages:

A lawyer's work has the same value as a labourer's inasmuch as all have the same right of earning their livelihood from their work (The Story of my Experiments with Truth, 1940). To bring this [equal distribution] ideal into being the entire social order has got to be reconstructed. A society based on non-violence cannot nurture any other ideal. (Bose, 1972, p. 50.)

He also advocated the concept of Trusteeship of poverty:

The rich man will be left in possession of his wealth, of which he will use what he reasonably requires for his personal needs and will act as a trustee for the remainder to be used for the society. In this argument, honesty on the part of the trustee is assumed. . . . Personally I do not believe in inherited riches. The well-to-do should educate and bring up their children so that they may learn how to be independent. . . . If the trusteeship idea catches, philanthropy as we know it, will disappear. . . . If the people did not behave like trustees, the state will, as a matter of fact, take away those things. (Bose, 1972, p. 51.)

The tragedy is that successive governments in India, even though swearing by Gandhi day in and day out, have accepted only the less significant elements of the Gandhian model—elements which, when converted into a plan of action, would create no other impacts than the ones we have seen during the past three decades. There is recognition that centre-down planning alone will not remove poverty, economic stagnation, or social instabilities. It is also recognized that the bottom-up process of planning must be launched now. But the political will to meet these imperatives is conspicuous by its absence.

REFERENCES

Agarwala, S. N. (1944) The Gandhian Plan (Bombay: Padma Publications).

Bannerjee, B. N. G. D. Parikh and V. M. Tarkunde (1944) People's Plan for the Economic Development of India. Indian Federation of Labour. (Delhi: Mukherjee Publications).

Bhalla, G. S. (1979) Spatial patterns of levels and growth of agricultural output in India. In R. P. Misra and K. V. Sundaram (eds.), Rural Area Development (New Delhi: Sterling).

Bose, N. K. (1968) Selections from Gandhi (Ahmedabad: Navjivan Publishing House), No. 268.

Bose, N. K. (1972) Studies in Gandhism (Ahmedabad: Navjivan Publishing House). 'Planning for the rich,' (1978) Economic and Political Weekly, August 1978 (Bombay).

Gandhi, M. K. (1924) Young India, 30 November (Delhi).

Gandhi, M. K. (1926) *Young India*, 7 October (Delhi).
Gandhi, M. K. (1934) *Harijan*, 2 November (Delhi).
Gandhi, M. K. (1942) *Harijan*, 26 July (Delhi).
Gandhi, M. K. (1952) *Rebuilding our Villages* (Ahmedabad: Navjivan Publishing House).
Gruchman, B. (1978) *State of Art of the Methods of Planning for Regional Development. An International Survey Report*, UNCRD (Nagoya, Japan).
Hanson, A. H. (1966) *The Process of Planning* (Oxford: Oxford University Press).
Hanumantha Rao, C. H. (1978) Urban versus rural or rich versus poor (Review article on M. Lipton, *Why Poor People Stay Poor – Urban Bias in World Development*), *Economic and Political Weekly*, 7 October.
Kurien, C. T. (1977) *Poverty, Planning and Social Transformation* (Delhi: Allied).
Mehta, A. (1978) Panchayat Raj, *The Hindu* (Madras), 31 August.
Misra, R. P., K. V. Sundaram and V. L. S. Prakasa Rao (1974) *Regional Development Planning in India* (Delhi: Vikas).
Misra, R. P. and K. V. Sundaram (1979) *Rural Area Development* (New Delhi: Sterling).
Planning Commission, Government of India (1978) *Report of the Working Group on Block Level Planning* (Varanasi: Sarva Seva Sangh Prakasan).
Planning Commission, Government of India (1973) *Challenge of Poverty and the Gandhian Answer* (Varanasi: Sarva Seva Sangh Prakasan).
Shah, K. T. (1947) (ed.), *Report of the National Planning Committee* (Bombay: Vora and Co.).
Singh, T. (1974) *India's Development Experience* (London: Macmillan).
Sundaram, K. V. (1978) Some recent trends in regional development planning in India. In Misra, R. P. *et al.* (eds.), *Regional Planning and National Development* (New Delhi: Vikas).
Sundaram, K. V. (1977) *Urban and Regional Planning in India* (New Delhi: Vikas).
Sundaram, K. V. (n.d.) *Territorial Division of Labour for Optimal Resource Use—The Indian Experience* (mimeo).
Thakurdan, P. *et al.* (1944) *A Plan of Economic Development for India* (Bombay).
Visvesvaraya, M. (1936) *Planned Economy for India* (Bangalore: Bangalore Press).

b. Africa

Chapter 11

Nigeria: The Need to Modify Centre-Down Development Planning

Michael Olanrewaju Filani

INTRODUCTION

One of the major characteristics of developing countries is the failure of their planning efforts, past and present, to fulfil stated objectives and bring about desired development. This failure has necessitated an increasing demand for a critical reappraisal of previous plans, in order to direct future efforts to more meaningful achievements. The increasing demand for appraisal can be ascribed to two main factors: First, there is a growing realization of the increasing disparity in socioeconomic activities between various geographical regions that make up these countries, and the income inequality amongst their people. Secondly, there is a general awareness that economic growth or the increase of national income per capita does not automatically constitute development in the sense of improving living conditions for the poorest groups in the society.

In Nigeria, oil exploitation and the attendant boom provide a good illustrative example. Under the impact of oil explorations, the Nigerian economy has grown in terms of accelerated gross national product, but it has not developed. The majority of the people are generally, not better off in the provision of and accessibility to the basic necessities of life including food, water, shelter, and employment. Yet according to Seers (1972), the provision of these basic necessities, and equality of access to them, constitute the only true universal in terms of development.

In Nigeria, as is the case with many developing countries, past development plans have been directed towards improving the country's performance on one or more economic indicators. Growth in aggregate product (GDP) has up until now been used as a proxy for development. However, the lumpiness in the GDP does not reflect the sharp spatial disparities that characterize the country's economic space. The use of the GDP as a development indicator has only given the erroneous impression of equal distribution of the fruits of development in spatial terms and amongst the different social strata in the country.

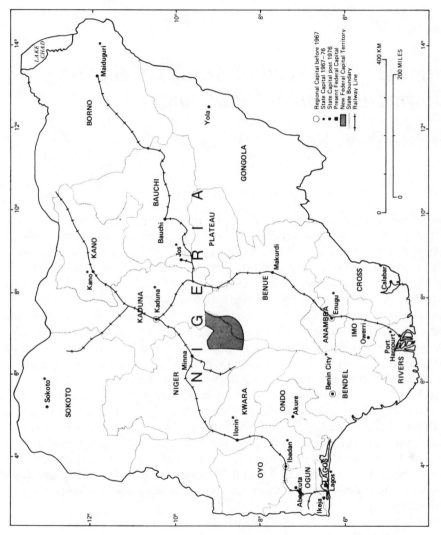

Figure 11.1 The States of Nigeria

In several respects, the Nigerian case typifies Lo and Salih's Model 1, described as resource-rich, open, dependent economy in an earlier chapter of this book (Chapter 5). Nigeria is richly endowed with agricultural and mineral resources, some of which are still relatively untapped. As a former British colony, its development planning strategies have been essentially capitalist market-oriented, concentrating mainly on the production of agricultural crops for export and the import-substitution approach to industrial development. While the technological know-how remains relatively low, the country's dependency bonds to the advanced industrial West have been further tightened and its peripheral position in the capitalist economic order consolidated. In the process, rural–urban migration has been heightened, leaving a vast majority of the country's population virtually poor.

The basic theme in this chapter, therefore, is that there is need for a new orientation in the country's planning ideology and strategies in order to ameliorate the present deficiencies, thereby improving the living standards of the country's poorest majority. After examining the various facets and consequences of past development planning strategies, the author goes on to suggest various ways through which past failures can be remedied.

PLANNING EXPERIENCE AND STRATEGIES IN NIGERIA

The Colonial Era

The history of development planning in Nigeria dates back to 1946 when as a result of the Colonial Development and Welfare Act of 1945, the Colonial Administration was requested to submit a ten-year development plan to guide the allocation of colonial development and welfare funds. This resulted in the preparation of the first development plan for the country, entitled 'A Ten-Year Plan of Development and Welfare for Nigeria, 1946' (Nigeria, 1946). This plan envisaged a capital expenditure programme of about ₦ 110 million for various activities over the ten-year period 1946–56. Out of this amount only ₦ 46 million was to come from the Colonial Development Funds.

The plan contained a list of projects considered desirable for the smooth functioning of the colonial administration and the projects were neither co-ordinated nor related to any overall economic goal (Central Planning Office, 1978). The main emphasis was on the development of a limited range of export crops needed to feed British industries and for the building-up of transport and communication systems to facilitate both the transportation of these crops to the sea ports of Lagos and Port Harcourt, and the easy movement of colonial administrators.

The main strategy for industrial development then was that of export promotion involving only first-stage processing of raw materials. This was done mainly to minimize transport costs and increase the over-all profitability of British industries. Industrial activity was characterized by saw-milling sheet-rubber manufacture, tin-ore mining, cotton ginning, groundnut shelling and groundnut oil extraction, palm fruit processing with hand-presses, and later on the pioneer oil-extraction industries. Before 1945, the bulk of the finishing-type industrial activity in the country featured small and medium-scale manufacturing activities such as printing and baking required to meet the immediate needs of the colonial administrators (Filani and Onyemelukwe, 1977). The result was that of the 47 industrial establishments reported in the country's industrial directory in 1945, 21 were purely of the processing type, 15 of the remaining 26 finishing establishments were either print-shops or bakeries (see Table 11.1).

In the early 1950s and towards the end of the colonial era, emphasis shifted from export-promotion to import-substitution for industrial development. By this strategy, virtually every input into the industrial process was imported. These inputs include the machinery, equipment, capital, technical know-how, and very often the semi-processed materials needed to produce the finished goods. In essence, all that was done in Nigeria was to put the 'finishing touches' to goods already produced elsewhere, or as Lewis described it, the strategy was no more than 'gathering other people's furniture' (Lewis, 1971). The change in orientation from export-promotion to import-substitution of manufactured

Table 11.1   *Industrial establishments in Nigeria by type in 1945*

Industrial type*	Total number †	Product type	
		Semi-processed	Finished
Meat processing	1	1	—
Saw-milling	1	1	—
Fruit processing	1	—	1
Vegetable and animal oil processing	3	2	1
Bakery	6	—	6
Food processing	1	1	—
Lime and block	1	—	1
Soft Drinks	1	—	1
Cotton ginning	10	10	—
Wearing apparel	1	—	1
Tannery	3	3	—
Drugs and medicine	1	—	1
Furniture and fixtures	2	—	2
Printing and publishing	9	—	9
Soap making	1	—	1
Coal products	1	—	1
Rubber processing	3	3	—
Wood carving	1	—	1
Total	47	21	26

Source: Federal Ministry of Industries (1971, *Industrial Directory* (Lagos).
* Excluding mining and service establishments.
† Excluding establishments whose start-up years and/or size (employment) were not indicated.

consumer goods was very rapid: up to 1958 about 50 per cent of the contribution of the manufacturing sector came from the semi-processing of raw materials for export; yet barely eight years after political independence, these export commodities accounted for less than 25 per cent of the value of total output in manufacturing (Mabogunje, 1977a). The remaining 75 per cent was made up of output from such import-substituting industries as textiles, shoes, beer, soft drinks, soap and detergents, cement, building construction materials, and metal products.

The transport systems developed during the colonial era helped to foster colonial administrative and economic control of the country. In this regard the role of rail transport was of particular significance. The first rail-line started from Lagos in 1895 and reached Kano in 1912, thus linking the south-western cocoa-producing areas with the north-central cotton- and groundnut-producing areas of the country. In 1913 the eastern line started from Port Harcourt which was founded as a port to export the coal of the Udi Hills as well as the tin and columbite from the Jos Plateau areas. Ten years later this line reached Kaduna where it joined the western main line. By 1927 Jos was connected with Kaduna and a branch line was constructed from Kano to Nguru. In two years, when another branch line from Zaria to Kaura Namoda was completed, a new spatial integration had been imposed on the country.

The effect of this new spatial integration on production and export trade was spectacular. Since the rail-lines cut across all the country's major ecological zones, many different parts, especially those areas producing main export commodities, were brought within easy economic reach of Europe. The coast-to-interior transport route system fostered the development of a highly monetized exchange economy and stimulated cash-crop production. As Harrison-Church pointed out, in the 25 years before the Second World War the railways in Nigeria allowed and partly caused the export of groundnuts to increase 200 times, and that of cocoa 30 times (Harrison-Church, 1949). Groundnut exports from the Kano region rose from less than 2000 tons in 1911, a year before the arrival of the railway, to about 20,000 tons in 1913, an increase of about 1000 per cent, a year after the railway reached Kano (Mabogunje, 1968).

With respect to roads, the colonial administration's stated policy was to develop them as feeders to the railways, Consequently, road development was so slow and so poor that by 1953 there were only 44,800 kilometres of roads out of which 17 per cent were described as federal roads and 20 per cent as regional. The remaining 63 per cent were local roads and tracks, mostly non-motorable, especially during the rainy seasons. The general lack of concern for road development was also manifested in the fact that at about the same time there were fewer than 22,000 vehicles in the country (Ayeni, 1978).

The institutional machinery for planning during the colonial era was a simple one geared towards the concentration of power at the centre, and mainly in the hands of the few colonial officials. At the apex of the machinery was the Legislative Council (later the Parliament) whose responsibility was to approve the plans formulated and prepared under the general direction of a Central Development

Board made up of a small hardcore of senior colonial government officials. Below these were Area Development Committees set up in each of the then three groups of provinces to advise the Central Development Board (Central Planning Office, 1978). These committees were in turn assisted by advisory bodies set up in each province and each division.

The main strategies of development during the colonial era in terms of export promotion, raw material valorization, and later import-substitution of manufactured consumer goods, together with transport development, emphasize the deliberate policy of exploitation of the country's natural resources at minimum cost by the colonial administration. Thus, the imposition of a colonial spatial order on the existing spatial system represented 'perhaps the greatest disequilibrating mechanisms' that have shaped the spatial patterns of the country's subsequent economic development (Ayeni, 1978). For one thing, the capital-cum port cities of the era 'had the largest and most skilled populations, the best infrastructures, the biggest concentration of the ruling elite and the highest potential for local entrepreneurship' (Daly, 1977, p. 4).

As industrialization in the cities progressed, the spatial contrasts of the dual economies became more prominent. This is particularly so because, as will be shown later, the agricultural population received little or no benefits in the form of Hirschman's 'trickle down effects'. Instead, contrasts between rural incomes and level of services and those of the cities were magnified. Also the spread of education drew many young and able-bodied citizens off the land to urban centres 'where unemployment, underemployment and urban poverty exploded' (Daly, 1977, p. 4)

Post-Independence Experience

Constitutional developments in the country by the mid-1950s brought the 1946–56 Colonial Plan to a premature end. A federal system of government was adopted in 1955 and each of the then three regional governments (East, West, and North) drew up separate development plans. Thus, the 1955–60 Economic Development Plan was launched and its life-span was later extended to 1962. These plans were, in the strict sense, no plan at all in that the schemes proposed were no more than an expansion of existing normal departmental activities (Adeniyi, 1978). Each of the programmes of the various governments had different goals and were lacking in coordination and over-all national or regional priorities. There was an increasing emphasis on sectoral growth of the economy with little or no regard for the spatial consequences of development projects.

Before the expiration of the 1955–62 plan some institutional changes had to be made to reflect the country's sociopolitical development and to facilitate the planning process. The whole country became politically independent in 1960 although the three regional governments had enjoyed self-government prior to this period. A Federal Ministry of Economic Development was created with corresponding ministries in the then Eastern and Western Regions (Central

Planning Office, 1978). In the Northern Region planning was the responsibility of the Ministry of Finance. A Joint Planning Committee was set up at the federal level to coordinate both the Federal and Regional Governments' planning efforts and to formulate rational goals and objectives of development for the country. The Joint Planning Committee was, in turn, responsible to the National Economic Council whose membership consisted of the Prime Minister, the Regional Ministers of Trade, Finance, Economic Planning, Agriculture, and Natural Resources. These bodies provided a framework for generating ideas from the people through the political forum to the officials. It was also through these bodies that plans were supposed to be coordinated and integrated among the divergent regions in order to achieve national goals and objectives.

Since independence, Nigeria has adopted the Keynesian type macroeconomic planning strategy leading to the production of periodic five-year development plans (Mabogunje, 1977a). As stated by Mabogunje, the main characteristic of this strategy is the reliance on 'monetary and fiscal policies to generate the appropriate stream of total spending so as to assure steady growth with full employment and no inflation (Mabogunje, 1977a p. 19). Thus so far three development plans have been drawn up in the country; the First National Development Plan 1962–68, the Second National Development Plan 1970–74 and the current Third National Development Plan, 1975–80.

The First National Development Plan (NDP) aimed at achieving the following main objectives (Federal Republic of Nigeria, 1962):

(a) to achieve a compound growth rate of 4 per cent per annum in the Gross Domestic Product;
(b) to save and invest 15 per cent of the GDP and at the same time raise per-capita consumption by about 1 per cent per annum;
(c) to achieve self-sustaining growth not later than by the end of the Third and Fourth National Development Plans;
(d) to develop as rapidly as possible opportunities in education, health, and employment and to improve access to these opportunities for all citizens; and
(e) to achieve a modernised economy consistent with the democratic, political, and social aspirations of the people, including the achievement of more equitable distribution of income both among people and among different regions.

The plan, launched in June 1962, had a structure similar to the 1955–62 programme in the sense that it embodied three separate Regional Governments' programmes with considerable overlapping of projects. Further political developments leading to the creation of the Mid-Western Region from the West in 1963 again warranted the drawing-up of a separate plan for the new region covering the remaining four-year period. However, the acceptance of the above common objectives and targets constituted a major difference from the previous plan, since these gave the First National Plan a semblance of national character.

Total planned expenditure amounted to ₦2366 million out of which 50 per cent was to be financed by external loans. The main emphasis was again on sectoral planning and sectoral growth of the economy. Although the desire to have equitable distribution of the fruits of development was expressed in the plan, subsequent investments portrayed little or no appreciation of the spatial effects and regional disparities in the country's developmental thrusts.

Investment in industry, particularly in import-substitution goods and in the manufacture of raw materials for export, was again encouraged. The lack of success there is revealed by the fact that in 1969, one year after the terminal period of the plan, about 75 per cent of the country's imports were still made up of manufactured goods, machinery, transport equipment and chemicals (Daly, 1978). Also the total actual disbursement throughout the six-year plan period amounted to ₦1073 million or only 45 per cent of the original planned expenditure (Central Planning Office, 1978).

Three main reasons had been adduced for the failure of the plan to achieve stated objectives. Firstly, the over-dependence on external finance for major projects which did not fully materialize. (For example, in the first two years of plan implementation only 12 per cent of total disbursement came from external sources which were originally supposed to cater for half of total investments.) Secondly, the political crisis and instability of 1964–66 and the subsequent civil war which lasted from 1967 to 1970 disrupted the smooth implementation of the programmes. Lastly, there was an acute shortage of executive capacity not only for implementing projects but also for supervising and monitoring progress made in them.

These problems notwithstanding, some measure of success was achieved. Mining became the fastest-growing sector with its share of the GDP increasing from 1.9 per cent (1962–63) to about 3.4 per cent (1966–67). Manufacturing also increased its share from 5.3 per cent to 7 per cent during the same period. Major projects completed during the plan period include the Port Harcourt oil refinery, the Nigerian Security printing and minting factory, the multi-million Naira Niger Dam at Kainji, the Bacita sugar mill, the Niger Bridge, the Eko Bridge, various port extensions and some trunk roads all over the country. Unfortunately, however, the iron and steel project, which was planned to be the cornerstone for national economic developmental take-off, never got beyond the drawing board.

Most of the projects completed between 1962 and 1968 were again urban-based and the rural sector was used to support a process of industrialization which had limited effects on rural inhabitants. Emphasis was placed on import-substitution industries such as textiles, beer, and soft drinks, but the 'trickle down effects' of these industries were so limited that by 1969 many arguments were raised in favour of intermediate and capital goods industries (Asiodu, 1969).

Although the first NDP was scheduled to end in March 1968 the civil war made it virtually impossible for the Second National Development Plan to be prepared until 1970. Political developments again necessitated further restructuring of the institutional machinery for planning during the second plan period. The most significant of these developments was the break-up of the country into 12 states compared with the former four regions of the 1962–68 plan period. There was therefore need for greater coordination of planning efforts 'in order to achieve a more cohesive and integrated national plan than hitherto' (Central Planning Office, 1978, p. 10). The Joint Planning Committee was changed to a Joint Planning Board with an enlarged membership. A National Economic Advisory Council (NEAC) was established, with its membership drawn largely from the private sector and including representatives from the country's chambers of commerce, the trade unions and the universities. The main function of the Council was to advise the Federal Government on planning economic and such other related matters that might help in harmonizing the public and private sectors of the economy. The former Economic Planning Unit in the Ministry of Economic Development was expanded and reorganised into a Central Planning Office (CPO) staffed with a core of professional planners. CPO functions as a planning agency for the Federal Government, coordinating the federal and state government plans. A Conference of Commissioners responsible for economic development in the 12 states and at the federal level was also created to review periodically economic matters of common interest to the country.

All these bodies participated in the formulation and implementation of the 1970–74 Second National Development Plan which was launched in 1970. The plan itself differed from its predecessors in many respects. It was the first to be formulated and prepared wholly by Nigerians. It was bigger in financial size and more diversified in its project composition. It also represented the first truly national and fully integrated plan in which the country's economy was viewed as a single organic unit. Above all, the plan contained five national objectives representing the first attempt to express the social and economic philosophy underlying the country's planning strategies. These five principal objectives were meant to establish Nigeria firmly as:

   (i)   a united, strong and self-reliant nation;
  (ii)   a great and dynamic economy;
 (iii)   a just and egalitarian society;
 (iv)   a land of bright and full opportunities for all citizens; and
   v)   a free and democratic society.

(Federal Republic of Nigeria, 1970, p. 32.)

These objectives were to be viewed as the ultimate or long-term goals of the country's development efforts in the bid to reduce 'inequalities in interpersonal incomes and promoting balanced development among the various communities in the different geographical areas in the country' (Federal Republic of Nigeria, 1970, p. 33).

The Plan envisaged a total capital expenditure of ₦3.2 billion (billion $= 10^9$) with the highest order of priorities given to agriculture, industry, transportation, and manpower development. In the second order of priority were such social services and utilities as electricity, communications, and water supplies. An average growth rate of 6.6 per cent per annum in the GDP was set as the minimum target.

Plan performance in terms of investments was much better than during the first plan period. Total actual capital investment amounted to ₦2.2 billion or about 66 per cent of the projected outlay. The GDP recorded an annual growth rate of 8.2 per cent. This was largely due to the performance of the mining sector which contributed 45 per cent of the GDP at the end of the plan period, constituting about three times the projected figure. Petroleum earnings, which by 1975 accounted for 95 per cent of the country's total export earnings, fostered industrialization between 1970 and 1974. However, lack of success of the plan, especially in the rural sector, is revealed by the fact that agricultural productivity not only fell in absolute terms, its share of the GDP declined from 36 per cent (1970–71) to 23 per cent in 1974–75. Also by 1975, importation of manufactured goods, machinery, transport equipment, and chemicals still constituted about 86 per cent of the total import bill (Daly, 1978). The low priority given to provision of job opportunities in the rural areas escalated rural–urban migration during the plan period.

The 1975–80 Third National Development Plan was launched in March 1975. The plan, which had an initial investment proposal of ₦33.8 billion, was in 1976 revised up to ₦43.3 billion after the creation of seven additional states making a total of 19 states in the country. The third plan represents the most ambitious one ever undertaken in Nigeria. In monetary terms, it is about 19 times the size of the first plan and 13 times that of the second. Even though sectoral planning still dominates the plan, it differs from its predecessors because of its emphasis on 'balanced development'. It states in part that 'a situation where some parts of the country are experiencing rapid economic growth while other parts are lagging behind can no longer be tolerated' (Federal Republic of Nigeria, 1975, 30). Consequently the size and distribution of both federal and state government's programmes are so structured as to 'generate growth simultaneously in all geographical areas of the country'. The plan also realizes that agriculture was, and would continue to be, the principal employer of Nigerians and that any improvement in the quality of rural life depends on the development and modernization of the agricultural sector. Thus allocation of capital expenditure to agriculture which was ₦215 million for 1970–74 rose to ₦1400 million for the 1975–80 plan period. In general, a successful implementation of the plan is expected to result in significant increase in per capita income and consequently in the standard of living of the populace.

Consequences of Past Planning Strategies

The various strategies of development adopted in Nigeria from the colonial era up to now have been categorized into three by Mabogunje. According to him the strategies are:

(a) concentration on an export agricultural production so as to generate enough foreign exchange reserves to purchase much needed imports;

(b) embarking on an import-substitution industrial strategy so as to incubate local processes which would eventually bring about basic socio-economic transformation; and

(c) attempting a Keynesian type macro-economic planning whereby public investment is used in a way to maximise capital output ratio in the economy as a whole (Mabogunje, 1977a).

Mabogunje went on to stress that while these three strategies have not necessarily been adopted independently of one another, there were periods when one of them was given greater prominence than others. From the above discussion of Nigeria's planning experience, it is clear that the first two strategies predominated during the colonial era while the post-independence period has been characterized by the third strategy, although also with strong emphasis on the earlier ones.

With these strategies of development Nigeria is today faced with the paradox that as the country has grown wealthier with respect to phenomenal increases in its GNP, the plight of the masses has become more precarious, while a small minority of its population have become extremely rich. In other words all that has been achieved so far is a situation in which social and spatial inequalities of income and access to opportunities have been intensified.

Such spatial inequalities of access to opportunities between urban centres and rural areas can be illustrated with the situation in Kano State. Kano State is made up of the area which used to be known as Kano Province, before the creation of states in 1967. The state had a population of 5.8 million in 1963 in an area of 43,053 square kilometres, with a density of about 135 persons per square kilometre. It is the major groundnut producer in the country and the richest amongst the six northern states. Yet it is one of the most backward in terms of educational advancement in the country. One major factor responsible for this backwardness is the extreme concentration of virtually all urban facilities in its capital city of Kano (see Table 11.2). Although historically Kano City had been very important, its domineering position was further reinforced in the colonial period by a transportation system which focused all routes on it, leaving the district headquarters unconnected with each other except through Kano. Thus over the years Kano City has developed as a primate city within its geographic region. The city's population of about 300,000 (1963) constituted only 5.1 per cent of the state's total, yet almost all the state's social and economic activities are concentrated in it. Table 11.2 summarizes the position as of 1971. The result of such concentration, as McDonnell (1964) emphasized, is that 'metropolitan Kano has swallowed many of the older functions of the district towns, as population has grown, as markets have expanded under the influence of the railway, and as its administrative, intellectual, and educational centres have become more specialized (McDonnell, 1964, p. 369). The concentration of activities in Kano city has had not only a dampening effect on the growth of

Table 11.2 *Contrasts between rural and metropolitian development, Kano State, 1971*

Item	Metropolitan	Other areas
1. Area	c. 30 sq. miles	+ 16,000 sq. miles
2. Population	c. 300,000	+ 5.3 million
3. Industries (operating)	+ 71 (capital investment ₦15 million)	2 (capital investment ₦300,000)
4. Industries (proposed)	79 (capital investment ₦30 million)	Nil
5. Wholesale establishments/ distribution forms	105	8
6. Departmental stores	7	—
7. Licensed buying agents	47	16
8. Insurance companies	11	—
9. Banks	11	1
10. Post office/Sub-post office	4	1
11. Post-primary institutions	16 + 2 post-secondary	11
12. Hospitals	4 including 1 specialist + various private clinics	3
13. Hotels	39 + 1 international	20
14. Electricity	Available throughout	2 centres
15. Pipeborne water	Available c. 2.5 million (gallons daily)	4 piped semi-urban 100,000 gallons
16. High courts	2	0
17. Cinema houses	7	2

Sources: McDonell, G. (1964) 'The dynamics of geographic change: the case of Kano, *Annals of the Association of American Geographers*, 54, 370 and Kumuyi J. (1973) *The Spatial Interaction* . . ., p. 124.

smaller urban centres around it, but has also aggravated the increasing migration of people to the city from its rural hinterland.

The strategies have also helped further in tightening the country's dependency bonds to the technologically advanced industrialized West and in consolidating the country's peripheral position in the capitalist economic order. The capitalist production structures, built up from the colonial era to the present, are controlled and regulated from Western industrial countries. This is done through international trade agreements which continue to keep Nigeria as a field of profitable exploitation by the large multinational corporations.

During the colonial era, concentration on the production of a limited range of export crops largely to feed British industries represented a continuation of the role assigned to peripheral regions by international capitalism (Mabogunje, 1977a; Hinderink and Sterkenburg, 1978). The capitalist production structure so built up was controlled and regulated from Western industrial countries directly through export production, and indirectly through price and monetary policies. Thus the colonial administration's direct investments in the agricultural sector and other indirect investments on infrastructure and social services were devoted to increasing output of export agricultural commodities.

Notably, the quantity of agricultural production grew by about 30 per cent between 1950 and 1960 while the GDP increased by approximately 4 per cent per annum. However, the increase in the value of export crops then was due more to fortuitous circumstances than to real changes over time in the volume of output (Mabogunje, 1977b). The growth in the value of the country's export crops was made possible especially because of the great demand and therefore rising prices for the crops after the Second World War. Using cocoa as an example, Oluwasanmi (1966) showed that in 1940 when the export of raw cocoa beans totalled 89,000 tons, the realized value was £1.5 million (₦3.00 million). In 1950, however, when export of cocoa rose by only 12.3 per cent to 100,000 tons, the realized value jumped to £18.9 million (₦37.8 million), amounting to a rise in value of approximately 916 per cent within a decade.

The phenomenal increase in the value of cocoa exports reflected in the growth of national income, but did not necessarily imply development in the rural areas or in standard of living of the rural producers of the crop. This was more so because Nigerian farmers received on the average much less than the world market actually paid for their agricultural products. Even though Marketing Boards were set up in 1948, among other things to encourage high productivity as well as to enhance the level of income for Nigerian farmers, it soon b - came obvious that these Boards were mainly instruments of economic exploitation (Filani and Onyemelukwe, 1978). As shown in Table 11.3, cocoa farmers received on the average 49.2 per cent of world prices for the period 1948–52

Table 11.3 *Percentages of actual world prices received by Nigerian farmers for major agricultural products, 1948–57*

Year	Cocoa	Palm oil	Palm kernel	Groundnut
1948	32.3	—	—	—
1949	98.4	55.8	100.0	39.6
1950	50.0	59.7	38.2	32.3
1951	46.9	59.1	40.0	22.3
1952	68.5	72.6	65.5	45.7
Mean 1948–52	49.2	61.8	60.9	35.1
1953	70.8	93.5	53.1	43.3
1954	43.3	71.4	64.1	45.5
1955	67.5	52.4	59.6	50.0
1956	96.2	46.2	58.4	46.1
1957	72.1	46.7	60.7	54.0
Mean 1953–57	70.0	62.0	59.2	47.8

Source: Olatunbosun, D. and Olayide, S.O. (1974) Effects of the Marketing Board on output and income of primary producers, in Onitiri, H.M.A. and Olatunbosun, D. (eds.), *The Marketing Board System* (Ibadan: Nigerian Institute of Social and Economic Research).

while palm oil, palm kernel and groundnut producers got averages of 61.8, 60.9. and 35.1 per cent respectively of world prices during the same period. In the following five-year period (1953–57) the average percentage increased for cocoa (70 per cent), palm oil (62 per cent), and groundnuts (47.8 per cent) while that of palm kernel dropped slightly to 59.2 per cent. In all, only in 1949 and 1956 did cocoa farmers receive more than 90 per cent of world prices paid for their products, while in 1949 palm kernel producers got the equivalent of world prices. In most other years less than two-thirds of the actual world prices went to the producers of commodities (Table 11.3).

Although it is sometimes claimed that differences between world prices and the amount paid to farmers were held back for development projects, these projects were largely concentrated in urban areas, fostering spatial disparities in development at the expense of the rural majority in the country. Today, after almost two decades of political independence, the pattern of investment allocation to development projects tends to reinforce the past predominance of urban centres. For example, of the total planned capital investment of ₦3192 million in the private and public sectors during 1970–74, only ₦526 million or 16.6 per cent was directly or indirectly invested in rural areas. This urban/rural imbalance is portrayed clearly for some selected sectors shown in Table 11.4. For instance, in industry, social welfare and town and country planning, less than 10 per cent of the total planned investment in each was allocated to rural areas. In all other sectors, with the exception of education, the rural areas shared less than 20 per cent of the investment allocations (Table 11.4).

Apart from non-realization of their fair share of world prices, the production

Table 11.4   *Urban/rural investment allocations in selected sectors of the Nigerian economy, 1970–74 Development Plan (all governments)*

Item	Total planned capital investment (₦ million)	Urban-based investment		Rural-based investment	
		₦ million	%	₦ million	%
Industry	172.2	155.4	91.2	16.8	9.8
Electricity	90.6	80.6	89.0	10.0	11.0
Water and sewage	103.4	84.4	71.6	19.0	18.4
Town and country planning	38.2	36.0	94.3	2.2	5.7
Education	277.8	196.8	70.9	81.0	29.1
Health	107.6	90.4	84.0	17.2	16.0
Social welfare	24.0	22.0	91.7	2.0	8.3
Agriculture and others	746.2	615.4	82.5	130.8	17.5
Total	1560.0	1281.0	82.1	279.0	17.9

Sources: Aluko, S. A. (1973) Industry in the rural setting (Ibadan, Proceedings of the 1972 Annual Conference of the Nigerian Economic Society), p. 217; and Federal Republic of Nigeria (1970), *Second National Development Plan* . . ., pp. 137–78 and 212–65.

techniques and organization of farmers remained within the traditional systems of farming with complete ignorance of modern techniques of cultivation and processing of these crops. As Mabogunje rightly observed, 'to concentrate on just output without any real concern about the farmer, about the means, the techniques and the organisation which he brings into this production is to miss the whole point of what development is all about' (Mabogunje, 1977a, p. 14).

The strategy of export-promotion and import-substitution for industrial development is subject to a number of defects. For example the fact that virtually every input into the industrial process is imported, has deprived Nigeria of any serious opportunity to explore, evaluate, and develop its own internal resources. With very little reliance on domestic sources of input, the linkage effects of the strategy remain tenuous (Lewis, 1971). The weak linkage between agriculture and industry thus provides one of the major explanations for the drift from rural to urban areas. Furthermore the import-substitution strategy which dominated development thinking in the 1950s and 1960s has not resulted in any fundamental change in indigenous technology. Rather it has underplayed the issue of 'technological transfer'. Apart from destroying the basis of local technology, this strategy has rendered the population in developing countries technologically impotent to meet the challenges of their own environment (Mabogunje, 1977a).

The spatial implications of the import-substitution strategy have been spectacular. Since many import-substitution industries were established without regard for the principles of comparative advantage, they are 'often high cost, tariff-protected and monopolistic enterprises' (Daly, 1977, p. 2). The heightened dependence on importation emphasized port location for the national space economy. Consequently an increasing proportion of industrial enterprises found port cities, particularly Lagos and Port Harcourt, the realistic least-cost location. The initial advantages enjoyed by these two cities and some regional capitals such as Enugu, Kaduna, and Ibadan and such other transportation nodes as Kano, Benin City, and Ilorin, have been reinforced. Table 11.5 shows the pattern of industrial activities in selected urban centres of the country as calculated from 1970 Industrial Surveys by the Federal Office of statistics. For example, the 12 major cities accounted for about 77 per cent of all industrial establishments in the country, 87 per cent of total industrial employment, 90 per cent of all industrial wages and salaries, and 93 per cent of gross industrial output. Metropolitan Lagos alone accounted for more than half of total industrial wages and salaries, about three-fifths of gross output, 49 per cent of the total industrial employment, and 38 per cent of total industrial establishments. Such a pattern of industrial concentration is repeated at the level of individual states. For example in Kano and Rivers States, the capitals (Kano and Port Harcourt respectively) contained all the industrial establishments surveyed (Adeniyi, 1978).

The implications for such a concentration of industrial activities and other development projects in a few urban centres are quite obvious, in a country where only 19.1 per cent of the total population live in urban areas (defined

Table 11.5   *Percentage share of industrial activities by some major centres in Nigeria, 1970*

Urban centre	No. of establishments	Total employed	Total wages and salaries	Gross output
Lagos Metropolitan Area	38.12	49.19	53.99	58.15
Ibadan	8.10	5.25	4.30	3.59
Kano	7.82	8.77	6.64	10.51
Benin City	4.97	2.23	1.15	0.56
Sapele/Ughelli	3.55	6.93	5.87	2.17
Kaduna	3.27	12.36	11.34	11.35
Jos	2.70	1.58	1.09	3.42
Abeokuta	2.56	1.67	1.86	1.63
Aba	2.27	0.69	0.66	0.82
Port Harcourt	1.28	0.93	2.25	1.84
Ilorin	1.13	0.67	0.50	0.65
Onitsha	1.13	0.25	0.06	0.02
Total	76.90	86.52	89.71	93.08
Rest of the country	23.10	13.48	10.29	6.92
	100.00	100.00	100.00	100.00

Source: Adapted from Adeniyi, E. O. (1978), p. 407

as places with at least 20,000 inhabitants) while the remaining 80.9 per cent live in the rural areas (Mabogunje, 1974). For one thing, the concentration has so increased the tempo of rural–urban migration that the few cities have grown tremendously in size and population. For instance, as shown in Table 11.6, with the exception of Ibadan and Zaria, the major cities in Nigeria grew at an annual rate of more than 7 per cent between 1952 and 1963. In fact, Kaduna and Port Harcourt grew at more than 10 per cent per annum, and Lagos at more than 8 per cent. Metropolitan Lagos, which had about 1.4 million people in 1963, had reached the two-to-three million mark by 1976 (Ayeni, 1976).

The rapid increase in the population of these urban centres has been due mainly to the influx of migrants from rural areas. The concentration of social and economic activities in the cities has generated the expectation of an escape from the drudgery and dreariness of rural life, and accelerated the massive migration from rural areas. This situation has been aggravated by the educational system largely patterned after that of the British, bearing little relationship to the real needs of the country. The system, while emphasizing academic rather than practical aspects of education, has taken children out of the workforce at periods when their help is critical to the farmers. The farmers themselves have become conscious of the limited opportunities which farming offers for attaining the 'good things' of life. Viewing the cities, therefore, as the primary sources of such things, they send their children to school not to become better

Table 11.6   *Population growth rates of major cities in Nigeria between 1952 and 1963*

City	1952	1963	Average annual growth rate (%)
Kaduna	42,647	129,133	10.6
Port Harcourt	59,548	179,563	10.5
Lagos	267,407	665,246	8.6
Kano	131,316	295,432	7.6
Aba	58,251	131,003	7.6
Enugu	63,212	138,457	7.4
Onitsha	77,087	163,032	7.0
Zaria	92,434	166,170	5.5
Ibadan	460,206	635,011	3.0

Source: Green and Milone (1972), p. 7.

farmers but to provide them with an escape route from traditional society (Damachi, 1973). For example, in his survey of the socioeconomic basis of rural–urban migration in Western Nigeria, Olusanya (1969) found that about 89 per cent of the parents interviewed thought it was a good thing for their children to migrate into the cities, while 71 per cent of the children interviewed expected to take up some other occupation than farming. In 1977, a similar survey was carried out in Ibarapa Division of Oyo State, where 415 farmers were asked how many of their children were or would be farmers. Over-all 61 per cent said none, 7 per cent said all, 21 per cent said some, and the rest were not sure (Williams and Filani, 1977). Despite a general feeling that there were likely to be social and economic problems in towns for young migrants, village dwellers continue to be overwhelmingly in favour of young persons moving to towns and cities. This is because there are simply no good opportunities for making a decent livelihood in rural areas.

The Keynesian-type macroeconomic planning adopted since independence suffers also from many limitations. Mabogunje (1977a) pointed out three of the main criticisms of this strategy, basing his arguments on the series of voluminous appendices to Myrdal's (1968) study on *Asian Drama: an Inquiry into the Poverty of Nations:* first, that development plans, being nothing more than fiscal budgets for public investments, avoid planning the whole economy, and to that extent lack operational control. As such little attention is paid to planning taxes and other measures to finance expenditures. Since attention is directed towards investments, the public budget, which constitutes the hardcore of the plans, becomes directed mainly towards accounting for public expenditures. Consequently plan-fulfilment is discussed in terms of how far various authorities have succeeded in spending the money allotted to them, and not in terms of concrete physical achievements, or how far the plans have succeeded in overcoming the real limiting factors in the operation of development policies.

Second, the strategy is inappropriate to the prevailing socioeconomic structure of Nigerian society. The Nigerian economy has a dualistic structure with a significant part still regarded as subsistent or traditional and largely outside the money market. That the development plan strategy relies heavily on monetary and fiscal policies to generate responses in the economy demonstrates its inappropriateness for this type of economy. The problem is further compounded by the fact that the goals and objectives of planning schemes in these five-year periodic plans have generally been set by bureaucrats and professionals, largely urban-based, who have managed to impose their ideas and value-judgments on the rest of the people. In essence, the population being planned for have not been taking any active part in the planning process. The professional planners have always assumed the role of technical experts capable of providing an objective assessment of the needs of the people. Quite often, especially in rural areas, these needs as perceived by planners and policy-makers are at variance with those of rural inhabitants, hence the apparent failure of projects meant for improving their conditions (Filani, 1977; Baldwin, 1957).

The third criticism of the development plan strategy relates to the fact that conditions for development are either assumed to exist in a *ceteris paribus* clause in which other things are held equal, or are assumed to be capable of being created automatically. This according to Myrdal (1968), presupposes that attitudes, institutions, and techniques in an underdeveloped country will automatically adapt to the development process 'under the impact of increased investment and without intentional and direct policy intervention by government' (Mabogunje, 1977a, p. 22).

Apart from Mabogunje's three main criticisms of the Keynesian macroeconomic planning strategy, the Nigerian situation is further compounded by plan indiscipline coupled with the series of changes in the country's political structure and planning machinery during plan implementation periods. The degree of plan indiscipline can be illustrated by the speed at which new projects are introduced mid-stream and the scope of existing ones varied without regard for rules and procedures laid down for effecting such changes. While it may be argued that plans must be flexible enough to accommodate unforeseen contingencies, the basic priorities of a plan document cannot be changed mid-stream without distorting the goals and pattern of resource allocation. For example, in 1977 alone the sum of over ₦ 600 million or about 12 per cent of the Federal Government capital expenditure went to projects which were not originally included in the 1975–80 plan (Central Planning Office, 1978). The distortions caused by these new projects have created many financing difficulties for the current plan. Thus the arbitrariness and ease with which previous plans have been revised and new projects introduced tend to cast some doubts on the commitment of policy-makers to development planning in the country.

In effect, the planning strategy adopted in Nigeria since political independence has so far perpetuated the spatial disparities in development which developed during the colonial administration. Income inequalities among the people have been widened and the large majority of rural dwellers still wallow in abject poverty.

The Need for New Orientation in Planning Strategy

The major weaknesses in Nigeria's development planning efforts from the colonial era to the present can be ascribed to the fact that planning has always been done 'from above'. In essence, the country has adopted the 'centre-down approach' to development planning. This approach has been characterized by:

(a) excessive centralization of the planning machinery in government ministries, providing little or no scope for popular participation;
(b) concentration of development efforts on some selected sectors of the economy without any appropriate linkage among the sectors and with the rest of the economy; and
(c) concentration of socioeconomic activities in the main urban centres hoping that this world generate the 'trickle down' effects that might improve conditions for the virtually neglected rural majority.

If the centre-down approach or planning 'from above' has failed to bring about the desired development, it is now necessary for Nigeria to fashion new development goals and objectives and reorient its planning strategy in a way more consistent with the country's sociocultural and economic structures and natural resource endowments. In doing this, the term 'development' itself must be interpreted to encompass much more than a rapid rate of growth in the GNP. Development must be seen as a process of increasing the quality of life of the great mass of the people through equitable distribution of growth, irrespective of the people's social strata and the territoral unit from which they come. As Mabogunje aptly put it, development 'is essentially a human issue, concerned with mobilizing communities and the whole society to engage in the task of self-improvement with the resources available to it' (Mabogunje, 1977a, p. 26). In order to accomplish this task of mobilization, changes have to occur in the structure of Nigerian society in such a way as to lead 'to a more equitable distribution of the social product over all groups of the population' (Hinderink and Sterkenburg, 1976, p. 14). Such changes would be directed towards the promotion of equal access to the material and non-material resources that are at the country's disposal.

The new orientation in strategy has to lay more emphasis on planning 'from below' rather than 'from above'. Planning 'from below' would concentrate on the society itself and on the vital importance of transforming those basic structural elements which have hitherto impaired the capability of Nigerians to respond effectively to new economic and cultural changes. The main objective of such transformation would then be to create a new society motivated by new value systems and cultural orientation. For this to happen, the people have to be involved both at the decision-making and implementation stages of planning. In effect, what is being called for is a strategy that is characterized by planning for and with the people.

In planning 'from below' attention must be paid to the spatial aspects of investments. Recent developments in the country seem to show an awareness for the need to reorganize the country spatially for purposes of effective planning. In terms of government functions, the creation of 19 states further the principle of decentralization. By bringing government nearer to the people, the creation of states has initiated the process of spreading development to more areas, especially the state capitals and divisional headquarters. The establishment of a new national capital 50 kilometres south of Abuja near the confluence of the rivers Niger and Benue may have significant spatial implications for subsequent economic development in the country (Figure 11.1).

Also the recent comprehensive local government reform in the country has led to the establishment of 320 Local Government Councils. This reform reflects a growing awareness of the importance of organizing society at levels lower than state level as a means of achieving an effective contact with the reality of conditions in which the majority of Nigerians live. There is therefore a need to capitalize on this reform to plan the country's development from the grass-roots. Effective planning can be most successful at this local government level where information on societal needs and available resources is more reliable than at higher levels of government. Each local government area can then become the basic planning unit at which the people can be effectively mobilized for the transformation of rural areas. To achieve the required transformation, there is a crying need for extensive and detailed maps showing resources distribution of each local government area. Such maps will then show the development potential of each area for purposes of national planning (Aboyade, 1968). Through a combination of cartographical and statistical analysis, valuable data would be available on both the human and environmental factors that define locational characteristics of each area. Such data would make local leaders aware of the developmental challenges of their areas, and at the same time enhance the level of rational decision-making at both state and federal levels with respect to location of investment projects.

Local governments should also be involved in the task of fomulating and implementing development plans. The federal government would then provide the leadership, financial, and technical assistance, while the state governments coordinate local government plans and assist in the provision of adequate infrastructure and additional resources to supplement local efforts. Local-interest groups can be easily galvanized and aroused to make contribution in cash and kind in order to improve the quality of life in their areas, especially as they would be the immediate beneficiaries of such local developments.

The spatial implications of such a planning strategy would transform the pattern of growth more effiectively. Emphasis would be placed on rural communities and the reorganization of fragmented farmland into large compact units in order to mechanize production and increase productivity. Investments made in towns and villages will be the kind that would not only help supply the kinds of support systems needed to solve problems of lack of facilities and welfare activities such as schools, hospitals, water supplies and electricity, but would also help to utilize the resources of rural areas. Industrial development would rely mainly on lower levels of technical and capital input and would be related to the products of the agricultural community. Emphasis has to be laid on the interrelationship between agriculture and industry. A basic transformation has to occur in the composition of industrial production with greater emphasis put on the internal production of most of the requisite inputs into agricultural activities. The only way by which the level of agricultural productivity can be raised is to develop industries that can internally produce fertilizers, simple mechanical implements of various kinds, and other inputs necessary in agriculture. It is pertinent to note that Nigeria or indeed any other developing country cannot be said to be on the path to achieving real development until rural problems are solved. This contention, therefore, re-emphasizes the views expressed by Friedmann and Douglass (1978, p. 3). After reviewing the planning experiences of six Asian countries, the authors concluded that 'the urgency of the present situation calls for heightened political commitment to an inward-looking and rural based strategy of national development' in order to overcome 'the contradictions of dualistic dependency' of the various economies.

Finally, given the existing sociopolitical environment in Nigeria, the planning strategy has to be a combination of planning 'from above' and 'from below' or a combination of 'centre-down' and 'bottom-up' approaches to planning. In this, the various local governments would serve as planning and executing agencies at the local levels, the state and federal governments would coordinate their activities, provide financial and technical assistance and general operational directives. In adopting this strategy, a deliberate attempt must be made to increasingly integrate a high degree of physical and regional planning into the country's existing preoccupation with sectoral planning of the economy. It is through such a concerted effort that the desired 'just and egalitarian society' so vividly espoused in the nation's Second National Development Plan can be truly achieved.

REFERENCES

Aboyade, O. (1968) Industrial location and development policy: the Nigerian case, *Nigerian Journal of Economic and Social Studies*, 10, 275–302.

Adeniyi, E. O. (1978) Regional planning. In Oguntoyinbo J. S., Areola, O. and Filani, M. (eds.), *The Geography of Nigerian Development* (Ibadan: Heinemann Educational Books (Nigeria) Ltd), pp. 401–10.

Asiodu, P. C. (1969) Planning for further industrial development in Nigeria In Ayida, A. A. and Onitiri, H.M.A. (eds.), *Reconstruction and Development in Nigeria* (Ibadan; Oxford University Press).

Ayeni, M. A. O. (1976) *A Regional Plan for Lagos State: Planning Lagos Metropolitan Area. Final Report* (University of Ibadan: Planning Studies Programme).

Ayeni, M. A. O. (1978) Patterns, processes and problems of urban development. In Oguntoyinbo, J. S., Areola, O. and Filani, M. (eds.), *The Geography of Nigerian Development* (Ibadan, Heinemann Educational Books (Nigeria) Ltd), pp. 156–74.

Baldwin, K. S. (1957) *The Niger Agricultural Project: An Experiment in African Development* (Cambridge, Mass.: Harvard University Press).

Central Planning Office (1978) *Review of Nigeria's Planning Experience* (Paper presented at the National Workshop on Planning Strategy for the 1980s; Nigerian Institute of Social and Economic Research).

Daly, M. (1977) *Development Planning in Nigeria* (mimeo, University of Ibadan, Planning Studies Programme).

Damachi, U. G. (1973) Manpower in Nigeria. In Damachi, U. G. and Seibel, H. D. (eds.), *Social Change and Economic Development*, pp. 81–97 (New York: Praeger).

Federal Republic of Nigeria (1962) *First National Development Plan, 1962–68* (Lagos: Federal Ministry of Information).

Federal Republic of Nigeria (1970) *Second National Development Plan, 1970–74* (Lagos: Federal Ministry of Information).

Federal Republic of Nigeria (1975) *Third National Development Plan 1975–80* (Lagos: Federal Ministry of Information).

Filani, M. O. (1977) Some fundamental issues in planning for rural development. In Adejuyigbe, O. and Helleiner, F. M. (eds.), *Environmental and Spatial Factors in Rural Development in Nigeria* (Proceedings of the 20th Annual General Conference of the Nigerian Geographical Association, University of Ife), pp. 6–14.

Filani, M. O. and Onyemelukwe, J. O. C. (1977) *A Geographical Analysis of Development Planning Problems in Developing Countries* (Mimeo, Department of Geography, University of Ibadan).

Friedmann, J. and Douglass, M. (1978) Agropolitan development: Towards a new strategy for regional planning in Asia. In Lo, F.C. and Salih, K. *Growth Pole Strategy and Regional Development Policy*, pp. 163–92 (Oxford: Pergamon Press).

Green, L. and V. Milone (1972) *Urbanization in Nigeria: A Planning Commentary*, pp. 7–8. (New York: Ford Foundation).

Harrison–Church, R. J. (1949), The evolution of railways in French and British West Africa, *Congress Internationale de Geographie*, 16 (4), 95–114.

Hinderink, J. and J. J. Sterkenburg (1978) Spatial inequality in underdeveloped countries and the role of government policy, *Ti jdschrift voor Econ. en Soc. Geografie*, 69, 5–15.

Kumuyi, A. J. (1973) *The Spatial Interaction of Central Places in Kano State of Nigeria* (Ibadan: University of Ibadan, Ph.D. Thesis).

Lewis, O. (1971) An appraisal of industrial development programme, *Quarterly Journal of Administration*, V (3).

Mabogunje, A. L. (1968) *Urbanization in Nigeria* (London: University of London Press).

Mabogunje, A. L. (1974) *Cities and Social Order*, pp. 35–36 (Ibadan: University of Ibadan Press).

Mabogunje, A. L. (1977a) *On Developing and Development* (Ibadan: University Lecture Series).

Mabogunje, A. L. (1977b) Development of the world production and trade up to the year 2000—a third world view, *GeoJournal*, 1 (3), 7–14.

McDonell, G. (1964), The dynamics of geographic change: the case of Kano, *Annals of the Association of American Geographers*, vol. 54.

Myrdal G. (1968) *Asian Drama: An Inquiry into the Poverty of Nations*, (New York: Pantheon).

Nigeria (1946) *Ten-Year Plan of Development and Welfare for Nigeria, 1946–56* (Lagos).

Olusanya, P. O. (1969) *Socio-Economic Aspects of Rural–Urban Migration in Western Nigeria* (Ibadan: Nigerian Institute of Social and Economic Research).

Oluwasanmi, H. A. (1966) *Agriculture and Nigerian Economic Development* pp. 118–19 (Ibadan: Ibadan University Press).

Seers, D. (1972) What are we trying to measure. In Baster, N. (ed.), *Measuring Development* pp. 21–36 (London: Cass).

Williams, V. W. and M. O. Filani (1977) Life on the land—problems in Nigeria, *Earth and Mineral Sciences*, **47** (1) 1–8. (The Pennsylvania State University).

Chapter 12

Ivory Coast: An Adaptive Centre-Down Approach in Transition

M. PENOUIL

The Ivory Coast is a very special case in Africa. After the Second World War it became one of the most prosperous French colonies, and contributed a large share to the budget of French West Africa. Since independence, the country has enjoyed an exceptionally high and steady growth rate. Other countries certainly have recorded comparable growth rates (8.3 per cent per year on average from 1960 to 1975), but in most such cases this has been based on petroleum or mineral exports, which is not the case with the Ivory Coast (see Tables 12.1 and 12.2).

This rapid growth has been accompanied by profound structural change, within the context of an ideology of explicit state capitalism. Some analysts have detected what they describe as a 'liberal option', but the term is certainly incorrect, because state interventions are numerous and fundamental. It is, however, clear that the Ivory Coast is an open market economy, and has chosen to ally itself politically with the Western world. Nevertheless there are probably few other African countries where the effectiveness of government economic policy has been as great. It is tempting to see in this the success of liberalism and capitalism. Certain authors have in fact laid great stress on 'the Miracle of the Ivory Coast', while others relentlessly criticize or express serious reservations regarding the future of this experiment. Regional or social inequalities, strikes, financial dependence on other countries, indebtedness, the absence of basic industries, etc. . . . are elements passed over in silence by the former, and by contrast commented upon by the latter who blame them on the socio-political choices of the country (Amin, 1967).

The Ivory Coast is clearly an example of 'development from above', as described by Hansen in chapter 1. It is true that this development is fuelled by a small number of activities generally aimed towards exports or the substitution of imports. Certainly contributions of capital from foreign sources play an essential role in financing development, and at any moment massive departures of capital could compromise the country's economic equilibrium. It is equally

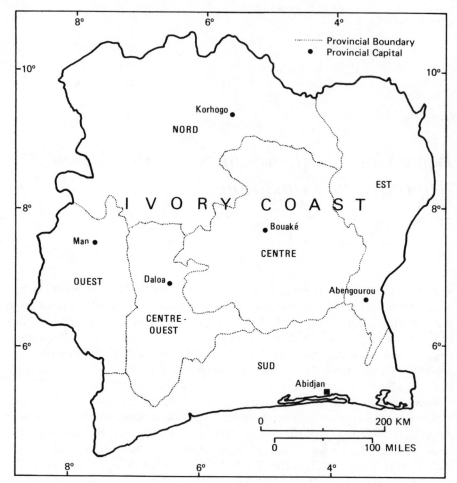

Figure 12.1 Provinces of the Ivory Coast

obvious that economic activity is most strongly concentrated in certain regions, and that the Ivory Coast is a classic example of Stöhr's centre-down, outward looking paradigm. (see Chapter 2).

However, specific case studies rarely fit general paradigms exactly. The Ivory Coast situation is perceptibly different from that of countries of Latin America or Asia. Lo and Salih (Chapter 5) describe four models, none of which in reality corresponds to the Ivory Coast situation. Several essential facts explain these differences: a relatively low population density which allows for considerable future agricultural expansion, and the actual character of the agricultural export sector, which is not always clearly differentiated from industrial activities. The rural world in this nation, as we will attempt to demonstrate, contains characteristics of both 'development from above' and 'development from

Table 12.1 *Annual rate of growth of the economy (in constant francs)*

Year	GDP* (%)	Import (%)	Export (%)	GFCF (%)**
1960				
1961				
1962				
1963	+ 10.2	+ 10.5	+ 9.1	+ 15.3
1964				
1965				
1966	+ 7.8	+ 8.6	+ 5.3	+ 3.4
1967	+ 0.6	+ 9.1	− 1.6	+ 3.1
1968	+ 17.2	+ 11.9	+ 26.9	+ 17.1
1969	+ 4.4	+ 7.9	+ 0.8	+ 7.6
1970	+ 8.2	+ 16.8	+ 7.3	+ 25.1
1971	+ 8.5	+ 0.9	+ 5.3	+ 3.4
1972	+ 6.8	− 0.9	+ 12.6	− 3.8
1973	+ 5.8	+ 21.7	+ 1	+ 25.6
1974	+ 6.0	+ 1.4	+ 10.3	− 1.3
1975	+ 6.9	− 6.4	− 7.6	+ 18.6

*Gross Domestic Product
**Gross Fixed Capital Formation
Source: Ministère du Plan de Côte d'Ivoire (1977).

Table 12.2 *Over-all development of the Ivory Coast economy since 1960 (at current values)*

Year	Gross domestic product (millions of francs CFA)	Population (in thousands)	Gross domestic product per capita (francs CFA)
1960	142,615	3735	38,183
1961	161,422	3840	42,037
1962	168,350	3945	42,674
1963	197,810	4050	48,482
1964	239,675	4165	57,343
1965	239,586	4300	55,718
1966	257,975	4430	58,234
1967	275,681	4560	60,456
1968	326,463	4765	68,514
1969	365,368	4940	74,002
1970	415,325	5115	81,138
1971	440,074	5264	83,600
1972	472,480	5060	79,275
1973	565,267	6200	91,172
1974	738,661	6460	114,544
1975	813,718	6720	121,036

Source: Ministère du Plan de Côte d'Ivoire (1977).

below'. It is this balance between the two strategies, it will be argued, which explains the relative success of the Ivorean experiment.

This chapter will reflect on the Ivory Coast's economic development with special attention to the development strategy chosen without stressing its ideological content. Socialist or capitalist options do not make very much sense in Black Africa. The Congo or Guinea are primarily underdeveloped, rather than socialist, countries. The Ivory Coast is primarily a developing nation rather than a capitalist economy. If a certain level of development is not achieved, governments have only a very imperfect command of the evolution of society. On the other hand, we believe that the interventions of the state can draw inspiration from certain development models, more particularly from plans of polarized growth or of balanced growth. It is from these perspectives that the experience of the Ivory Coast will be examined.

A basic observation underlying this examination must first, however, be made. Plans for balanced growth or polarized growth are only plans; and political intentions as outlined in official speeches are often very far removed from observed reality.

It is essential therefore to describe the reality of the Ivory Coast development process. But an objective interpretation of the facts is a delicate matter, and development problems can only be interpreted and dealt with over a period of time. Observable facts do not allow us to rely too easily on pre-established schemes; often it happens that apparently divergent development approaches unite in the reality of development. The Ivorean success is the result of a great experiment, and a historic combination of strategies (Penouil, 1971). One strategy may be preferable to another, depending upon the stage of development attained. The Ivory Coast demonstrates that in the first phases of development polarization is difficult and dangerous. It can only have a chance of success when a minimum infrastructure has been created and a minimum level of revenue attained. Once polarization is begun it can only bear fruit if its negative effects are neutralized by other development processes, which achieve a sometimes precarious social and economic equilibrium. These secondary processes can appear in a 'natural' way, without conscious political decision, but to assure permanency of development they must be integrated into the national strategy. One of these compensatory development processes seems to have come from economic activities springing not from the modern economy, but from profound social changes in the organization of production since the 1950s. In the urban environment these activities are in what the ILO calls the 'informal' or non-structured sector. To integrate the organization of production into a rural environment, we prefer to use the phrase 'transitional activities'. *It is believed that any strategy for development must bring together action on traditional activities, transitional activities, and modern activities, as it must combine actions relevant to balanced growth planning and polarized growth planning.* The major error of the last few decades has been that of presenting, as exclusive, actions that must be viewed as complementary. The art of development policy consists of balancing these different components in terms of development

Table 12.3   *Gross domestic product (distribution and development by sector) (in billions of francs CFA, current)*

Year	Primary	%	Secondary	%	Tertiary	%	Total
1960	60.1	46.8	19.9	15.2	49.6	36	130.5
1961	62.3		24.8		60.1		147.2
1962	60.2		27.5		63.4		151.1
1963	74.4		30.2		74.4		179.0
1964	87.2		36.		93.6		216.8
1965	84.4	39.4	40.6	19.0	89.	41.6	214.0
1966	88.1		50.4		94.2		232.7
1967	36.1		55.2		106.		197.3
1968	100.7		62.4		133.		296.1
1969	106.5		73.6		149.2		329.3
1970	112.6	30.2	89.0	23.8	171.9	46.0	379.5
1971	116.6	29.6	101.3	25.8	175.7	44.6	393.6
1972	125.1	29.4	111.5	26.2	186.6	41.4	425.2
1973	159.1	31.2	121.5	23.8	229.2	45.	509.8
1974	190.0	28.8	163.1	23.5	321.0	47.7	674.1
1975	233.6	31.8	181.7	24.7	319.9	43.5	735.2

*Figures in this table include indirect taxes.
Source: Ministère du Plan en Côte d'Ivoire (1977).

achieved and present potential. This is what we call the *Strategy of Over-All, Progressive Development.* The Ivorean success stems from the fact that coherent choices, adapted to the situation of the moment, have been put into action.

This study will begin by examining how in the course of Ivory Coast history there has been a successful combining of the strategies of balanced growth and polarized growth.

In a second section we will examine the current situation to show how in fact growth comes about by the simultaneous evolution of modern functions, traditional activities, and transitional activities, and how a rational policy could accelerate this growth.

THE COMBINATION OF DEVELOPMENT STRATEGIES IN THE ECONOMIC HISTORY OF THE IVORY COAST

The Ivory Coast is a young country. It is a pure creation of French colonial policy which established a united administrative structure there, where before only assorted ethnic groups had existed. The borders of the new colony, however, fragmented some ethnic groups; and important portions found themselves in Upper Volta, Mali, and Ghana. In the beginning there was no cultural, ethnic, geographical, or economic unity; more than 50 ethnic groups spoke 100 different dialects, and population was less than 2 million in 1920. Indeed the Ivory Coast seemed to be one of those African countries where there would be serious problems of tribalism. Achievement of national integration must have seemed all the more difficult because the new country was going to include a large number of foreigners coming mostly from Upper Volta. The bonds between these peoples, in larger part animists in the south, Muslims in the north, some matrilineal, others patrilineal, exposed for decades to European influence, or hidden away in secluded forest regions, must have seemed tenuous.

The potential for development was equally very uneven. The south is a region of forest and heavy rainfall (up to 2.5 metres of precipitation per year). The north is a region of savannahs with a semi-arid climate; the forest disappears, except for a few trees along riverbanks, and agricultural life depends on the amount of rainfall. Therefore the south can raise crops of coffee, cocoa, or tropical fruits. It is an area of palm trees, coconuts, rubber trees, and a great variety of crops (cassava, yams, etc.). The north is the domain of cattle ranching (often mediocre), of millet, and of cotton and rice when conditions are favourable. Thus one area has a production which assures adequate sustenance and permits creation of an important flow of exports; in the other, production gives only a meagre sustenance with very few exports. It is not surprising that with such a starting point marked inequalities still exist today, and it would surely be dangerous to attribute these solely to the chosen development approach. It is probable that whatever the political trends, the constraints of the human environment and the natural potential would have been a determining influence. If one is aware of the great difficulties which have existed since the beginning, the current achievements are remarkable.

Ivory Coast's development seems to have taken place in four phases, the last of which has scarcely begun. In these various stages, varied development strategies have been undertaken. At the outset there was a simple hunter–gatherer economy (1860–1930). The second phase rested on a mechanism of balanced growth, limited however to one section of the territory (1930–52). In the third phase a plan for polarized growth prevailed but without real industrial power (1952–70). The last and current phase seems to be the undertaking of a total, generalized growth plan.

The Hunter-Gatherer Economy

This form of economy existed before the Ivory Coast became a French colony. By the second half of the nineteenth century, various trading posts had been established by the French, English and Dutch, whose ships would put into port at various points on the coast to buy local products, in particular products of native palm groves (Bretignère, 1931). The most important incursion for France was carried out in the east of the country by a merchant from La Rochelle (Verdier, 1895). This man, who came to Africa as an employee, set up in business for himself in the area of Assinie near the mouth of the Comoe, and close by the present Ghanaian border. Verdier played an important role in the establishment of the Ivory Coast. During and after the War of 1870 he secured the representation of French interests in the region. Agreements entered into by him with chiefs of the region permitted the legal foundation of the French presence there, in the face of British claims. And finally one of his employees, Treich-

Laplene, carried out the exploration of the eastern part of the Ivory Coast and established a liaison with the Binger expedition coming from the north. In 1892 the boundaries of the French territory were specified, and in 1893 the Ivory Coast became a colony, with Binger the first Governor.

In the years that followed a series of events decisive to subsequent development of the territory took place.

The first involved the setting-up of a new kind of production framework. Verdier had completed his commercial activities by creation of a coffee plantation at Elima. If in the first years production was very low, it nonetheless foreshadowed the agricultural future of the Ivory Coast. Meanwhile, exploitation of forest resources from a 'hunter–gatherer' point of view was not abandoned, and with this in mind Verdier obtained in 1893 a grant of some 5,500,000 hectares of land in the Ivory Coast. One could foresee an economic system like that of the Belgian Congo or French Equatorial Africa—a system implying exploitation of certain forest resources (rubber, for example), and doubtless a concentration of new cultures in large-scale European plantations. Binger's action led to the buying-back of granted lands, thus leaving the door open to the subsequent rise of small native plantations, whose development began after the First World War and expanded after the Second World War.

The second important event was the development of infrastructure. In those days, the administrative and commercial centre of the Ivory Coast was located on an offshore sand bar not far from Assinie at Grand Bassam. On this inhospitable coast the sand bar was an obstacle to commerical development. Construction of a wharf at Grand Bassam was an important step forward in opening up the country to external commerce (1897), but development remained difficult, and at other points along the coast a lack of installations rendered any commercial activity precarious.

Finally, for strategic as well as economic reasons the construction of a north–south railway was undertaken, starting from the coast and going towards Upper Volta and Niger. It appeared that the rational departure point was the principal port of the territory, Grand Bassam, but under pressure from an officer of the French Navy another starting point was chosen, the small, unimportant village of Abidjan. Justification for the choice was a fundamental one: the existence of a deep ocean trench in the immediate vicinity. From there it would be possible to pierce the offshore bar and permit creation of a perfectly sheltered port on the Ebrié lagoon, accessible to ships of considerable tonnage, and with channel drainage assured in a natural fashion by currents created by the great depth of surrounding waters. The building of the railway would prove a long and exacting task but from this time on one of the essential elements of development strategy was in place: on the one hand the site which would later become the development pole of the country was chosen, the site of a village open both towards the world and towards the north and accessible to neighbouring countries without reference to any theoretical polarized growth plan—and with good reason. Moreover, the railway signified a quest for a firm interrelationship between south and north in West Africa. Certainly when the choice was made, illusions still existed about great riches in the continent's interior; but desire to proceed with development of the area's resources was unquestionable.

Nevertheless, until the 1930s export capacity remained limited, both because of the absence of important mineral resources and the lack of development of local agriculture.

Balanced Growth

This second stage was inaugurated by the actions of various territorial governors, who worked to develop cocoa and coffee plantations. The outlines of this new kind of cultivation gradually took shape, and in spite of hostility of local peoples, the benefits of the new production were gradually recognized, and were solidified by expansion of the export trade. Revenues, although small, permitted the setting-up of a trading type economy, which doubtless exploited the local people but which also opened the door to consumption of new products. Also— and this was crucial—a dual production structure was set in place with, on the one hand, vast European plantations, mostly one-crop operations, and on the other, native plantations. Expansion of native-grown coffee took place only after 1939 (1939: 18,000 tons, and in 1959: 104,000 tons). The development of cocoa began at an earlier date: 1000 tons in 1920, 55,000 tons in 1930, rising to 75,000 tons in 1955. These native plantations were the start of the evolution, and the starting point of the transitional sector.

To begin with the native plantations were a combination of traditional subsistence agriculture and new production methods. Burnt-out patches of land cleared in forest areas were the beginning. The planter tilled the soil, sometimes under the eagle eye of an administrator, cultivating coffee or cacao trees alongside the yam and cassava plants which provided the family's food. Quality of production was questionable for a long time, and commercialization brought with it many problems. Nonetheless, these African planters have become a new social class. This gradually evolved, first with increasing use of salaried labourers to ensure certain operations were carried out, especially at harvest time. Use of manpower from Upper Volta also created a kind of rural capitalism, where ownership of land guaranteed unquestioned social status and a sure income. Then this social group, especially after the Second World War, organized to defend its own interests—at one and the same time against the administration (demanding an end to forced labour) and against the abuses of the commercial system. In this way the coffee planters' association formed a springboard to the future, by means of which Houphouet Boigny started his political career.

To what extent is it possible then to speak of balanced growth? The Ivory Coast does not offer a spectacle of diversified development spread out over the whole country, but certainly activities were created in response to demand generated by a rise in agricultural income. Here we find the chief feature of the theory of balanced growth. It would be fallacy to pretend, in a country on the eve of development, that a large variety of activities could develop simultaneously. Above all, then, balanced growth produces enterprises aimed at generating revenues, to justify creation of industries to take the place of imports, and this is progressively what has happened. Significantly, the principal industry of the country created about 1930 was a textile mill (Establishments Gonfreville), located beyond the coastal zone at Bouaké. The industrialization process, however, was very slow, though relatively general, until the Second World War. The following text from the Government of the AOF illustrates the situation:

> The Ivory Coast is a French West African colony where European industry is most advanced: one can enumerate two brick-works, one at Mossou near Bassam belonging to the Catholic Mission and the other near Dabou. There are seventeen sawmills, one of which is under the direction of the railways. Seventeen enterprises are devoted to cultivation of cocoa and nineteen to coffee, these products being mechanically processed on the spot. Factories for the picking and milling of cotton number five. Two plants handle kapok, three sisal hemp. Moreover three companies are in the business of milling rice. One plant makes banana flakes. There are five ice manufacturers, two at Abidjan, one at Bassam, one at Bobo Dioulasso and one at Sassandra. One of the two plants at Abidjan specializes in industrial refrigeration: it has refrigeration equipment for conserving food products. Four electricity plants are operating in Abidjan (two plants), at Bassam and at Badikaha. Four oil manufactures are installed at Dedougou, Badikaha and Sassandra (which has two). Two soap factories are operating at Badikaha and Abidjan. At Dabakala a factory distills essences for perfume (sage, lemon, aromatic grasses, etc.). Abidjan has two printing plants. Three carpentry and cabinet-makers shops exist at Bassam, Abidjan and Badikaha. Seven workshops for mechanical repairs (garages) are at Bassam, two at Abidjan, two at Sassandra, one at Badikaha and Dedougou. Besides all these industries one could mention the cloth mill of the Catholic Mission at Ouagadougou which makes thick-pile carpets. An ivory carving industry is installed at Grand Bassam. One could mention besides, a cooperage and a manufactury of oil barrels at Badikaha. Enterprises connected with the building industry are five in number, grouped together at Abidjan. Two harbour transport firms (four motorbikes and three boats), one at Bassam, the other at Lahou. (Gouvernement General de l'AOF—1937.)

This lengthy list, which nonetheless omits the important Gonfreville plant at Bouaké, demonstrates the absence of large industry and the relative dispersal of small industry.

This came about because the Ivory Coast had as yet no true urban structure. Commercial decisions established the capital at Grand Bassam; an unhealthy climate brought about its transfer to the other bank of the Ebrié lagoon at Bingerville. But for the creation of a large industrial city the new location was no better than the old. Thus above all it was with the founding of the capital at Abidjan in 1934 that the first beginnings of real industrialization saw the light of day. Abidjan had only 1400 inhabitants in 1912; 9000 in 1926. It reached a population of 22,000 in 1939 and by 1955 had grown to 125,000. If the Second World War hastened the growth of certain enterprises, it was the opening of the Vridi Canal and creation of the port of Abidjan which brought about true industrialization.

This period also saw the strengthening of administrative and commercial structures in the country. The contribution of Syrian and Lebanese merchants is obvious at the trading-post level. They contributed to the spread of a money economy, at least in the south. The aftermath of war saw an acceleration of this, notably with the founding of the first banking establishments. This gradually maturing process thus prepared the way for the polarized growth stage.

The Polarized Growth Period

Creation of the port of Abidjan completely changed the fundamental plan for Ivory Coast development. The setting-up of new installations, however, was almost coincidental with the country's independence, thus permitting the start of a redesigning of structures.

In the 20 years between 1952 and the early 1970s the country's growth has forged ahead. From 1955 to 1970 traffic handled by the port of Abidjan has multiplied five times. Growth in primary products is three times its 1955 level and per-capita productivity has doubled (see Table 12.4).

This growth is that of the Abidjan core. Abidjan's expansion is obvious from its population growth, from 124,000 in 1955 to over a million in 1977.

Significant also is the gravitational pull which the Abidjan core exerts on the surrounding area. The population is made up of a large majority of migrants, coming sometimes from other parts of the country, sometimes from neighbouring states: 71.7 per cent the former and 26.6 per cent the latter. As the city expands her field of attraction has grown too, now reaching out to all West Africa, including Anglophone territories. This expansion has in part limited

Table 12.4   *Total traffic for the Port of Abidjan (in tons)*

Year	Imports	Exports	Total	Annual growth (%)
1955	547,594	399,421	947,015	—
1960	757,112	1,009,597	1,776,709	+ 26
1961	1,057,538	1,320,827	2,378,365	+ 35
1962	1,070,360	1,432,744	2,503,106	+ 5
1963	1,127,222	1,760,062	2,887,284	+ 15
1964	1,327,320	2,059,381	3,387,701	+ 17
1965	1,412,278	2,199,908	3,612,186	+ 6.6
1966	1,683,241	2,389,970	4,073,211	+ 12
1967	1,664,237	2,481,429	4,145,666	+ 1.7
1968	1,968,612	2,776,397	4,745,009	+ 14
1969	2,003,672	3,163,993	5,167,669	+ 8.9
1970	2,335,100	2,745,591	5,080,691	− 1.6
1971	2,616,370	2,726,343	5,342,713	+ 5.3
1972	2,965,027	2,959,987	5,925,000	+ 10.6
1973	3,468,232	3,091,445	6,559,677	+ 10.7
1974	3,558,984	3,039,296	6,597,331	+ 0.5
1975	3,442,976	2,618,724	6,061,250	− 8
1976	4,650,000	3,430,000	7,610,000	+ 23.7
1977	4,729,000	3,137,000	7,863,000	+ 3.3

Source: Ministère du Plan de Côte d'Ivoire (1977)

that of secondary centres. Bouaké, Man, Daloa, Korhogo and Gagnoa have all had much lower growth and have not been able to follow in the wake of their leading city.

Abidjan thus has become the true symbol of the new nation, not only because of successful growth but also because here is found a happy synthesis of Ivorean and other African peoples. The confining of many of the country's activities to Abidjan, and its choice as economic and administrative capital, illustrate once more the existence of a centralized development process oriented towards the outside world, a point criticized by both Stöhr and Hansen in earlier chapters. The following points should however be considered:

Ivorean development could only take place through agricultural activity and exports. At the outset the food consumption structure was very different in the north from that of the south. Millet and corn predominated in the north; while cassava, bananas, and citrus were grown in the south. No industrialization process could be guaranteed. Thus the choice of a port was a logical step. Choosing a city in the country's centre (Bouaké) would have required a burden: some communication network with a north–south axis and a Bouaké–Abidjan segment, which would have changed nothing. Anyway the Abidjan–Niger axis was built more for political than for economic reasons.

Finally we must establish that Abidjan's growth was in large measure determined by the growth of production in the country as a whole. Indeed it is equally linked to the industrialization process. Industrialization developed because of the creation of a thriving industrial base. By examining official figures we can establish that no one part of the country attained more than 15 per cent of the whole. The most important component in the industrial sector, textiles and clothing, produces slightly less by value than the export of coffee or cocoa. The polarization on Abidjan then comes about from industrial concentration. It is from growth of demand, ultimately, that industry draws its dynamism and its new outlets. Agricultural development and growth of the revenues it distributes is now the basis for development strategy, aided by a government practising a policy of high prices to producers compared with those paid in neighbouring countries. Here the Ivory Coast is moving away from the 'development from above' plan, for there is a simultaneous polarization and diversification of enterprises. This is the important point, for polarization is often linked to a thriving industry which animates the centre where it is located. When this happens the centre expands by diversification and the grouping together of enterprises, often resource-based. This strategy of development resembles, at least in part, the propositions of Stöhr, in Chapter 2.

If then there is polarization in Abidjan, it comes about less from the creation of a crowded intersectoral nucleus, than by organization of a well-distributed urban area endowed with important administrative, cultural, commercial, and financial infrastructures. Thus 65 per cent of the Ivory Coast's total business dealings are realized at Abidjan with the major portion of business transacted outside Abidjan being in sectors requiring decentralization, such as the lumber industry (saw-mills), or agriculture (e.g. cotton picking, oil extraction).

Table 12.5 *Per capita agricultural income by region (in francs CFA)*

South	17,400
East	18,200
Centre-west	17,300
Centre	14,000
North	3,600
West	5,700
South-west	5,700

Source: Planning Ministry, Policy of Regional Development and Over-all Development, July 1970.

Abidjan's economic growth thus has caused an aggravation of interregional inequalities. Table 12.5 furnishes data on income inequalities between regions, and we see the seriousness of the problem.

This inequality is reflected in social structures. In the city of Abidjan inequalities are obvious between an Ivory Coast middle class based on administration and politics and more and more involved in business, and a working class having difficulty finding steady work outside the informal sector, and made up largely of foreigners. In the agricultural south the inequity is between owners of land and the salaried labourer, especially as the Abidjanois are beginning to buy up and acquire agricultural lands in their territory. In the north inequality is still more flagarant between a small minority of officials and tradesmen and the great mass of peasants. This inequality seems to be growing.

The 20 per cent of the population earning lowest incomes received 8 per cent of national income in 1959, and 3.9 per cent in 1970. The 20 per cent with highest incomes received 55 per cent of total revenue in 1959, and 57.2 per cent in 1970. The poorest 60 per cent of the population have seen their share dwindle from 30 to 22 per cent between these two dates (Lecaillon and Germidis, 1977).

Polarized development, 'from above', doubtless augmented these inequalities. But these figures need interpretation. For one thing, data on national book-keeping has greatly undervalued payments 'in kind' which mostly accrue to the lowest income earners.

Many other African countries suffer from much greater inequality in distribution of incomes than the Ivory Coast. The 20 per cent earning highest incomes in Gabon receive 71 per cent of revenues, in Sierra Leone 64.1 per cent, in Senegal 64 per cent, in Nigeria 60.9 per cent and in Zambia 57.1 per cent.

If the lowest earners' share of income has declined, it is that share pertaining to the 7th and 8th deciles which has grown the most: from 15 per cent in 1959 to 20.9 per cent in 1970. The share of 5th and 6th deciles on the one hand, and the 9th and 10th on the other, varies little. The lowest incomes are those of rural foreign labourers (from Upper Volta). The highest incomes are often earned by Europeans. The Ivory Coast economy paradoxically functions like a 'centre' in its relations with those on its borders. Today this is more important than effects connected with its peripheral character in regard to developed countries; in distribution of revenue at least.

Finally one must mention, with Lecaillon and Germidis, that all nominal incomes are growing; though we are faced with an unequal division of the effects of growth, each inhabitant benefits to a degree. In this redistribution the chief beneficiaries are members of the middle class, officials and planters, rather than the highest-income group. This growth presents a certain danger. It can only be pursued by maintaining a large flow of exports destined to cover the considerable cost of the country's external indebtedness which has in the past financed new investments. Pursuit of growth also supposes revenues being distributed to a growing portion of the population which would assure, by growth of final demand, new openings for industrial trade and creation of new industries.

Towards a Universal Growth Plan

The development strategy of the Ivory Coast since the 1970s seems to follow a coherent universal plan. It involves a rise in agricultural production in the already developed zones, but also a reproduction of this model in other areas. Growth has been pursued and accelerated in favourable circumstances (the rise in coffee prices, for example).

This direction was foreshadowed by the palm-oil plan which permitted a diversification of production in the forest zone of the south. But above all the government has attempted to revive the experience of the east in the south-west. Thus creation of the port of San Pedro seems destined to play in the south west the role filled in the east by Abidjan. Also the expansion of rubber and palm plantations here would ensure cultivation of an important forest region, bringing relief to the over-exploited south-east. This conforms to the logic of 'development from above' based on capitalist uniformity.

Projects are also under way with the objective of assuring a growth of rural incomes of central and northern areas—expansion of rice crops (from 250,000 tons in 1965 this had reached 460,000 tons by 1975), and chiefly the sugar plan, which may appear very ambitious but which provides for cultivation of regions very underdeveloped before. Finally, the north is more interested in the rise of cotton (from 12,000 tons of cotton seed in 1965, production grew to 60,000 tons in 1975, and 75,000 tons in 1977), and the first results of stock-breeding. The Ivory Coast seems to be returning to the policy which caused its initial success, the recognition of the essential role of agriculture.

We are also seeing some decentralization of industrial plants, e.g. in the textile sector with creation of a spinning–weaving mill at Dimbokro (Utexi) and at Agboville (Cotivo). These small towns are of course situated within Abidjan's zones of influence, but the effort to decentralize is nonetheless remarkable for the size of factories created. Similarily the agricultural projects already begun will have important repercussions on the determination of locations for agricultural and food plants in the north and west.

Table 12.6   *Number of maternity and hospital beds by administrative region in 1969 and in 1975*

Region	Maternity beds					Hospital beds					Total beds				
	1969 Beds	1975 Beds	1969 Inhab. per bed	1975 Inhab. per bed	% incr. or decr.	1969 Beds	1975 Beds	1969 Inhab. per bed	1975 Inhab. per bed	% incr. or decr.	1969 Beds	1975 Beds	1969 Inhab. per bed	1975 Inhab. per bed	% incr. or decr.
West	158	252	3898	2798	+28	317	381	1945	1850	+5	475	633	1298	1113	+14
North	169	287	3357	2117	+37	321	354	1767	1716	+3	490	641	1158	948	+18
Centre	605	759	2127	1046	+4	1186	1331	1085	1166	−7	1791	2090	718	743	+3
West-Centre	217	283	2637	2713	−3	476	466	1202	1647	−37	693	749	825	1025	−24
East	68	92	3636	3485	+4	112	85	2208	3772	−71	180	177	1373	1811	−32
South-west	78	111	1551	1654	−6	112	211	1080	870	+19	190	322	636	570	+10
South, not including Abidjan	700	874	1330	1496	−12	728	895	1299	1461	−14	1428	1769	652	739	−13
Abidjan	369	590	1322	1694	−28	1021	1473	512	678	−32	1417	2063	369	484	−31
South including Abidjan	1096	1464	1339	1576	−18	1749	2368	839	974	−16	2845	3832	515	602	−17
Total excluding Abidjan	1995	2658	2196	2048	+6	3252	3723	1341	1462	−9	5247	6381	835	853	−2
Total including Abidjan	2391	3248	2051	1984	+3	4273	5196	1147	1240	−9	6664	8444	736	763	−4

Source: Ministry of Public Health, SAPS. Planning Ministry, DED.

This industrial rebalancing will be strengthened if two important projects see the light of day and enjoy the success anticipated: first, development of the pulp and paper industry using forest resources, particularly in the south-west (planting in the region of San Pedro); and second, a plan to mine the iron ore deposits at Bongolo in the Man region. This plan could completely alter the economy of the south-west by establishment of a micrometallurgy plant at San Pedro, although the international context for this is unfavourable at present.

Evolution of industry implies a modification of the urban structure, and some centres are undergoing rapid expansion: San Pedro, a tiny village in 1965, had a population in 1975 exceeding 30,000; Yamossoukro grew in the same period from 8,000 to 35,000 inhabitants. But above all one sees the growth of towns which could tomorrow become large cities, when redistribution is accomplished. Bouaké has reached a population of 175,000; Man, Korhogo, and Daloa have all reached 50,000; besides, new centres of growth in an area a hundred or so kilometres around Abidjan, could slow down the growth of the capital Dimbokro 30,000 population, Agboville (27,000), Divo (37,000), and Dabou (25,000) are examples of this. This evolution in urban structure has brought with it a new balance in the redistribution of social infrastructures. Table 12.6 illustrates this evolution in the realm of health care.

Further industrialization remains dependent on the country's resources. Here also signs of decentralization exist. Already profits have accrued from the construction of the Cossou dam, indicative of events which could improve the energy supply across the whole country. The Taabo Dam on the Bandama and the Buyo on the Sassandra would permit, according to plan, the tripling of electrical energy production between 1975 and 1980, while assuring a supply to the centre and south-west. Thus the willingness to incorporate a balanced spatial equilibrium seems clearly demonstrated.

However, the model of polarized growth has not been completely abandoned. A fundamental role has devolved upon Abidjan, the principal centre for establishment of new industry. This role can only be understood if one analyses carefully the functioning of the Ivory Coast economy and inquiries into the method by which the evolution of traditional, modern and transitional activities are going forward today.

THE REALITY OF IVORY COAST DEVELOPMENT: A COMPLEX AFFAIR

Over and above actions of governments, development comes about by the real evolution of economic and social structures, and these structures do not evolve in an automatic or similar manner. Any given society is composed of parts turned towards the past, solidified in custom; elements partially evolving, but incapable of attaining a complete mutation; and of dynamic organizations which foreshadow its future shape. The life of a society is the product of interaction between these differing elements. In Ivory Coast society we see a combination of these diverse elements. Taking into account the direction of this study, it is clearly demonstrable that political stability and the unity of the country have been assured due to the intelligent action of President Houphouet-Boigny, who was able at one and the same time to secure the support of the traditional chiefdoms and to neutralize it by enacting new powers to preserve the country's equilibrium—a complex task. Finally it is important to demonstrate the significance of traditional forces in the various domains, especially in regard to the problems of the rights of people and their welfare. The institution of marriage, the dowry, the inheritance, could not simply be altered by legal dictum. Their roots went too deep. In all rural areas the organization of work is influenced by ancient custom, especially in the division of labour between men and women, and between age groups. Certain elements of these older cultures can slow down development; but equally surely, respect for ancient customs has helped these people avoid the shocks and antagonisms other countries have suffered.

A great diversity of organizational forms has been retained in the Ivory Coast in order to assure administration of the modern economy, which was formed by the placing together of foreign enterprises, with foreign capital and direction (an example is the textile sector in Gonfreville, or the extraction of oil in Blohorn) and many small enterprises financed by foreign or national funding (particularly in commerce); and the union of national and foreign moneys in the realms of banking, agriculture, and industry. The Ivory Coast has been able to develop diversified enterprises adapted to local demand, with much better success than her neighbours, although the latter are often better endowed by nature. One might be tempted to attribute this success only to the modern sector, with its skilled policy of substitution for imports, opening up of markets, and assuring contributions of capital. But this view would neglect the destiny of the major portion of the people, who labour outside the scope of recently developed enterprises. Certain people in critical analysis of Ivory Coast development maintain it is merely a façade, and the lot of the great portion of people has not improved; that the infrastructure has stayed the same so that one cannot speak of true development. The object of this study is to show that the Ivory Coast deserves neither excessive tributes, nor abuse from the critics. Many transformations are coming about slowly, and are only partially successful. But the total population is pledged to them. The modern sector is the result of a rapid mutation in the economic structure. Such a basic change can only take place in the entire social body, and this slow transition is well under way. The role of the transitional activities involved will be described.

1. Transitional Activities in the Ivory Coast

What is meant by 'transitional activities'? Recent publications of the International Labour Office (ILO) have accented the role of the non-structured 'informal sector', but specific descriptions are difficult. Transitional activities differ from the modern sector. Production in non-agricultural enterprises is carried out in the urban milieu, and the features most often noted are:

(1) family ownership of enterprises;
(2) small scale of operations;
(3) labour-intensive technology;
(4) skills acquired outside the formal school system;
(5) unregulated and competitive markets.

These enterprises cover a wide domain, from small traditional cottage industries with low productivity and limited value on the social scale such as shoeshine or cigarette vendors, to repair services affiliated with modern production. The work of the ILO is concerned with this sector as it forms a kind of regulator for employment in towns where wage-earners form only a fraction of the available active population.

Here the preference is to put the problem differently. It is unrealistic to draw a picture of the Ivory Coast as having two sectors, a modern and a traditional one, when the majority of the populace is engaged in enterprises relating to neither one of these. Instead it must be shown how traditional organization forms have been abandoned to benefit new activities springing from needs in the modern economy, with the understanding that most of these small producers cannot create enterprises which have the modern sector's standards of efficiency and organization.

These transitional activities fit as comfortably into the agricultural domain as into industry and the service sector. Between the exclusively family-type agricultural operation (home-grown food for home consumption) and the agriculture of the huge capitalistic plantations, a dynamic intermediary organization has sprung up. At the centre of this expansion is found a juxtaposition of products for export and foodstuffs for home consumption. Here again we find a certain overlapping of strategies 'from above' and 'from below'.

Another fundamental point: these activities can only exist because of introduction of the modern sector into the traditional economy. Activities of transition respond to two types of demand: one, to assure a flow of agriculture exports under conditions of low capital utilization and low wages to labourers; the second, to respond to demands for goods and services arising from the modern economy. More rarely—but the cases are interesting—transitional production activities compete with goods offered by modern industries: for example, in sectors with a structure of small craftsmen producing various goods such as clothing, furniture, pottery, etc. Small tailor shops compete nicely with more elegant modern-sector establishments.

A third large group contains diverse activities of repair work, a flourishing segment in the towns of the Third World. Abidjan is amply furnished with 'radio doctors', motorcycle surgeons', garage mechanics, electricians, watch-repairers, all charged with satisfying the needs of a not-too-demanding clientele.

Thus transition-type enterprises are very varied but have common features. They could not exist without the desire for imitation, or the demand which comes from the presence of modern enterprises. They are characterized by an organization which is neither that of traditional society nor of modern production. The agricultural sector also has a dualist structure. Small planters ensure their subsistence with food crops generally looked after by women (cassava, banana trees, etc.). Commercial crops like coffee and cocoa are often worked by men with the help of salaried foreign labour, generally from Upper Volta. In the crafts sector, production is achieved with very low capital investment: non-existent headquarters, or perhaps one small room; plant consisting of a few tools; absence of stock; production done on demand, the first materials purchased as needed or supplied by the client. Salaried labourers are few, it is often a family enterprise, and earnings are low. Tiny quantities are sold. It is difficult to form an exact estimate of the place of these transitional activities in the Ivory Coast economy. But it is estimated that nearly 80 per cent of the employees in agriculture come from the transitional sector, and at least 50 per cent in second- or third-rank occupations. Its importance varies in urban centres according to the centre's size, being predominant in small and middle-sized towns and less important in large cities.

There is no indication of a decline of these activities over the last few years. The growth of Abidjan has not been followed by a parallel growth of modern employment. In the 'Bidonvilles', which have multiplied in the last ten years, the number of salaried employees entering the modern sector has been low. Only transitional occupations have ensured a precarious existence. In rural areas the money economy has progressed considerably but transitional occupations remain predominant.

The importance of these enterprises in the economic life of the country is thus well demonstrated; but it is crucial to clarify their role in the over-all process of development.

2. The Role of Transitional Enterprises in the Development of the Ivory Coast

To understand the importance of these enterprises, it is necessary first to stress their diversity and ability to evolve. Ivory Coast agriculture is organized essentially in a dualistic way, with foodstuffs raised for home consumption, and commercial production for export. Yet to an observer of the rural countryside, there have been marked changes in the last 20 years. From 1950 to 1965 agriculture in the south still reflected the ancient practices of the hunter–gatherer economy. On one parcel of land, mixed inextricably, would be found coffee plants and banana trees, cassava, corn, and various other edible plants, all sprinkled amongst the last remains of the great tropical forest. Today the rural landscape is evolving. The needs of towns have caused veritable fields of produce to appear. The quality of coffee and cocoa is improving and the density of planting increasing. Certainly coffee plantations like those of Latin America are still in the future, but transformations are obvious in comparison with other African countries. The agriculture sector's internal dualism separates the case of the Ivory Coast from the analyses of Stöhr which draw conclusions from one

setback in the financing of commercial operations by underdeveloped economies. Ivory Coast agriculture is responding excellently to national needs. Since
1973 exports as a group have grown by 172 per cent. Imports in volume have
grown by 10 per cent while the population has grown by 17 per cent.

This dynamism is met with again in the urban sector of transition. If technology is limited, it sometimes approaches that of modern enterprise: transport
businesses, long-distance hauling, tailors, to name a few. Such businesses have
the best opportunity of becoming integrated into the modern sector.

Two essential aspects identify the position of transitional activities. Firstly,
these transitional activities have provided the conditions under which modern
capitalistic activities have evolved and flourished. Secondly, these activities
help cover the essential needs of the populace due to a genuine, independently
functioning interchange with modern economy activities.

Transitional Activities Make Possible the Creation of Productive Capitalistic Structures

The economic theory of underdevelopment is concerned with the labour factor
in the development process, following Lewis (1954). This theory is still, however,
imperfect and unrealistic, and does not clarify how surplus manual labourers
can exist in an urban milieu where job creation is limited. In the case of Abidjan,
the modern sector can only give employment to one in ten. But it would be
false to affirm that there are 300,000–400,000 unemployed there. That might
be true if one considered only salaried labour, but socioeconomic reality is
very different. The notion of disguised unemployment too, in the traditional
theory, seems dangerous and ambiguous. The explanation for it only considers development as evolving in the Western model, where marginal work
productivity could appear low, even negative, as could the productivity of the
agricultural worker, if one referred chiefly to saleable productivity. In reality
one portion of economic activity of the Ivory Coast is outside the market or
takes place in a different market from the one for modern goods.

Because one portion of the population lives outside the modern economy,
urban growth can be pursued in spite of the relatively low growth of salaried
employment. Existence of informal enterprises permit high rates of urban
growth. Certainly, and we return to this point, the income from such activities
is not great. If the head man of an enterprise earns a rate above minimum wage,
salaries are less for the others, and perhaps nill for an apprentice. But if, contrary to the suggestions of Todaro (1969), hope of finding salaried work in the
modern sector is faint, here the apprentice can reasonably look forward to becoming salaried, or even creating his own business (Mettelin and Schaudell,
1978; Lachaud, 1979).

It would be tempting then to see the informal sector a reservoir of labourers
from which modern sector contractors would be able to draw. Research,
however, reveals that workers in these sectors generally wish to continue working there. It is thus not certain that mobility towards modern wage-earning

jobs would be great. It is more probable that these enterprises are part of a system of family solidarity, if sometimes simplistic. True, members of a tribe can gather together in a town with their 'brothers' hoping to find employment, but this is often a case of sharing out the tasks of a certain job, or contributing certain resources. Often from such poorly paid transition enterprises, meagre earnings are obtained. If these activities flow directly from modern enterprises and a huge urban expansion, it is still difficult to see in this a case where outside capital is creating a reserve of urban labourers. Until recently, government policy has been hostile to the creation of this type of operation, seeing in it competition with the modern sector, and fearing a large financial drain. But these transitional enterprises are the result of adaptation of the traditional rural milieu to the impact and attraction of urban life. They help solve employment problems which without them would be insoluble. And they make a minor contribution to the modern economy at the working level.

Transitional Enterprises and the Creation of an Autonomous Linkage Connecting Markets for Services and Products

Here we formulate an important hypothesis, of the existence of an autonomous liaison between transitional and modern economies. A simple idea: transitional activities furnish goods and services which come from needs created by the modern economy; but they market them at relatively low prices, leaving the producer only a tiny remuneration with which he can only buy goods produced at the same level. This is the case with agricultural products for export. Documentation is scarce but research done in Abidjan (Mettelin and Schaudell, 1978; Lachaud, 1977; and Lubell, 1976) seems to confirm the hypothesis. Listing of average incomes is not significant, due to irregularity of earnings. During the enquiry for example, 16 per cent of auto mechanics interviewed indicated they had had no repairs to carry out during the preceding week. As to their clientele, it was mostly households and enterprises which they serviced regularly or which were in the modern sector.

Research on urban markets supplied comparable results. Enterprises remain small, prices are difficult to evaluate and compare with prices in the modern sector. The conditioning is very different, sales often being by the piece or in a little pile, never by the kilo. Thus prices vary greatly according to the time of day (lowering of prices at end of market time), or according to the competition. Earnings are generally small, below 30,000 francs CFA per month in three-quarter of the cases. Quality and quantity sold differ. In certain market-places, the presence of a European clientele can modify the usual mechanisms and bring to that transitional market the practices of the modern economy; for example at Abidjan the Plateau Market receives a European clientele of some importance, and prices here differ from those charged in other quarters such as Adjame or Treichville. But this is an instance of transitional enterprise responding to varying situations.

In all it seems possible to affirm that there indeed exists a self-sustaining autonomous linkage relating to transitional activities, made up of various elements: small markets, very low profits, modest mark-up, no accumulation of real capital to modernize the enterprise; a mostly local clientele with low buying power and not exacting about the quality of goods bought.

In this way the informal sector enterprises supply chiefly either clients who also come from the transitional sector, urban or rural, or minimum wage-earners incapable of buying in the big modern establishments.

In closing, it should be remembered that the transitional sector is not homogeneous, and conclusions presented here could change. In addition it is one of the characteristics of this sector to offer the whole gamut of production units, traditional to modern. This is why these have been referred to as *transitional activities*, rather than as a *transitional sector*, because the word 'sector' implies a homogeneity that does not exist here.

The second comment concerns relations between modern and transitional activities. In speaking of a specific linkage, the connections are sometimes underestimated. But transitional enterprises do purchase in the modern sector, although not in a large way. Productive capital is tiny with goods often bought on sale; there is frequent use of sale materials in repair shops; foodstuffs are bought nearby from other producers, truckers, or farmers. If the relationship exists it is more on the level of the dynamic of needs: individuals wanting to obtain superior-quality merchandise when income allows. Finally there could be a relationship at the purchasing level, where people connected with the

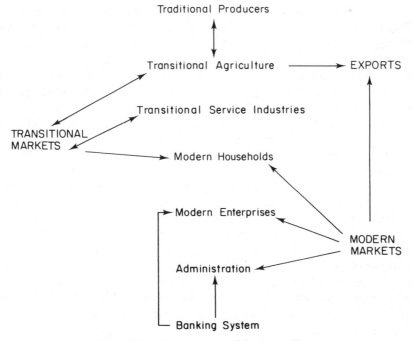

Figure 12.2 The economy of the Ivory Coast

modern sector due to the nature of their work become buyers in transitional markets, thus posing a problem of competition between the two. One could sum up the functioning of an economy like the Ivory Coast as shown in Figure 12.2.

In sum this outline demonstrates the limitations of policies which only consider the growth of the modern economy and ignore that of the essential transitional sector, even though the revenue flow is sluggish from one to the other.

This outline also demonstrates that development in the Ivory Coast nicely combines strategies 'from above' and 'from below'.

Stöhr (Chapter 2 of this book) lists three fundamental traits in characterizing the latter:

(1) full utilization of local resources;
(2) the satisfying of essential needs;
(3) the integration of economic patterns.

We have tried to demonstrate that these three objectives have been—at least partially, and progressively—achieved here, in a complex economic structure but integrated nonetheless into national and local realities.

3. Towards an Over-All Development Policy

For a long time Ivory Coast policy has not favoured transitional enterprises. A certain hostility has existed among planners encountering enterprises not integrated with ease into a capitalistic development plan. In the agricultural sector the importance of local plantations was largely recognized, but more for political reasons than for the evolving of a true development strategy. For a long time prices paid to producers were relatively low; this allowed the Bureau of Stabilization for coffee and cocoa to make substantial profits, which financed numerous public investments. Paradoxically, economic policy has been more favourable to large modern units, since independence, than it was under French colonial rule. But for several years now a timid revolution has been taking shape. First a substantial rise in prices has been given to producers. This could be explained by world price rises and internal price hikes for industrial products, but it also demonstrates a desire to improve planters' living conditions. In the industrial domain, a certain evolution was connected to the policy of 'localization'. Such a policy had limits, and enterprises directed by Ivory Coast natives other than public ones have been rare. The 'localization' of enterprises then changed to intervening on behalf of small enterprises. In this spirit the Office of Promotion of Ivory Coast Enterprise (OPEI) was created. This organization aims at supplying financial and technical assistance to new enterprises with Ivory Coast capital. It has acted to aid new businesses created by Ivory Coast natives of middle class rather than those in the transitional sector, with enterprises directed by managers chosen by them. One would hope this bureau's scope would evolve, and planning services might become more interested in transitional activities. In any case this demonstrates the difficulties met with in assisting transitional activities. The change to modern forms of this kind of enterprise is a delicate operation since, in evolving, the enterprise is cut off from its linkages and its markets, without necessarily achieving a reasonable state of competition *vis-à-vis* the modern sector. Modernization, in fact, has repercussions on costs, and consequently on prices. Evolution of transitional enterprise will doubtless come about less by internal evolution than by creation of new, better-equipped units with better-trained workmen, using improved business and management practices. These new units will then take the place of former enterprises whose production and management methods could not ensure their survival.

Thus the three chief aspects of a policy favouring transitional enterprises can be outlined as follows:

(1) A certain degree financial assistance will be necessary to permit certain transformations or certain creations. Organizations like the OPEI could respond to this need, being in a position to better define those sectors where such interventions would be timely.

(2) Such creative action would allow this educative movement to go beyond the existing training-on-the-job apprentice system.

(3) Any movement in favour of commercialization would have to assist informal enterprises to escape the limits of a market which too often is simply reduced to a section of the big city, or a few villages in rural areas. Formation of marketing cooperatives could, for example, facilitate sales across the whole territory, and even explore activities where production requires considerable manpower—for instance the clothing trade where transitional enterprise already competes easily with the modern sector.

It would be most valuable to install such a three-fold policy in view of the logical over-all development the Ivory Coast has already achieved.

In conclusion, the Ivory Coast's development seems rich with lessons for everyone, not because of the apparent miracle that one sees there, but because behind this success—still imperfect and tenuous—profound change and development can be discerned. This has come about less by the erection of great blocks of flats in the capital city, than by a slow but profound alteration of the whole economy. This change is still strongly influenced by the outside world, even though internal forces have played a determining role. Connections with external powers are obviously indispensable for a small country which is not too well endowed with mineral resources and must acquire, from outside sources, part of its energy and primary materials and almost all its equipment. The polarization of Abidjan, so often cited as an example in a critical vein, or contrariwise to underscore the results of growth, is such a recent event that it is only part of the explanation of the historical evolution of the Ivory Coast economy in the past three-quarters of a century. But this polarization was made possible by transformations which affected all the south of the country, especially the rural economy. It only succeeds at the present time because special structures assure the functioning not only of the urban economy, but of the whole country's economic system. These special structures are the role of transitional activities which has been examined here at some length. The Ivory Coast's success is indeed the reflection of an over-all development strategy. This should not be considered as a permanent fixture. On the contrary, the logic of Ivory Coast strategy is that of a perpetual rethinking concerning the country's equilibrium, always with a view to creating a more progressive and dynamic society.

REFERENCES

Amin, S. (1967) *Le Développement Economique de la Côte d'Ivoire* (Edition de Minuit).

Bretignère, A. (1931) *Au Temps Héroïque de la Côte d'Ivoire* (Editions Pierre Roger).

Gouvernement Général de l'AOF (1937) *La Côte d'Ivoire* (L'Agence économique de l'AOF).

International Labour Office (1976) *L'emploi, la Croissance et les Besoins* Essentiels.

Lachaud, J. P. (1976) *Travail et Développement: Concepts et Mesure* (Bordeaux: Université de Bordeaux I).

Lachaud, J. P. (1977) *Contribution à l'Etude du Secteur Informel en Côte d'Ivoire. Le Cas du Secteur de l'Habillement à Abidjan* (Bordeaux: Université de Bordeaux I).

Lecaillon, J. and D. Germidis (1977) *Inégalité des Revenus et Développement Economiques: Cameroun, Côte d' Ivoire Madagascar, Sénégal* (Presses Universitaties de France).

Lewis, W. A. (1954) Economic development with unlimited supplies of labor, *Manchester School of Economic and Social Studies*, Vol. 22

Lubell, H., H. Joshi and J. Mouly (1976) *Abidjan: Urbanisation et Emploi en Côte d'Ivoire* (Bureau International du Travail).

Mettelin, P. and S. Schaudel (1978) *Les Activités de Transition et le Secteur Informel à Abidjan* (Bordeaux: Centre d'Etudes d'Afrique Noire).

Ministére du Plan de Còte d'Ivoire (1977) *La Côte d'Ivoire en chiffres, édition 1977–78* (Abidjan)

Penouil, M. (1971) Le miracle ivoiren ou l'application réaliste de théories irréalistes. *Année Africaine* (Pedone).

Penouil, M. (1972) Growth poles in underdeveloped regions and countries, *Growth Poles and Regional Policies* Eds. Kuklinski and Petrella (Mouton).

Penouil, M. (1975) L'Afrique de l'Ouest: uniformité et diversité: vers un modèle dualiste de développement, *Monde en développement*.

Todaro, F. (1969) A model of labor migration and urban unemployment in less developed countries, *The American Economic Review*, **59** (1).

Verdier, A. (1895) *Trente Cinq Ans de Lutte aux Colonies*. (Paris: Léon Chailley.)

Development from Above or Below?
Edited by W. B. Stöhr and D. R. Fraser Taylor
© 1981 John Wiley & Sons Ltd.

Chapter 13

Tanzania: Socialist Ideology, Bureaucratic Reality, and Development from Below

JAN LUNDQVIST*

VISION AND REALITY

Tanzania, in terms of official assistance granted and over-all interest, is one of the most positively perceived Third World countries. A general view seems to be that Tanzania is seriously committed to attacking problems connected with poverty and underdevelopment. This view is not so much based on hard facts as on the content of the speeches and writings of President Julius Nyerere. President Nyerere has the ability to present an inspiring vision of what a nation and a people should be. The credibility of these ideas is ensured by some of his policies, especially his rejection of an affluent life-style. In contrast most Third World leaders are often regarded as political 'light-weights', marionettes expressing platitudes which seldom seem to affect their own position, or style of life.

But people in the long run cannot live only on visions. For the people concerned, and for the granting agencies, real progress is called for; and this requires not only ideology but also political action. The discussion in this chapter focuses on the relationship between political documents, their implementation, and the effects on people's situations in the various regions and localities of Tanzania. An assessment is made of the ways and extent to which policies have been either 'centre-down' or 'bottom-up' as outlined by Stöhr in Chapter 2 of this volume.

A note of clarification of the concepts of 'centre-down' and 'bottom-up' is warranted. It is true, as argued by Stöhr (Chapter 2 of this book), that the 'from above' strategy builds on a monolithic and uniform concept of development. On a national scale this implies, amongst other things, that a universal type of development will spread out from one or more centres to the rest of the

*The author would like to acknowledge the help of Rune Skarstein in preparing this chapter, which corresponds to the state of about mid-1979.

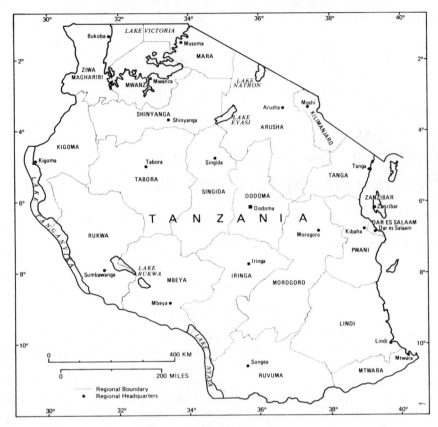

Figure 13.1 Regions of Tanzania

country, or put another way, from the core to the periphery. A spatial system
is created where '...a core dominates some of the vital decisions of populations
in other areas' (see Chapter 1 of this volume). In the 'centre-down' philosophy
there is no logical contradiction in this relationship, due to the monolithic
concept of development.

In the 'bottom-up' strategy the spatial system is different. Sub-national areas
should have the right to determine, for example, how to use the surplus generated, what technology to employ, and how to organize services for basic territorial
needs (see Chapter 2 of this volume). There is thus a potential conflict of interest and power between national and local/regional levels, and also between the
regions.

It is important to point out two problems in distinguishing between these
two concepts. First, there is always a 'centre-down' component, due to the
simple fact that government institutions exist; and a strong centre does not
necessarily exclude the possibility of a 'bottom-up,' type of development. On
the contrary, a committed and strong leadership is a prerequisite for realization

of a 'bottom-up' type of development. This is illustrated in the case of China, which has often been put forward as an example of the 'bottom-up' strategy (see Chapter 6 of this volume).

Secondly, the 'bottom' is a very vague concept. One would imagine that a 'bottom' exists in any society, and thereby in any strategy. But it can clearly have different meanings. In the 'bottom-up' strategy it is equated with groups, territorial units, and perhaps even with 'the masses'. It is thus a concept with a communal, collective meaning. In the words of Mao Tse-Tung: 'Now we are promoting construction and the mass movement. One needs a bit of "from top to bottom" such as government directives and orders, regulations and systems, etc., but in the main the masses must do things themselves' (Green, 1977, p. 7).

In the 'centre-down' strategy it is rather individual persons who are at the bottom end of the chain through which innovations diffuse, i.e., innovations spread through the urban system, or through social networks.

The distinction between the two interpretations of what is the 'bottom' has implications for the institutional framework of plan formulation and implementation. In most developing countries there are obviously few well-formed groups; and weak groups seem to require more support and organization than weak individuals.

PHASES OF POLITICAL ORIENTATION AND ACTION IN TANZANIA

There have been four clearly distinguishable phases in the transformation of Tanzania. The first phase was the six-year period after Independence, 1961–66. During this time Tanzania practised a strategy that was rather conventional and conformed with advice from industrialized countries. It was concerned with growth and transformation of the various sectors of the economy and did not, for example, consider regional patterns of development. It was supposed that mechanization and industrialization were to be furnished through private foreign and official assistance.

The second phase covered the period from the Arusha Declaration in 1967–72. The aim to restructure the Tanzanian economic and social structure in a socialist direction was now openly stated, and steps were taken to put African socialism into being. In concrete terms the plan was mainly based on two indigenous resources: land and labour. Foreign assistance would not be a requisite. The ideology underlying this policy, including the idea of self-reliance and voluntary arrangements at the village level, quickly gained considerable good-will, both abroad and within the country.

In the third phase, after 1973, the voluntary option for village development had virtually disappeared. Time had run short for results through education and good example. Compulsory action from the centre was introduced on a national scale to accelerate transformation of the settlement and land-husbandry system. At the same time ideological aspects became cloudy, and the transformation was confined to physical change. The external political orientation did not change, however. In concrete terms, the new policy first of all meant 'villagization', i.e. a movement of the scattered population into nucleated settlements; but there was no direct interference in the social system.

There are now hints of a fourth phase. Industry is to be upgraded and in the Third Five-Year Development Plan (TFYP), will receive a large share of government moneys. (TFYP was approved by the National Assembly in April 1978.) In agriculture an increase in productivity will be attempted requiring new technology, chiefly ox-ploughs. Changes have also been indicated to strengthen the role of the individual in the economy. This will probably mean a geographically uneven development, as there are not sufficient resources to implement such a design across the whole country. Also it will probably mean a strengthening of foreign assistance on a less discriminatory basis.

To help visualize these four phases, they are indicated in Table 13.1, which gives their position in terms of 'internal' and 'external' orientation of development policy. The fourth phase is identified by rudimentary signs of change in the last years. An alternative development is therefore possible. It would seem, however, that only assistance without strings, which was considered useful the development of Tanzania, has been welcomed. It is, however, important to stress, especially in regard to the latter point, that even though actual policies of Tanzania can be divided into periods with rather diverging trends, the vision drawn by President Nyerere has not been abandoned. But he admits it will take time to see it realized. 'In 1967 a group of the youth . . . asked me how long it would take Tanzania to become socialist. I thought thirty years. I was wrong. . . I am now sure that it will take us much longer!' (Nyerere, 1977, I).

Clearly, realization of his vision will take time. A central issue is the role of bureaucracy, which has the task of making the vision come true. But there is reason to believe that a large number of people in the bureaucracy

Table 13.1   *The four phases of Tanzanian development since 1961*

		External orientation	
		Open and positive attitude towards foreign assistance	Critical attitude towards foreign assistance
Internal Orientation	Development 'from above'	Phase I (Phase IV)	Phase III
	Development 'from below'		Phase II

have either not understood the political goals or are, for one reason or another, unable or unwilling to work towards their implementation.

It is therefore not so much a question of whether the Tanzanian people have the kind of leader they deserve and need, but rather whether the leader has the kind of bureaucracy he deserves and needs. It is consequently important for our analysis to distinguish between the strategy presented in political documents, and the strategy implemented.

PHASE I: 1961–66

From Neglected Colonial Territory to Ambitious Nation

At Independence in 1961, Tanzania was amongst the poorest nations in the world in terms of traditional criteria: per capita production for the market economy, and consumption. Potential for development had been considered low and, as a consequence, the colonial power had not interfered much in its internal policy or social structure, as had happened in other African countries. However, foreign domination was clear enough in terms of organization and direction of the economy. About 85 per cent of exports consisted of raw materials, and external trade contributed about 50 per cent to GDP. Physical and social infrastructure were concentrated in towns and in export or cash-crop-producing areas. The remainder of agriculture and the rural population were largely left to themselves, or were called upon as a labour reservoir for labour-demanding estates. Industry in the modern sense was tiny, contributing less than 5 per cent to GDP.

In the middle of what is here called Phase I, the First Five-Year Plan (1964–69) (FFYP) was brought forth. It was the first national plan, in the sense that the previous plan had been based on an IBRD report prepared before Independence. It was thus a document of great importance, and received considerable praise when presented. The President himself saw it as '... realistic and viable ... as vital to the future of this Union [Tanganyika and Zanzibar] as the attainment of independence itself' (Presidential Address to Parliament, 12 May, 1964, in United Republic of Tanganyika and Zanzibar 1964, vol. I, p. vii).

Before dealing with the actual situation and how it changed, some of the major plan objectives should be outlined:

In the agricultural and rural sector a 'transformation approach' was to be implemented by creating 74 village settlement schemes with about 250 family units in each. Through a complementary 'improvement approach', extension facilities were supposed to stimulate so-called progressive farmers. The agricultural sector would in any case decrease in relative importance: its contribution to GDP was expected to fall from about 58 per cent in 1960–62, to 48 per cent in 1970, to 37 per cent in 1980. Subsistence agriculture would decrease at about the same pace, from a 32 per cent contribution to GDP in 1960–62, to 22 per cent in 1970, down to 14 per cent by 1980 United Republic of Tanganyika and Zanzibar 1964. vol. I, pp. 21–22).

The role of the state was to furnish the country with an improved infrastructure, which was neglected and in bad condition. This, together with other favorable conditions, was supposed to smooth the way for private, mainly foreign investments in industry and trade. Almost 80 per cent of the budget for the FFYP was expected to come from abroad.

An accelerated growth rate from 5 per cent to 6.7 per cent was expected, in order to reach long-term goals set for 1980.

Rural and Regional Achievements

As indicated, the rural sector was clearly divided into two systems as far as crop orientation was concerned: export crops (mainly intended for industrial processing), and food crops for local consumption. Sisal, cotton, coffee, tea, and cashew nuts were essential cash or export crops accounting for about 75 per cent of total net monetary output in agriculture (United Republic of Tanganyika and Zanzibar, 1964, vol. I, p. 20).

Regional distribution of these crops was uneven, with 4 out of 17 regions producing some 60 per cent of the total. Export and cash-crop-producing areas stretched in a belt in the north along the border of Kenya, Lake Victoria, and Uganda. The central and western parts of the country were virtually without these important crops, and agriculture in the south was of minor importance (Jensen, 1968).

Crop orientation and regional distribution at Independence was quite concentrated, a condition which became even more so in the following years, as a natural result of plan intentions that: '. . . self-sufficiency of the family agricultural unit will be discouraged and specialization in one or two crops urged', (United Republic of Tanganyika and Zanzibar, 1964, vol. I, p. 20). During the first five years of Independence there was rapid growth in volume and value of major crops with the exception of sisal. The value of cotton, coffee, and cashew nuts more than doubled between 1961 and 1966 (United Republic of Tanzania, 1969, p. 14).

Monetary income naturally increased in regions producing these crops compared to the other regions where staple crops were grown, but only a small amount circulated through the market system, with the remainder used for subsistence consumption. Value of total marketed production per capita as a result showed large variations between regions. In 1966, when this colonial type of orientation of the economy probably reached its climax, Dodoma in central Tanzania marketed a production equivalent to Sh. 28 per inhabitant, while Kilimanjaro near the Kenyan border marketed production worth Sh. 258 per inhabitant (Jensen, 1968). Value of production marketed from other regions varied between these two extremes, with high-income regions in the north clearly predominating.

Continued emphasis on the same agricultural crops was successful in that production increased, and by and large so did value. But it was not successful in terms of area and labour productivity. Cotton for instance had been encouraged in Sukumaland since the War, and average acreage under cotton per household had increased from less than 1 to $3\frac{1}{2}$ acres. This expansion had been partly at the expense of food crops, and had been made possible by extension into bordering areas with less population. But since labour productivity and crop husbandry did not appreciably improve, growers did not benefit to the extent implied by production figures. Rather it was the traders and transporters who benefited (Mapolu and Phillipsson, 1976, p. 51). It was, literally speaking, regional growth, not regional development.

Relative growth and wealth in cash and export-producing areas at independence was in this way stimulated by FFYP policy. In addition to crop orientation, the policy of settlement schemes contributed to a concentration of rural development resources which meant a split in rural policy as a whole. The Ministry of Agriculture, Forests, and Wildlife, for example, had advocated a policy supporting individual farmers. It had responsibility for carrying out the 'improvement approach', and was allocated about $6 million to do so. In comparison, the newly created Ministry of Lands, Settlement, and Water Development was entrusted with the responsibility of implementing the 74 village settlement schemes (amongst other things), and was allocated not less than £20.8 million (United Republic of Tanganyika and Zanzibar, 1964, vol. I, pp. 2 and 27).

By the end of 1965 about 20 settlement schemes were established in the central and south-central parts of the country. Hopes for transformation of the rural sector through these schemes, however, soon faded. Besides a heavy overcapitalization, the schemes were haunted by what seemed to be the usual problems of such undertakings: 'unsuitable' settlers, insufficient planning, and problems with infrastructure, especially water.

At the same time, concentration of rural investment resources in these small 'islands' of development left most peasants and the traditional agricultural ways largely isolated. New social and regional disparities arose, and because of these frustrating developments the village settlement scheme was abandoned in 1966.

Matters worsened due to abruptly falling export prices for the important cash crops. Export earning from sisal were reduced by more than 50 per cent between 1963 and 1967 and besides the disruptive consequences on the national economy, effects were severe on a regional and local level, since geographical locations of estates were heavily concentrated. In the two most important sisal-growing areas, Tanga in the north-east and Morogoro about 200 kilometres west of Dar es Salaam, employment (and earnings) declined from 160,000 in 1964 to less than 110,000 in 1968 (United Republic of Tanzania, 1972, p. 39).

Coffee and cotton, two of the other prime export earners, also faced problems due to quota agreements in the case of coffee, and competition from synthetics in the case of cotton. In regional terms peripheral areas in the north, south, and west, together with areas south of Lake Victoria, were affected.

Foreign Reliance and Social Growth

The fall in total export earnings was serious, as imports had been greatly encouraged. Further strain on the economy resulted when a number of projects with anticipated foreign investment never started, or came to a halt. The plan was in fact an elaborate wishing-list, a desire for a number of foreign-financed projects, based on only a few concrete offers. As a result Tanzania had to proceed on an *ad hoc* basis, proposing sometimes alternate projects to those in the plan.

But the most serious assault on the FFYP and its philosophy, followed since independence, came about when political strings attached to foreign aid became obvious. Thus Tanzania rejected West German aid, conditional as it was on there being no East German representation in mainland Tanzania. A big sugar project in Kilombero, 300 kilometres south-west of Dar es Salaam, was affected. At the same time in 1965–66, Tanzania almost broke off relations with the Commonwealth as a result of Great Britain's stand on Rhodesia.

These events meant that about half the anticipated funds never materialized, and the results became clear in terms of economic performance. A tricky situation had also evolved in social formation and regional stratification. The settlement schemes, and support of 'progressive farmers' had caused problems. And a more than fourfold increase in Tanzanian civil servants—from 1170 to 4937 between 1961 and 1967 (United Republic of Tanzania, 1969, p. 24)— created a new class structure in the urban system, (see Shivji, 1976, for a discussion of class formation in Tanzania). Tanzania was thus changing rapidly in a number of respects.

Comment

Half-way through the FFYP period it was clear that the chosen strategy was unsuccessful. The main objectives as listed above were not achieved. It was basically a 'centre-down' strategy with capital, technology, and expertise expected largely to come from abroad. But the centre failed to supply enough input. Some inputs which actually arrived were inefficiently used, with limited spread effects. This was the case with the settlement schemes. Growth was patchy and slow; but more importantly, development in the country was influenced more by external factors than by Tanzanian authorities. The signal that a new phase was imminent came with the Arusha Declaration in 1967.

PHASE II: 1967–72

Socialist Perspective and Consolidation of Bureaucracy

The Arusha Declaration is probably best known in conjunction with the concept of self-reliance. According to common interpretation, the Declaration was an act of necessity, a logical consequence of frustrations over reliance on foreign aid as experienced in the FFYP. Self-reliance is thus often viewed favourably in the light of foreign experience. But it is obvious that the internal situation played an important role, at least in the timing of the presentation. 'The President was to say, in fact, that had he waited another eighteen months to take the Arusha initiatives it would have been impossible to assault these various citadels of privilege' (Saul, 1971–72, p. 12).

Privilege indeed was directly attacked: leaders were forbidden to have more that one salaried job, or to own more than one house, etc. Strengthening of central and party control indirectly meant the government had now created a framework within which equality and participation of a large section of the population were feasible.

Nationalization of banks, certain industries, and estates, shortly after presentation of the Arusha Declaration, was an expression of the policy of self-reliance. And since self-reliance and socialism were presented as being closely related, nationalization was looked upon from this dual perspective, at least by foreign owners. It has been argued that nationalization served primarily to strengthen the bureaucracy; but whatever the motive it is clear that changes which took place after the Arusha Declaration created a situation where a 'bottom-up' strategy could be implemented.

A second Five Year Plan, 1969–74 (SFYP) translated some of the Arusha Declaration intentions into planning guidelines. Rural development had top priority. Major emphasis was put on Ujamaa development: 'modern traditionalism', and other forms of cooperative productive activities were to be expanded. SFYP also advocated a spread in urban development to divert expansion from Dar es Salaam into nine regional centres. Industrial employment was to increase by about 50 per cent from 1969 to 1975. Some of the most interesting changes in independent Tanzania occurred during implementation of the SFYP. But a continuous development strategy consistent with documented policy and principles of the SFYP started before 1969, and was halted before the end of SFYP (1974), as will be discussed below.

Rural and Regional Development

Ujamaa: Modern Traditionalism, or the Building of Socialism

The policy of Ujamaa in rural development has aroused great interest far beyond the boundaries of Tanzania. It is the central and therefore the critical focus of Tanzanian policy. Presentation of the concept has been so frequent that its content has not been questioned. It was launched as a twentieth century version of traditional African village life. A great deal of initiative and self-reliance was to be given to the people.

> The first essential for successful Ujamaa is that it should come from the people, rather than being imposed from above.... It is vital that, whatever encouragement TANU [Tanganyika National Union], Government or parastatals give to this type of group, they must not try to run it; they must help the people to run it themselves. (United Republic of Tanzania, 1969, pp. 27–28.)

More specifically the Ujamaa idea includes communal arrangements in ownership, work effort, and the sharing of benefits.

However, as Mushi has pointed out, 'Ujamaa was possible only at the primary family and, in some cases, the extended family levels. What existed beyond these levels was ujima rather than ujamaa' (Mushi, 1971, p. 22). 'Ujima' means that the communal arrangements are concentrated in work efforts during certain peak periods, but does not include communal ownership. The obvious implication here is that Ujamaa does not mean return to a continuity temporarily upset by colonialism, but rather that Ujamaa on a larger scale had to be built.

Besides the intention to give people a vital say in initiation and management of the villages, a frontal approach was decided upon in that rural areas everywhere would be transformed simultaneously. This soon proved impossible. Obvious regional differences in response appeared from the start. In 1969 the Mtwara region along the border with Mozambique had one-third of all villages in the country. At the other extreme, half the country had only 15 per cent of registered villages (see Table 13.2). This uneven distribution lessened but then stabilized, and for four years about one-fifth of villages were concentrated in one region, Mtwara, and one-fifth were spread out in nine regions with low Ujamaa village density. The total number of villages increased rapidly from 1969 to 1972, but there was stagnation in most regions between 1972 and 1973. Figures for 1974 (given in United Republic of Tanzania, 1975) show a reduction from 5628 to 5008 villages between 1973 and 1974. The reason given was an amalgamation of smaller villagers; but figures for 1974 seem uncertain, and since 'operation village' was introduced at the end of 1973, they are not included below.

Actual development was thus not successful from the point of view of mobilizing the whole country at the same time, nor was it successful from a 'people's participation' viewpoint. Response was particularly low in more affluent regions such as Kilimanjaro, Arusha, and West Lake in the north and north-west, with only 1–3 per cent of population living in registered villages by 1973. There are reasons for the low and geographically uneven number of Ujamaa villages. First, the figures not only reflect voluntary response but they highlight the fact that, to some extent, force was used. In Mtwara region with the largest number

Table 13.2   *Concentration of Ujamaa villages, by region*

	1969		1970		1971		1972		1973	
High concentration										
One region	264	(33)	465	(24)	748	(17)	1088	(20)	1103	(20)
Two regions	412	(51)	815	(42)	1399	(31)	1801	(32)	1815	(32)
Three regions	472	(58)	1100	(56)	1991	(45)	2431	(44)	2477	(44)
Low concentration										
50 per cent										
of all Regions	125	(15)	242	(12)	822	(18)	1105	(20)	1163	(21)
Total:	809	(100)	1956	(100)	4463	(100)	5546	(100)	5628	(100)

Note: figures in parentheses are percentages.
Source: figures compiled from United Republic of Tanzania, 1975.
Synthesis: The first part of the Table shows the number of Ujamaa villages in one, two and three of the regions having most of the villages, e.g. 44% of all villages are concentrated in three regions in 1973. At the other end (second part of Table) there were only 21% of the villages in nine out of the eighteen regions in 1973.

of villages, this was motivated by the conflict with the Portuguese over Mozambique. But by the end of the 1960s compulsion was clearly not authorized by the president (in a speech in 1968, the President re-emphasized that Ujamaa should be voluntary; see Nyerere 1973). It was therefore only some energetic souls dealing with Ujamaa development too forcefully, who rapidly raised the number of villages in their regions.

The growing number of villages between 1970 and 1972 resulted from a massive campaign and a partially new strategy, usually called an 'operation'. Using the excuse of unforeseeable floods in the Rufiji Valley about 150 kilometres south of Dar es Salaam, the government in 1969 had already arranged new sites for settlements. A more or less *de facto* compulsory movement of people took place. New 'operations' followed in 1970 and until 1973, on a regional basis, first in Dodoma in central Tanzania and then in Kigoma.

No obvious reasons for the stagnation in growth of villages appear. It was probably the combined effect of frustrating experiences and incompetent bureaucracy. There are few good examples of villages built and maintained by villagers themselves, and some of those which could have been labelled 'good' as judged by intentions of the SFYP and ideological documents, were unscrupulously harassed. Best known is the Ruvuma Development Association (RDA), formed by 16 villages. In terms of both cash and food-crop production and other activities, they exemplified the intentions expressed by President Nyerere and the SFYP (Coulson, 1977). RDA was, however, considered a threat, both by expatriates advocating planned settlement schemes and by bureaucrats; and in 1969 the President dissolved it. According to Coulson (1977, p. 91) he was not prepared to accept slow speed of development where small groups of committed individuals formed villages, nor did he accept a differentiation between committed socialists and the rest of the people.

Bureaucrats—because of educational background, experience, and expectations—were ill-equipped to deal with the complexities of rural development. They had little or no training in agricultural development, and to vindicate their role they devoted more and more effort in areas where they could compete with the peasants who they considered stubborn and resistant. Consequently there was a '. . . growing commandist trend in the bureaucratic implementation of the Ujamaa policy' (Mapolu and Phillipsson, 1976, p. 46).

Actually the bureaucracy had involved themselves in a number of ways: in planning, decision-making, organization, and evaluation of villages (as described by Boeson *et al.*, 1977, for West Lake region). The involvement not only frustrated village initiative but it became costly, and the results frustrated good agricultural practices because of a preference by bureaucracy for technical input. Mapolu and Phillipsson (1976) describe how tractors were generously supplied to cotton cultivators in Sukumaland. The result was that peasants cultivated larger acreages of cotton than they could weed and harvest with available labour, and in the end more than half the crop was left in the fields to rot.

Bureaucratic insufficiencies are adequately summarized by Hydén (1975):

A main reason why the socialist results of the Ujamaa programme so far have been meagre is that any revolutionary strategy is ultimately a class strategy and not a development strategy that can be bolstered by, for example, bureaucratic control and technical assistance inputs. (Quoted from Boesen *et al.*, 1977, p. 16.)

Finally, special steps were taken to increase ways for regions to stimulate development of Ujamaa. A regional development fund was established in 1968 which allocated one million shillings to each region for projects not exceeding Sh. 50,000 each. This of course also increased the power of regional authorities. According to Collins (1976, p. 29), during its first years of operation, allocation of the Regional Development Fund, however, went largely into infrastructure. Thus 47.3 per cent of total funds between 1967 and 1970 went to roads, bridges, and wells. A large portion of this benefited Ujamaa villages but it is remarkable that direct Ujamaa aid over the same period was not more than 2.5 per cent.

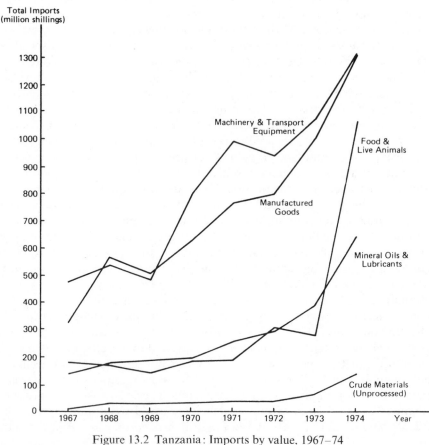

Figure 13.2 Tanzania: Imports by value, 1967–74

Source: Diagramme based on figures in Nnunduma (1977).

Effects of Ujamaa on Production and Regional Income

In the light of the above circumstances production declined, at least during a period after the movement to Ujamaa villages, partially due to crops left behind and partially to other labour demands at the new location (Omari, 1976, p. 135). But national agricultural growth has been declining for a long time. Low performance in this sector has caused great concern to authorities, not only because it raises questions about the policy's relevancy, but also because food imports have increased since the 1960s, placing heavy burdens on foreign currency reserves (see Figure 13.2).

Besides the problem with over-all production, evidence shows that the regional income gap for smallholders increased during the first half of the 1970s. There are relatively wealthy farmers, especially in northern regions with a favourable climate. These are areas where missionaries have in the past encouraged growth of commerical crops. Geographically it is interesting that these areas with relative land scarcity have a fairly high income level on average— though there are certainly poor people here as well—whereas regions with land abundantly available have relatively small high-income groups.

Historical and geographical factors are not confined within regional boundaries. Considerable intraregional differences exist. On the local level socio-political aspects also contribute to revenue variations. Raikes (1978) describes how cooperatives and access to other official sources have been mechanisms also causing intraregional differences; but generally wealthy farmers have not been members of Ujamaa villages.

Decentralization of Administration and Organization of Parastatal Sector

Priority to rural development also implied changes in the administration of planning. One step was the establishment of the Regional Development Fund mentioned above. A further step to strengthen regions' involvement in planning their own development came in 1972 with so-called decentralization. This policy called for the upgrading to rank of Minister of Regional Commissioners whose position had been ambiguous; appointment of Regional and District Development Directors; and decentralization of 40 per cent of the development budget under direction of the Regional Development Director and his staff.

As a logical consequence, regional and lower government levels then formulated development plans to be sent to the Prime Minister's office for review and coordination. At the same time local governments and town councils which had been inherited from before Independence, were abolished. The steps taken can be interpreted in different ways. Reorganization could imply a more effective administration; alternatively, it could form a basis for carrying out political ideals; or it could fulfil both ambitions. In light of the Tanzanian ideology as expressed by President Nyerere, one would expect reforms to guarantee implementation of socialist principles in planning, i.e. the involvement of people in planning at the local level and above. Ironically these same proposals regarding the carrying-out of reforms were largely based on recommendations from consulting firms not famed for their socialist bias, namely McKinsey of New York.

The results can be assessed in two ways, each one having two opposing poles: (1) expediency *vs.* ideological conformity; or (2) bureaucratic control *vs.* popular participation.

As mentioned, previously the Regional Development Fund was largely spent on infrastructure. There was considerable popular demand for RDF but many applications had to be rejected as calling for social service schemes exceeding administrative capacity (Collins, 1976). Fund allocation seems thus to have been biased in favour of projects that were easy to identify, did not require complicated judgments, and could be carried out with or without popular support or assistance. Another interpretation of allocation of the Fund is that it reflects the difficulty villagers and non-bureaucrats have, articulating projects and forcing them through. There is a similarity with the discussion on Ujamaa above.

These difficulties are not confined to working relations between the regional, district, and local village levels. Similar weaknesses occur between regions/ districts and the national level. Discussing the decentralization measures of 1972, Green remarks:

> The 1972–74 results indicate clear progress... with one glaring exception. Serious regional proposals for the 1975–80 Plan were to have reached the Prime Minister's office for review and co-ordination work to commence early in mid-1974. By late 1974 many regions and districts (by no means all) seemed not to have started, yet alone completed, a serious planning exercise. (Green, 1974, p. 38.)

Regional proposals for the 1975–80 plan were supposed to be worked out by teams of expatriates from various countries and international agencies. Thus Sweden for example, was involved in two regions, Norway in one, etc. Recruitment of these teams was sometimes not done early enough to present a plan document, and there were coordinating problems.

Important also was the unfulfilled intention to decentralize 40 per cent of the development budget to regions. According to the rules, the Prime Minister's office (PMO) had the right to eliminate projects proposed by regions. This arrangement could work on the principle that one central body must coordinate regional proposals, but it is obvious that the cuts were much greater:

> Data from Rukwa Region indicate that the original submission of their 1976/77 plan stood at shs. 48,095 mil. This is what represented the views of the people of Rukwa since it had been approved by both government organs and Party organs in the region. Pressure from the centre, i.e. PMO and Treasury, caused the regional authorities to cut down the budget halfway to shs. 24,447 mil. This submission was not even accepted for it had to be cut down further to shs. 15,373 mil. Ultimately the approved regional budget was 15.18 mil. (Shao, 1977, p. 42.)

According to the same source the region's share of the total plan resources was only 10.74 per cent for 1976–77 (Shao, 1977, p. 41).

Clearly, decentralization succeeded in terms of central bureaucratic control but failed in ideological conformity, popular participation, and expediency. Coulson (1977, pp. 93–94) summarizes decentralization thus: ' ... it was a decentralization of the civil service: for elected local government was abolished ... the Party strengthened the salaried Party officials at the expense of elected representatives ... '.

Changes in the parastatal sector were similar to those above in strengthening central bureaucratic control. Through nationalization of a number of major enterprises in 1967, the National Development Corporation (NDC), the main parastatal organization, found its duties considerably increased. Its evolving role from 1965 to 1967 is illustrated by the stock of fixed investment it controled: in 1965, Sh. 123 million, which in 1967 had doubled, to 243 million (Packard, 1971–72).

Growth in the role played by the NDC was, however, not coupled with a change in obligations, which were to carry on 'business ... at a net profit, ... to be self-financing. It did this very well at first due to windfall gains from nationalization, foreign management and capitalist work customs.' (Packard, 1971–72 p. 64). The potential danger was that NDC was living a life of its own, out of phase with the rest of the economy and the intentions of stated policy. It therefore was reorganized into three parastatals, each responsible to a Ministry. They in turn were to be coordinated through the two central Ministries, of Finance (Treasury) and Planning (Devplan).

Steps were also taken to change the base below the mentioned three-level structure. In the President's Circular Letter No. 1, 1970, the initiation of Workers' Councils was announced, which were supposed to have an advisory function. Results were discouraging in both economic and workers' participation aspects. A growing number of companies showed losses, with the exception of the four big branches: breweries, cement, tobacco, and diamonds; and a positive role for the Workers Council was not achieved because of a lack of education and of clearly-defined aims (Hydén, 1972).

Dealing with the changes in the parastatal sector, Packard (1971–72, p. 66) notes:

'As we have seen, of course, the directive on Workers' Councils partially fills the gap of socialist 'control' over the parastatal organizations. But the re-organization of 1969 is really concerned with the planning system *from the centre;* and in this sense it lays little emphasis on the socialist content of the proposed planning system.

Comment

The most obvious changes between 1967 and 1973 were a strengthening of central control in both rural and urban/industrial sectors. As mentioned at the beginning, this could very well have been conducive to implementation of a 'bottom-up' development. But central control was not imposed to ensure popular participation in planning and project implementation. Instead it became an umbrella for bureaucratic interference. Besides a deceleration of the growth rate in the agricultural sector during Phase II there were increasing inter- and intraregional differences in terms of income and response to Ujamaa policy.

PHASE III: 1973–1978

Villagization By Order

One of the final moves in a primarily voluntary Ujamaa was the compulsory movement to development villages. A nation-wide 'Operation Village' was decided upon at the sixteenth TANU Biennial Conference in 1973. By the end of 1976 over thirteen million, that is the whole rural population, were reportedly living in such villages (Nyerere, 1977, p. 41). The policy which started out under the banner of socialist development was more and more being carried out in pragmatic terms, i.e. as a necessary step to better living conditions for the people. It is important therefore to assess in what ways the potential for improved living conditions has been increased.

As noted above, production decreased—particularly in the initial phase of settlement in the new place. This is not surprising, but more disturbing in the long run is the effect of lack of selectivity in the choosing of sites.

It seems that access to and from towns and authorities played a central role. As a result new settlements were located near roads. As Moore comments: (1976, p. 5:16), 'One wonders again whether this was for the officials' convenience rather than the farmers' needs.' In some cases this had disastrous results for agricultural progress.

> In certain regions, among them the West Lake Region, the existence of roads has been the key determinant for the location of villages. In the Karagwe District this is a disastrous choice. Here the roads are mainly located at the top of the mountains where the soils are thin and sandy, excellent for roads, but unsuitable for cultivation. The traditional villages are located in the valley bottoms where the fertile soil has been washed down and where the drinking water is available. (Boesen *et al.*, 1977, p. 172.)

Production in sites with poor soil could to some extent be compensated for by better access to input factors, markets, and more rational work organization. It seems, however, that only road locations and access to markets have improved. Work organization in development villages seems to be irrational, and has not contributed to mobilization of resources (Mapolu and Phillipsson, 1976). The shift into new villages has also caused disruption in traditional social values, which has been counterproductive to production efforts as well as having negative ecological consequences (Mascarenhas, 1977).

Input factors cost money; and neither government nor the peasants themselves can finance substantial input factors on a national scale. Coulson (1977, p. 94) asks, 'Why should the Government create impossible demands for services (water supplies, schools, dispensaries, agricultural extension and famine relief) which it must have known it could not fulfil?'

Furthermore, the argument for villagization was not very relevant as a prerequisite for viable units to supply the (limited) amount of input factors. The problem was not only a dispersed population, but in fact there was also a concentration problem. About 35 per cent of the rural population was living on a mere 5.5 per cent of the land area. This allowed for cultivation an over-all average of only 1.2 acres per person (Nnunduma, 1977, p. 84). People in such densely populated areas could have been efficiently supplied with inputs (Coulson, 1977) and at the same time production could have continued as before.

CRISIS AND REGIONAL DISPARITIES

The new strategy direction was put on trial almost at once. In 1973–74, changes in the international economic system struck Tanzania. A manifold increase in imports prices, especially for energy, came at a time when Tanzania was facing one of its severest droughts and food imports were crucial.

The change came suddenly. Figure 13.2 shows development of costs for selected imports over the period 1967–74, and it indicates a continuous increase in import of manufactured goods, machinery, and transport equipment. Decline in imports for machinery and transport equipment in 1971–72 reflects restrictions in saloon-car purchases. The most surprising feature is that predominantly agricultural Tanzania in 1974 had to import food and livestock for amounts not significantly less than moneys spent on industrial goods. The import bill for food and livestock not only represented a rapid increase in quantity ordered, but was also due to last-minute ordering (Green, 1977). The drought, of course, not only affected food production, but that of export or cash crops as well. Cotton production was most severely hit and showed a decrease in export from an index of 145 in 1972 to 85 in 1975. Coffee declined slightly in 1974 but recovered in 1975 (Nnunduma, 1977, p. 99). Existing regional disparities were intensified by the crisis. In coffee-producing areas in the extreme north, south, and west, decline in production was balanced by increased prices on world markets, which enabled peasants to slightly increase their incomes. Poorer regions, mainly in the central, southern, and western parts of the country, raising mostly food crops, were less fortunate. On average here earnings were about 15 per cent less in 1975 than in 1969.

In urban areas the crisis hit low-income groups harder than higher-income groups because of their different consumption patterns. While middle-grade civil servants in Dar es Salaam had a rise in price index for goods and services of about 51 per cent between March 1974 and March 1975, the increase in consumer prices for minimum-wage earners in the same period and district rose by 73 per cent (Nnunduma, 1977).

The crisis also affected industrial production with higher prices, import restrictions, and scarcity of raw materials.

Comment

It was possible to implement 'Operation Village' over a short period, partially because of the integrated spatial administrative system, and in spite of passive or negative response from large numbers of the peasants. But movement to new land caused disturbance in production, and further increases are to some extent hampered by bad site selections. Another shift to new locations has taken place; and this could perhaps be done systematically, but economic and social costs would be heavy. On the other hand an increase in production, especially in volume marketed, is essential in financing of centrally based services which are supposed to be supplied to villages.

PHASE IV: 1978–

A New Perspective in Planning

The second and third development strategy phases discussed herein are distinguished by the important ideological principles which have been formulated and abandoned. Phase four is not so easily distinguishable. However, formulations in the Third Five-Year Plan (TFYP), and comments by officials make it clear that important changes can be expected. The most important speak of a change of view in relative importance of the sectors. Industry and urban development will be upgraded at the expense of agricultural or rural progress.

But important changes are also announced for the rural sector. Until realization of 'Operation Village' in 1976, efforts had concentrated on institutional aspects and the settlement pattern; scant attention had been paid to producer price policies; but steps to stimulate rural growth by other means were finally taken. By May–November 1974, substantial price increases had been announced. There have been new increases since then. For the 1978–79 season price increases were announced in December 1977 (*Daily News*, 31 December 1977). This was done early so that farmers could respond in the coming season. Increases were for cash crops only, prices for food crops remaining stable. Cash-crop production is clearly a real concern: '... our cash-crop production continues to decrease from year to year, although due to favourable prices for some of them, the monetary value of these crops has not decreased' (United Republic Tanzania, 1978b, p. 11).

For the immediate future other important changes have also been indicated. In his Christmas Message for 1977 Prime Minister E. Sokoine announced the appointment of 4000 village managers (*Daily News*, 29 December 1977), for whom Sh. 30 million were to be allocated in the 1978–79 budget (United Republic of Tanzania, 1978b). The same article states that 'the role of the village manager and his relationship *vis-à-vis* the elected village government has yet to be spent out', and 'The unsettled issue is whether there should be spontaneous or induced development from the peasantry collectivised in some 7,500 registered villages '

There has also been an announcement that TFYP will place special emphasis on ox ploughing, to raise area productivity. Until the end of 1977 increase in production was due to cultivation of new land (*Daily News*, 27 December 1977).

In the TFYP for Economic and Social Development, 1 July 1976 to 30 June 1981, the role of agriculture for the long-term industrial plan (1975–95) is described as follows (p. 43):

> ...the success of the strategy depends heavily upon success in agricultural production and mineral exploration and exploitation, because the basis of the strategy is the utilisation of local resources to the greatest extent feasible in all industries ... Agriculture will have the following role to play:
> (i) to supply sufficient quantities of local food crops and industrial raw materials;
> (ii) to produce more food, investable surpluses and export crops to earn the foreign exchange needed to support industrialization.

In view of the agriculture sector's performance so far, these are indeed high expectations.

There is a renewed concern about industrial development and the making of new contacts with foreign interests, although one hears few official comments. In the section on industries in the TFYP, nothing is said about the role of foreign investments. Many writers, especially in the press, have interpreted the new policy as proof that Tanzania has left the rural sector to itself and/or that foreign dominance over the Tanzanian economy will increase and eradicate what remains of the policy of self-reliance and striving towards the social organization of production. This is a static interpretation, and is essentially incorrect. As Mapolu and Phillipsson (1976, p. 43) point out in respect to the Chinese Revolution: 'Either the inexistance of an industrial base leads to the stagnation of productive forces in agriculture or—more often—to a heavy *de facto* reliance on imported technology and expertise which reinforces the links with imperialism.'

CONCLUSIONS

Tanzanian policy and development have been sources of great interest to both academics and officialdom everywhere. Praise for Tanzanian ideology and its implementation has been expressed in a growing number of articles since the end of the 1960s. Many articles still appear but sceptical comments, often from a Marxist perspective, are now more common. There is frustration both concerning the way leftist intellectuals have been treated in Tanzania (including students at the University of Dar es Salaam—e.g. students who protested against a raise in pay for civil servants in March 1978, were harassed by police 'on orders from higher levels', see *African Political Economy*, No. 10, 1977); and also concerning the power concentrated within the bureaucracy and its inability to enter into dialogue with the masses of the people. Writers point, as well, to lack of initiative, the growing influence of the political party, and inefficiency in administration.

Many of these criticisms can of course be brought against a number of countries. The most serious one is concerning the relation between political intention, implementation, and the results achieved. Production results in agriculture have officially been described as poor, and development according to the principles of Ujamaa is also poor. A local self-reliant Ujamaa policy could have avoided some of the problems created through central control. But a very important prerequisite for making such a policy real is that there must be local resources of such quality and tradition as to make choices of technology, input provisions, etc., possible. Tanzania still has some way to go in this respect, and its industrial base is still not sufficient to cater to the mechanization of the countryside.

Continuous and increasing regional and local differences in income and poor response to central policies, are indicators that the 'frontal' development approach has been ineffective. This failure in development has occurred even though considerable resources and thought have been devoted to promote certain policies by government, the party, and the civil service.

Nevertheless, comprehensive political, physical, and institutional changes have taken place within a short time-period in Tanzania. Few other countries have exceeded Tanzania's pace, and results of these changes are still not fully known. In this perspective the years ahead will be as interesting as those years when the ideology was evolving, and should be studied accordingly. Unfortunately, it appears that the absorbing interest in Tanzania as a model of development was concerned more with the model itself than with results obtained from its implementation in Tanzania.

REFERENCES

Boesen, J., B. D. Madsen and T. Moody (1977) *Ujamaa-Socialism from Above* (Uppsala: SIAS).

Collins, P. D. (1976) Decentralization and local administration for development in Tanzania. IDS, *Discussion Paper No. 94*, Sussex (mimeo).

Coulson, A. (1977) Agricultural policies in mainland Tanzania, *Review of African Political Economy*, 10, 74–100.

Green, R. H. (1974) Towards Ujamaa and Kujitegemea: income distribution and absolute poverty eradication, aspects of the Tanzania transition to socialism. IDS, *Discussion Paper No. 66*, Sussex (mimeo).

Green, R. H. (1977) Towards socialism and self-reliance. Tanzania striving for sustained transition projected. *Research Report* No. 38 (Uppsala: SIAS).

Hydén, G. (1972) *Socialism och samhallsutveckling i Tanzania. En studie i teori och praktik* (Lund: Cavefors).

Hydén, G. (1975) Ujamaa, villagization and rural development in Tanzania, *ODI Review*, 1.

Jensen, S. (1968) Regional economic atlas mainland Tanzania. *Research Paper* No. 1, BRALUP (Dar es Salaam: University College) (mimeo).

Mapolu, H. and G. Phillipsson (1976) Agricultural cooperation and the development of the productive forces: some lessons from Tanzania, *Africa Development*, 1, 42–58.

Mascarenhas, A. (1977) Resettlement and desertification: the Wagogo of Dodoma District, Tanzania, *Economic Geography*, 53, 376–80.

Moore, J. (1976) Villages and villagisation in Mwanza Region, *Mwanza Integrated Regional Planning Project*, vol. III, Chap. 5 (Stockholm).

Mushi, S. S. (1971) Ujamaa: modernization by traditionalism, *Taamuli*, 1 (2), 13–29 (Dar es Salaam: University of Dar es Salaam).

Nnunduma, J. (1977) Economic analysis of the Tanzanian experiment (Dar es Salaam) (mimeo).

Nyerere, J. (1973) *Freedom and Development. A Selection from Writings and Speeches 1968 to 1973* (Oxford: Oxford University Press).

Nyerere, J. (1977) *The Arusha Declaration Ten Years After* (Dar es Salaam: Government Printer).

Omari, C. K. (1976) *Strategy for Rural Development: Tanzania Experience* (Dar es Salaam: East African Literature Bureau).

Packard, P. C. (1971–72) Management and Control of Parastatal Organizations, *Development and Change*, III (3), 62–76. (The Hague: Institute of Development Studies)

Raikes, P. (1978) Rural differentiation and class formation in Tanzania, *Journal of Peasant Studies*, 5 (3), 3–25.

Saul, J. S. (1971–72) Planning for socialism in Tanzania: the socio-political context, *Development and Change* III (3), 3–25. (The Hague: Institute of Development Studies).

Shao, A. A. (1977) *The mechanism for approving projects in the annual development plans* (Mzumbe, Tanzania: Institute of Development Management) (mimeo).

Shivji, I. G. (1976) *Class Struggles in Tanzania* (New York: Monthly Review Press).

Tanzania students protest politicians' spoils (1977) *Review of African Political Economy*, 10, 101–105. (Memorandum of Students, 5 March, 1978).

The United Republic of Tanganyika and Zanzibar (1964), *Tanganyika Five-Year Plan for Economic and Social Development, 1st July 1964 to 30th June 1969*, vols. I and II (Dar es Salaam).

The United Republic of Tanzania (1966) *Background to the Budget. An Economic Survey* (Dar es Salaam).

The United Republic of Tanzania (1969) *Second Five Year Plan for Economic and Social Development, 1st July, 1969 to 30th June 1974*, vol. 1: *General Analysis* (Dar es Salaam).

The United Republic of Tanzania (1972) *The Economic Survey 1971–72* (Dar es Salaam).

The United Republic of Tanzania (1975) *The Economic Survey 1973–74* (Dar es Salaam).

The United Republic of Tanzania (1978a) *Third Five Year Plan for Social and Economic Development* (published in Swahili 1978, now translated into English for publication).

The United Republic of Tanzania (1978b) *Speech by the Minister for Finance and Planning, Introducing the Estimates of Public Revenue and Expenditure for 1978–79 and the Development Plan 1978–79* (Dar es Salaam).

The following articles from *Daily News* of Tanzania:

The year 1977 that was: on Agriculture, 27 December, 1977.

1977 the year that was: on National Economy, 29 December, 1977.

Crop prices for next season up, 31 December, 1977.

Development from Above or Below?
Edited by W. B. Stöhr and D. R. Fraser Taylor
© 1981 John Wiley & Sons Ltd.

Chapter 14

Algeria: Centre-Down Development, State Capitalism, and Emergent Decentralization

KEITH SUTTON

> *In this writer's view, it can safely be said that Algeria today is consolidating the ground for its social and economic development more effectively and hopefully than any other Arab country.*
>
> (Sayigh, 1978, p. 521)

The undoubted growth of the Algerian economy in recent years understandably evokes eulogies such as that above. More practically, this improved economic standing has imbued the World Bank and other international loan agencies with the confidence to provide funds for many recent development projects, especially in the industrial and hydrocarbon sectors. However, this euphoria has been tempered by recent self-criticism within official Algerian circles and by observers concerned over the less successful translation of this economic growth into economic and social development. (Chaliand and Minces, 1972; Lazreg, 1976; Sutton, 1976).

It could be argued that such caveats are hypercritical when one considers the recent historical background to Algerian economic growth. The country's route to independence was long and cruel, with a guerrilla war from 1954 until 1962 during which heavy losses were inflicted on both sides, but particularly on the Muslim population which suffered many thousands of casualties and the displacement from their homes and crops of up to half the rural inhabitants. Independence in 1962 brought the exodus of their own volition of most of the one million French citizens, with the result that the functioning of many sectors of the economy faltered as basic skills were lacking and the bottom fell out of the market for consumer goods and services. It has been calculated that between 1959 and 1963 Gross Domestic Product (GDP) at current prices fell by about 28 per cent. In constant prices it fell even more, and Amin considers that by 1963 the real value of production had dropped 35 per cent below the 1960 level (Norbye,

1969; Amin, 1970; Knapp, 1977). Within agriculture, favourable climatic conditions aided cereal production but severe drops were recorded in the production of wine, vegetables and industrial crops. The partial and later near-complete loss of the French market for Algerian wine exports was especially severe. The production of manufacturing industry in current prices fell 20 per cent, 1959–63, and the drop in investments was even more abrupt, 82 per cent between 1961 and 1963, at current prices (Amin, 1970). Only oil and gas production increased at a rapid rate, 1960–64, and growing oil exports enabled the country to overcome the economic trauma of independence, and the subsequent setback associated with Algeria's determined quest for economic independence through the nationalization of mines, hydrocarbon assets, and commercial interests. It is in this context that recent strong economic performance should be gauged, as well as by international comparisons within the Third World.

Before focusing on the post-independence economy, it is necessary to stress the inequalities of the dual economy which prevailed in colonial times. In 1958 the average income per capita in Algeria was equivalent to £90 at current prices, which was about five times greater than in India or about twice that of Tunisia. However, this relative prosperity only benefited a small minority. Estimates for 1954 contrasted an average income of 360,000 old francs per head (£365) for the European population, with a mere 29,000 old francs (£29.50) for the Moslems. These French Algerians thus had an income one-third higher than the concurrent average in France. Further, in agriculture per capita income was only about £23, and thus the agricultural population in Algeria was little better off than people in the poorest countries in the world (Norbye, 1969). These marked social disparities would have found strong regional expression because of the concentration of the European population and the industrial sector in the larger cities and the richer coastal plains of the northern Tell. Interior and eastern regions predominantly based on traditional peasant agriculture lagged well behind in terms of income levels and standard of living. Thus there was a strong correlation between income per head and proportion of non-Moslems in the different departments. Income per head ranged from £245 in Algiers department to £25 in Batna (Norbye, 1969).

THE CENTRE-DOWN APPROACH

After an experimental flirtation with self-management and workers' control as an approach to running industry and agriculture the newly independent Algerian government pursued a policy of centralization. State capitalism and more rigid centralism through increased bureaucratic dominance were substituted for the more localized decision-making procedures and worker and peasant mobilization approaches which had stemmed from both ideology and spontaneity. Foreign-owned businesses and industries were nationalized leading up to the take-over of remaining French oil interests in 1971. If, for a while, this may have broken the diffusion of innovations through the multi-national firm structure, central to some of the centre-down literature (see

Chapter 2), the system was re-established by large state companies calling in international expertise.

While stressing that 'the policies of the 1965–71 period were thoroughly state capitalist, and the single-minded pursuit of centralism and industrialization by such means has left a legacy of rural stagnation . . .', Nellis cautiously suggests that a recent re-emphasis on popular mobilization and participation allows a re-evalution of earlier strategies (Nellis, 1977, p. 533). A reversion to centralization was a calculated risk to create the strong economic and political base on which later to build the more decentralized and participatory economy and society which had always been desired. This hypothesis will receive support from this chapter reviewing Algerian economic strategy and regional development policies. It will be argued that centre-down approaches, while achieving growth, have had a limited developmental impact, and so significant beginnings have been made to promote decentralization allowing opportunities for a bottom-up strategy.

In an interim situation, however, the continued central role of state companies and of industry-orientated national plans lend support to alternative explanations condemning decentralization efforts as cosmetic, and participatory socialist statements from government ministers as a smoke-screen behind which a meritocratic bourgeoisie has been strengthening its control. According to Lazreg, the state apparatus 'is used by the technocratic and the administrative bourgeoisie, along with the petty-bourgeoisie, as a means to reproduce their conditions of existence' (Lazreg, 1976, p.176). Certainly by 1972 state companies controlled 90 per cent of the industrial sector and employed 70 per cent of the industrial work-force (Economist Intelligence Unit,1977).

The leading state company, Sonatrach, reflects the financial dominance of the oil and gas sector, particularly since the 1973–74 oil price rises. Undoubtedly, the fundamental role played by this capital-intensive and high-technology industry has favoured the centralizing forces. By 1974 hydrocarbons accounted for 41.5 per cent of GDP. Oil exports were then 92.5 per cent of total exports. Estimates of likely government revenues for 1977 are that hydrocarbons will account for 69 per cent of them (Economist Intelligence Unit, 1977; Balcet and Nancy, 1975). Obviously the motive force behind recent growth in GDP and in investment rates (see Table 14.1) has been this important oil sector. If recently oil production has only increased slowly, the OPEC action in raising prices has ensured the financing of Algeria's ambitious industrialization programmes, including hydrocarbon-based industries. Furthermore, natural gas has yet to realize its full potential. By the early 1980s gas exports are expected to be similar to oil exports in energy equivalent, and gas will account for much of the expected growth in Algeria's hydrocarbon revenues to an estimated $10 billion (Sutton, 1978b).

Hydrocarbon industries have been central to Algeria's general strategy for economic development which has given priority to industrial over agricultural investment and, within industry, to creating basic industries with later multiplier effects promoting secondary industries. This approach can be traced back to

Table 14.1   GDP* growth, per capita GDP, and re-investment rates in Algeria

Year	Annual growth in GDP (%)	Ratio of gross domestic investment to GDP (%)	GDP per capita (current dinars)
1963	− 1.5	21.8	—
1964	9.0	11.7	1138
1965	11.2	9.9	1182
1966	− 0.1	11.2	1193
1967	11.1	19.8	1239
1968	10.5	25.1	1385
1969	9.6	29.8	1474
1970	6.8	36.2	1598
1971	2.6	36.2	1591
1972	16.6	40.1	1794
1973	8.8	42.6	2036
1974	51.1	36.5	2979
1975	10.9	50.7	3206
1976	20.3	48.7	3704

Source: Sayigh, 1978 (1963–73 statistics); International Financial Statistics,
    XXXI (10), October 1978 (1974–76 statistics and GDP/per capita statistics).
*Gross Domestic Product.

the French Constantine Plan of 1959, which initiated several major projects in
the iron and steel and hydrocarbon industries. The strategy was then reformulated for a newly independent situation by the French adviser G. Destanne de
Bernis, who advocated the establishment of industries industrialisantes (Destanne de Bernis, 1971). Arguing a centre-down approach, he considered that
reasonable living standards could not be provided from the limited agricultural
resources and so industrial employment must be created. In turn industry would
supply fertilizers and machinery for agriculture, as well as forming a market
for the resulting increased agricultural output. Heavy basic industries are thus
being developed to exploit Algeria's raw materials. The strategy is to create
an iron and steel industry, to generate an important amount of electricity,
to orientate petrochemicals towards agricultural needs, and to establish a major
cement-manufacturing capacity. Much of this had been achieved by the late
1970s, and this permitted the initiation of mechanical and electrical construction
industries more orientated towards consumer goods. Critics of this industries
industrialisantes strategy maintain that it leaves the bulk of the population
marginal to the economy (Chaliand and Minces, 1972).

Some reorientation of political and financial interest towards the rural
sector has tempered these criticisms, but industrialization seems likely to
remain the foundation of Algerian policy. Thus, after having absorbed 48.4 per
cent of the funds of the First Four Year Plan (1970–73), industry remains central
to the Second Plan accounting for 43.5 per cent of its funds (Balcet and Nancy,

1974). While the Second Four Year Plan (1974–77) has a more diversified industrialization policy, the further development of the hydrocarbon industry and the enlargement of Algeria's steel-manufacturing capacity remain central. Plans are under way for a second larger steel-works at Mactar, near Oran, with a 10 million-ton capacity. Diversification is seen in the vastly increased range of industrial products forecast for 1977 (République Algérienne Démocratique et Populaire, 1974).

The achievements and on-going projects in Algeria's heavy industrial capacity exceed its likely domestic consumption, especially of steel and petrochemicals. While it is advantageous for a country to benefit from the value-added that comes from processing natural resources hitherto exported as raw materials, the harsh realities of a world trade system warped by tariff and quota restrictions and suffering from excess productive capacity in sectors such as steelmaking, threaten to pose export problems for Algeria's burgeoning industries.

The range of industries and their production levels are impressive, but employment effects are more limited. By 1976, industrial employment had reached 323,097, plus another 195,947 employed in the building and public works sector. As Table 14.2 indicates, the basic industrial sectors of iron and steel and hydrocarbons are still not providing employment concomitant with their share of investment funds. However, significant growth in employment took place between 1972 and 1976. Just as the First Four Year Plan fell short

Table 14.2   *Growth of industrial employment by sectors, 1967, 1972, and 1976*

Sectors	1967	1972	1976	Percentage growth 1972–76
1. Extractive industries	12,129	15,585	19,156	23
2. Oil and natural gas	7,276	23,310	62,007	166
3. Food and drink industries	20,177	32,231	42,836	33
4. Textile industries	7,659	27,477	33,988	24
5. Leather and shoe industries	7,342	7,355	7,215	− 2
6. Chemical industries	6,305	8,578	14,531	69
7. Construction materials	7,474	12,014	22,078	84
8. Iron and steel	3,952	13,163	28,766	119
9. Production and transformation of metals	15,325	24,804	52,502	112
10. Wood, cork, and furniture industries	7,764	7,733	11,878	54
11. Paper and printing industries	5,931	5,733	10,823	89
12. Other manufacturing industries	1,205	4,799	6,385	33
13. Building and public works	62,450	131,723	195,947	49
14. Electricity, gas, water, and health services	5,759	6,474	10,932	69
Total	170,748	320,981	519,044	62

Sources: Secrétariat d'Etat au Plan, 1972, p. 17; Balcet and Nancy, 1975, p. 509; Secrétariat d'Etat au Plan, 1977, p. 7.

of its objective of 170,000 new industrial jobs, the early results of the Second Plan suggest a similar failure to achieve targets of 85,000 additional jobs in industry and 138,000 in building and public works. Meanwhile the male work-force will increase by about 470,000 in the period 1973–77. Between 1973 and 1980, an expansion in the work-force of 890,000 men and 110,000 women is anticipated. Clearly, industrialization is not seen as a solution to Algeria's unemployment problem until well into the 1980s.

Although an agent for centralization, Algerian industry in the mid-1960s had pretensions to decentralization through self-management or *autogestion*. With the growth of state companies this novel approach was swamped by an emphasis on technocratic values and economic growth. Self-management took a stronger hold on agriculture, where French farms were mostly transformed into *autogestion* estates 1962–63. Initially, abandoned estates were occupied by landless agricultural workers to preserve both crops and jobs. This spontaneous situation was regularized in 1963 and the remaining French land was expropriated and reorganized into some 3000 self-managed units (King, 1977). Following the 1963 creation of the *Office National de la Réforme Agraire*, which named an official director to work in each unit parallel to the elected management committees, popular socialism gave way to state socialism. 'La bataille autour de la gestion populaire se solda par la victoire du pouvoir central' (Hermassi, 1975, p. 213). The *autogestion* sector declined into a mixture of centrally controlled state farms and unregulated cooperatives. Centralization and closer bureaucratic control replaced what had been largely theoretical decentralized and localized decision-making. The government argued that efficiency criteria demanded this central control and that production losses should be minimized while an improved basis for a more efficient form of participatory socialism was being established. Criticisms of policies of centralism and state capitalism were difficult to refute. Only recently have government statements been re-evaluated at their face value rather than interpreted with cynicism (Nellis, 1977). Since June 1975 *autogestion* estates have been granted more freedom of decision-making. The state directors have been removed as has the centralized accounting system. The spirit of decentralization accompanying the *Révolution Agraire* has now encompassed the earlier established self-managed sector (Ollivier, 1975), but this followed a decade of domination by a technocratic centre-down approach to its production, marketing, and supply requirements.

REGIONAL DEVELOPMENT AND PLANNING

Whereas national development plans, operational since 1967, have been dominated by industry, they have been accompanied by regional development plans, each lasting three or four years. However national rather than regional policy dominated into the 1970s.

The initial three-year plan, 1967–69, achieved 82 per cent of its target with investments of 9.1 milliard dinars, largely in preparatory infrastructure and training. With increased capital supplies from higher oil prices, and aided by world inflation, the 1970–73 plan over-achieved its investment target of 27.7 milliard dinars. Eventual investment reached at least 34.2 milliard dinars (Balcet and Nancy, 1974) and perhaps as much as 40.5 milliard dinars (Sayigh, 1978). By comparison, each regional programme had funds of only 0.6 to 0.8 milliard dinars in the 1960s and 0.8 to 1.0 milliard in the 1970s. It should be emphasized that these increased national investments were in the industrial and infra-structural sectors and, to a lesser degree, in agriculture. Housing, education, and social sectors fared less well. Thus industry absorbed 48.4 per cent and infrastructure 14.6 per cent of the 34.2 milliard dinars invested in 1970–73. It can be argued that these more productive investments permitted a higher rate of re-investment which by 1973 stood at 40 per cent. This high level of re-investment was at the expense of improved social, housing, and consumer levels for all social groups and regions, while the undoubted economic growth achiev-ed through this sacrifice has been limited to certain regions and to the minority of the population employed in the secondary and tertiary sectors. The three privileged poles of development, Algiers, Oran, and the Skikda–Annaba–Constantine triangle, between them took 75 per cent of investments and received more than 60 per cent of jobs created (Jacquemot and Nancy, 1973). Further-more, investment-based indices are illusory as the 19 per cent increase over planned investments did not correspond with a quickened completion of projects; rather the reverse. Planners had to admit that industrial projects in iron and steel and mechanical engineering had only achieved 50 per cent of their objective after three of the four years of the plan. So one of the first tasks of the Second Four Year Plan (1974–77) was to make up these gaps in completion. The real success rate, 1970–73, in terms of physical rather than financial achieve-ments, was only 70–80 per cent (Duteil, 1973).

In terms of employment, this industry-dominated national plan also fell short of its admittedly low target. Until the 1980s industry has been given a relatively limited role in the creation of jobs in comparison with its large share of invest-ment funds. Industrial employment data for 1973 have been analysed to evaluate the degree of spread of economic development (Sutton, 1976). The northern coastlands are dominant to the disadvantage of the interior High Plains and High Steppe but the oil and gas field regions of the eastern Sahara also stand out. Established urban-industrial concentrations such as Algiers, Oran, Constantine, and Annaba maintain their prominence, but their neighbouring *dairate* also now have substantial industrial work-forces. Location quotients in Table 14.3 emphasize this maldistribution. Thus the *wilayas* of Algiers, Oran, Annaba, and Tlemcen dominate, with a better than proportional position also being achieved by Oasis, the *wilaya* containing the major oil and gas production.

Thus in industrial terms, although the government has promoted regional development programmes and has tried to divert some private investment away from Greater Algiers the strength of polarization forces has continued. The new coastal industrial poles associated with the oil and gas pipeline terminals at Skikda and Arzew have only strengthened coast—interior disparities, and the Arzew growth pole really only reinforces the established industrial centre of

Table 14.3　*Industrial location quotients, 1974*

Wilaya	Industry and building and public works	Industry
Algiers	2.67	3.04
Annaba	0.92	1.25
Aurès	0.49	0.25
Constantine	0.55	0.56
El Asnam	0.25	0.28
Médéa	0.42	0.14
Mostaganem	0.31	0.33
Oasis	1.90	2.17
Oran	1.88	1.78
Såida	0.22	0.33
Saoura	0.73	0.22
Sétif	0.37	0.33
Tiaret	0.20	0.22
Tizi Ouzou	0.88	0.45
Tlemcen	1.33	1.10

Source: Secrétariat d'Etat au Plan, 1976, p. 69; Commissariat National au Recensement de la Population, 1967.
Statistical base: 1974 employment: 1966 population.
Location quotient has been calculated as follows:

$$\frac{\dfrac{\text{Number employed in industry in } wilaya}{\text{Number employed in industry in Algeria}} \times 100}{\dfrac{\text{Population of } wilaya}{\text{Population of Algeria}} \times 100}$$

Oran. Thus 1973 employment data suggest that Greater Algiers and the coastal regions remain dominant despite the undoubted spread of the fact of industrial employment further afield in the interior. Despite close government control over a centrally planned economy, Algerian experience follows that elsewhere in achieving more limited spread effects than expected (see Chapter 1 of this volume).

The results of the 1977 population census provide some supporting evidence of a degree of 'spread' in that, contrary to expectations, Algiers and the largest metropolises grew more slowly than the Algerian urban population as a whole. Corrected census figures give a population of 1.523 million for Algiers which suggests that it grew at the same rate, 1966–77, as the total urban population. There seems to have been a slowing-down of the capital's population growth rate from about 1971–72 which contradicts the arguments stressing centralization, and other large cities share in this paradox (Prenant, 1978). Constantine, Oran, and Annaba grew between 53 per cent and 72 per cent, less than the urban population growth as a whole which was 75 per cent. Instead, 1966–77 saw the demographic expansion of the smaller urban centres where some industrialization and the provision of commercial, social, and administrative services resulted in more complete urban functions. Prenant interpreted these trends as indicative of the parallel movement towards the increased centralization of decision-making and the decentralization of activities. The growing role of state companies with their headquarters in Algiers appears to be centralization, yet increasingly their production units are being established in regional urban centres, so aiding decentralization. The particular case of the capital, Algiers, can be seen to be partly illusory in that its relatively slow population growth has been accompanied by the spilling-over of demographic vitality and industrial expansion into the adjacent Mitidja Plain where the formerly intensive 'colonial' agricultural system has regressed in face of industrial competition for space and manpower (Mutin, 1977).

As a palliative measure the industry-dominated plans have been accompanied by regional development plans. Initially, these 'special programmes' covered the complete wilayas* of Oasis (1966), Aurès (1968), Tizi Ouzou (1968), Titteri (1969), Tlemcen (1970), Sétif (1970), Saïda (1971), and El-Asnam (1972) (see Figure 14.1). More recently, only the more lagging regions of wilayas have been designated in programmes for parts of Constantine (1973) and Annaba (1973). That for Annaba concentrates on 35 communes suffering from a peripheral location against the Tunisian border. The city of Annaba was excluded as it was already benefiting from national industrialization policies (Mignon, 1974).

In 1978 five of the new smaller Saharan wilayas, Béchar, Adrar, Laghouat, Ouargla, and Tamanrasset, were granted regional development plans (Prenant, 1978). As well as aiming to lessen regional disparities in infrastructural, social, and educational provision, these 'special programmes' also promote specific agricultural and industrial developments appropriate to the region's potentiality. In introducing the 1970 programme for Sétif, it was claimed that 'by its policy of regional equilibrium the revolutionary government stresses its determination to abolish regional inequalities and to re-establish social justice' (Ministry of Information and Culture, 1970, p. 12). It has been estimated that between 15 and 20 milliard 'current' dinars have been allocated to regional development plans in the decade up to 1978 (Prenant, 1978).

Industrial employment data for 1973 have demonstrated the effectiveness of the earlier programmes. Thus the 1968 programme for the Aurès has resulted in the employment of 895 persons in a large textile plant at Batna. A further 9358 persons were employed in a large building and public works sector in the wilaya and success in setting up industrial establishments in tanning, quarrying, clothing, and wood and furniture manufacturing is in evidence for several towns (Sutton, 1976). Batna now contains several schools and colleges to redress the low 39 per cent scholarization rate recorded in 1968. Thus this policy of special programmes amounts to a government attempt to promote spread effects away

* The territorial administrative hierarchy has been re-organised and renamed since independence. The French system of 3 large départements in Northern Algeria, divided into arrondissements and then into a complex mixture of settler-dominated and native communes, was soon replaced by a more straightforward hierarchy of 15 départements, 82 arrondissements, and 676 communes. These were arabised into wilaya, daïra, and beladia, respectively. Then in 1975 the number of wilayat (plural) was increased to 31 and of daïrat to 153, a measure taken to strengthen decentralisation which served to multiply the tertiary activities in a larger number of administrative centres.

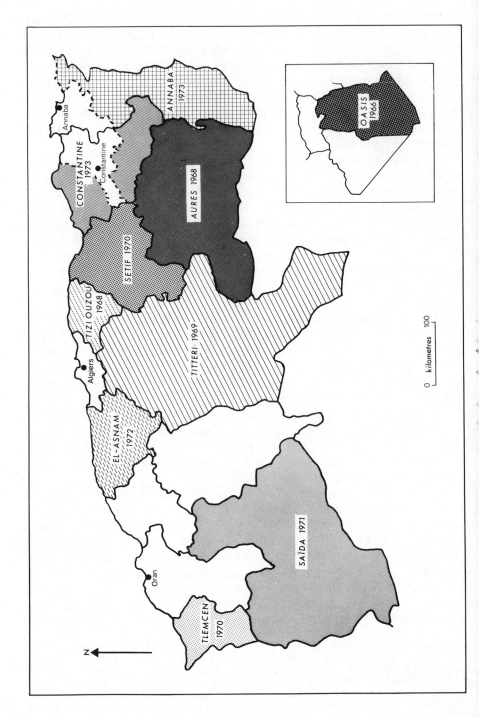

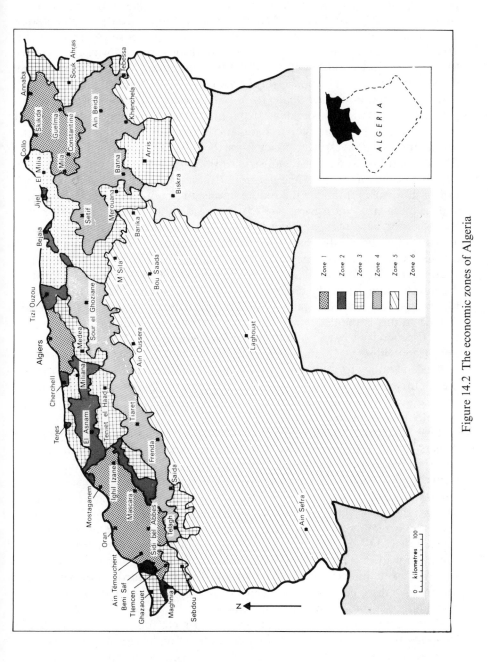

Figure 14.2 The economic zones of Algeria

from national development poles. However, in a review of the mechanics of preparing these programmes, Mignon emphasizes that officials from the *Secrétariat d'Etat au Plan* in Algiers are decidedly in control. While some consultation takes place with local interests who put forward projects, the opinions of technicians and administrators of the *wilaya* authority are preferred to those of elected representatives of the communes. Mignon suggests that *déconcentration* rather than *décentralisation* is in operation (Mignon, 1974).

In the Second Four Year Plan (1974–77) a greater realization of what the regional problems were, was evident. In population terms the plan aims to stabilize the interior mountain and steppe zones and to limit further migration to coastal towns. Industrial investment policy is partly linked to these zonal objectives in that it will (a) pursue intense industrialization in the strongly urbanized zones with the exception of the Algiers region, (b) establish national-scale industries in inland towns, (c) favour the mountainous and high plains zones for small- and medium-sized industries, and (d) give priority to desert and steppe zone locations for industries enjoying low transport inputs. Relatively, though, it appears that the established urban coastal centres, except for Algiers, will continue to be favoured (Jacquemot and Nancy, 1973).

However, the Second Plan's regional policy is wider than just industrial decentralization. From the outset it calls for further studies of both the tendency to concentrate on the coastal zone and the problem of promoting the interior. Linked to this is the desire to limit the growth of major cities and encourage that of secondary towns, notably in regions destined to be new zones of attraction (République Algérienne Démocratique et Populaire, 1974). To assist these studies and policy formulations, the Second Plan offers a new zonal division of the country into six economic regions (see Figure 14.2). Some of the characteristics, opportunities, and constraints of these regions could form the basis of a spatially more sensitive regional policy.

Zone 1 comprises the more developed urbanized area focused on the poles of Algiers, Oran, and the Annaba–Constantine–Skikda triangle. It concentrates Algeria's modern economic activities but also many of its urban housing problems. Living standards are twice as high as the national average, and 4 or 5 times those of the mountain and steppe zones. Zone 1 has benefited from the heavy-industry orientation of earlier development plans.

Zone 2 is regarded as a 'transition zone', comprising the interior valleys of the Chéliff and Soummam and other small coastal plains and *piedmont* regions. Natural constraints are few and modern agriculture and self-management units are significant. Indices of employment and living standards approximate the national average, development potentiality is high, and already in evidence is some industrialization and irrigation expansion.

Zone 3 covers the main mountain regions which, together with the steppe, compose the major regional problem. In relation to resources it is over-populated despite strong out-migration. Low living standards, below-average industrialization and an under-representation of modern agriculture all reflect and exacerbate the zone's unfavourable physical conditions.

Zone 4 is fairly extensive, being composed of the east–west alignment of High Plains. Despite reasonable ecological conditions for agriculture, population densities are low. Socioeconomic indices are close to the national average, and together with Zone 2 it is seen as a potential reception area for inflows of population and economic growth.

Zone 5 is more extensive, encompassing the Steppe and Saharan Atlas, and represents a major problem region. Low population densities are still too high for the poor resource base. Semi-aridity is a paramount problem and future out-migration appears inevitable.

Zone 6 is basically the Saharan south plus an extension northwards in Eastern Algeria. Relatively high urbanization has enabled social advances to be made, and hydrocarbons and tourism have stimulated some development (République Algérienne Démocratique et Populaire, 1974).

In comparison with the administrative units (*wilayas*) employed in early regional development plans, these zones have a greater ecological and socioeconomic homogeneity. The Second Plan puts them forward for discussion and further study, but does make some regional policy implications. Zones 2 and 4 are considered likely reception areas for 'surplus' population from the more problematic Zones 3 and 5, and as such should be favoured in regional development policies. Coastal Zone I will inevitably receive further population and economic development, but attempts should be made to lessen the polarization trends prevalent hitherto.

A major policy development contained in the Second Plan with these aims in mind is the *Plan Communal*. This adds a certain devolution of decision-making and management to local communities as opposed to the centre-down form of regional development measures practised before. The *plan communal* groups together in a coherent body all the activities of a local nature proposed, undertaken, and managed by commune authorities and yet financed from central government funds. It is intended that this new instrument of policy shall become 'une des dimensions importantes du système de planification et de sa démocratisation' (République Algérienne Démocratique et Populaire, 1974, p. 175). Both development and decisions about it should thus be diffused much more widely, as the content of policies is decided after consultation with the *assemblées populaires* in each commune. In many urban communes these *plans communaux* will form part of a wider scheme for the modernization of 33 towns. In less favoured rural areas each *plan communal* is expected to play a part in a wider policy of income redistribution through the raising of rural income levels by the establishment of new economic activity. Thus a 'Special Programme for the 200 Poorest Communes' has been elaborated, linked with social programmes of redistribution to help families with annual consumption per capita of below 500 dinars. At the broadest level these *plans communaux*, both urban and rural, aim to retain in decent living conditions the maximum number of people in the countryside and interior towns, and to organize the out-migration of surplus population to the more favoured zones (République Algérienne Démocratique et Populaire, 1974; Jacquemot and Nancy, 1973).

This recent departure in regional development planning increases the import-ance of the commune authorities, the *assemblées populaires communales*. Obviously their technical ability to implement plans must be supplemented from central government and *wilaya* resources but their range of decision-making is widened. This localization of developmental planning started with the imple-mentation of the Agrarian Reform measures, many of which were administered on the basis of local decisions. It is proposed now to turn to the problem of rural development, seen by some as central to Algeria's development problems, be-fore returning to the role of commune-based authorities in an embryonic bottom-up strategy now evolving in the course of the re-evaluation of the country's over-all approach.

THE CENTRAL PROBLEM OF RURAL DEVELOPMENT

Despite the dominace of industrialization in national plans and the fundamental role of oil exports, the agricultural sector of the economy and the rural com-ponent of the population remain very important. Indeed in a longer-term per-spective the problem of rural development can be regarded as central to the social and spatial objectives implicit in the achievement of economic develop-ment. Although agriculture only contributed 7.3 per cent of GDP in 1974, in employment and residence terms the rural-agricultural areas play a greater role. In 1973, out of an active labour force of 2.5 million, 1.3 million worked in agriculture but of these, 40–45 per cent were underemployed (Economist Intelli-gence Unit, 1977). Another 1973 estimate was that 61 per cent of the population were rural and this would increase to a rural population of 10.4 million by 1980, still 55 per cent of the total (République Algérienne Démocratique et Populaire, 1974). However, its social importance is not matched by agriculture's economic contribution. Low productivity and near-stagnation have typified both tradi-tional and modern sectors of agriculture. Agricultural value added in 1971 (averaged over 1969–73) stood at 2255 million dinars, only a slight increase over 1965 (averaged over 1963–67) when it was 2205 million dinars (constant prices). The decline in viticultural production accounts for much of this poor perfor-mance in the *autogestion* sector, where other crops such as cereals, market garden produce, and Mediterranean fruits together achieved a 24 per cent mo-netary increase in production from 1965 to 1971 (Balcet and Nancy, 1975). Consequently Algeria has become a seriously food-deficient nation, and recently up to a quarter of Algeria's oil revenues have been required to pay for basic food-stuff imports (Nellis, 1977). Not only is agriculture failing to produce sufficient raw materials for Algeria's developing industries or enough foodstuffs for its rapidly expanding population, it is also failing to generate an adequate demand for industrial products. Thus the 'industrializing industries' strategy is threatened by a weak market for those products geared towards agriculture. It has been suggested that realization of this weakening of the internal dynamism of an agriculture industry-based development strategy prompted pressure for a regeneration of Algerian agriculture, with the social upheaval of agrarian reform as a major impetus for change (Balcet and Nancy, 1975).

Only in 1971 was the reform of the traditional sector of Algerian agriculture commenced. Here about 900,000 peasant families, with inadequate holdings or landless, formed the theoretical beneficiaries of agrarian reform. Briefly, the Agrarian Revolution consisted of a phased redistribution of land and livestock, the recipients or which were organized into various production cooperatives. These were then grouped with pre-existing *autogestion* estates and private farms into service cooperatives (CAPCS). Also a peasants' union (UNPA) and a programme of new 'socialist villages' were established (Sutton, 1974; Ministère de l'Information et de la Culture, 1972). Redistribution has proceeded in three phases. The First Phase, 1972–73, concerned state and communal land and that belonging to religious endowments. The Second Phase, 1973–75, expropriated excess land from private landowners and all land from absentees. Detailed landholding limits were established according to type and potential of land use. From 1975 the Third Phase reorganized communal and private land and livestock in the *pastoral* regions. A Fourth Phase is to affect the forestry and alfalfa zones. As well as restructuring the agricultural economy this reform has two other primary objectives; firstly, to change rural ways of life so that they approach urban standards, thus diminishing rural out-migration; and secondly, to increase control over the factors of production so as to better fulfil domestic needs for agricultural and food products (Balcet and Nancy, 1975).

A 'land bank', the *Fonds National de la Révolution Agraire,* was set up, through which passed land in the process of expropriation and redistribution. By April 1977, 1,109,944 hectares had been redistributed, together with 824, 651 palm trees and 158,614 sheep (Banque Nationale d'Algérie, 1977). Regional statistics show a predominance of interior High Plains *wilayat* in terms of land redistribution (Sutton, 1978a). Thus already agrarian reform is partly redressing the Tell–interior imbalance now interpreted by the Second Four Year Plan as central to regional development policy.

An important supporting measure has been the all-embracing service cooperative (CAPCS). As well as providing services for cultivation, such as seed and machinery, marketing farm output, and purchasing inputs, each CAPCS functions as an intermediary for state assistance. The aim of establishing a CAPCS in each rural commune was achieved by 1977 when 654 were recorded (Banque Nationale d'Algérie, 1977).

Although the reform programme is incomplete many criticisms have already been levelled at it, and these particularly centre on its limited impact in terms of area and people involved. The 1972–73 agricultural census showed that 3,903,285 hectares in holdings of above 10 hectares could be partly expropriated as excess land (Secrétariat d'Etat au Plan, 1975). Initial estimates of land to be redistributed during the second phase of the reform were about 1 million hectares. The census reduced this to 640,000 hectares but only 500,000 hectares were actually expropriated and by late 1975 only 366,000 hectares had been redistributed. The initial shortfall was due to false declarations, manipulation of the limits on size of holdings, joint possession, etc. Further shortfalls resulted from successful appeals by landowners or, quite simply, from the *de facto* recovery of land by the former owners. The final discrepancy between land expropriated and redistributed stemmed from the acquisition of wasteland, abandonment by recipients of land allocated to them, regional shortage of recipients, and especially, administrative delays (Guichaoua, 1977). Patently the 1977 results fall well behind expectations. Evidence of evasion is such that Abdi concludes that 'large landed property-holders would have thus for the most part escaped the nationalisation measures' (Abdi, 1975, p.37).

The expropriation of land held by absentee owners especially was averted, in part by weak legislation. Thus, parents and children of victims of the War of Independence were excluded from expropriation measures and so the vast majority of absentees were not affected. Also the presence of any member of the family on the farm holding when the census of ownership was carried out sufficed to avoid the absentee label (The two aspects..., 1978). Further criticism centres on the favouring of traditional crops by the production cooperatives and on the increased mechanization of reform land, with much machinery remaining under-used. Production criticisms merge into political criticisms about the hierarchical organization of cooperatives, the relative lack of democratic decision-making, and the malfunctioning of the peasants' union. Even President Boumedienne 'has continually criticised certain civil servants who are said, in alliance with the landed class, to have attempted to hinder the reform' (Algeria looks . . ., p. 92). Finally, a major reservation concerns the limited social impact of the reform. The second phase, which concerned privately held land, only reduced it by 10 per cent and only affected 1 per cent of private landowners. Of the 900,000 peasant families, all potential beneficiaries, only 83,813 had received land by 1977, and indeed eventual aims only envisaged 150,000 recipients. This could be increased when the third phase of reform redistributes livestock and pasture resources. Previously half the flock was owned by only 5 per cent of the 170,000 graziers, a minority who also monopolized the communal pastures and exploited their shepherds through severe share-cropping contracts.

Despite limited preliminary results the Agrarian Revolution is on-going, and the expropriation of excess and absentee land could theoretically be re-enacted. Meanwhile cooperatives and other institutional bodies are gaining in importance. While admitting that many remain unaffected by reform, Ollivier considers the Agrarian Revolution to be central to Algerian development strategy, in particular the way strategy has evolved in the 1970s (Ollivier, 1975). He links the Agrarian Revolution and the 1954–61 War of Liberation in that, until 1971, economic policy was largely an extension of that war to gain economic independence. Then in 1971 a more revolutionary agrarian policy became central to the national development strategy. Central to Ollivier's interpretation are the reform cooperatives which, through their modest size and freedom of organization, give scope for initiative and local responsibility and form part of a wider decentralization of management in the whole agricultural sector. Thus the *autogestion* estates became much freer after 1975 when state directors' positions were eliminated, and state marketing and supply organizations replaced by cooperative bodies elected by producers. These are the CAPCS at the commune level, and *Coopératives d'approvisionnement en fruits et legumes* at the *wilaya* level. Ollivier evokes the CAPCS as the main instrument of rural development. It integrates the formerly separate sectors of agriculture in that they are all represented on its administrative council including private landowners. Thus there exists for the first time at the level of the commune, an instrument for planning rural development, entirely controlled and managed by agricultural producers. Large landowning interests are now alienated from the governing FLN party, and new broader organizations such as the UNPA and the APCE reflect a new political orientation. Furthermore, the whole reform operation has followed a more decentralized procedure through the APCE. In this way the Agrarian Revolution has become the decisive political battle-ground to consolidate Algeria's chosen development strategy (Ollivier, 1975).

The political significance of agrarian reform will be pursued later. However, it should not disguise the fact that the direct social and spatial impact of land redistribution has been limited. This criticism could be muted if the reform were widened into a labour mobilization programme aimed at tackling the central problem of rural development. Ollivier advocates the intensification of agriculture by replacing some *autogestion* estates by more efficient cooperatives. He points out that reform production cooperatives on poorer land are already creating more jobs per hectare than the *autogestion* estates (Ollivier, 1975). A long-established advocate of labour mobilization or human resource investment, Tiano has always linked it with agrarian reform. Labour-intensive rural renovation projects should be pursued, with priority given to projects capable of creating more permanent jobs. Thus the construction of earthworks to conserve soil requires 180 man-days per hectare and will result in 30 man-days' permanent employment per hectare, as the hill-slope can then be properly cultivated. Similarly, a minor irrigation work requires 450–550 man-days per hectare in preparation and results in 20–105 man-days of permanent employment (Tiano, 1967, 1972). Tiano estimated that Algeria offered 1200 million days' work for rural renovation, 650 million for irrigation improvement, plus many more for road construction, urban improvement, etc. Mobilization would be less expensive if the state reduced the minimum wage and paid it in kind, using food aid to replace wages. In the mid-1960s Tiano calculated that if Algeria devoted 500 million dinars a year to such a programme the possible schemes would take 15 years to complete and would require 80 million days' work each year. Work-days in agriculture would have been increased from 200 million to 520 million a year, thus providing more than one million new jobs, and moreover, agricultural production would have quadrupled (Tiano, 1967, p. 59). The annual investment of 500 million dinars compares well with investments annually channelled into industry. Such rural mobilization must be accompanied by agrarian reform so as to interest those involved. Permanent jobs and land would be allocated by lottery so there would always be some hope and incentive for those mobilized.

A limited amount of mobilization has taken place. National service has provided manpower to construct a trans-Sahara road. During the Second Plan national servicemen will participate in the *Barrage Vert* programme to restrain the northward spread of arid erosion through the use of a linear forested zone. Furthermore, a parallel *service civil*, of both sexes, will assist regional development by enlisting technical personnel for civilian rather than military duties (République Algérienne Démocratique et Populaire, 1974). Vacation-time mobilization of students has been used for campaigns of publicity and explanation in connection with agrarian reform. So far, however, mobilization has stemmed from central government, and it remains to be seen whether the new local organizations such as cooperatives and the APCs can promote similar local projects.

SOCIAL AND SPATIAL DISPARITIES IN LIVING LEVELS

Before focusing on new policy orientations it is necessary to evaluate the progress made in reducing regional disparities and improving the living standard of the average Algerian, and more particularly of the poorer section of the community. Data deficiencies, particularly in terms of regional incomes, and the lack (at time of writing) of results of the socioeconomic section of the 1977 population census, make this a difficult task.

One rough measure of aggregate progress in national growth is the per capita GDP, which increased slowly until 1971 and then accelerated and maintained its momentum thereafter. The slow growth from 1138 current dinars per head in 1964, to 1591 current dinars per head in 1971 would have been partly negated by inflationary pressures and changes in the international value of the dinar. Table 14.1 shows the parallel rise in rate of reinvestment as an austerity programme restricted the proportion of the growing GDP available for consumption. While post-1971 inflation and exchange-rate value of the dinar have worsened, the more rapid growth in GDP per capita up to 3740 current dinars in 1976 represents undoubted national economic progress to which the rise in oil prices contributed considerably. However, such an index is a measure of the national economy and surrogate data will have to suffice in an appraisal of social and regional differences.

Data for 1974, based on an annual survey of the industrial and much of the service sector, suggest an annual average male income of 11,976 dinars. The highest-paid group, the *cadres et techniciens supérieurs*, received 25,440 dinars or 112 per cent more than the average, and the lowest-paid group, unqualified personnel, were paid 6876 dinars a year or 43 per cent below the average. The ratio of average income of the lowest-paid and the highest-paid groups was 1 : 3.7. This social discrepancy compares with the twelve-fold difference between Europeans and non-Europeans prior to independence. However, 1974 data exclude civil servants and the agricultural sector, the two ends of the social spectrum. Lower-paid agricultural workers have benefited from the recent extension to them of national minimum-wage legislation as well as family allowances and other welfare benefits. The SNMG (*Salaire national minimum garanti*) in agriculture was raised to 15.30 dinars per day (3978 dinars per year) in 1976. A higher SNMG applied to the non-agricultural sector amounting to 4992 dinars per year in 1976 (Secrétariat d'Etat au Plan, 1976). Family allowances would supplement these minima. This limited attempt at social equalization also included ceilings on incomes of the highest-paid civil servants, but the scope for evasion by means of fringe benefits was considerable.

A major problem with this minimum-wage legislation is that it could not apply to the hundreds of thousands of peasants cultivating their own microholdings, nor could it apply to the unemployed. When these underprivileged groups are considered, social income disparities remain high.

An assessment of regional disparities in levels of living is equally difficult. Regional income data are not available, but when social disparities are tied to the restricted spread of industrial employment and the high proportion of the population involved in agriculture in the interior *wilayas*, the likelihood is of continued strong regional differences. Some comparisons with Tiano's earlier figures are possible. In 1954 the ratio between departments with the lowest average income (Oasis and Saoura) and that with the highest (Algiers) was was 1 : 9.7. If the Saharan departments are excluded it was 1 : 7.1 (Tiano, 1967, p. 121). In the absence of more recent regional income data, surrogate variables have to be used. Coast–interior disparities in industrial employment in the early 1970s suggest that the spread of industrial employment should not obscure the continued strength of regional contrasts. An alternative social surrogate quoted by Tiano for 1963, i.e. the number of hospital beds per 10,000 inhabitants, produced a ratio of 1 : 5 for Northern Algeria. A ratio of 1 : 7.2 can be calculated for 1974, but with a 1966 population base. This suggests some divergence, but the fact that it employs the new smaller *wilayas*, which automatically enhance regional discrepancies, and that 1966–74 population growth will not have been even, renders straight comparisons extremely difficult. A calculation for 1972 using the old, larger *wilayas*, produced a ratio of 1 : 3.8, suggesting convergence in terms of regional provision of hospital beds (Secrétariat d'Etat au Plan, 1975). Scholastic indices also suggest progress from a ratio of 1 : 3.25 in 1966 to one of 1 : 1.75 in 1975. Certainly the rapid extension of education facilities since independence and the expansion of medical infrastructure before and after the enactment of free medical care in 1973, reflect governmental desire for greater social and regional egalitarianism. Whereas some progress towards these goals can be shown in social terms, in the field of income and employment opportunities marked disparities continue to give cause for concern.

AN EVOLVING BOTTOM-UP STRATEGY

Evidence has been presented in this chapter to support the contention that a shift in development policy has taken place since 1971. Regional development programmes and an increased emphasis on agrarian reform and rural development do not constitute a total paradigm change to a strategy of agropolitan development. However, it is contended that a bottom-up strategy of development is evolving in Algeria.

Many shortcomings of centralized development strategies based on heavy industrialization alluded to by Friedmann and Douglass have emerged. Dualistic economic structures have prevailed; hyper-urbanization threatens Algiers; urban and rural unemployment remain intolerably high; undoubted income inequality exists despite theoretical salary ceilings; and domestic food production increasingly fails to meet the country's needs (Friedmann and Douglass, 1978). Elements of the alternative 'strategy of accelerated rural development' can be detected. Whereas the new reform villages (Sutton and Lawless, 1978) do not amount to an agropolis, they certainly represent an increased social investment in rural districts, which can help reduce rural–urban migration. Organizational forms are now present which in part meet the idea of agropolitan districts. The service cooperatives group together all rural social strata, not just the recipients of reform land, and this concept is being applied to the extensive pastoral areas in the third phase of the Agrarian Revolution. The reduction of differences between rural and urban incomes through diversifying rural employment to include non-agricultural activities has to some extent flowed from regional development programmes, is advocated in zonal policies of the Second Plan, and could be implemented by service cooperatives and the APC authorities. A degree of local power of decision-making is emerging whereby commune bodies represented by the APC are encouraged to promote local development programmes as part of a *plan communal*.

At the national level there is some indication that the Third Four Year Plan (1978–81) will shift the focus of investment away from the industrial sector and will pay more attention to problems of rural development ('Algeria looks..., p. 92). Thus certain criteria of an agropolitan policy (Friedmann and Douglass, 1978), can be tentatively met in the Algerian case study. Perhaps a better case can be made for an evolving bottom-up strategy by re-emphasizing some of these beginnings separately rather than forcing them into the model suggested by Friedmann and Douglass. As Stöhr states, development 'from below' has to be specific to the country's sociocultural, historical, and institutional conditions. Many of his essential components of a strategy 'from below' are partially and increasingly in evidence in Algeria. Land reform, better structures for communal decision-making, priority to projects for basic needs rather than export-base production, and other policies favouring peripheral and rural areas can be seen, if only in embryonic form, to be entering Algerian development strategy (see Chapter 2 of this volume). Adoption of the whole package of measures is obviously a long way off and is not really being contemplated in any co-ordinated fashion at present. After all, the financial dominance of the oil and gas industry, and the focus of investment on industry, have not been lessened; merely supplemented by new policy orientations. Social welfare indices together with legislation setting national minimum-wage levels and programmes for regional economic development suggest that the government's strategy includes the promotion of disadvantaged sections of Algerian society and the lagging regions. In the absence of firm data showing achievements towards this end, interim evaluation must rely on the interpretation of certain new policy orientations as marking a start in the shift of overall development strategy.

Two major innovations have been the *Assemblées Populaires Communales* and the CAPCs with their devolved powers of deciding and initiating policy. The APCs, enlarged into *Assemblées Populaires Communales Elargies*, played a principal role in putting agrarian reform into operation. They carried out and verified a census of private land from which expropriation decisions could be made. At the same time they drew up a list of potential recipients of land. Within each commune the APCE was left to decide, within zonal guidelines, on the precise limits to be adopted above which land was expropriated and redistributed. Furthermore, the relative success of these elected consultative bodies, one APC in each of the 676 communes, has prompted the return of one-party democracy in the *wilayat* and eventually, in 1977, an elected national assembly (Nellis, 1977). Is this a case of diffusion from the periphery to the centre?

Once agrarian reform was under way the APC was paralleled by another local body, the CAPCS, again at the commune level. All sections of agricultural production are represented in the administrative council of each CAPCS, and as well as providing input and output services it functions as an intermediary for state assistance and as a watchdog to ensure adherence to the texts of the reform. The potential central role in rural development which the CAPCS can play has been reviewed earlier. Ollivier elaborates on several possibilities which have still to be tried (Ollivier, 1975). It can organize a system of cultivation contracts whereby part of each holding can be devoted to crops of national importance, part to cover local needs, and part left to the producer's initiative. It can co-ordinate those rural activities which operate on a scale larger than an individual holding, such as the promotion of irrigation, reafforestation, transport, storage, publicity, and training. The CAPCS can develop cooperation beyond the level of the commune as certain development options require larger, regional-scale organizations. It can promote links with other economic sectors, especially with agricultural processing and food industries.

While these two local bodies provide an organized outlet for local initiative, their wider participation in rural development is encouraged by the policy of *plans communaux* established by the Second Plan. These *plans* could provide the means to master urban growth problems and to put rural promotion policies into operation. Some of the details of this new approach, the *plans communaux*, have been discussed above. Perhaps the significant diffusion of decision-making to the local level should be stressed. As the Second Plan states, this will be 'un facteur de mobilisation des capacités tant humaines que matérielles' (République Algérienne Démocratique et Populaire, 1974, p. 176).

These beginnings of an alternative bottom-up strategy can also be interpreted as amongst the first signs of a wider political decentralization. The re-interpretation by Nellis of Algeria's self-proclaimed but often criticized status as a socialist state lays stress on the evidence of mobilization and decentralization in political, management, and economic development fields. 'Efforts to "decentralise, deconcentrate and democratise" the political system ... have resulted in the election of consultative "popular assemblies" in the 676 *communes* and 15 *départements* (Nellis, 1977, p. 535). Parallel to the Agrarian Revolution and commencing also in 1971 has been the extension of *la gestion socialiste* in the industrial and service sectors. From 1974 this approach, involving workers in the management of their own productive efforts, has been extended through the state companies. Permanent workers in each 'unit' elect an *assemblée des travailleurs* which in theory has considerable consultative and co-managerial powers. Hopefully this will mobilize workers by permanently associating them with decision-making. Duties as well as rights are involved in this participation, which by mid-1976 was being operated in some 40 enterprises, totalling 550 units and 150,000 workers. State enterprises outside of *la gestion socialiste*, including Sonatrach, were being prepared for this approach, as were social services and the traditional public administration sector (Nellis, 1977).

An even larger participation and mobilization exercise was the 1976 debate over *la charte nationale*. Discussion over the proposed national charter was encouraged at all levels from local and employment units to national conventions. It certainly indicated the degree of dissatisfaction with the inefficiencies and privileges of the system of administration, and indicated the public wish to play a greater part in decision-making. Much of the criticism was of the growing technocracy rather than leadership. An interesting link between these political events and the rural development programme is suggested by Ollivier. He stresses the political role of the Agrarian Revolution and considers that its political effects have probably been greater than its other results so far, in that the successful national debate on *la charte nationale* in May–June 1976 took place in a more favourable climate, thanks to the dynamism created by agrarian reform (Ollivier, 1975).

Perhaps in regional development terms one would not go as far as Nellis in his reassessment of the degree of commitment of Algerian politicians to socialist policies. He reflects that 'though there is always something suspicious in any argument which insists on taking steps to the rear in order to ensure future progress, this seems to have been the Algerian experience' (Nellis, 1977, p. 533). Centralized development of a heavy industrial base has always been seen as a development *sine qua non* rather than as a retreat into a familiar, safe paradigm. Nevertheless, a more decentralized approach is emerging, with greater emphasis on rural development, and could evolve into a bottom-up strategy. This interpretation finds support in the exhortation of the Second Plan that the planning system should be progressively supported by processes of democratization. Furthermore, progress in decentralization and in the struggle against the concentration of decision-making should permit the enlargement of the body of effective participants in planning. 'Cet enrichissement dans l'application de la décentralisation doit avoir pour but ... d'assurer une plus grande diffusion des actions et des fruits de développement à travers l'ensemble du pays' (République Algérienne Démocratique et Populaire, 1974, p. 243). Whether this amounts to the participation of Algeria in the general reassessment of development strategies detected by Brookfield in the mid-1970s is debatable (Brookfield, 1978). He considers that this re-thinking involves three major issues: firstly, the growing concern that income and assets should be redistributed in favour of the poor; secondly, revitalization of the debate over urban or rural bias in development strategies; and thirdly, the questioning of conventional regional development policies which remain too closely linked with non-metropolitan urban development. As a result of widespread disenchantment with growth-based strategies fundamental dogmas are being challenged, including the prior role of industrial growth and the desirability of shifting people out of agriculture into the modern sector. An emergent doctrine of 'self-reliant development' is increasingly advocated, with emphasis on internal trade, a questioning of the need for high technology, and an adherence to the arguments of Schumacher. It all focuses on a new priority for rural development.

In their outline of a new agropolitan paradigm, Friedmann and Douglass also urge that priority be given to rural development, and that planning for it be 'decentralised, participatory, and deeply immersed in the particulars of local settings' (Friedmann and Douglass, 1978, p. 167). To achieve this, rural planning should be ecologically specific with substantial control over development priorities and practice exercised by the people of local districts. Some institutions for this more devolved approach are emerging in Algeria, and increasingly the political will to initiate local development policies is becoming evident at commune level and is being encouraged at government level. An indication that this is more than just the beginning of a bottom-up strategy will be when local and regional bodies go beyond the promotion of rural development, and start to question the continued central role of industrialization in national economic strategy.

REFERENCES

Abdi, N. (1975) Réforme agraire en Algérie, *Maghreb-Machrek*, 69, 34–40.
Algeria looks to agriculture to solve social problems, *The Middle East*, 45 (July 1978), 92.
Amin, S. (1970) *The Maghreb in the Modern World* (Penguin).
Balcet, G. and M. Nancy (1974) Chronique économique—Algérie, *Annuaire de l'Afrique du Nord*, XIII, 373–95.
Balcet, G. and M. Nancy (1975) Chronique économique—Algérie, *Annuaire de l'Afrique du Nord*, XIV, 497–524.
Banque Nationale d'Algérie (1977) Direction du Financement de l'Agriculture, *Révolution agraire—situation générale et par Agence au 30 avril 1977*.
Brookfield, H. (1978) Third World development, *Progress in Human Geography*, 2 (1), 120–32.
Chaliand, G. and J. Minces (1972) *L'Algérie indépendante (Bilan d'une révolution nationale)* (Paris: Maspéro).
Commissariat National au Recensement de la Population, (1967) *Recensement Général*

de la Population 1966. Situation de Résidence par Commune. Ministére des Finances et du Plan (Algiers).

Destanne de Bernis, G. (1971) Deux stratégies pour l'industrialisation du Tiers Monde. Les industries industrialisantes et les options algériennes, *Revue Tiers Monde*, **XII** (47), 545–63.

Direction des Statistiques (1973) Fichier des grandes unités. Liste des établissements classés d'après leur localisation géographique-Tirage de janvier 1973 (Unpubl. mss. Algiers).

Duteil, M. (1973), L'Algérie d'un plan à l'autre, *Jeune Afrique*, 669, 60–70; 670, 55–69.

Economist Intelligence Unit (1977) *Quarterly Economic Review of Algeria*, Annual supplement 1977, 1–28.

Friedmann, J. and M. Douglass (1978) Agropolitan development. Towards a new strategy for regional planning in Asia, Lo, F. and K. Salih (eds.), *Growth Pole Strategy and Regional Development Policy*, pp. 163–192 (Pergamon Press).

Guichaoua, A. (1977), La Réforme agraire algérienne. Portée et limites. Politique agricole et transformations sociales, *Revue Tiers-Monde*, **XVIII** (71), 583–601.

Hermassi, E. (1975), Etat et société au Maghreb—etude comparative, Editions Anthropos. (Paris).

Jacquemot, P. and M. Nancy (1973) Chronique économique—Algérie, *Annuaire de l'Afrique du Nord*, XII, 535–58.

King, R. (1977) *Land Reform, A World Survey* (Bell).

Knapp, W. (1977), *North West Africa. A Political and Economic Survey* (Oxford University Press).

Lazreg, M. (1976) *The Emergence of Classes in Algeria* (Westview Press).

Mignon, J. M. (1974) Le plan communal algérien; l'expérience des programmes Spéciaux de Constantine et d'Annaba, *Revue Tiers Monde*, XV (58), 389–96.

Ministère de l' Information et de la Culture (1972) *La Charte de la Révolution Agraire.*

Ministry of Information and Culture (1970), *The Faces of Algeria—the Wilaya of Sétif.*

Mutin, G. (1977), Développement et maîtrise de l'espace en Mitidja, *Revue de Géographie de Lyon*, 52 (1), 6–33.

Nellis, J. R. (1977) Socialist management in Algeria, *The Journal of Modern African Studies*, 15 (4), 529–54.

Norbye, O. (1969) The Economy of Algeria. In Robson, P. and D. A. Lury, (eds.), *The Economies of Africa*, pp. 471–521 (George Allen & Unwin).

Ollivier, M. (1975) Place de la révolution agraire dans la stratégie Algérienne de développement, *Annuaire de l'Afrique du Nord*, XIV, 91–114.

Prenant, A. (1978) Centralisation de la décision à Alger. Décentralisation de l'exécution en Algérie, La mutation des functions capitales d'Alger, *Revue française d'études politiques méditerranéennes*, 3 (2–3), 30–31, 128–50.

République Algérienne Démocratique et Populaire (1974) *IIe. Plan Quadriennal 1974–77. Rapport Général.*

Sayigh, Y.A. (1978) *The Economies of the Arab World* (Croom Helm).

Secrétariat d'Etat au Plan (1972) *La Situation de l'Emploi et des Salaires en 1972.*

Secrétariat d'Etat au Plan, (1975) *Annuaire Statistique de l'Algérie, 1974.*

Secrétariat d'Etat au Plan (1976) *Annuaire Statistique de l'Algérie, 1975.*

Secrétariat d'Etat au Plan (1977) *Les résultats de l'enquête: emploi et salaires de 1976.*

Sutton, K. (1974) Agrarian reform in Algeria—the conversion of projects into action, *Afrika Spectrum*, 9 (1), 50–68.

Sutton, K. (1976) Industrialisation and regional development in a centrally-planned economy—the case of Algeria, *Tijdschrift voor Economische en Sociale Geografie*, 67 (2), 83–94.

Sutton, K. (1978a) The progress of Algeria's agrarian reform and its settlement implications, *The Maghreb Review*, 3 (5–6), 10–16.

Sutton, K. (1978b) The Algerian natural gas industry, *The Maghreb Review*, 3 (9), 14–20.

Sutton, K. and R.I. Lawless (1978) Population regrouping in Algeria—traumatic change and the rural settlement pattern, *Transactions of the Institute of British Geographers*, 3 (3), 331–50.

The two aspects of Algerian socialism, *Civilisations*, XXVIII (1–2) (1978), 2–30.

Tiano, A. (1967), *Le Maghreb entre les Mythes* (Presses Universitaires de France).

Tiano, A. (1972), Human resources investment and employment policy in the Maghreb, *International Labour Review*, 105 (2), 109–33.

c. Latin America

Development from Above or Below?
Edited by W. B. Stöhr and D. R. Fraser Taylor
© 1981 John Wiley & Sons Ltd.

Chapter 15

Brazil: Economic Efficiency and the Disintegration of Peripheral Regions

PAULO ROBERTO HADDAD

INTRODUCTION

A recent and comprehensive study of regional development inequalities in Brazil (Universidade Federal de Pernambuco—Pimes, 1977), prepared from the most up-to-date statistical data, concludes that regional inequalities clearly still persist in the post-war period, when compared with international standards (Williamson, 1965) (see Table 15.1). Although it is difficult to reproduce all the conclusions presented in that study, the most relevant social and economic indicators showed that Brazilian regional problems are closely related to social issues and require a political solution, with emphasis on equity criteria.

Since the 1950s, the federal government has paid special attention to the problems of inequality in regional development, by preparing a number of national development plans. Table 15.2 clearly indicates the growth of the regional element in national development plans since 1963, emphasizing information on strategies, objectives, and instruments of spatial policy adopted (Cintra and Haddad, 1978b). These policies were intensified after the development phase in the country which consolidated a process of import-substitution providing Brazil in the early 1960s with one of the most advanced and diversified industrial structures amongst developing countries. This industrial structure was concentrated mainly in the metropolitan areas of Rio de Janeiro and São Paulo, so that spatial policies evolved in the midst of centre – periphery relations between developed and underdeveloped regions. This point was stressed over and above the unequal relations existing between production structures, since spatial policies were developed in the time after the 1964 military coup, which restricted civil rights in the country for over a decade and also, from a political viewpoint, intensely concentrated decision-making powers *vis-à-vis* the nation's development alternatives.

As a result of these two historical factors, a basic characteristic of Brazilian spatial development policies has been their uniform pattern in certain compo-

Figure 15.1 Regions of Brazil, 1960/70

Table 15.1 *Index of inequality—$V_w{}^*$, Brazil, 1950–70*

Year	20 States	5 Macro-regions	North/South
1950	0.65	0.50	0.45
1951	0.65	0.50	0.45
1952	0.67	0.51	0.46
1953	0.65	0.51	0.47
1954	0.66	0.51	0.46
1955	0.65	0.50	0.46
1956	0.63	0.49	0.43
1957	0.62	0.47	0.42
1958	0.63	0.49	0.43
1959	0.59	0.44	0.40
1960	0.58	0.43	0.38
1961	0.58	0.43	-.39
1962	0.54	0.41	0.38
1963	0.59	0.44	0.38
1964	0.54	0.42	0.37
1965	0.53	0.41	0.36
1966	0.55	0.44	0.38
1967	0.54	0.43	0.37
1968	0.55	0.44	0.37
1969	0.56	0.45	0.39
1970	0.57	0.46	0.40

Source: Universidade Federal de Pernambuco, Pimes, 1977.

$$*V_w = \frac{\sqrt{\Sigma_i(y_i - y)^2 \cdot (f_i/n)}}{\bar{y}}$$

where y_i = Product per capita: region i;
$\bar{y}$ = Product per capita: Brazil;
f_i = Population: region i;
n = Population: Brazil.

nents (i.e., centralized decisions at federal government level, adoption of a polarized development model from the expansion of economic activities in industrial areas, a monolithic conception of development, etc.) all of which fit properly into the centre-down development paradigm, as described by Hansen and Stöhr (Chapters 1 and 2 of this volume).

Further emphasis has been given to the role of spatial planning in the development process of the country: policy instruments have diversified from the isolated use of fiscal incentives to attract private investors to depressed areas, towards more integrated development programmes to exploit natural resources in vital areas of the Amazon, or the nation's mid-west. Objectives have been broadened from an initial concern with social strains in depressed areas, to a better articulation of sectoral policies and those of an urban-regional nature; and finally, political priority given to spatial policy objectives in recent years indicates that they are considered critical for the country's development. Why then are the results of this emphasis seemingly only mediocre in their impact on the poorest in our population, in the disadvantaged regions?

Table 15.2 *Spatial policies in national development plans*

National plan	Strategies of regional development	Priority given to objectives of spatial policies	Major policy instruments	Socioeconomic impact on peripheral areas
Plano Trienal 1963–65	Promotion of development in less developed areas. Spatial dispersion of government efforts (although special emphasis was given the north-east)	Adequate to weak (sociopolitical tensions)	Transfer of public investments into socioeconomic infrastructure. Capital incentives and technology transfer to peripheral areas	This period of the plan characterized by national political crisis and instability in implementing different economic plans which implied a low profile for spatial policies
Plano de Ação Econômica do Governo 1964–66	Promotion of development in less-developed areas with concentration on the north-east	Adequate to weak (sociopolitical tensions)	Transfer of public investments into socioeconomic infra-structure. Capital incentives and technology transfer to peripheral areas	Main emphasis was in implementation of micro-economic policies of stabilization and growth; spatial policies had lower priority and were not intensive
Plano Estratégico de Desenvolvimento 1968–70	Promotion of development in less-developed areas through expansion of economic frontier by incorporation of new resources and creation of regional poles. Consolidation of development process in the centre-south	Required for the country's development process	Transfer of public investments into socioeconomic infra-structure. Capital incentives and technology transfer, tax transfer from the federal government to states and municipalities	Main emphasis was directed towards attaining a high growth rate for GNP. Modern industries moved to the north-east, bringing high productivity levels but minor employment benefits

Metes e Bases 1970–72 and II Plano Nacional de Desenvolvimento 1972–74	Promotion of development in less-developed areas through expansion of economic frontier by incorporation of new resources and creation of regional poles Consolidation of development process in the centre-south	Required for country's development process	Transfer of public investment into socioeconomic infrastructure Large expansion of transport and communication networks from core regions to peripheral areas Tax transfer from federal government to states and municipalities Capital incentives and technology transfer	Greater emphasis given by plan to expand resource regions, increased accessibility to national and international markets but at the same time increased reliance on external production factors and public funds from federal government
II Plano Nacional de Desenvolvimento 1975–79	Consolidation of development process of industrial core region with higher emphasis on urban development policies Expansion of resource regions and creation of agro-mineral growth poles Promotion of development in depressed regions	Critical for the country's development process	Growth centre policies Promotion of modern industries in peripheral areas Capital incentives and technology transfer	With the slowing down of growth rate of GNP a lower priority was given to special regional programmes of federal government for depressed areas and resource regions; a major effort is being directed towards the large metropolitan areas.

Source: Cintra and Haddad, 1978b.

CONCENTRATION OF ECONOMIC POLICY AND POWER

Some Reasons for Past Centre-Down Policies

Although information about regional economies in the last century in Brazil is scarce, statistical evidence in the 1872 national census indicates that there was no major divergence between levels of development of the north-east and south in that period. But this situation altered towards the end of the century. At the start of the twentieth century the south was ahead of the north-east in pace of industrialization and expanding investment in infrastructure. In 1900, for example, the per-capita railroad mileage open for traffic in the south was some twenty times greater than that in the north-east (Denslow, 1978).

After the 1929 economic crisis, a plan for import-substitution was adopted in Brazil, based upon industrialization as an alternative to the primary commodity export scheme which had been in existence since colonial years. This process created a strong geographical concentration in major production areas in the south, chiefly in the Rio-São Paulo axis. Investment in the urban and communications infrastructures required for the coffee economy, as well as expansion of regional markets created by high productivity and the favourable income distribution profile of the coffee crop, were the main causes of the historical pattern of uneven regional development in the country.

Although regional inequality was still a problem, there was a trend, from 1950 to 1965, towards a levelling-off in per capita income differences among the states and regions (see Table 15.1). After the economic stagnation of the 1962–67 period resulting from the decline in import-substitution, the level of inequality increased again until 1970, the last year for which regional statistics are available.

During these two periods, convergent or divergent patterns of regional inequalities could be explained away by historical factors, which is not the objective of this chapter; but some of the elements of the national planning system will be analysed which in recent years have strongly influenced spatial policies in the country. It should be noted that these policies are considered to be increasingly centralizing and concentrating resource mechanisms which favour the interests of public and private organizations of the core regions.

After the 1964 military coup, the problem for economic planners in Brazil was to bring about the rapid recovery of an economy which had recently been operating far below its great potential. This recovery, which got off to an impressive start after 1967, came about by the prudent application of conventional fiscal, exchange, and monetary policies. The generalized trend was towards reinforcement of centralization—both *vertically* (*inter*governmental), and *horizontally* (in *intra*governmental relations). This centralization took place irrespective of the possible neutrality or selectivity of the respective policy instruments.

This option originated in the political rationale of the 1964 coup. The absence of any tradition of macroeconomic policies in the public bureaucracy, combined with the necessity to pull the nation's economy out of stagnation as soon as possible, did not permit the adoption of a decentralization which might have led to the slow process of relaxing institutional rigidity. But a scarcity of human resources with experience in such policies demanded that the available personnel be concentrated in a few of the most crucial organizations in the official set-up, and these became actual decision-making centres. Finally, since conflicts between the objectives and goals of the various proposed policies could not be resolved between social groups and regions without threat to the new political regime, it was necessary to train technical staff of make decisions on what the stability of the economic system would be in the long run (Hadded, 1978). What are the implications, for Brazil, of this power concentration, in terms of the formulation and implementation of spatial development policies?

Intergovernmental Relations

In the area of *vertical* concentration of decision-making power, the federal government has progressively curtailed the powers of states and municipalities to deal independently with economic policy. In the last few years the whole spectrum of tax legislation has been under control of the federal government; states do not even set the rates for their own taxes. As a result the trend has been towards the federal government taking an increasing share of total taxes paid: in 1930 the Union held 51 per cent of total public revenue, states held 37 per cent, and municipalities 12 per cent. By 1975, this had changed to 73 per cent, 24 per cent, and 3 per cent, respectively.

Tax reform in 1967 established a system of transfer to states and municipalities to offset concentration of resources at the federal level. About 50 per cent of such transfers were, however, allocated to capital investments in specific programmes, and were rigidly portioned out to various sectors, thus controlling the autonomy of the states and municipalities in terms of expenditure. In a strategic move to avoid such control devices, the states began to resort to internal and foreign debt to finance investment programmes. In 1972 their total debt was 39 per cent over total revenue. That figure jumped to 57 per cent in 1975. More recently, the debt capacity of states and municipalities has come under the control of federal legislation (Siqueira, 1978; Weiterschan, 1978).

As a consequence of these limitations, there has been a considerable change in planning at the state level (urban planning issues will not be considered here) in favour of a process simplifying ordinary planning procedures and discarding the stages of definition of objectives, goal specification, and selection of policy instruments, while emphasizing diagnosis and control. Planning is then considered as the maximum development of the state's latent negotiating capability, for which the reception of roles' (where states are briefed on guidelines as established by the central government) becomes the dominant stage (Boisier *et al.*, 1972; Cintra and Haddad, 1978a).

The loss of decision-making power by states provides them with only limited participation (even smaller for municipalities) in the formulating of spatial development policies, and drives the states into a struggle for resources for funds and programmes defined by central planning agencies. Besides, when they have to formulate guidelines for programmes involving strong conflicts of interest between states, these agencies tend to become inefficient. The Second National Development Plan, 1975–79, for example, emphasizes as one of its main objectives the definition of an industrial decentralization policy in the central-southern region so as to avoid too great a concentration of vigorous activity around the Greater São Paulo area. This is a typical example of a problem which concerns various states and regions, the solution to which can only be worked out by economic policies controlled by the federal government (e.g. fiscal incentives, official loans at subsidized interest, tariff exemptions for capital goods imports, etc.), whose policies are basically manipulated to reach certain over-all or sectoral goals, proposed by national planning without any adequate differentiation being made amongst states of the central-southern regions. Therefore, in the past few years, these states have internalized the social costs of the industrial decentralization process. In order to increase their bargaining power to attract private investment, the states must bear the burden of expenditures for the economic infrastructure, thus sacrificing funds which should go into basic social sectors. Financial participation in investment and the granting of tax exemptions (presently under federal control) also puts a heavy burden on future budgets through expansion of non-compressible expenses, and because of the low elasticity of tax revenue in relation to income.

The general conclusion one arrives at, from this experience of loss of decision-making autonomy by public agencies and institutions at the sub-national level, is that development values, styles, and patterns must emerge from the central federal orbit, and permeate the regions, states, and municipalities through a hierarchy established in various development policies, all formulated and implemented at different government levels. This hierarchy is solidified by the concentration of political power and resources, and through policy instruments in those government agencies which attend to the interests of the core regions. Thus, development programmes and projects which might answer the basic needs of regional and local populations are lost in the endless maze of public bureaucracy, because they are not charted along that critical path of, what in the current development model is labelled 'of national interest'.

Intragovernmental Relations

If it is believed that planning processes can be implemented at various levels—national, regional, state, local—and under many approaches—global, sectoral, spatial—then it is wise to question the relative position that the many agencies of spatial development policies hold in the bureaucratic set-up of the public sector. When the Brazilian post-war social and economic planning experience is examined, it is evident (see Table 15.2) that in the first plans of the federal government there were generic references to regional development in the country. In the present decade, planning increased the analytical focus on regional issues, trying to connect goals of sectoral growth with programmes for regional development. This introduced the urban dimension to the scope of regional planning. In a word, they provided the problem with an appropriate analytical approach.

Appraisal of the role that spatial development policies play in the national or state planning process cannot, however, be made from explicit statements of intention in government documents or in guidelines of public institutions. The appraisal must start with an analysis of the decision-making processes occurring within the shifting forces of political organizations. In the Brazilian experience, this argument can be illustrated by the decline of SUDENE's role as a regional planning agency. It had responsibility for development of the north-east and was created in 1959 with wide supervisory powers to coordinate and control preparation and execution of development programmes which were accountable to other federal bodies. SUDENE's first task was to make the sectoral operation of the various Ministries and public agencies conditional upon a five-year regional master plan; and second, it was to demand greater participation of the elite and local leadership in their decision-making so that this decision-making reflected the expectations of the region's various social levels in its development. SUDENE was also to be responsible for administering a broad set of economic stimuli such as fiscal incentives, import-duty exemption, total or partial income tax exemption, etc., aimed at attracting private investors towards the region.

In practice, however, the legal scope of SUDENE was greatly reduced (Goodman and Albuquerque, 1974). In its planning, SUDENE depended greatly upon the initiative and cooperation of various sectoral agencies of the federal government operating in the area, and it was directed and influenced by the political and institutional stability (or lack of it) within these agencies. Also, budgetary resources controlled by SUDENE were limited and unstable, thus allowing a degree of outside control over the effective implementation of programmes and projects which regional interests considered priorities.

There were two main reasons for this. Firstly, there are always tremendous political–institutional obstacles to the coordination by any superintendency or planning secretary post, because in their relationship with other sectors of federal or state administration, these sectors strive to preserve their decision-making 'territory' and to refute any and all attempts by planners to interfere in their various sectoral development programmes. Secondly, national development plans in general have many objectives which by their nature conflict, yet which must be attained simultaneously in an allotted time-period. Yet it was not established precisely what the willingness would be to sacrifice, within a given system of values, one objective in favour of another. The result of this lack of any trade-off possibility in the planning process meant that in the dynamics of decision-making, conflicts of objectives would only be resolved by interest-group pressures. And historically, the representatives of populations of depressed areas are the ones who have the least bargaining power and input in constructing workable programmes of which they themselves are the targets.

These remarks concerning the phasing-out of SUDENE as the supervisory and planning agency of the north-east in favour of a new order—which again strengthened federal government agencies in the area—also illustrate the over-all reasons for the failure of other regional planning agencies at federal and state levels, which were created with the political intention to become legitimate centres for aiding local low-income groups. However, the limited participation of these groups could be enlarged, if strategies which make up these policies had different built-in mechanisms for personal and regional income distribution. This could be very advantageous. The content of such policies should therefore be assessed.

SOCIAL VALUES AND CONTENT OF SPATIAL DEVELOPMENT POLICIES

One of the main characteristics of the 'old paradigm' which has guided actions of policy-makers in recent years in Brazil is the supposed difference between the positive, 'what is', and the normative, 'what should be'. In accordance with this paradigm, the triumph of economics as a science comes at the moment it manages to get rid of value-judgments, beliefs, and ideologies, and starts solving problems by means of empirical verifications through the method of scientific explanation. The analytical tool economists have available to deal with typical problems of the public sector is the *welfare theory*, concerned with deriving normative criteria for the optimal allocation of scarce resources from simplified hypotheses about individual consumer and enterprise behaviour. This theory, however, was developed *assuming the absence of power in individual relations*, which renders most recommendations of the theory extremely naïve from a political–organizational viewpoint.

That trend in economics which attempts to establish its autonomy as a discipline is relatively recent, and only dates back to the last century. For classical economists, economics meant political economy: allocation of scarce resources that could be applied in alternative ways to different uses, as well as distribution of the results of these allocations amongst individuals and social groups, decisions which by their very nature are extremely political (Boulding, 1970).

As a consequence of isolating power relations from their analytical system, economists tend to split decisions into 'economic' and 'political'—with economic decisions allocating scarce resources, and political decisions redistributing the results of such allocations. In the Brazilian development experience, for example, governmental effort was directed in the past 15 years towards guaranteeing economic efficiency, leaving redistributive policies to execute, outside the productive system, all corrections deemed necessary to soften social costs connected with the growing expansion of GDP. The basic hypothesis of this strategy, which did not work in practice, is that any unfavourable effects upon income and property distribution can be compensated for, so that those who lose will be rewarded by those who gain a larger share of an increasing product. Balancing this development process 'from above' (see Chapter 2 in this volume), it has been found, through various social and economic indicators that official policy, in timidly enforcing such compensation schemes, has allowed certain growth mechanisms to become institutionalized, with benefits going only to a minority stratum of the Brazilian population, which has had a highly concentrating effect.

To illustrate this, some aspects of the objectives of redistributive justice should be examined in the recent experience of Brazilian regional planning.

GROWTH AND EQUITY IN THE DEPRESSED PERIPHERY

Major factors of regional development which have controlled recent public policy in Brazil assume that redistribution and equity objectives are subordinate to the accelerated expansion of per capita regional income, since such strategies can be achieved by indirect methods, through priority allocations to objectives. In increasing the income and tax base of a regional economy, it would be easier for the regional power to manipulate greater resources, which would allow for solutions 'from above' to filter down to the social and economic problems of the poorest in the population.

An assessment of this strategy can be made through social and economic indicators starting from the recent experience in the Brazilian north-east, where a per capita product growth rate above national average was keenly pursued. It can be seen that:

> the fundamental characteristic of this process is the selective integration of a few urban sectors and restricted social groups in isolated regional centres into the production system dominated by São Paulo and Rio de Janeiro; the dualistic nature of the north-eastern economy was aggravated by recent policies; urban–rural income differentials were enlarged, and income distribution disparities increased markedly in the last decade. (Goodman, 1975.)

To demonstrate the frailty of the per capita income growth rate as a dominating criterion for attainment of regional development, let us examine the results of two studies which used non-monetary indicators (mortality levels and employment) to verify the impact of a strategy which assumed spill-over effects derived from efficiency-oriented policies, to be helpful in the achieving of social justice and a better quality of life.

From research on the demographic evolution of the north-east compared with Brazil as a whole, in the period from 1940 to 1970, the following conclusions can be mentioned (Carvalho and Wood, 1978):

(1) The central north-east life-expectancy was about 44.2 years in the decade 1960–70, just a bit above the most recent estimates for Africa. At the same time, in the southern region of Brazil (see Table 15.3), life-expectancy was 61.9 years;

(2) The index of dispersion, in calculating regional and rural life-expectancy, using for criteria the relative percentages in regional populations for 1940, 1950, and 1970, indicates that regional inequalities grew between 1930 and 1950, but declined in the last decade;

Table 15.3   *Life-expectancy at birth, by region, 1930–40 to
1960–70 (years of age)*

Region	1930–40	1940–50	1960–70
Amazonia	39.8	42.7	54.2
North-east/north	40.0	43.7	50.4
North-east/centre	34.7	34.0	44.2
North-east/south	39.3	39.2	49.7
Minas	43.0	46.1	55.4
Rio	44.5	48.7	57.0
São Paulo	42.7	49.4	58.2
Paranà	43.9	45.9	56.6
South	51.0	55.3	61.9
Central-west	46.9	49.8	57.5
BRAZIL	41.2	43.6	53.4
Range	16.3	21.3	17.7
Index of dispersion	3.86	5.84	4.88

Source: Carvalho and Wood, 1978.
Note: Estimates for the period 1950–60 are not available as the 1960
census was not published in its entirety.

(3) Estimates by income group indicate that average life-expectancy levels
of high-income families greatly exceed those of poorer families (see Table
15.4). If disparities between high-income and low-income families for
various regions are examined, it is observed that difference is again greatest
for the north-east centre region.

At the same time, studies on family and personal income distribution in
Brazil indicate that the incidence of absolute poverty is greater in the north-
east, both as to interregional distribution of the poor and from the viewpoint
of high frequency of the poor in intraregional distribution. However, in the
last few years the performance of the north-eastern economy has been satisfactory, measured by growth rate of per capita product, which has frequently
been above the national average. It is therefore wise to query, what is the most
adequate approach in dealing with problems of the north-eastern population:
a regional economic approach, or a social approach?

Several social scientists, dissatisfied with the perverse results of recent
development policies upon income and wealth distribution in the north-east,
question whether these policies implemented on behalf of the region's 'interests',
do not conceal instead the spatial dimensions of the class conflict which is under
way at the level of income growth throughout the economy. They also advocate
changing the use of 'regions' as the cut-off point for economic policies, arguing
instead in favour of social development programmes directly aimed at benefiting low-income groups, chiefly in depressed areas. These groups could eventual-

Table 15.4   *Life-expectancy at birth, by region and by household income, 1970 (years of age)*

Region	Total	Household income (cruzeiros)				
		(1)	(2)	(3)	(4)	(4) − (1)
		1–150	151–300	301–500	501+	
Amazonia	54.2	53.4	53.9	54.8	58.2	4.8
North-east/north	50.4	50.0	50.8	52.7	55.7	5.7
North-east/centre	44.2	42.8	46.1	50.3	54.4	11.6
North-east/south	49.7	48.9	50.3	51.9	54.9	6.0
Minas	55.4	53.8	55.4	55.6	62.3	8.5
Rio	57.0	54.1	54.8	57.6	62.1	8.0
São Paulo	58.2	54.7	56.1	58.7	63.9	9.2
Paranà	56.6	54.8	56.5	59.3	63.7	8.9
South	61.9	60.5	61.2	63.4	66.9	6.4
Central-west	57.5	56.5	57.1	58.2	63.3	6.8
BRAZIL	53.4	49.9	54.5	57.6	62.0	12.1

Source: Carvalho and Wood, 1978.

ly be favoured by the spill-over effects of present development strategies, but such benefits would not occur until a time so far in the future as to be politically intolerable.

Publication of the results of the 1970 industrial census facilitated a study of the regional structure of industrial employment in the Brazilian economy and its evolution since 1950. In a study by Haddad (Haddad, 1977b) personnel employed in industry in the years 1950, 1960, and 1970 were analysed. Nineteen industrial sectors were studied, in 25 states and territories, over a period when the country was putting great stress on a policy of import-substitution and expansion of the economy.

In analysing the structure of industrial employment, it can be seen that in 1950 there was a high degree of spatial concentration. 81 per cent of total industrial employment was located in only six states: São Paulo, Guanabara, Minas Gerais, Rio de Janeiro, Rio Grande do Sul, and Pernambuco; and moreover, the Rio–São Paulo axis contained 57 per cent of that total. Results also indicated that these six states managed to attract 80 per cent of national industrial employment growth between 1950 and 1970. Thus the high concentration was maintained, and within the six most developed states, São Paulo alone absorbed 56 per cent of new industrial jobs in the national economy. And this state of São Paulo began the decade with almost 49 per cent of the industrial employment of Brazil as a whole.

The position of the north-eastern economy should be outlined here. Industrialization in the north-east advanced at the slowest pace of any region in the country: in 1950 it had some 16.8 per cent of total industrial employment. This share declined to 10.3 per cent by 1970. The north-east would have had to start the 1970s with 174,000 more jobs than its recorded total, to hold the same relative position it held in 1950 *vis-à-vis* the rest of Brazil, equivalent to an extra 63 per cent added to its industrial employment level for 1970.

To identify components of regional growth, a method of analysis was employed (see Table 15.5) with a rationale evolved from simple empirical evidence, i.e., growth of employment is greater in certain industries than in others, and greater in some regions than in others. Therefore a particular region can show a faster economic growth than the average of a group of regions, either because there are more dynamic sectors (structural components) in its industrial mix, or because that region has a growing participation in the national employment in individual sectors (differential component).

In determining the difference between structural and differential components, this method permits identification of the different forces bearing upon regional growth. The structural component tells us that a few sectors in the national

Table 15.5 *Patterns of regional growth of industrial employment in Brazil, 1950–70*

State	Total net shift		Differential shift		Structural shift	
	Absolute value	Percentage value	Absolute value	Percentage value	Absolute value	Percentage value
Acre	341	0.10	369	0.15	− 28	− 0.02
Amazonas	2,202	0.66	1,020	0.40	1,181	0.83
Pará	2,146	0.65	1,988	0.79	158	0.11
Amapá	763	0.23	941	0.37	− 177	− 0.12
Rondônia	1,086	0.33	1,107	0.44	− 20	− 0.01
Roraima	− 197	− 0.06	− 124	− 0.05	− 73	− 0.05
Maranhão*	− 7,063	− 2.12	1,889	− 0.75	− 5,174	− 3.63
Piauí*	2,487	0.75	3,043	1.20	− 556	− 0.39
Ceará*	1,140	0.34	6,831	2.70	− 5,692	− 3.99
Rio Grande do Norte*	− 8,083	− 2.43	− 3,516	− 1.39	− 4,567	− 3.20
Paraíba	− 26,952	− 8.10	− 13,699	− 5.41	− 13,253	− 9.29
Pernambuco*	− 77,759	− 23.38	− 34,122	− 13.48	− 43,637	− 30.58
Alagoas*	− 24,369	− 7.33	− 10,866	− 4.29	− 13,504	− 9.46
Bahia*	− 13,155	− 3.96	1,167	0.46	− 14,322	− 10.04
Sergipe*	− 20,642	− 6.21	− 10,937	− 4.32	− 9,705	− 6.80
Minas Gerais	− 27,856	− 8.38	− 4,791	− 1.89	− 23,065	− 16.17
Espírito Santo	6,280	1.89	8,268	3.27	− 1,988	− 1.39
Rio de Janeiro	− 26,661	− 8.02	− 42,733	− 16.88	16,062	11.26
Guanabara	− 91,618	− 27.55	− 120,226	− 47.49	28,609	20.05
São Paulo	236,004	70.96	142,760	56.39	93,244	65.35
Paraná	41,630	12.52	40,571	16.03	1,059	0.74
Rio Grande do Sul	− 8,229	− 2.47	− 10,265	− 4.05	2,036	1.43
Santa Catarina	19,616	5.90	26,021	10.28	− 6,405	− 4.49
Goiás	13,165	3.96	13,668	5.40	− 503	− 0.35
Mato Grosso	5,731	1.72	5,406	2.14	326	− 0.23

* North-eastern states.
Source: Haddad, 1977b.

development process grow more rapidly than the others, accounted for by certain factors such as technological innovation, etc. Since any of the specialized regions (São Paulo, for example) in the more vigorous sectors of the nation's economy will achieve positive structural variation in employment terms, it is crucial to research the possibilities for each region, in order to choose with care the best possible locations for industries or businesses being located there.

But regional growth must not be examined solely on this basis. The different regional sectors present varying productive capacities and performance capabilities. Therefore a given region (Paraná, for example) can develop more quickly than others, since it has shown the capability of attracting a growing percentage of activities or companies, even though it may belong to a slow-growth sector at the national level. Major forces which bring about these readjustments are almost always of a locational nature, i.e., variations in transportation costs, specific fiscal stimuli for certain areas, differentials in relative prices of inputs amongst regions, etc. It is therefore necessary to study the locational advantages of each region in order to pinpoint possible advantages of slow-growth sectors at the national level, as well as to clarify the reasons for the favourable performance of these sectors in the over-all picture.

The difference between the effective growth of each region and its hypothetical estimated growth (employing national rate of employment growth for this region) is calculated by two factors, one structural and one differential: or, total net shift = structural shift + differential shift; thus,

$$\left(\sum_i Ei^1 j - \sum_i Ei^0 j\right) - \sum_i Ei^0 j(r_{tt} - 1) =$$
$$\sum_i Ei^0 j(r_{it} - r_{tt}) + \sum_i Ei^0 j(r_{ij} - r_{it})$$

where:

Eij = employment in sector i of region j;
$\sum_i Ei^1 j$ = employment in all sectors of region j in 1970;
$\sum_i Ei^0 j$ = employment in all sectors of region j in 1950;
r_{tt} = national rate of employment growth;
r_{it} = national rate of employment growth in sector i;
r_{ij} = rate of employment growth in sector i of region j.

This method is simply a set of indicators and not a model, and does not explain the trends shown in its results. It is, however, a valuable instrument in the exploratory stage of studies about patterns of the spatial concentration of economic activities. Although the method can be more revealing when we have a higher degree of sectoral disaggregation, it is obvious from Table 15.5 that the position of the north-eastern states is precarious (they are responsible for almost half of the total negative shift), and that from the six states with the greatest share of total industrial employment, only São Paulo indicated a rate of growth above the national average, with positive results for both differential and structural components. To illustrate, we can interpret, from Table 15.5, the results for

Minas Gerais: this state would have to begin the present decade with 27,856 more jobs than the recorded figure, in order to hold the same relative position it held in 1950. This was because the industrial structure of the state at the start of this period was composed mainly of traditional sectors (it had a negative structural component equal to 23,065 jobs, or 16.7 per cent of the total negative structural variation); and because the productive sectors had a relative performance in Minas Gerais which was less favourable than the other states, on average. (Negative differential component equal to 4791 jobs, or 1.89 per cent of total negative differential variation).

More recent data (Universidade Federal de Pernambuco, Pimes, 1977) regarding the situation of underemployment and under-remuneration in the country, indicate the seriousness of the problem in the north-east. Giving as a definition of *visible underemployment* 'the population who worked part-time but would like to work full time, besides those people who are occasionally working part time for economic reasons', information from the National Household Survey of the Census Bureau (FIBGE) allows calculation of the different rates of visible underemployment (total, urban, and rural) for 1973. The north-east shows the highest rates (8.7, 9.2, and 8.0 per cent respectively) amongst all regions when compared with values reflecting average levels for the country (6.1, 4.9, and 5.0 per cent). If the under-remuneration of up to half a minimum wage (see Table 15.6) is taken as a defining level of disguised underemployment (which figure would contain many of those obviously underemployed), it can be seen that while the average of employees paid this amount was rather high (19.5 per cent in 1972, and 14.8 per cent in 1973), the position of the north-east (Region V) appears to be dramatic (46.2 per cent in 1972, 36.9 per cent in 1973), since the minimum wage can be considered a value which already indicates absolute poverty. Information contained in Table 15.6 indicates that these interregional differences also occur for self-employed workers, and show even less favourable values.

The employment issue in the north-east is closely connected to policies of industrial expansion adopted for the region at the start of the 1970s. Concern with attracting private investors to generate employment opportunities for existing labour and demand for an economic infrastructure to be constructed in the region, inspired planners to work out legislation allowing the north-east region to use profits originating from an up-to-50-per cent income tax deduction for companies, and also giving exemption from import duty for the importation of equipment, financial support to projects with negative rates of real interest, etc. The end-result was a decreased cost of capital (compared with its opportunity cost) to entrepreneurs who base decisions upon market price in deciding upon the size and concentration of investment projects.

The current wage rate, on the other hand, legally determined by minimum subsistence level, does not reflect the relative scarcity of labour in this region where unemployment or disguised underemployment prevails. If therefore a 'social' price system had been developed in order to select and assess projects for the region, employment here would have been much higher, because more

Table 15.6 *Monthly income levels of employed and self-employed workers by region, 1972 and 1973*

Income classes in relation to legal minimum wage	Cumulative frequency and distribution (in percentage points)													
	Region I		Region II		Region III		Region IV		Region V		Region VI		Total	
	1972	1973	1972	1973	1972	1973	1972	1973	1972	1973	1972	1973	1972	1973
Employed workers														
up to $\frac{1}{4}$	1.3	0.9	2.1	1.3	2.5	1.8	9.3	5.5	16.0	11.7	0.8	0.3	6.2	4.2
$\frac{1}{4}$ to $\frac{1}{2}$	5.1	4.1	8.4	6.2	10.6	8.8	28.5	18.7	46.2	36.9	2.2	1.3	19.5	14.8
$\frac{1}{2}$ to 1	16.0	24.5	29.0	29.4	44.2	40.2	55.3	56.3	75.0	71.1	8.9	13.1	43.3	43.5
1 to 2	61.8	58.6	66.3	62.4	79.2	75.5	85.3	81.7	90.1	88.2	53.6	47.3	75.5	72.5
2 to 3	79.7	75.7	82.0	77.9	89.2	86.1	92.2	90.0	93.9	93.3	72.2	68.1	89.6	84.1
3 to 5	89.0	88.0	91.2	89.9	94.7	93.2	96.5	95.3	96.7	96.2	83.4	80.9	93.3	92.3
TOTAL	100.0	100.0	100.0	100.0	100.0	100.0	100.0	100.0	100.0	100.0	100.0	100.0	100.0	100.0
Self-employed workers														
up to $\frac{1}{4}$	7.8	8.7	4.7	2.7	8.9	8.6	20.0	15.0	32.8	30.9	9.0	3.6	18.5	16.3
$\frac{1}{4}$ to $\frac{1}{2}$	18.8	23.5	11.9	10.4	19.7	19.7	35.4	32.3	53.3	50.8	22.0	14.3	32.9	31.2
$\frac{1}{2}$ to 1	37.8	43.4	26.2	24.1	39.6	40.0	59.0	56.2	77.6	76.4	40.2	33.3	53.6	52.7
1 to 2	69.5	71.5	56.8	50.2	71.8	66.4	83.6	81.4	93.0	92.4	70.9	63.3	78.0	75.7
2 to 3	83.6	81.4	74.2	62.7	85.0	76.3	90.9	87.5	.97.0	95.6	83.8	74.0	87.9	83.0
3 to 5	91.4	91.5	95.3	93.0	93.2	89.7	95.7	94.3	98.5	98.4	91.3	89.0	93.6	92.6
TOTAL	100.0	100.0	100.0	100.0	100.0	100.0	100.0	100.0	100.0	100.0	100.0	100.0	100.0	100.0

Source: Universidade Federal de Pernambuco, Pimes, 1977.
Note: Region I: Rio de Janeiro; Region II: São Paulo; Region III: Paraná, Santa Catarina & Rio Grande do Sul; Region IV: Minas Gerais & Espírito Santo; Region V: Maranhao, Ceará, Rio Grande do Norte, Paraíba, Pernambuco, Alagoas, Sergipe, Bahia; Region VI: Distrito Federal.

labour could be employed per unit of investment, since the social cost of labour is lower than market wage rates. In a research project using the following formula for computing the social cost of labour (SCL):

$$SCL = c - \frac{1}{So}(c - m),$$

where c = consumption level of the urban worker measured at social cost,
      m = marginal productivity of labour in agriculture,
      So = present value of one unit of investment, in consumption terms,

it was concluded (Bacha *et al.*, 1974) that within the many hypotheses for estimating values, the variation in social cost of labour in the north-east varies from 50 to 60 per cent of private costs. If these values were adopted in the decision-making process for projects in the region, reduction in the relative price of labour could stimulate a higher volume of current production, the easing of the unemployment problem, and an improvement in personal income distribution. As alternatives, research suggests, at the implementation level, the reduction or elimination of costs of social insurance contributions for the labour force, or the granting of direct subsidies for the employment of labour.

NATURAL RESOURCES AND REGIONAL GROWTH

Inequality in industrial development policies for depressed areas is also revealed in programmes and projects for exploitation of the natural resource base (both agricultural and mineral) in peripheral regions of the country. Often these resources form the necessary base for starting the development process in a particular area, aided usually by product demands from other regions or from abroad.

In the present sociopolitical climate where, in a spatial context, centre–peripheral relations prevail between developed and underdeveloped areas of the country, those regional economies whose export base comes from resources only, become extremely vulnerable and unstable. This is because the dominating centre, by investment in transportation and communications in the periphery, creates a relationship of domination over the periphery which is characteristic of an internal colonialism. In such areas where a high resource potential is identified, large agricultural and mining projects are created, financed, and controlled by private capital from centrally located companies, aiming to generate, at low cost, an export surplus of raw materials and food products which is required to support the industrialization process in other regions, or to solve acute problems in the balance of payments.

It is not at all certain when the pattern of regional development in Brazil is seen in historical perspective, that the major projects mentioned above will succeed in ensuring stable development conditions in regions where they are installed. The exporting of natural resources only provides a starting point for the initial take-off in any regional economic development. In order for such exports to be able to continue this development, a strong dynamism is necessary, coupled with the capacity for dispersing this dynamism to other productive sectors of the region. In different periods of the country's history, there have been regions where growth occurred for a few decades through the extraction or cultivation of primary products; but the socioeconomic structures needed for sustaining the development process were never created.

This is an interesting point to illustrate by giving examples where the central power of the public administration channelled forces towards the strengthening of mechanisms in order to dominate peripheral areas. This was almost always in an area where conflicts had occurred between regional interest and the objectives of macroeconomic stabilization and growth policies. It is interesting to note contradictions which have occurred between the intentions of planners trying to carry out development processes 'without devastating the nation's natural resources' (objective of the second NDP) and the over-all strategy prevailing in current economic development planning in Brazil. It can be reported, for example, that the federal government policies for the Amazon over the last decade included an ambitious road-building programme combined with colonization programmes along the main routes, research concerning natural resources to clarify uncertainties about the region's development potential, and a variety of fiscal incentives to attract private investment for the agricultural, cattle-raising, forestry, and mining sectors (Haddad, 1977a; Mahar, 1978).

Although implementation of such policies is still in embryo stage, the first studies indicate that if recent development of a socioeconomic infrastructure in the Amazon has reduced its isolation from the rest of the country, and has introduced consideration of new investment alternatives, on the one hand, then on the other hand the strategy of accelerated occupation and heavy exploitation of natural resources through great mining projects has brought about heavy destruction of the local flora and fauna.

The planners' great desire to establish an optimal rate of natural-resource exploitation in the country, with social welfare improvement in mind, requires a broader investigation of conditions for the proper use of these resources. Social rates of discount could be introduced in specific projects, and adequate cycles of resource use so as to minimize environmental pollution in the interest of future generations, and so on. The broader implications of this discussion lead us to a redefinition of the development process in ecological terms.

A narrower view would be for Brazil to instigate a natural-resources policy with aims which were not merely reflections of those interested only in growth in the economy, the group which has had all along a dominating position in decision-making in the planning process.

It is now unfortunately a probability, with current difficulties in balance of payments (growing foreign debt, excessive growth of imports, overvalued exchange rate, etc.), that the country will exploit its renewable and non-renewable natural-resource potential at an undesirable rate in order to accelerate export growth, as exports have become a critical factor in present development planning.

Several recent studies indicate the way in which regional settlement and colonization policies for the Amazon and the centre-west are developing the land through their special programmes (Poloamazônia, Polocentro, Prodepan, etc.), organized by the federal government. These programmes again emphasize efficient production as the dominating objective for small farmers. Official colonization projects have given a major role to food-processing companies to take advantage of potential growth in these regions, 'mainly to obtain a significant contribution for the growth of the Gross National Product'. Inspired by fiscal incentives, subsidized rural credit, and a high intensity of public capital formation (mainly in transportation and communications infrastructure), the major food-producing companies have begun cattle-raising over huge tracts of land, leaving large areas unexplored for other speculative possibilities. The result is frequently the destruction of the small producer, an increase in concentration of ownership of real estate, job eradication, and an accumulation of benefits to the upper decile of rural income earners (Goodman, 1978; Wood and Schmink, 1978).

This problem is illustrated in quantitative terms through the visible implications of the extensive cattle-breeding in the Amazon region (Mahar, 1978):

(1) At subsidized prices made possible by fiscal incentives and easy credit, where less than 350 concerns submitted cattle-raising projects to the SUDAM (Superintendency for Development of the Amazon Region), these establishments quickly acquired 800,000 square kilometres of Brazilian agricultural frontier land (an area approximately ten times the size of Austria);

(2) A frontier strategy of extensive cattle-raising, and such a concentrated pattern of land-ownership, contribute only in a very minor way to rural employment or to permanent human settlement in the area. To illustrate the potential impact of a possible alternative strategy, Mahar has calculated the possibilities of additional employment which could result from an intensive pattern of land use, using estimates based upon average land/labour ratios observed in the different states of the Amazon. Results of calculations by projects indicate that the potential employment figure (68,112) is more than four times that created by cattle-raising projects (15,468). This clearly demonstrates the heavy opportunity cost of cattle development in the region.

THE EMERGENCE OF A NEW PARADIGM?

The problems which have to be faced by policy-makers in Brazil over the next decade are no longer problems of bridging the economic gap between effective and potential GNP, or those related to new challenges of an anti-cyclic nature. They are the more complex issues of elimination of absolute poverty in urban and rural areas, identification of economically feasible energy alternatives, the restructuring of supply to provide for greater production of public and wage goods, and maintaining the per capita GNP growth rate and of price stability in the face of adverse developments in the international economy.

Providing solutions to these new challenges (the problems themselves are old, but have only now gained political momentum) cannot be accomplished under the aegis of a decision-making organization that concentrates all its power in the core regions, depending on technocracy and public institutions to solve the problems. Consider, for example, urban and rural poverty in its various aspects (poor nutrition, illiteracy, underemployment, poor housing conditions, etc.). It is obvious that we have barely scratched the surface of these insistent problems, if concentration continues to be on conventional methods like those applied so far, i.e., the organizing of a group or a new institution to deal with the matter, or the establishing of budgetary resources to finance necessary actions, or as happens frequently, a fine-sounding propaganda scheme is launched to make public opinion feel that 'problems are being solved'. What always happens is that these resources never meet the proportions or seriousness of problems; and small attention is given to assessment of effectiveness or efficiency of expenditures made. The end-result is that no substantive improvement occurs.

A recurring theme in Brazilian spatial planning over the past 20 years has been great rhetoric on social justice, which appears in all official programmes and projects. Implementation of programmes and projects, however, has generated some adverse results in terms of redistribution and social equity for low-income peripheral area populations. On the one hand, spatial policies were explicitly aimed at reducing differences between regions in terms of quality of life; while on the other hand, macroeconomic stabilization and growth policies, because they were not tailored to each region's needs, produced unfavourable results for development in peripheral areas, based as they were on criteria of efficiency in the private economy, and not on the social cost of the region's development.

Even in the current development stage of Brazilian capitalism, there are significant forces and potential resources available in peripheral areas which, if tapped, could help solve some of the problems. Such solutions would be less elegant perhaps, but certainly more adequate than those monotonously proposed by unimaginative politicians and technicians in government bureaucracies. Use of non-conventional sources of energy, a realistic housing policy for low-income groups, stimuli to help put proper technologies into place, a plan for integrated rural development, defence of consumers, and environmental preservation programmes: these are some of the themes of a new paradigm of development which are being thrust forward for government attention by political pressures. But they will only become reality if, in their evolution, they involve initiatives and greater participation from those social groups 'from below' which are at present outside the bureaucratic establishment.

REFERENCES

Bacha, E. L., A. B. de Araujo, M. da Mata, and R. L. Modenesi (1974) *Análise Governamental de Projetos de Investimentos no Brasil* (Rio de Janeiro: IPEA IMPES).

This Brazilian experience fits quite well with the expectation that comes from an assessment of the centre-down paradigm when approached from the standpoint of international comparison; for instance in W. Stöhr and F. Tödtling (1978).

Boisier, S., A. T. M. Sizva and C. A. Lodder (1972) *Analysis del sistema de planeamiento estadual en Brasil* (Rio de Janeiro: CEPAL).

Building, K. E. (1970) Economics as a political science, *Economics as a Science*, pp. 77–96 (New York: McGraw-Hill).

Carvalho, J. A. M. de and C. H. Wood (1978) Mortality, income distribution and rural–urban residence in Brazil. Paper presented in Symposium on Socio-Economic Change in Brazil (Madison:University of Wisconsin, May).

Cintra, A. O. and P. R. Haddad (eds.) (1978a) Aspectos da factibilidade do planejamento estadual no Brasil. *Dilemas do planejamento urbano e regional no Brasil* pp. 143–65. (Rio de Janeiro: Zahar).

Cintra, A. O. and P. R. Haddad (eds.) (1978b) *Dilemas do planejamento urbano e regional no Brasil* (Rio de Janeiro: Zahar).

Denslow, D. (1978) As exportações e a origem do padrão de industrialização regional do Brasil. In Baer, W. *et al.* (eds), *Dimensões do desenvolvimento econômico brasileiro* (Rio de Janeiro : Campus).

Goodman, D. E. (1975) O modelo econômico brasileiro e os mercados de trabalho: uma perspectiva regional, *Pesquisa e Planejamento Econômico*, 5 (1), 89–116.

Goodman, D. E. (1978) Expansão de fronteira e colonização rural: recente politica de desenvolvimento do Centro-Oeste do Brasil. In Baer, W. (eds.), *Dimensões do desenvolvimento econômico brasileiro.* (P. P. Geiger and P. H. Haddad, Rio de Janeiro: Campus).

Goodman, D. E. and R. C. de Albuquerque (1974) As Politicas de desenvolvimento regional. In Goodman, D. E. and R. C. de Albuquerque, (eds.), *Incentivos á industrialização e desenvolvimento do Nordeste*, pp. 149–94 (Rio de Janeiro: IPEA/INPES).

Haddad, P. R. (1977a) Natural resources and regional development: lessons from the Brazilian experience. Paper presented at Fifth World Congress of International Economic Association (Tokyo, August–September).

Haddad, P. R. (1977b) Padrões regionais de crescimento do emprego industrial de 1950 a 1970, *Revista Brasileira de Geografia*, 39 (1), 3–45.

Haddad, P. R. (1978) Os economistas e a concentração de poder—notas preliminares, Fundação João Pinheiro, Beto Horizonte, *Análise e conjuntura*, 8 (3), 2–7.

Mahar, D. J. (1978) *Desenvolvimento Economico da Amazônia* (Rio de Janeiro: IPEA).

Siqueira, P. P. de (1978) Sistema tributário nacional e planejamento estadual. In Cintra, A. O. and P. R. Haddad (eds.), *Dilemas do Planejamento urbano e regional no Brasil*, pp. 156–65 (Rio de Janeiro: Zahar).

Stöhr, W. and F. Tödtling, (1978) An evaluation of regional policies—experiences in market and mixed economies. In Hansen, N. M. (ed.), *Human Settlement Systems* (Cambridge, Mass. : Ballinger Co.).

Universidade Federal de Pernambuco, Pimes (1977) *Estado atual e evolução recente das desigualdades regionais no desenvolvimento brasileiro* (Recife) 3 v.

Weiterschan, H. M. (1978) *Politica econômica ao nivel estadual no Brasil: avaliação da experiência recente a proposição de alternativas* (Belo Horizonte: Fundação João Pinheiro) 2 v.

Williamson, J. (1965) Regional inequality and the process of national development—a description of the patterns, *Economic Development and Cultural Change*, 13 (4).

Wood, C. H. and M. Schmink (1978) Blaming the victim: small farmer production in the Amazon colonization project. Paper presented at Annual Meeting of American Association for the Advancement of Science (Washington).

Development from Above or Below?
Edited by W. B. Stöhr and D. R. Fraser Taylor
© 1981 John Wiley & Sons Ltd.

Chapter 16

Chile: Continuity and Change—Variations of Centre-Down Strategies under Different Political Regimes

SERGIO BOISIER

INTRODUCTION

Purpose of the Study

The present study on Chile analyses this country's experience in the use of regional (sub-national) development policies in helping to solve some basic problems, particularly in relation to the poorest social stratum of population. Chile represents a highly attractive case study in the context of use of regional development policies, due partly to the fact that regional policies have been institutionalized for over a decade, and partly to the important political and economic changes that have taken place in Chile. The study covers the period 1965–78, during which time three ways of dealing with regional development problems may be clearly identified.

DESCRIPTION OF THE COUNTRY

It is a well-known fact that regional development policies are not independent of the spatial features of a country, so that the main characteristics defining Chile's spatial organization must be described, albeit briefly.

Chile's natural resources are distributed broadly throughout the length of its territory according to a specific and diversified pattern. Metallic and non-metallic minerals are found in the north, the best soil resources occur in the central zone, while forest and energy resources are located in the south. Fishery resources are considerable and highly varied along the whole length of the country's extensive coast.

The urban population, in spite of this broad distribution of natural resources, however, is heavily concentrated in a typical primacy structure.

Figure 16.1 Planning regions of Chile, 1970

In 1970, 43.6 per cent of the nation's urban population lived in the capital city, Santiago, and only 16.6 per cent in the next two cities following in size.

The territorial distribution of economic activity also shows a considerable degree of spatial concentration, whatever the territorial division used as a reference. Thus for example, the so-called metropolitan region which generated 43.6 per cent of Chile's GDP (gross domestic product) in 1974, comprises only 2.27 per cent of the whole territory, yet includes 37.1 per cent of the population. Chile's total GDP in 1978 was estimated at US $12 billion and its per capita income at about US $1,100.

From the standpoint of spatial organization, Chile presents a clear case of geographical concentration of population and production, territorial centralization of the administrative system, and centre–periphery relationships in its spatial–economic system.

System of Regional Administration

The system of government and internal administration of Chile is of a unitary nature. The country is currently divided into 13 regions (one of which is the metropolitan region) and 40 provinces. In accordance with changes introduced by the present government, the structure of the system of government and territorial administration (regional, provincial, and municipal) is based on the establishment of three vertical subsystems: (1) the hierarchical subsystem, whose components are the President of the Republic, the Minister of the Interior, the regional intendentes, the provincial governors, and mayors; (2) the technical subsystem, composed of the National Planning Office (ODEPLAN), the Regional Planning and Coordination Offices (SERPLAC), the Provincial Planning and Coordination Offices, and the Municipal Planning and Coordination Offices; and (3) the participative subsystem, consisting of a national development council, regional development councils, provincial advisory committees, and municipal development councils. Some of these bodies are still in the formulative stage.

Background of Regional Development Policies in Chile

The conditions under which the Chilean nation developed (a permanent state of war for over 300 years owing to indigenous resistance, and external wars with bordering countries), in addition to geographical, social, and cultural differentiation, led to the consolidation of an extremely centralized system of administration.

There is nation-wide consensus that centralism as thus defined is a defect which needs to be corrected, although such consensus is based not on a thorough analysis of the situation but rather on some relatively vague recriminations against the growth of the capital city in contrast with the poverty and lack of means of the other provinces. This consensus seems to have been reached in the second half, or more precisely in the last third of the nineteenth century, and has become stronger since the first quarter of the present century. (Sanhueza, 1977, p. 1.)

This anti-capital feeling and the consensus referred to above, combined with some situations that are described later in this study, constitute the initial factor which subsequently led to the establishment of regional policies.

The central government's need to consolidate national sovereignty in areas remote from the seat of government (particularly when some of those areas had been conquered in wars with bordering countries), and the no-less-important need to satisfy the most active regional groups (although regionalism has found little political expression during the twentieth century) in exchange for permanently necessary political support, led to the establishment of an inorganic regional development legislation, which Stöhr has called 'implicit regionalisms' in the case of Chile (Stöhr, 1976).

In 1960, after the strong earthquakes and tidal waves which destroyed part of the South of Chile, the executive was accorded extraordinary legislative powers. The result was the restructuring of the Ministry of Economic Affairs (which became the Ministry of Economic Affairs, Development, and Reconstruction), a considerable impetus to economic planning through the principal governmental development agency, the Corporation for the Development of Production (CORFO); and even more important, the setting-up in each province of a provincial development committee headed by the provincial authority and composed of government officials and representatives of the private sector. The Decree/Law establishing these Committees laid down as their basic objective: 'incorporating the provinces in the study of regional development plans and in the supervision of the implementation of annual programmes for the investment of government resources'. For a variety of reasons (such as the multiplicity of institutions, lack of government technical support), these committees, with very few exceptions, had no effect whatsoever; and three years after their establishment they had practically disappeared.

In short, throughout this century, but in particular during the 15 years between 1950 and 1965, a number of independent measures adopted by successive governments—sometimes in response to political pressures generated in the periphery, sometimes as autonomous action by the centre—helped to lay the basis for the systematic institutionalization of regional planning in the second half of the 1960s.

PHASE ONE: REGIONAL PLANNING AND SOCIAL MODERNIZATION UNDER THE CHRISTIAN-DEMOCRATIC FREI GOVERNMENT
1964 to 1970

The Socio-political Context

It is a well-known fact that Chilean society has been developing in an uneven manner during the present century. This unevenness, which has been very marked in terms of sociopolitical and economic development, gradually led to a situation which John Friedmann has so aptly called 'the crisis of inclusion'

(Friedmann and Lackington, 1966). In response to this situation in which the economic development and urbanization process in Chile was incapable of meeting growing social demands, there began to take shape a new structure of political forces, which advocated radical changes in the life of the nation, and proposed new national objectives. The struggle for power amongst the new political forces was temporarily resolved in 1964 in favour of a combination of the centre (on social-democratic lines) headed by the Christian Democrats.

The general objectives, both economic and social, upheld by the government of the time were expressed in the following manner in an official strategic document (ODEPLAN, 1968):

   (1) to speed up the rate of growth of gross domestic product;
   (2) to achieve a redistribution of income and property in favour of the under-privileged groups;
   (3) to attain full employment of the labour force;
   (4) to gradually arrest the inflationary process;
   (5) to improve the country's economic sovereignty;
   (6) to provide a level of basic education for the school-age population, increase school attendance at the secondary level, and furnish technical training facilities for workers;
   (7) to improve the population's state of health;
   (8) to promote the construction of dwellings to keep pace with population growth, and improve the standard of housing;
   (9) to institute an efficient system of justice within reach of the whole population;
  (10) to promote organization and participation of the community in development activities;
  (11) to improve relations between workers and employers;
  (12) to integrate the various national sectors organically into the formulation and implementation of development plans.

This list of objectives was far from constituting a traditional collection of programmed good wishes. Indeed, Chilean society began to be shaken by a series of drastic economic and social government measures. That the government elected in 1964 was seriously proposing to modernize Chile's economic and social structure is beyond doubt, if some of the main government actions of that period are reviewed. (To illustrate this: during the period 1964–70 a reform was introduced in the Constitution concerning ownership rights, a process of agrarian reform was carried out, a reform of the educational system was initiated, steps were taken to stimulate and give impetus to rural trade unionism, the tax system was modified to include a property tax, the planning function was introduced at highest government level, laws were passed and action taken to encourage establishment of intermediary organizations (neighbourhood committees, mothers' centres, etc.), Latin American integration was promoted in particular through the Cartagene Agreement, copper production was 'Chileanized', a Ministry of Housing and Urbanization was instituted, vigorous impetus was given the development of heavy industry, etc.). Thus the broad set of objectives listed above derived from the integrality or integrational approach that characterized interpretation of Chilean society by the intellectuals and politicians of the governing party of that time. A publication by a respected economist, Jorge Ahumada, *La Crisis Integral de Chile* (Chile's Integral Crisis) (1966) exerted an enormous influence on the Christian Democrat intelligentsia.

But to the question whether the prevailing structure of Chile permitted these manifold objectives to be tackled simultaneously or with any chance of success, the answer is No. This explains, from one point of view, the emergence of regionalization and regional policies in Chile.

> The attainment of these goals, however worthwhile they may be in themselves, is contingent upon the existence of a broad consensus with respect to the means to be employed and a thorough-going mobilization of the population around the objectives of national development. Yet neither consensus nor mobilization is likely to materialize so long as the country is internally divided, with rural areas set off against the city and, within their own communities, the urban sub-proletariat against the organized working and middle-class groups. A unifying framework for policy and program was therefore sought that would establish the basis for both consensus and mobilization in the society, be more responsive to the new demands, and permit the linkage of local energies to national power. This unifying framework we may call *the urban-regional frame* for national development. (Friedmann, 1969, p. 14.)

Thus what Friedmann has called 'the urban-regional frame' appeared as a means of achieving the general aim of social modernization, an idea which developed on fertile ground: the anti-capital or anti-centralist feeling of the majority of the population.

The Ideological and Doctrinal Background of Regional Planning in the Period 1964–70

From a strategic and tactical point of view, introduction of the regional dimension into the country's organization seemed more than justified by the above reasoning. Nevertheless the idea of regional development, or rather the institutionalization of regional planning as a public function, also responded to deeper factors than simple tactical considerations.

One of the central values of the government party's political philosophy, Christian Humanism, was solidarity. Solidarity appeared as the answer to the concept of struggle, whether individual as in the liberal model, or class struggle, as in the Marxist model. Solidarity was expressed in programme form in the social integration postulate (Palma, 1976). The operability of this postulate in practice involved setting up participation machinery and procedures. It is no accident that in the first strategic document on regional planning it should be affirmed that 'the region thus appears as an instrument of action for the development policy and as an instrument of participation for the individual, who is the object and the subject of planning' (ODEPLAN, 1968, p. 35). The same document suggested national integration (physical, economic, and sociopolitical) as the central objective of regional development.

The concept of the state as promoter and representative of the common good (another value of the Christian-humanist philosophy), combined with the integraism of both humanistic and technocratic origin in their vision of Chilean society, led the governing coalition to decisively support planning—first as an instrument for the management of public affairs and social mobilization, but later, owing to political opposition, only as a somewhat neutral and technocratic instrument of management.

As one analyst says,

> The delays and adoption of safeguards in the approval [of the law establishing the planning system] stemmed from the search for a compromise that would guarantee that the governing party would not insist on extending the planning area beyond the strictly technical sphere. (Palma, 1976, p. 52).

Formulation of Regional Development Policy

Notwithstanding the integral approach which characterized the socioeconomic interpretation of intellectuals and politicians of the governing party of the time, the fact is that the planning body as set up—the National Planning Office (ODEPLAN)—was composed of a group of specialists with the most technocratic and economic approach in the new government. Of course, this situation was a challenge to the new-born group of regional planners who, in order to win intellectual respectability, had to show that they were as technocratic as, or even more technocratic than, their colleagues responsible for over-all planning. Another contributing factor were the foreign experts, especially hired to advise the group of regional planners. The two most important advisers wrote:

> Experts were imported to bear sacred truths acquired at the fount of knowledge in distant countries overseas. Their message was at first not questioned. But what served these experts as truth was, indeed, little more than the generally accepted body of doctrine which informs nearly all regional analysis. (Friedmann and Stöhr, 1967, p. 27).

Analysis of regional development policy for this period could begin with a description of the guidance system. This system related to the level at which regional development programme objectives and criteria were defined. Generally speaking, it was the level where authority for the development policy was delegated to the various regional population groups (Stöhr, 1975). From that point of view, Chile's regional development policy over the period 1964–70 consisted of a system of national guidance, which was compatible, moreover, with the orientation 'from the centre down' which—from a different angle—typified the regional development policy under study.

An essential substantive characteristic of the regional development policy of the time lay in its primary orientation towards geographical areas, as subjects of development, in preference to an alternative orientation towards people. In terms of categories used by Cumberland in his study on regional planning in the United States (Cumberland, 1972), Chile's regional development policy gave preference to 'aid to places' over 'aid to people'.

This approach was manifest from many angles. One was the regionalization of the country, a task which consumed an excessive amount of time, effort, and capacity for political struggle, as though the question of defining regions were an end in itself.

Another point was the government's evident preoccupation with the amount invested in each region (as a criterion for assessing results achieved), without questioning, for example, the employment effect of such investment (Achurra, 1972; ODEPLAN, 1970). Yet another point was the emphasis on reducing the disequilibrium between the average per capita income of the regions as an objective *per se*, ignoring first the dynamic relation between interregional disequilibria and the intraregional disequilibria of income; and secondly, the relation between regional disequilibria and the functional and/or personal disparities in distribution of income.

There is no doubt that interregional income disequilibria were and are as large in Chile as in many other Latin American countries. Table 16.1 shows a proxy measure of regional income, i.e. per capita regional geographical product (RGP).

Table 16.1   *Selected regional social indicators*

Region	Index of per capita RGP* (1970)	Rate of infant mortality (1968)	Rate of elementary school attendance (1969)	Regional distribution of housing deficit (percentages) 1969
I. Tarapacá	123.2	65.9	97.4	4.8
II. Antofagasta	241.5	85.3	86.0	2.5
III. Atacama/Coquimbo	97.7	89.3	84.2	10.2
IV. Valparaiso/Aconcagua	101.3	60.2	87.7	6.7
Metropolitan Zone	123.1	56.6	79.3	31.5
V. O'Higgins/Colchagua	92.0	79.9	78.4	4.6
VI. Maule/Talca/Linares/Curicó	62.7	92.1	76.8	6.4
VII. Bío-Bío/Ñuble/Concepcion/ Arauco/Malleco	69.3	111.9	83.0	23.0
VIII. Cautín	44.5	131.1	92.3	2.4
IX. Valdivia/Osorno	66.1	124.8	89.5	4.1
X. Llanquihue/Chiloé/Aysén	62.1	117.4	81.8	3.4
XI. Magallanes	194.1	50.9	93.5	0.3
Average or Total	100.0	83.4	82.6	100.0

Source: Achurra, 1972.
*Regional Geographic Product.

A second substantive characteristic of the regional development policy of the time was its conception within the context of the 'centre-down' paradigm developed by Hansen (see Chapter 1). According to Hansen, the essential position within this paradigm is that development, whether spontaneous or induced, starts in a relatively few dynamic sectors and geographical clusters and will (or should) spread to the rest of the spatial system in the course of time. All the theorizing and practice of growth poles fits into this scheme.

If a single element were to be used to characterize the theoretical foundation of Chile's regional development policy in this period, it would be the growth pole concept. In fact, starting with the very definition of planning regions, the growth pole concept has become the main rationalizing element of regional development policy, despite the fact that the concept was misinterpreted in theory and used in a completely inefficient manner as a policy instrument. A thorough analysis of this situation was made by the author for the United Nations Research Institute for Social Development (Boisier, 1971). It is interesting to note that this determined use of a growth pole strategy does not—as might have been supposed—derive from an explicit theoretical interpretation of the structure and spatial functioning of the Chilean economy (in fact, such an interpretation is still lacking today), but simply from the need to 'use what was to hand' in terms of regional development instruments within the framework of a neocapitalist system.

In line with the foregoing observations, the regional development policy of the period contains some of the most important biases attributed to general regional development policies in vogue as from the 1960s (Stöhr and Tödtling, 1977): notably, urban and industrial, with large-scale answers to almost any type of regional problem, based essentially on highly centralized organizations and market mechanisms.

A third characteristic of regional policy during this phase was its economic bias, the reduction of all phenomena to objectives (and obviously to policies and policy instruments) contained within the context of regional economic variables and problems. Not only does this appear very clear upon examination of regional development policy components, but it is also evident from observing the complete absence of professionals in the fields of sociology, anthropology, political science, and administration, in the technical group concerned with formulation and implementation of regional policy. In fact, if such professionals did exist, their efforts were completely and absolutely subordinated to, and limited by, those specialists trained in engineering and economic disciplines.

This economic bias in regional policy relegated the population, the social groups, and in the last instance, individuals—the final subject of development—to a merely subsidiary position. Regional policy did not primarily seek to improve living conditions of the poorest population groups, but aimed rather at achieving the greatest economic efficiency compatible with the deconcentration

Table 16.2 *Percentage of poor by region (1970 Census)*

Regions (North-south)	Total population, a	Total number of poor, b	Percentage, c $\frac{b}{a}$. 100
I. Tarapacá	176,180	38,301	21.7
II. Antofagasta	251,655	49,409	19.6
III. Atacama/Coquimbo	493,504	140,528	28.5
IV. Valparaíso/Aconcagua	909,545	152,675	16.8
V. O'Higgins/Colchagua	479,670	108,976	22.7
VI. Maule/Talca/Linares/Curicó	624,209	153,440	24.6
VII. Bío-Bío/Ñuble/Concepción/ Arauco/Malleco	1,451,919	335,550	23.1
VIII. Cautín	420,620	115,431	27.4
IX. Valdivia/Osorno	435,540	87,512	20.1
X. Llanquihue/Chiloe/Aysén	361,288	78,790	21.8
XI. Magallanes	91,625	8,653	9.4
Metropolitan Zone (Santiago)	3,425,943	647,139	18.8
TOTAL	9,121,698	1,916,404	21.0

Source: *Mapa de extrema probreza*, ODEPLAN—Instituto de Economia, Universidad Catolica de Chile, Santiago, Chile, 1974.

and decentralization pursued—so much so that of the eight provinces located at the bottom of a scale representing a compound socioeconomic indicator (see Boisier, 1974a), only two are in a policy priority region. In contrast, of the seven provinces in the best positions on the same scale, four constituted a part or all of the priority regions.

In a subsequent study an attempt was made to identify the location of that portion of the population which, according to criteria used in the same study, were living under conditions of extreme poverty. It is interesting to compare the regional results of the study, shown in Table 16.2, with the *de facto* priorities of regional policy in this phase.

The three regions with the highest percentage of poverty (Regions III, VI and VIII) were by no means high-priority regions, while regions with greater priority in regional policy (Regions VII, II, IV and I) in general showed poverty indexes below national average. Even in absolute terms, the poor living in priority regions numbered only 575,935 persons, which is something like 30 per cent of the total indicated in the study. The essentially economic content of regional policy is therefore clear.

The Specific Content of Regional Policy

Even at the risk of presenting a distorted view because of space limitations, it is necessary to describe some of the specific components of regional policy in the period 1964–70. The typical division of subjects in development plans will be used for the purpose.

1. Regional analysis

The analysis of the spatial structure and functioning of the Chilean economy which serves as a basis for policy formulation is more descriptive than interpretive in nature, and centres on a study of the state of disorganization—physical, economic, and social—of the country. Officially the interregional analysis was accompanied by a standardized analysis of each region with accent on identification of the region's main problems, growth potential, and role within the national development scheme. The regionalization of the whole country represented a basic task in which a wide range of criteria were used. This process has been described by Stöhr (1976), so it will not be referred to here except to note the considerable political problems deriving from regional definitions.

2. Regional objectives

The objectives established at the regional level were mainly associated with attainment of adequate levels of physical integration, economic integration, and sociopolitical integration, and also with reduction of interregional income disparities and, as a historic constant of Chile's development, the economic and demographic consolidation of the most remote regions of the country. As specific strategies were prepared for each region, these objectives were reduced merely to economic growth targets.

3. Regional strategy

The regional development strategy designed to attain the above objectives was based, as already stated, on the growth pole concept. To that end, various kinds of poles in the national territory were identified and assigned a degree of priority. Those defined were: a national development pole (Santiago), three multi-regional development poles (Valparaíso, Concepción, and Antofagasta), ten regional development poles (Arica, Punta Arenas, La Serena-Coquimbo, Valdivia, Talca, Temuco, Puerto Montt, Iquique, Rancagua and Osorno), and 20 intraregional development centres consisting of towns of over 20,000 inhabitants.

Regional development strategy defines actions of 'control' over the centre or metropolitan region, and of 'development' in priority peripheral regions. Regional priorities and appropriate action for groups of regions are defined in the following terms: (i) spatial rationalization and control policy (metropolitan region); (ii) reconversion and polarization policy (Regions IV and VII); (iii) diversification and integration policy (Regions I, II and XI in the extreme north and extreme south); and (iv) support and complementarity policy (Regions III, V, VI, VIII, IX and X) (ODEPLAN, 1970).

4. Policy instruments

Some of the main regional development policy instruments used at the time were:

(a) the establishment and strengthening of regional development corpora-
tions: Committee for the Progress of Arica, Magallanes Corporation,
Northern CORFO Institute, Chiloé-CORFO Institute, Aysén-CORFO
Institute, Iquique and Pisagua Programming Committee;

(b) establishment of regional councils of intendentes (the Regional Council
of Intendentes of the Bio-Bio Region VII, and Regional Council of
Intendentes of the Maule Region VI);

(c) regional sectoral committees to coordinate the regional work of agencies
with common interests in any sector, for example the agricultural sector;

(d) the industrial location policy, which was the most important and successful regional policy instrument and operated by means of customs duties
on capital equipment and tax concessions, and the provision of infrastructure;

(e) regionalization of the public sector capital budget with a view to more
efficient allocation of public sector resources.

PHASE TWO: REGIONAL PLANNING AND SOCIAL MOBILIZATION UNDER THE SOCIALIST ALLENDE GOVERNMENT, 1970–1973

The Ideological and Political Context: Transition to Socialism

The 1964–70 social modernization period was directed and controlled by a
single highly ideology-oriented political party which was in this way the creator
of an essentially exclusive political project. Partly due to this characteristic—
which makes it difficult to form political alliances—and partly due to the
aggravating factor of a slow-down in economic growth towards the end of the
period, the 'crisis of inclusion' noted by Friedmann became more acute, partially
creating conditions that opened the way for triumph of a Marxist coalition in
the 1970 presidential elections.

The governing coalition during this phase (Popular Unity), was structured
and dominated by the two most important Marxist parties in Chile, the communist party and the socialist party. The ideological framework which determined the government's policies was therefore always crystal-clear. The Basic
Programme of the Popular Unity government stated as follows in 1969: 'The
only genuinely popular course, and therefore the fundamental task which the
Government of the People has before it, is to finish with the domination of the
imperialists, the monopolies, and the land-owning oligarchy, and embark on
the construction of socialism in Chile.'

Thus began the transition-to-socialism period, through a political tactic
called 'the Chilean way to socialism'. This attempt took place in a political
medium characterized by the presence of three political blocs of similar size,
which in view of the drastic project of social change proposed, necessarily

converted the government bloc into a minority bloc. 'Consequently, the first and fundamental problem that arose following Salvador Allende's triumph was that of power, or even more specifically, that of how to use the executive power to launch a series of changes that would lead to the conquest of total power' (Arriagada, 1978, p. 66).

This problem was tackled by the government from a purely political approach and from an economic approach. From the latter viewpoint, the government's efforts were directed at creating an area of social property in the economy.

> The Popular Government's first step, which is a determinant of any other result that may be desired in the economic, social and cultural field, is the materialization of changes in the structure of ownership, intended to eliminate the imperialist, monopolist, financial and land-owing power. Thus a strong public and mixed ownership area will be set up with the workers participating in its management, which will permit the complete diversion of the major part of production to satisfying the needs of the broad masses, and at the same time a change in the capitalist production relationships that prevailed in those production sectors. (ODEPLAN, 1971, p. 27.)

The creation of an 'area of social property' had, of course, distinct territorial and regional connotations in view of the uneven geographical distribution of economic activity. Furthermore, the establishment of the area of social property was considered, at least at the outset, as a regional development policy instrument. In fact, an official document states:

> To consolidate the Social Property Area in a specific geographical context means, therefore, to ensure in their full magnitude the production, investment, supply, distribution and marketing process whose materialization and requirements differ according to the sector and region concerned. (ODEPLAN, 1972, p. 28.)

The Regional Planning Concept in this Phase

Economic planning is of the essence in a socialist economy, as it constitutes the main mechanism for allocation of resources. It is therefore perfectly logical that the government should wish to consolidate a comprehensive planning system, including regional planning at a high level. (The system organized in 1971 included a National Development Council, the National Planning Office with its regional and sectoral offices, the Budget Office of the Ministry of Finance, the National Statistical Institute, the National Institute for Technological Research and Standardization, the Natural Resources Institute, the National Scientific and Technological Research Commission, and all the regional development bodies.) As stated by the Minister–Director of Planning (for the first time the government gave ministerial status to the Director of the National Planning Office in 1971), 'The Popular Government attaches great importance to decentralization and regional planning' (Martner, 1972, p. 21). On this point the government merely institutionalized a concept already widely supported by the population. Even the electoral triumph of Salvador Allende was generated in the provinces, not the capital. (In the province of Santiago Mr. Allende obtained 34.5 per cent of the votes, while in the whole country he obtained 36.3 per cent.) Political logic thus demanded that the provincial discontent implicit in the voice of the electorate should be given due attention.

Nevertheless, the outstanding fact from the standpoint of this analysis is that the Popular Unity government inherited the regional planning function in a two-fold sense, which would ultimately mean the maintenance of most important aspects of the former approach. First, the government inherited an organization-al structure in the field of regional development which nothing suggested needed modifying, and which expanded, but did not change; secondly, it also inherited a technical team which grouped together nearly all the specialists trained in the previous phase, that is, nearly all the Chilean experts able to contribute know-ledge and experience in this field. Regardless of whether or not it supported the new government, this technical team shared the same view of regional planning, so that most of the biases and characteristics of the work done in the previous phase were transferred to this one. Over and above initial statements, the documents on regional planning describing the work to be done in this period began to repeat subjects, approaches, and proposals essentially similar to those of the previous phase.

The following general development objectives were laid down by government for the period 1971–76:

(1) to achieve greater economic independence through incorporation of natural resources and basic wealth in resources belonging to the nation;
(2) to move from an exclusive economy to an economy of popular participa-tion through large-scale creation of new jobs;
(3) to improve distribution of income;
(4) to restructure the means of production in order to improve living levels of the masses;
(5) to increase material wealth and rechannel its use;
(6) to extend the area of social ownership and convert it to development leadership;
(7) to achieve a sustained national development, moving from a stagnant economy to a constantly expanding economy;
(8) to create a more balanced spatial economy.

To achieve these objectives, a strategy was proposed on the basis of the mass incorporation of the whole population into the process of change. Two develop-ments were to take place simultaneously: a substantial rise in employment, and a significant real increase in wages and salaries for the lower-income stratum. In other words, the aim was to increase the consumption of the population—a type of consumption which presumably would meet the following conditions: (i) designed to satisfy the basic needs of the bulk of the population; (ii) based on or supported by the organization of the masses to meet their own needs; (iii) based on the full utilization of natural resources and simple technologies employing considerable labour; and (iv) associated with a progressive levelling of incomes.

Such a strategy was surely of a mobilizing nature even when, as stated by one analyst, 'the whole mobilizing effort had an ambiguous destination: "popular power", "popular movement", "the workers"' (Palma, 1976); and it should have had profound regional impact, due to its association with exploitation of natural resources and the use of 'soft' technologies.

As Stöhr writes in Chapter 2, policy emphasis in a 'bottom-up' strategy is linked to: territorial, organized services for basic needs; rural and village development; labour-intensive activities; small and medium-size projects; technology which permits the full employment of regional human, natural, and institutional resources on a territorially integrated basis. If one compares these elements with those of the economic strategy of government as outlined in the previous paragraph, there exists a high degree of coincidence.

So the base was laid for the designing of regional strategies and policies within the 'bottom-up' model of development, resting firmly on the concept of self-reliance. This however did not happen, because: (i) regional planners were less conscious about alternative development strategies than their national counterparts, or they were well behind national planners in terms of political ideology; (ii) concern with a 'bottom-up' type of development is rather recent and there was not, and still is not, a well-structured theory available (as Stöhr, in Chapter 2, correctly points out); (iii) the whole political process was interrupted in 1973, which prevented development and implementation of emerging new ideas; and (iv) there were other side-effects of the over-all strategy, connected with the pattern of spatial distribution of population and economic activity, and the inertia existing in these processes.

To some specialists the contradictions between strategy as postulated and real conditions prevailing in the country were clear from the very inception of the government. For example, Geisse (1971) maintained that:

the deconcentration options based on the ODEPLAN model of the previous administration (1964–1970) might be thought to lose some of their validity *vis-à-vis* the new government's programmes: for example, for a determined redistribution of income and the considerable increase in demand through the consumption of the low-income sector, which might presuppose a regional redistribution in favour of the poorer regions.

This is not necessarily so:

(1) In the Santiago Metropolitan Area there is also a concentration of poverty, if not greater in proportion, yet substantially greater in number so that its demands are generally given preference. Therefore, it is quite likely that the effects of government policies aimed at providing work and shelter will also continue to be concentrated there.

(2) A major part of the greater demand will initially be covered by idle installed capacity, in view of the ample margins existing in nearly all branches of industry, a capacity which is also concentrated in the Metropolitan Area.

(3) Industrial investment will be directed, for the most part, at eliminating the bottlenecks in existing plants, almost all of which are located in Santiago.

(4) The best possibilities for agro-industrial development are also found in the regions within the immediate vicinity of Santiago". (Geisse, 1971, p. 260–261).

This is the general framework within which the regional planning concept in this phase of the study was inserted. Some of its most important characteristics are commented on below.

A distinctive feature of regional planning, despite its formal differences, during the period 1970–73 was its continuity. This is partly explained, as noted earlier, by the continued operation of the same technical team. It also derives partly from a simpler question related to the small degree of freedom offered by Chile's spatial and economic structure for the introduction of substantive change. This continuity is partly associated with the absence of a specific theory of regional development within socialist–Marxist thinking (or more specifically, absence of a theory for the transition to a period of socialism) for which regional development problems, like others, simply derive from the capitalist mode of production. This same fact helps explain a certain adherence by regional planners to the hypothesis of social and spatial symmetry, according to which solution of superstructural problems automatically brings with it the solution of territorial problems. Nevertheless, this was one of the several contradictions that characterize statements on regional development during this period. In fact, the social and spatial symmetry hypothesis appears implicit in the total theoretical denouncement of 'dependent capitalism', as the underlying reason for any regional development problems. To quote a senior official of the régime. 'The present national form of this structure in Chile is the result of past dependent capitalist development, and the regional problems observable today are the result of the contradictions to which it gave rise' (Bedrack, 1974, p. 19). This oversimplified view is not shared, however, by the principal technical experts of the national planning office, some of whom wrote:

> it should be noted that, even when the inherited spatial structure is mainly the result of the development of a capitalist economy, it is not possible, in practice, to accept a simple mechanical concept and affirm that merely the construction of socialism will lead to a more adequate territorial structure. (J. Cavada *et al.*, 1972, p. 4.)

In view of this characteristic of continuity—notwithstanding certain changes which were introduced and which will be analysed later in the study—the concept of regional planning which prevailed in this phase contained, as noted, biases similar to those in the previous phase. Economics still dominated the regional approach, and individuals and social organizations were still absent, as subjects, from regional planning.

Regional policy guidance was definitely intended to be a national, centralized type of guidance.

> There is a certain conviction that regional planning is only possible in a system of centralized planning, since only thus is it possible to understand and act in a region conceived in the national context with a role assigned to it from the centre. (J. Cavada *et al.*, 1972, p. 4.)

How this centralization was to be made consistent with the strengthening of democracy, was something that was never seriously discussed.

Perhaps the most innovative factor appearing in the regional planning approach in this phase was the replacement of growth poles as a strategy basis, by the concept of 'integrated areas'. A more thorough analysis reveals, however, that beyond the attack on the traditional polarized development strategy, a more sophisticated concept of polarization was used which was on the one hand already evident in the last technical documents of the previous phase (ODEPLAN, 1970) and which, on the other hand, stemmed directly from proposals aimed not at discarding a polarized development strategy, but adapting it to conditions in Latin America (Boisier, 1974b). This fact is very clear in Bedrack's writings on the spatial development strategy of the period 1970–73 (Bedrack, 1974).

Another innovation relates to the introduction of new geographical categories of analysis, but not of management. Thus the idea of a macro-zone as a supra-regional category (Macrozona Norte, Macrozona Central, Macrozona Central-Sur, Macrozona Sur, and Macrozona Sur-Austral); and the concept of Basic Territorial Unit as the minimum information and planning level (Babarovic and Wellhoff, 1972).

Noteworthy in this period are both the expansion of the regional planning system to all regions (through establishment of regional planning offices in each), and the effort involved in preparing development plans in each region. These plans were basically print-outs of investment projects, however.

The Specific Content of Regional Policy, 1970–73

As for the previous period, it is necessary to give a brief account of the specific content of regional policy in its principal elements.

1. Regional analysis

In essence, the regional analysis which was to serve as basis for regional policy in this phase coincides with the analysis made in the period 1964–70. What should be noted, however, was the attempt to accentuate the interpretive character of the analysis, even though this was done solely by total adherence to the dependent capitalist development theory, without questioning the validity of that theoretical approach as a category of sub-national analysis. Moreover, as happened in any case with most official statements on the state of Chilean society during this period, the regional analysis took the form of a severe social criticism.

2. Regional objectives

The following basic regional development objectives were established:

   (i) to re-adapt the spatial structure of the country in accordance with requirements of the proposed production process, and at the same time to
transform the transport system;

  (ii) to decentralize population growth, by substantially altering trends of
migratory flows;

 (iii) to create in each region sources of productive labour that would help
increase production and raise household income;

 (iv) to mobilize idle or under-utilized resources;

  (v) to introduce changes in the regional structure during the six-year
period, basically through accelerated development of industry;

 (vi) to increase each region's average productivity, and to expand all those
activities with high social productivity;

(vii) to create a social infrastructure capable of ensuring the whole population the right to health, education, recreation, and housing;

(viii) to institute the mechanisms to promote cultural development in every
region;

 (ix) to link participation of the people with all aspects of the process of
constructing the new society;

  (x) to obtain a harmonious, proportional, and integrated regional development in the whole of the national territory.

3. Regional strategy

The strategic thinking of this period did not manage to settle as clearly as in the
preceding period, due partly to interruption of the governmental period and the
priority given to the conjunctural political struggle, from 1972 onwards.

In any case, an approach that sought more equitable objectives at both the
interregional and the intraregional level came into force in this phase. The
national regional development strategy seems to have been based on six fundamental ideas: (1) the promotion and consolidation of medium-sized urban
centres (10,000–300,000 inhabitants); (2) the incorporation of new areas in the
production process; (3) the intensive exploitation of the coast and marine resources; (4) development and consolidation of border areas; (5) strengthening
of regional industrial development through industrial complexes; and (6)
priority in treatment of the most depressed areas, from an economic and social
viewpoint. (This last is perhaps the only really new element of the above
strategy. With a few modifications, the new regional division was essentially the
same as that prevailing in earlier phases.)

4. Regional policy instruments

The definition of regional development policy in this phase was not without a
certain amount of ambiguity. This to a certain extent explains the absence of
official statements in terms of regional policy instruments, which are only outlined in the following manner (Soms, 1972):

(i)   decentralization of the state-owned banking system and other non-banking credit institutions;
(ii) regionalization of the public sector capital budget;
(iii) establishment and rationalization of regional development corporations;
(iv) structuring of a regionalized marketing and distribution system;
(v)   generation and rational study of regional investment projects;
(vi) administration of regional enterprises.

This type of regional policy instrument needs no comment, but the lack of consistency between real functioning of an economic system, objectives proposed, and the instruments supposedly designed to achieve these objectives is only too glaring. A sort of wishful thinking is visible in the following statement by one of the experts associated with the planning office:

> The future orientation of regional development, the overcoming of the historical disequilibria between provinces as a result of the *modus operandi* of the capitalist system, the full use of idle resources, and the righting of century-long social injustices, depend on the efficiency and energy the People's Government may use to activate what are known generically as the instruments of regional economic control. (Soms, 1972, p. 75.)

It is not possible to evaluate the conception of regional planning at this point, at least in terms of specific actions and results. Like all government activities of the period, it was to come to a premature end, institutionally speaking, in 1973.

PHASE THREE: REGIONAL PLANNING AND SOCIAL DISARTICULATION UNDER THE MILITARY JUNTA, 1973–

The military government which assumed complete power in 1973 represented a drastic break with the development style the country had been following—with the natural differences proper to the various political groups which came into power—since the end of the 1920s. As has already been stated, this former mode of development had been marked by lack of synchronization between social and political development and economic growth. The break was total and radical, and once more in the history of Chile, a new government sought to create an irreversible social situation by disassembling the country's economic, political, social, trade union, and cultural structures, in order to build a new society.

The Ideology of the Military Government

The basic ideological guidelines of the military government were made explicit in a document entitled 'Declaration of Principles of the Government of Chile', published six months after the take-over.

In the general context of the doctrine of national security, the military junta defined its policy as anti-Marxist, nationalist, and authoritarian. At the same time, it proposed a completely liberal economic system, which included a gradual return to the private sector and denationalization of property (mining, agricultural, and especially industrial property), and the opening up of the economy to foreign trade; and it postulated freedom of prices so as to enable the market to operate as the main system for the allocation of resources. This liberal system is based at least on two principles relating to the role of the state: the subsidiarity principle (where only those functions which intermediate bodies and private agencies cannot carry out, devolve on the state), and the principle of the functional and territorial decentralization of state power.

Concept of Regional Planning During this Phase

Only the initial confusion produced by the violent break-up of the institutional system, and the immediate move by several groups to take up hasty positions, could provide a basis for the hope of maintaining and expanding the regional planning function within the new government system. As will become clear from events culminating in 1978, the basic inconsistency between the economic system now evolving and attempts at planning regional development were resolved in favour of the more powerful technical and power groups—who are opposed as a matter of principle to the very idea of planning.

Regional planning in this phase seemed linked to the ideas of national security and state decentralization. It is the National Planning Office which maintains the national security, while state decentralization is handled by the traditionally powerful Ministry of the Interior and the recently created National Commission for Administrative Reform (CONARA) and both implicitly support regional planning activities from an administrative point of view.

The point of departure for the initial treatment of regional problems in this phase lies in a (mimeograph) programming document prepared by a group of economists just before the overthrow of the Popular Unity government. This document was intended to serve as an economic policy programme for the military junta. It argues decisively in favour of functional decentralization as a form of combatting excessive 'statism'.

Since the document mentioned contained almost no reference to territorial decentralization, immediately following the change of government the regional planners of ODEPLAN prepared a confidential study for the military junta (*Restauración nacional y desarrollo regional. Bases para una política*; confidential, October 1973), in which an effort was made to reinstate the function of regional planning. It also attempted to make the most of the situation of total power, to bring about changes in the country's spatial development pattern. It is worth mentioning that this document was prepared by the same experts who had worked in the two previous political phases, a point which once again provides support for the phenomenon of essential continuity in the concept of regional policy. It is also important to note the influence which this document had on subsequent official declarations.

Within a short time, however, the first signs of a power struggle—over the handling of regional questions—appeared between the National Planning Office (attempting to defend its legal authority in this respect) and the Ministry of the Interior (which, realizing the potential source of political power which the regions represented, tried to take over administration of regional development), and lastly the National Commission for Administrative Reform, which considered regionalization as the most suitable method for self-legitimization. This struggle only served to help the most radical supporters of the over-all development model to gain control, in the end, of the regional planning apparatus.

In any case the military government committed itself initially to regional planning in a manifesto by the President of the Republic which maintained that:

> the modernization of the State of Chile, owing to its particular geographical conditions, requires a system permitting administratively and regionally decentralized development, so that co-ordination and participation of the regions in terms of national integration, security, socio-economic development and administration may materialize in the most perfect form possible. (CONARA, 1976, p. 33.)

With this act, the government dissolved the former political and administrative division of the country (25 provinces); and a new regional division was adopted. New provinces were identified and the structure of the system of regional, provincial, and communal government and administration was defined. (Actually, however, with some modifications the new regional division was essentially the same as that prevailing in earlier phases.)

The basis for the new regionalization is the search for national integration, the country's geopolitical requirements, and the need to limit overpopulation and growth in Santiago due to excessive spending on urban infrastructure. The same decree lays down the difference between pilot and non-pilot regions. The first pilot or priority areas are the two regions in the extreme north, the two in the extreme south, and the region of the Bio-Bio basin in the central part of the country. As may be seen again, the geopolitical criterion is a determining factor in setting regional priorities.

Obviously, the conception of regional planning during this phase is mainly contained within the broader concept of the doctrine of national security. This is evident both in the criteria and in the language used to set regional priorities. The territory is seen as a 'being', as a living organism, while pilot regions are defined, nerve centres and urban outports identified, national unity is postulated, etc. (CONARA, 1976).

A mixture of theoretical concepts, to some extent quite coherent but also to some extent internally inconsistent, are contained in the strategy. On the one hand, the centre–periphery nature of the structure and spatial functioning of the Chilean economy can be seen (although a purely economic interpretation of the centre–periphery relationship is used). As a result it is proposed to use poles or centres of growth as points for concentrating investment (the term 'nerve centre' is used); but at the same time these poles are considered as connecting points for different transport systems, something more often associated with the concept of central places than the concept of growth poles. On the other hand, the neoclassical approach reappears in terms of the stress laid on improving spatial mobility of resources and goods, including the operation of the market and price system, but there is some contradiction in the use of concepts like the centre–periphery model and development poles.

Despite numerous declarations in favour of decentralization and the fact that the regional development strategy states that:

> the greatest responsibility for implementation [of the strategy] devolves on the regions themselves, on the efficiency with which their regional governments act, and the effort which local entrepreneurs make in terms of capacity, initiative and response to official stimuli (ODEPLAN, 1975, p. 256).

It is obvious that as far as guidance of regional policy is concerned, this is again a national process, and a highly centralized one, this time for purely political reasons associated with the very nature of a *de facto* government. It cannot be ignored, however, that the regional authority set up in this phase (the Regional Intendant) has considerably more power and autonomy than any other regional authority had in the past. It is not irrelevant that in all cases he happens to be the civil as well as military regional authority.

Nonetheless, the process of transfer of power to regional authorities has taken place within a broader process of decentralization and reduction of the public sector of the economy. Although regional authorities now have more power, their field of control is smaller than before; the net result being a reduced capacity to make autonomous decisions.

The Specific Content of the Policy

Despite the growing disfunctionality which regional placing is acquiring *vis-à-vis* the problems of the Chilean economy and *vis-à-vis* the gradual consolidation of the over-all development model, it has been possible to propose certain lines of strategy during this period and even to put some policy instruments into operation.

1. The regional analysis

It has already been said that the Chilean economy has been over-analysed, not only nationally but regionally. A new diagnosis, however, is being made, stressing the descriptive aspect. Two relatively new elements have been introduced: a strong emphasis on analysis and the role of natural resources, and an evaluation of the geographical situation of Chile, a factor which is considered almost an additional resource.

2. Regional objectives

The proposed objectives of regional development for this phase involve: (i) the search for a balance between natural resources, geographical distribution of population, and national security; (ii) the need to direct the participation of the nation in defining its own destiny; (iii) the need to place the organization of geo-economic space at the service of national objectives; (iv) the need to guarantee equal opportunities to the entire population; and (v) the need to watch over the continuing existence of the territory as a living organism.

3. Regional development policies and policy instruments

Stress was placed on the regionalization (differentiation and priorities) of sectoral policies, particularly in sectors of direct production, rather than stressing regional policies as such. The main regional policy instrument brought into service during this phase is the National Regional Development Fund (NRDF). Set up in 1974, it is financed with a budgetary contribution equal to at least 5 per cent of total tax and tariff earnings, excluding the real estate contribution. The Fund is distributed amongst the regions in accordance with priorities of the strategy. In 1976 the Fund's resources amounted to rather more than $70 million, over which the regions have only limited autonomy once their share is fixed. Moreover, regions in the extreme north and south have continued to benefit from special régimes in taxes and tariffs, the latter not being very significant in an increasingly open economy. Mention should also be made of the use of incentives for employment in some regions, and of an official proposal to place a special location tax on activities in the metropolitan region (this last type of instrument under normal political conditions would be difficult to implement).

Conflict Between the Development Style and Regional Planning

Despite the formal importance which planning and the planning agencies have had in Chile, it is a fact that basically the economic policy has been directed by the government's financial agencies, even during the Popular Unity administration. This helps explain the continuing presence of conflict—sometimes visible, at other times invisible—between planners and those directly responsible for carrying out economic policy. During the present phase, the team responsible for economic policy (Ministry of Finance, Ministry of the Economy, and the Central Bank) took the National Planning Office under its control from the beginning, at least in regard to over-all planning. The conflict was therefore between regional and national planning. The nucleus of this conflict was the opposition between proposals for non-discriminating homogeneous economic policies with the same impersonal norm for everyone, upheld by national planners and economic policy authorities; and the need for territorial discrimination, or differentiated economic policies, the argument supported by regional planners. At a certain point the conflict spread and became confrontation between regional authorities and the President of the Republic himself, as depicted by the press during 1977.

As the national economic model begins to show apparent short-term successes (reduction of inflation, improvement of public finances, the bettering of the external trade situation, etc.), the relative power of the economic policy team has increased and has finally brought about the elimination of the function of regional planning in 1978, taking advantage of a period of internal reorganization of the National Planning Office. In the new system only the regional planning and coordination secretariats have been retained, as agencies for improving regional information (on resources and projects), so as to provide incentive for private investment. Nonetheless, all the administrative aspects of regionalization still remain; and the regions, as a geographical level of the administration of the state will be incorporated into the new constitution which is currently under study.

AN OVER-ALL APPRAISAL

After nearly 14 years of existence, the regional planning function has practically been eliminated in Chile. This can only be explained by a combination of facts: what regional planning was (its characteristics, its conception and instrumentalization), and what it was unable to be.

It is a fact that there never was any lasting political commitment to regional planning in Chile. Friedmann noted three conditions to be fulfilled if a regional development policy were to succeed in generating this type of commitment. First, the policy must be simple in conception and dramatic in its effect; second, the policy must be incorporated into new institutions capable of surviving in a medium characterized by frequent political change; and third, the policy must quickly produce visible results (Friedmann, 1973).

The foregoing analysis indicates that in the case of Chile these conditions were only partly fulfilled. It is true that the function of regional planning was incorporated into a 'new' institution which demonstrated its capacity for survival, but the explanation for this may be precisely the lack of real power of the National Planning Office. On the other hand owing to the always eminently technical orientation of policy formulation, it never invoked images to fire the popular imagination (like the building of Brasilia or the development of the Venezuelan Guyana, for example). Only at the beginning of the present phase did regional planners propose a project of this type (Proyecto Aysén-2000), aimed at mobilizing the country's creative energy towards colonizing and development of a vast rich region in the extreme south. No popular mass version of the regional development strategy, in any of its phases, was ever produced in Chile to inform the nation, or incorporated into the work of regional development. Lastly, in view of the typically economic objectives proposed during the three phases described, the fulfilment of the third condition laid down by Friedmann—i.e. the generation of visible results—could not be expected. When results did occur, they had strong sectoral implications (the outcome of the industrial location policy, for example).

In these conditions, regional planning and the planners themselves were not in a position to resist the onslaught of forces upholding the validity of an essentially liberal and neoclassical economic model. The result was therefore perfectly foreseeable, especially in Chile's current political context.

This inability to generate adequate political support is a direct result of the intrinsic features of the concept of regional planning from its beginnings. A historical analysis shows that the formulation of regional policy in Chile has been more of an intellectual exercise than a plan of action. It has also been more an independent construction by the centre, than a logical response to problems of the periphery. This partly explains the centralized form which regional planning always adopted. This centralization is also explained by the country's strong centralist tradition, a cultural element developed over hundreds of years during which the 'centre' and the state necessarily controlled the entire country.

Regional policy in Chile being what it was—an autonomous and centralized intellectual construction—it is fairly obvious that, leaving aside superficial differences manifested in each phase, it is a typical example of the 'centre-down' paradigm. As such it has always constituted an exercise in economics, the objectives of which had very little in common with the real living conditions of the poorest stratum of the population (but this argument should not be taken too far, since a regional policy is not exactly the same as a social development policy), and very little connection with any intermediate social organizations (macro- or micro-). In the circumstances, regional policy was also unable to arouse political support from the base which could have enabled it to counter the lack of institutional political support.

In brief, regional planning was a simultaneous phenomenon of continuity and change, of elitist conception and artificial application, which therefore stood apart from the country's real social problems. It will have to be rebuilt from its foundations.

REFERENCES

Achurra, M. (1972) Los desequilibrios regionales en Chile y algunas reflexiones sobre el proceso de concentración, *Revista Interamericana de Planificación*, **VI**, (23).

Ahumada, J. (1966) *La Crisis Integral de Chile* (Santiago de Chile: Editorial Universitaria).

Arriagada, G. (1978) La doble crisis de Septiembre de 1973, *Revista Mensaje*, 226 (Santiago).

Babarobic, I. y F. Wellhoff (1972) *Sistema urbano y polarización en Chile*,(Universidad de Chile: Centro de Planeamiento), (mimeo).

Bedrack, M. (1974) *La estrategia de desarrollo espacial en Chile (1970–1973)* (Buenos Aires: Ediciones SIAP-Planteos).

Boisier, S. (1971) *Polos de desarrollo: hipótesis y politicas en América Latina* (Geneva: United Nations Research Institute for Social Development).

Boisier, S. (1974a) Information systems for regional development in Chile, *Regional Information and Regional Planning* (The Hague: Mouton).

Boisier, S. (1974b) Industrialización, urbanización, polarización: Hacia un enfoque unificado, *Planificación regional y urbana en América Latina* (México: Editorial Siglo XXI).

Cavada, J., E. Marinovic y R. Rocha (1972) Conceptos generales sobre el desarrollo regional chileno en el proceso de transición, *Nueva Economia*, May–August (Santiago).

CONARA (Comisión Nacional de Reforma Administrativa) (1976) *Chile: hacia un nuevo destino* (Santiago), 33.

Cumberland, J. (1972) *Regional Development. Experiences and Prospects in the United States of America* (The Hague: Mouton).

Friedmann, J. (1969) Politicas urbanas y regionales para el desarrollo nacional en Chile: el desafio de la próxima década, *Chile: La Década del 70* (Santiago: Ford Foundation).

Friedmann, J. (1973) *Urbanization, Planning and National Development* (Beverley Hills: Sage Publications).

Friedmann, J. y T. Lackington (1966) *La hiperurbanización y el desarrollo nacional en Chile: Algunas hipótesis* (Santiago: Centro Interdisciplinario de Desarrollo Urbano y Regional, Universidad Católica de Chile).

Friedmann, J. and W. Stöhr (1967) The uses of regional science: policy planning in Chile, *Papers, Regional Science Association*, **XVIII**.

Geisse, G. (1971) Decentralización a partir de la actual concentración urbana y regional, *Chile: búsqueda de un nuevo socialismo* (Santiago: Universidad Católica de Chile).

Lo, Fu-Chen and K. Salih (1978) *Growth Pole Strategy and Regional Development Policy. Asian Experience and Alternative Approaches.* (Oxford: Pergamon Press).

Martner, G. (1972) La estructuración de un sistema nacional de planificación, *Revista Interamericana de Planificación*, **VI** (23).

ODEPLAN (Oficina de Planificación Nacional), (1968), *Politica de desarrollo nacional* (Santiago).

ODEPLAN (1970) *El desarrollo regional de Chile en la década 1970–1980* (Santiago) (mimeo).

ODEPLAN (1971) *Resumen del plan de la economia nacional 1971–1976*, Series **VI**, No. 3 (Santiago).

ODEPLAN (1972) El área social: un análisis regional, *Nueva Economia*, May-August (Santiago).

ODEPLAN (1975) *Estrategia nacional de desarrollo regional 1975–1990* (Santiago).

Palma, E. (1976) *Estado y planificacion: el caso de Chile.* (Santiago: ILPES), (mimeo).

Sanhueza, B. (1977) *Notas para perfilar una politica de desarrollo regional en Chile*, Document CPRD E/20 (Santiago: ILPES).

Soms, E. (1972) Los instrumentos de dirección económica regional, *Revista Interamericana de Planificación*, **VI** (23).

Stöhr, W. (1975) *Regional Development. Experiences and Prospects in Latin America* (The Hague: Mouton).

Stöhr, W. (1976) La definición de regiones con relación al desarrollo nacional y regional en América Latina, *Ensayos sobre planificacion regional del desarrollo* (Mexico City: Editorial Siglo XXI).

Stöhr, W. and F. Tödtling (1977) *Spatial Equity. Some Antitheses to Current Regional Development Doctrine, Papers, Regional Science Association*, 38.

Suarez, H. (1975) *Perspectiva histórica de la regionalización y sistemas de administración* (Santiago: Departamento de Administración, Universidad de Chile, Sede Occidente).

Chapter 17

Peru: Regional Planning 1968–77; Frustrated Bottom-Up Aspirations in a Technocratic Military Setting

Jos G. M. Hilhorst *

INTRODUCTION

The words appearing in the title of this chapter—however short the period since they were introduced—convey rather different notions, depending on the speaker. One's choice of content for the terms 'centre-down' and 'bottom-up' is determined either by the speaker's interpretation of what the main features of a given planning system were in the past, or by his concern as to what its main features should be in the future. Stöhr (Chapter 2) emphasizes that not only is the level of decision-making important when distinguishing between the two types of planning, but also the criteria used in decision-making.

In this chapter, only the level of decision-making will be observed; that is to say, 'centre-down' planning is understood to be the process of decision-making in which lower levels of the decision-making hierarchy are only involved insofar as they carry out decisions in accordance with the criteria set by the upper levels of the hierarchy. The reason for this divergence in method of approach from another chapter of this volume is because in true 'bottom-up' planning (i.e. when lower levels have autonomy), the lower levels are left free to choose their own criteria for decision-making. They are autonomous in setting their objectives. There is no doubt that in such a situation the need exists for a conflict-resolving mechanism, which would have as one function to bring some similarity and/or order of priority into the objectives of various regions. It would also determine on which issues regions could decide to apply 'selective spatial closure' (Stöhr and Tödling, 1977), especially when the process of development from below, as described in Chapter 2, reaches the point where extraregional markets become crucial to a region's development.

* The author wishes to thank Edgardo Quintanilla, Lincoln Villanueva, and Walter Stöhr for their valuable comments.

By development is meant the broadening of the scope of action available to an individual, a group, or the population of a region or country (Hilhorst, 1976). Regional development thus implies development of the various groups within that region's boundaries, and the shifting away or even removal of the constraints limiting them from reaching their objectives.

Whether one agrees or disagrees with the objectives or decision-making criteria of a country or region's decision-makers has a certain relevance from a moral point of view. However, the investigation as to whether development—as defined here—comes about by centre-down or bottom-up decision-making would seemingly be obstructed if additional conditions are introduced regarding the path taken or strategy adopted to produce the development desired. In this regard Stöhr's necessary condition (a certain crucial measure of regional autonomy) constitutes, for many, one of the objectives of regional development.

These introductory observations regarding the title of this chapter seem to be in order, first because they explain a certain deviation from the line indicated by Stöhr and secondly because some of the concepts set out in the Introduction will be used elsewhere in this chapter.

The following section deals with the nature of the regional problem in Peru at the beginning of 1968, the period under consideration. Then the framework of national objectives will be discussed, and a further section will outline policies for regional development. Finally an analysis of regional policy implementation will be presented, and a summing-up will provide some conclusions.

THE NATURE OF THE REGIONAL PROBLEM IN PERU, 1968

The take-over by the military in October 1968 inspired many Peruvians, especially young professionals and the military, with great enthusiasm. This is not to say that the new efforts at regional planning that were instituted did not take account of what had gone before, in the period after 1964 when Belaúnde was president. A number of Belaúnde's socioeconomic policies were closely concerned with the regional problems of Peru. He introduced the Agrarian Reform Law of 1964 in an attempt to better integrate the population, especially that of the Sierra (highland) region, into the national life. Of similar concern were his efforts towards construction of the *Carretera Marginal de la Selva*, a road to run parallel with the main axis of the Andean ranges but on their eastern slopes. This latter concern was related to the problem of spatial integration in the country, a problem that has not yet been satisfactorily solved.

Einaudi (1975) has observed that the sociological features of the Peruvian population and the spatial characteristics of the territory they live in make communication difficult, and this is turn limits effective government. Similarly, he says, given these conditions, Cuban-style revolution in Peru is not too probable. There is much truth in this observation, which was made at a time when the road network had improved considerably as compared to the situation in the late 1950s, when it was still impossible to travel by road from the northernmost coastal city of Tacna on the Chilean border.

Table 17.1   *Urban and rural population by natural zones: 1940, 1961, and 1990 (thousands)*

Natural zone	1940			1961			1990 (Projected)		
	Urban	Rural	Total	Urban	Rural	Total	Urban	Rural	Total
Costa	1,156	627	1,783	3,015	897	3,912	11,375	1,623	12,998
Sierra	940	3,065	4,005	1,446	3,681	5,127	3,209	5,096	8,305
Selva	126	294	420	301	570	871	936	1,435	2,371

Source: National Planning Institute (INP), 1971a, Appendix 1.

Traditionally, the country has been divided into three zones: the coast, mostly a desert area crossed by some 50 rivers that come down from the Andes and provide water for irrigation-agriculture; the Sierra, densely populated mountain ranges and valleys containing the largest part of the country's population and its poor; and third the Selva, which constitutes part of the Amazon rain forest (see Table 17.1). Before the Pan-American Highway—running from north to south along the coast—was finished in the late 1950s, the country was effectively divided into a number of subsystems that really only communicated by sea. Thus Chiclayo and Trujillo were the main centres on the Costa Norte, Arequipa was the main centre in the south, and Lima was the centre of the Costa Central and the Sierra Central.

The centuries-long relative isolation of the centres of these subsystems—which were established mainly by Spanish cultural influence—had created separate sociopolitical entities, with opposing interests. In the north the power base of what Peruvians call the *oligarquía* was located, concerned mainly with sugar production; in the Lima area the groups related to import–export trade, banking, and some mining and cotton interests; whereas in the south interests centred traditionally around trade in wool, and latterly in this century, around copper and gold mining; and the newly founded industries were concentrated here. Thus the population of the Sierra constituted a mere appendix to the life of the other subsystems (see Table 17.2). Their weak economic condition found a parallel in *res politica*: when elections were held, most of them were excluded due to illiteracy. The population of the Selva was hardly considered in policy-making, except perhaps for the inhabitants of Iquitos, capital of the department of Loreto, and a left-over from the formerly flourishing exploitation of natural rubber.

Belaúnde's problems in attacking the regional issue were complex. Although the Agrarian Reform Law was passed by parliament, its implementation was sabotaged by the voting-down of considerable parts of the required budget. The party that had advocated land reform since the 1930s, the APRA (*Alianza Popular Revolucionaria de America*), with its main base on the Costa Norte, was among those who crippled Belaúnde's efforts. Still, some advance was made, especially in the southern Sierra, before Belaúnde was removed from office in 1968.

Table 17.2   *Income per capita and population by department and natural zones*

Department	Income per capita (thousand soles, 1961)	Population in (thousands, 1972)	Costa	Sierra	Selva
Lima	30.2	3927.1	×		
Pasco[a]	15.1	d		×	
Ica	12.4	373.3	×		
Arequipa	12.0	561.3	×		
Lambayeque	11.3	533.3	×		
La Liberted	11.2	808.4	×		
Tacna	11.2	99.5	×		
Madre de Dios	11.1	25.2			×
Tumbes	10.9	79.3	×		
Coronel Portillo[b]	10.4	120.5			×
Loreto[c]	9.8	420.0	×		
Junin	9.7	905.4[e]		×	
Moquegua	9.6	78.0	×		
Piura	9.0	880.0	×		
Ancash	8.1	755.1	×		
Cajamarca	6.9	956.6		×	
San Martin	6.9	233.9			×
Huánuco	6.6	431.7		×	
Apurimac	6.3	321.1		×	
Amazonas	6.1	213.0		×	
Cuzco	5.5	751.5		×	
Huanvacelica	4.3	346.9		×	
Ayacucho	4.3	479.4		×	
Puno	4.1	813.2		×	

Source: Webb (1975) for income data; ONEC (1975) for population figures.
[a] Pasco is the Department where the mining operations of the Cerro de Pasco Corporation were carried on. The corporation was nationalized in 1974 (Kruyt and Vellinga, 1977).
[b] Province.
[c] Excluding the province of Coronel Portillo.
[d] See Junin.
[e] Includes the population of Pasco.

As a professor of architecture who taught in Lima before becoming president, Belaúnde was well aware of another problem of regional imbalance: the burgeoning growth of Lima-Callao, which almost trippled in size from 1940 to the time he came to office. The *barreadas* already covered an extensive portion of greater Lima and held about a quarter of its population (ONEC, 1975, p. 9). In 1940 Lima had a population of 911,000. By 1972 this had increased to 3.1 million (ONEC, 1975). By 1963, more than 60 per cent of industrial production was generated in Lima (INP, 1967), although it only had 20 per cent of the total population. As Boisier puts it:

This process of concentration in Lima tends to be enhanced on the basis of four factors: (i) the production structure is biased towards final consumer goods, with a strong market orientation; (ii) production consists mainly in simple transformation with a strong dependence, upon imported inputs; (iii) industrial infra-structure in the country's interior is limited, especially as far as energy is concerned; and (iv) the administration of taxes and credits is concentrated in Lima, which tends to favour industrial concentration there. (Boisier, 1972, p. 117, translated by the author.)

This process of economic concentration went hand in hand with one of centralization of decision-making. Lima became the centre, and almost nothing could be done elsewhere unless Lima had agreed. An important reaction to this over-centralization was the emergence of a number of so-called departmental development corporations that began to receive central government assistance from the National Fund for Economic Development after 1964. In Departments like Arequipa, Piura, and San Martin they fulfilled an important role, but mainly they served the interests of the local elite.

Although interregional imbalances decreased over the period 1946–66 (as Slater, 1975, has shown) at least in terms of certain indicators, there is no doubt the problem was as serious as before, because of the increasing numbers of people involved.

NATIONAL POLICIES

In November 1968, one month after the military take-over, the National Planning Institute (INP) published a document on the government's strategy for long-run national development (INP, 1968). The new government, calling itself revolutionary (although neither capitalist nor socialist), made its intentions clear to the country and to the world. The three main objectives were:

(a) to integrate the national population in order to achieve the full utilisation of the human resources as well as the basic potential of the country; (b) substantially improve the distribution of an income per capita that would be not less than double the present (1968) level; (c) mobilize the contribution of the external sector to the national development policy, reducing the present conditions of external dependence and vulnerability that characterize the economy (INP, 1968, p. 6).

The document went on to to say that in order to reach these objectives in a 20-year period, a number of structural reforms would be necessary, and the simultaneous creation of new development poles. The structural reforms would be related to agrarian reform, a new mining policy, a new industrial policy, a new structure for the public sector (including tax reform), and a new expenditure pattern for social services; while each of the four socioeconomic regions (Figure 17.1) would be favoured by the expansion of agriculture and mining, and the development of urban industrial poles.

In more concrete terms, the government intended to reduce underemployment by increasing the number of fully employed from 71.5 per cent of the labour force in 1971, to 83.2 per cent by 1975. Close to 1.2 million jobs were to be created over the 1971–75 period. This objective would be reached by making the economy grow at a rate of 7.5 per cent per annum on the basis of investments

Figure 17.1 Peruvian regionalization policy

that would occupy a 21.3 per cent share of GDP by 1975, starting with a corresponding figure of 12.8 per cent in 1970 (INP, 1971b, pp. 17–19). This was to be an effort that would bring about a yearly rate of growth of gross investment of almost 20 per cent over the period. Although private investment was expected to grow from 17.7 billion soles to 28.0 billion by 1975, public investment would have to give the most important boost: it would have to increase four times, from 10 to 40 billion soles, which would almost reverse the percentage distribution for public and private investment.

Public investment was to be: 25 per cent to manufacturing, almost 8 per cent to agriculture, and mining investments would amount to more than double the latter percentage (INP, 1971b, p. 20).

These medium-term objectives were cast in a framework of 15 generic objectives. The regional inequality issue is clear in a number of these objectives. It is treated explicitly in objectives No. 8 ('establishment of integrated propulsive industries that maximally utilize the country's natural resources; increase in the productivity of established industries in order to enter international and regional markets at competitive costs'), and No. 11 ('reduction of the disequilibrium in the distribution of the population over the national territory') (INP, 1971b, p. 15).

In other words, a growth pole policy was to be pursued, and the scale of propulsive industries had to be derived from international and regional (that is to say, ALALC and Andean) markets. The emphasis on natural resource development is understood as relating to mining and fishing, as per objective No. 14. Objective No. 11 related to a purely physical planning issue and was not very satisfactory, as the regional problem was then and still is essentially a social and political problem.

In order to meet the medium-term objectives, the government announced its intention to carry out a number of policies in relation to six issues: (i) what is called *Participación social*; (ii) the role of the state in the development process; (iii) the financing of development; (iv) the external sector; (v) science and technology; and (vi) integration (by which was meant the process of Latin American integration).

Concerning the first and second issues, the plan for 1971–75 was most specific: social participation (which was defined in terms of (a) training, social awareness, and organization by the majority; (b) activating systems of investment by labour-intensive means; (c) the stimulation of cooperative and self-operated enterprises; and (d) the stimulation of dialogue between the majority and national decision-makers) was to be achieved by means of agrarian reform, reform of the enterprise, educational reform, and the reduction of unemployment and underemployment. It was expected that agrarian reform, based on the new Law of 1969, would contribute to a better distribution of land tenure, income, and political power in favour of the majority of the *campesinos*. Cooperative forms of ownership of land would be the preferred framework, in carrying out reform. Reform of the enterprise would continue to be executed under the General Law of Industries, the Law on the Industrial Community, the

General Fisheries Law, and the General Mining Law—all these decrees passed in the 1970–71 period. These laws were expected to function as key elements in attaining the objectives of the revolution.

Educational reform was considered the most important task the government had set itself for the medium term. In 1971 it considered this the most difficult task.

It is worth while considering the background of the various reforms. For most members of the officer corps of the Peruvian armed forces, the value of education as a determinant of social mobility and development was evident. Most of them were witness to the effects of their own education upon the quality of organization of the institution they served in. Einaudi argues that the concerns of the military were:

> (i) institutional autonomy and survival; (ii) public order and the control of remote areas; (iii) foreign policy and boundary questions; and (iv) more recently, national development, including under that rubric education, industrialization, control of strategic materials (petroleum, tele-communications), and general national planning and support for central government authority and administration (1973, p. 309).

The last set of concerns was stimulated by their analysis of the causes of peasant unrest over the decade preceding their forceful intervention, as well as the causes of the guerrilla movement that had been active until 1965. The role of the CAEM (Academy for Higher Military Studies) has been discussed by many authors—Cotler (1975), Einaudi (1973, 1975), Lowenthal (1975), to name a few—and it is generally agreed that it was crucial in putting forward a doctrine on the causes of what was termed *seguridad nacional*, that became part of the intellectual baggage of most officers. In simple terms, the violence of the decade was considered to have been caused by extreme social and economic imbalance between the various groups of the population. In order to alleviate the violence it was necessary to alleviate inequality. To achieve this, it was necessary to redistribute property and income, and to provide the masses with more political power. Thus, the educational system must also be changed.

Class conflict in terms of Marxist theory does not play a role in this approach; rather it is man's individual misunderstanding of his social responsibility that drives him to treat his fellow men unjustly. In this sense, the Peruvian revolution was neither capitalist nor communist. Analysis of the problem was not based on Marxist principles; its solution could therefore not be phrased in such terms. The solution, which emphasized community of interest among workers and capitalists and enhanced the participation of workers in equity capital in industrial enterprises, could not be termed capitalist, since it denied supremacy of the individual and of private interests. Thus the way in which the agrarian reform was executed (and the General Laws on Fisheries and Mining, as well as the Law on the Industrial Community) was coherent and was aimed at legislating away the conflict of interest between labour and capital, between rich and poor.

In this context it becomes understandable that the military had to remove the *oligarquia*, since they more than any others behaved in a most capitalistic fashion. The ambivalent attitude of the military towards labour unions is also explained; they eventually would no longer have a role to play. This is explained by the objectives of the Law on the Industrial Community, which implied that workers in industry would eventually obtain 50 per cent of the votes due to their owning half the equity capital, which would make them in this way co-responsible for what went on in various enterprises and branches of industry. It also explains the remarkable role that SINAMOS (*Sistema Nacional de Apoyo a la Movilizacion Social*) had to play, and why this organization was so strongly opposed by a party like APRA that had acquired more and more bourgeois characteristics.

This brings us to the second issue, the role of the state in the development process. The military intended to reorganize the administration, but also wanted the state to take on heavier responsibility in the country's economy. This was done mainly by investment in mining and related industries, but also by nationalization of public utilities and the railways. The state also acquired the majority of banking interests, and eventually moved into distribution of major food products. In this context, it is interesting to note that the plan for 1971–75 says:

> the development of mining and of the related industries requires that in their vicinity be constructed a road network, railways, ports, electric power works and urban centres. As a consequence, the areas surrounding these production centres will constitute themselves as real foci of regional development. This requires co-ordinated action from the different ministries and public enterprises in order to avoid bottlenecks, to permit the diffusion of benefits and to attract private and complementary investment. Therefore, in the medium run, regional planning will have great importance. (INP, 1971b, p. 27.)

The government obviously thought at this stage that private enterprise would fall in with its objectives. Also it seems to have argued (like so many growth pole planners do and did) that there would be a considerable degree of automatic action on the part of entrepreneurs. The literature on this subject tells a different story.

The planning system that would implement these and other policies was headed by the Instituto Nacional de Planificación (INP). It maintained strong ties with the sectoral planning offices attached to each of the Ministries. Their programmes of work and the objectives for each Ministry were thus coordinated. Part of INP were the regional planning offices, initially one for each of the four regions. But from 1972 on their number increased, and now stands at nine (1978), since a number of subregional offices have been opened.

By 1975 the Peruvian Revolution was in serious trouble. Internal opposition from left and right was increasing. More and more opponents were shipped out of the country. National newspapers had been taken over by the government in June 1974, the independent press was banned, the army broke a strike of

Table 17.3   *Gross Domestic Product by type of expenditure, 1970–76 (at 1973 price rates, 10⁹ soles)*

	1970	1971	1972	1973	1974	1975	1976
1. Private consumption	240.8	251.6	260.7	273.8	290.7	299.8	314.8
2. General government consumption	38.7	41.2	42.8	46.1	51.0	56.8	60.1
3. Gross fixed capital formation	42.9	46.7	48.4	60.9	79.5	84.8	73.9
4. Variations in stocks	− 3.1	2.0	− 3.3	2.9	13.6	21.6	15.5
5. Final demand = 1 + 2 + 3 + 4	*319.3*	*341.4*	*348.6*	*383.7*	*434.8*	*463.1*	*464.3*
6. Exports of goods and services	62.7	59.6	65.0	53.1	50.0	48.1	48.6
7. Imports of goods and services	48.2	50.3	50.1	54.9	72.4	81.3	68.5
8. GDP = 5 + 6 − 7	*333.8*	*350.7*	*363.6*	*381.9*	*412.3*	*429.9*	*444.6*
9. Rate of growth		5.7	3.6	5.0	7.9	4.3	3.4

Source: INE, 1978.

policemen with tanks, and riots and looting occurred in the centre of Lima. The economy was in crisis and exports started to fall, while imports increased sharply. The balance of payments showed an enormous deficit. The rate of growth of GNP, which had been close to 5.5 per cent per annum over the period 1971–75, dropped sharply (INE, 1978). (See Table 17.3).

In August of 1975 President Velasco was replaced by General Morales Bermúdez, and a new phase began. The period of reform was over, and the government now devoted itself to crisis management. Whereas in the first phase an explicit attempt had been made to make the country more independent economically, the investment and nationalization policies, a slump in the fishing industry due to the disappearing anchovy, and a world economic crisis causing decrease in export demand, all made it necessary to let foreign financial interests acquire a new foothold in the country. New loans had to be contracted, and in September 1975 the sol was devalued by 16 per cent. In January and in June 1977 successive restrictive economic measures were announced. The effects of the measures on real wages were disastrous: by January 1975, average wages still stood 22 per cent higher than in 1968, but by December 1976 they had fallen to 25 per cent below the level they had been eight years earlier (*Lateinamerika Nachrichten*, 1977, p. 4).

REGIONAL DEVELOPMENT POLICIES

Figure 17.2 shows the four regions and the subregional divisions that accompanied the 1968 strategy document (INP, 1968). The long-run strategy had the following main characteristics: in the north (Region I), a growth pole would be established based on the non-metallic minerals of Sechura, the petroleum and gas of Talara, and the agricultural potential of the region. In the centre (Region II), Pucallpa was designated to become a pole, while in the south (Region III),

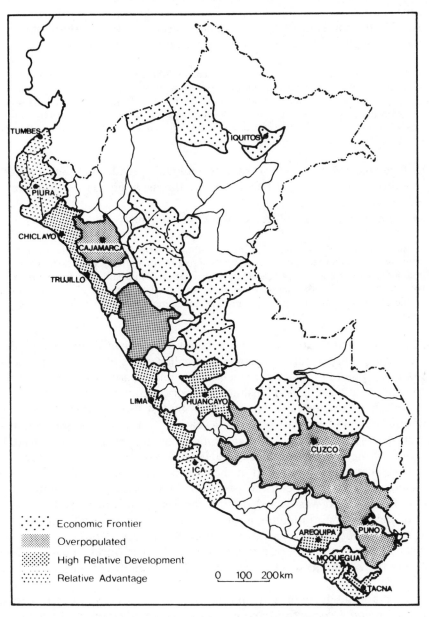

Figure 17.2 Areas of Concentrated Action in Peru

the main urban centres would be interconnected. For Region IV, the east, no specific ideas were put forth at the time, except that prospecting should be intensified.

By 1971 the government's plans had become clearer. Its long-term objectives in the area of physical planning and natural resources use were:

1. to integrate the national territory by means of infra-structure and communication systems that should permit the efficient development of activities and the free flow of goods and persons...; 2. to develop the territory so as to facilitate the individual in selecting a way and place to live...; 3. guarantee sufficient territory, useful and disposable for permitting human settlement that would satisfy the needs of the population. (INP, 1971b, p. 54.)

The first objective is directly related to regional specialization and economic integration with the Andean Pact countries, the last refers to infrastructure and communications.

In the medium run, six objectives were to be pursued:

(i) strengthening of economic links and of administrative organization between neighbouring complementary zones...; (ii) the creation of demand centres that counterbalance Greater Lima...; (iii) the institutionalization of mutual-aid forms of labour...; (iv) the broadening of social services to the marginalized groups...; (v) an increase in the amount of arable land...; and (vi) an increase in the non-rural population. (INP, 1971b, pp. 55–56.)

In order to reach these objectives the country was divided into five kinds of zones, which when grouped together were called zones of concentrated action (ZAC), (see Figure 17.2) with: (a) the zones of relative advantage (characterized by a concentration of mineral, fishing, and agricultural resources); (b) zones of relatively high development, characterized by fast urbanization, a good agricultural base, and a relatively well-developed transport and services infrastructure; (c) zones of population saturation, characterized by being part of the Sierra, with high population density and limited natural resources; (d) economic frontier zones, which are found in the east, but although far from centres of demand, have good agricultural potential; and (e) the Metropolitan Areas. Thus, Lima-Callao was excluded from the zones mentioned under (b). The rest of the country, containing approximately 20 per cent of the population, was to be excluded from special attention. Some areas, however, would still benefit from construction of infrastructure and other works, intended chiefly to serve the development of areas within the five kinds of zones listed (INP, 1971b, pp. 57–60).

It is important to note that highest investment priority was given to the *zonas de ventajas comparativas* (areas with relative advantages). In the north, fishing was to be promoted in Paita, agriculture in the water-rich valleys, petrochemicals and fertilizers in the area of Talara and Bayóvar, agro-industrial expansion in the twin cities of Piuara-Sullana and Chiclayo, pharmaceuticals in Chiclayo, and metal industries in Trujillo and Chimbote. The latter two urban zones were to act as counter-magnets to Lima-Callao, and they would be leaders with Piura-Sullana, Chiclayo, Cajamarca, Tarapoto and Huaraz functioning as subregional centres dispensing administrative and other services. The other urban centres of the north were assigned secondary roles.

In the central region, Greater Lima was to decentralize industrial expansion to cities like Huacho-Barranca, Chicha-Pisco, and Ilo. Huancayo and Ayacucho in the Sierra and Ica on the cost were to become regional centres.

In the south, Arequipa was to perform the role of counter-magnet by means of industrial diversification, while Puno, Cuzco, Juliaca, and Tacna must acquire more regional importance.

For the east, the only decision was that Iquitos had to be strengthened and it was not clear what this was to mean.

Thus the growth pole strategy, that was mentioned in the November 1968 document, was repeated in the medium-term plan. The term is not used in classification of zones, but there is no doubt this kind of strategy was pursued.

The 1971–75 plan also devoted attention to regionalization of the administration. The central government was to be reorganized and the administrative organization defined with a territorial criterion emphasizing the search for positive correlation between development and administrative regions. The main characteristics of the work to be done during this search were identified: urban hierarchies were to be determined, the important cities to be reinforced: those on the coast in order to integrate their hinterlands, those in the Sierra in order to break their dependence on the coastal cities. Coastal corridors had to be strengthened to promote creation of political–administrative units of groups of centres that had hitherto been semi-autonomous. Departmental boundaries were also to be adapted to new developments, and each department should have its own development authority which would direct and coordinate inter-sectoral and sectoral activity, all this in close relationship with the corresponding regional office of the INP. It would therefore be necessary to deconcentrate state developmental action in favour of the regions. This would be accompained by: (a) a redefinition of selection, allocation and administration, especially of financial resources; (b) the establishment of a common fund upon which the sectoral and intersectoral organizations of the region would draw in accordance with their budgets and operating capacity; and (c) the reallocating of the functions of the state to bring administration closer to the individual user of its services, his realities, and his problems.

The military attached great importance to planning. During the first phase, especially until 1974/75, the INP had considerable influence over formulation of plans and policies. It was keenly involved in the entire government planning effort, and was a key factor in approval of the bi-annual budgets. All investment projects of government and state enterprises required approval from the project division of the Institute. The INP formed the coordinating unit of the sectoral planning offices attached to each of the Ministries.

By 1972 it became clear that the regions were far too big for a large number of the planning projects. Therefore over the next two years subregional offices were established in Cuzco, Puno, Tacna, and Ayacucho. Each of these regional planning offices prepared sections to be included in the volume on regional development policies for the five-year plan.

The work of the regional planning offices was to be coordinated by the INP regional programming division, which disappeared in December 1974.

In 1972 SINAMOS (*Sistema Nacional de Apoyo a la Movilización Social*) was established, and it acquired all the staff and funds assigned to departmental development corporations in 1964. Since these corporations had been largely expressions of interests of bigger local industrialists and farmers, this or a similar move on the part of the government had been expected for some time. In some cases, as in Arequipa, it meant that planners lost contact with the businessmen of the area. Nevertheless, the existing special incentives for industrial location in Arequipa (exemptions from duties on certain imports) were continued. Similar advantages had been given to Iquitos, in an effort to offset high transportation costs to this isolated city. This author is not aware of other locational incentives. The leading principle was that the main industries would be located by the government in accordance with the objectives of regional development, and private enterprise would be 'automatically' attracted to these locations.

The policy of administrative reform that was to lead to decentralization was also to be regarded as an instrument of regional development. As will be seen, this policy has hardly taken effect as yet.

REGIONAL ASPECTS OF POLICY-MAKING

Agrarian Reform

With the advent of the new government, the agrarian reform that had begun under Belaúnde shifted its emphasis from the southern Sierra to the Costa Norte. The big sugar plantations were expropriated first and transformed into agricultural production cooperatives (CAPs). Here, as elsewhere, the emphasis of agrarian reform was not to subdivide existing production units, but to replace individual or family ownership by communal operations with permanent workers becoming the communal owners. On the cost they were mainly CAPs, whereas in the Sierra the form chosen was the agricultural society of social interests (SAIS). These normally centre around one or more formerly private large estates, and they sometimes include communally held land on the borders of these estates, the so-called *comunidades campesinas*. (The foregoing are the two main groupings.) In addition, however, the *comunidades* were reorganized into new forms, while certain groups of farmers were also established, called *grupos campesinos*.

The Ministry of Agriculture organized land reform on a regional basis: twelve zones were established, and each in turn subdivided into sectors. Smith (1976) in his description and analysis of agrarian reforms shows how this reform by gradual steps greatly increased state influence in the rural areas. In the first phase of agrarian reform for a given sector, a special management committee (*Comité Especial de Administración*, CEA) would be set up. It would be composed of 'two representatives of the Ministry of Agriculture (one of them to be the chairman), one from the Banco Agropecuario, one from the Banco Industrial, and two from the workers, elected in a general assembly' (Smith, 1976, p. 99), the 'workers' being the permanent workers on the expropriated land. In a second phase, one or more CAPs or SAISes would be created, depending on whether the area was on the coast or in the Sierra.

In a third phase, when it became clear that the CAPs and the SAISes behaved towards seasonal workers from neighbouring *minifundios* or *comunidades campesinas* in the same capitalistic manner as the former owners (or worse), the government decided to introduce PIARs (integrated rural development projects), which in most cases were to have an area of influence coinciding with sectors, as defined above. Each PIAR would include not only the CAPs or SAISes in the sector, but also the *comunidades campesinas*. In fact, the PIARs were planned as an instrument for the better distribution of income generated in the areas, while also serving as a basis for providing training to peasants as well as for provision of other services (Giles, 1974).

Conlin, who has been studying especially the participation aspect of the PIARs, concludes: 'Domination has changed its face. No longer is it the brute subjection of the peasant to the hacienda owner. . . . It is the subjection of the peasant to the bureaucratic machine, accomplished under the guise of participation' (Conlin, 1974, p. 165). Production planning at the level of the CAP or SAIS is mainly determined by the authority of the *ingeniero* from the Ministry. Likewise at the level of the central cooperative and the PIAR, the role of government representative is a decisive one.

Similarly, the representation of production units—via *ligas* (leagues) and federations—in the National Agrarian Confederation (CNA) was strongly influenced by the zonal offices of SINAMOS, who were responsible for carrying out the first specific objective of the five-year plan, 'to increase the effective participation of the national majorities . . . by means of its organization in intermediate institutions'. Many felt this form smacked unacceptably of corporatism (Cotler, 1975), while the APRA and other parties also criticized the government for introducing political organization in this manner.

This is not the only problem the government has had with agrarian reform. Another is of a more basic nature, and very little can be done about it: There is no way to distribute land so that each family can own the minimum required by the Agrarian Reform Law, even if it were held communally. Smith says: 'In 1972 there were 440,000 holdings of less than one hectare on 171,652 ha., and the number of holdings of one to five hectares. . .(stood at) 586,200 on 1,330,472 ha.' Thus although many acquired land during the process of agrarian reform after 1964 and especially since 1968, the total amount of land available is insufficient by far to provide every family—whether individually or in communal ownership—with enough hectares to satisfy the minimum plot size requirement of the 1969 Law.

This is especially relevant to the Sierra, particularly in the southern portions. The average farmer here did not benefit from reform, and poverty in these areas cannot be expected to lessen for many years to come. These areas will remain as sources of supply of unskilled migrants to the cities, certainly as long as government refrains from instituting a serious programme for opening up the jungle areas (Hilhorst, forthcoming). Meanwhile, most of the (modern sector) investment, including large-scale irrigation, has gone to the coast.

Fixed Public Investment

Thus we find that for the years 1971–72 fixed investment by state enterprises in the coastal zone, including Greater Lima, constituted 68 per cent of the total for this type of investment (based on Fitzgerald, 1976a, p. 89). Central government investment in the same area for this period amounted to 64 per cent of the total for the whole country. For the period 1973–74 the plan was less strongly oriented towards the coast; 54 per cent of all government investment was destined for this area, but 24 per cent of the total was considered 'national', that is, this percentage could not be allocated to a particular region (INP, 1973, pp. 11–74). This skewed spatial distribution of investment was due partly to the large irrigation works established close to Ecuador (the Chira Piura irrigation scheme), and partly to such major schemes as Tinajones (Costa Norte) and Majes-La Joya (Costa Sur). There were also the industrial projects: metal-working industries in Trujillo and Chimbote (assigned to Peru by the Andean Group), the extraction of phosphates close to Bayóvar (where the pipeline carrying oil from the Selva would also have its terminus), a fisheries complex in Piura (all these in Costa Norte); a similar complex in La Puntilla (Costa Central), the sodium carbonate factory at Haucho, zinc and copper refineries, at Lima and Ilo respectively; the paper mill at Paramonga as well as the Cerro Verde copper mine—some of these in the central region and some in the south, but all on the coast. Unfortunately no statistical data are available regarding the implementation of the plan for 1973–74. However most of these projects are known to be completed or in an advanced stage, and also in the period 1971–74, 65 per cent of public investment was concentrated on the coast (INP, 1976, p. 7). For the southern Sierra, however, by 1972 planning had started to implement a tourist circuit that would open up the treasures of Cuzco, Lake Titicaca, and of course Machu Pichu, to foreign travellers. This project has become quite a success as a foreign-exchange earner, but it means relatively little to the majority of the peasantry in the area who are still working land that is hardly sufficient to keep their families at a bare subsistence level. More than 63 per cent of the people here cannot satisfy even their basic needs (Couriel, 1978, p. 125).

Statistical data are available for the period 1975–76, and also for 1977. For the first two years, 71.5 per cent of public investment, and more than 80 per cent of all productive public investment excluding the oil pipeline, was located on the coast. Direct employment results from these investments were especially felt in Iquitos, Piura, Arequipa, and Lima, where 63.4 per cent of public investment was concentrated (INP, 1976, pp. 6–7). The projects that went on during this two-year period required more than three times the amounts budgeted in these years, to reach completion. In other words, the commitment to a continued spatial bias was built in for the period at least until 1981. The positive aspect of this programme is that only 18 per cent of the amount the government has committed itself to, will go to Lima–Huacho–Pisco, while more than one third goes to Trujillo–Chimbote.

For the year 1977, 60 per cent of government investment was planned to be spent in the coastal area, and only 14 per cent in the Sierra. For various reasons, implementation of the investment plan led to the location of 62.5 per cent of government investment on the coast, and 13.5 per cent in the Sierra (INP, 1978, p. 11).

Educational Reform

Educational reform based on the General Law on Education, of March 1972, was important in many ways. Two results should be mentioned: it caused considerable political unrest, especially because of a strike supported by the leftist teachers' organization, SUTEP; and it set an impressive example of reorganization in the government's administration. The more crucial aspects of education reform are that it did away with the traditional separation of vocational and general education; it accepted bilingualism (especially of relevance in the Quechua and Ayamara-speaking areas) and it promoted creation of a community educational system, from the base towards the top. In addition, the new law embraces the principle of lifelong or continuing education. Andrews Hay (1976) praises the reform, saying it takes into account the worldwide debate on educational requirements.

Reform is being implemented through a new organizational pattern of schools, the basic unit of which is the *nucleo educativo comunal* (community educational nucleus), the NEC. The NECs may include various schools, but one of them is assigned the role of *centro base* and is usually equipped with facilities that will be of service to the other schools in its designated area. All schools in such areas, including private ones, form part of the NEC.

These areas have been determined by the zonal directorates of the Ministry of Education, and each directorate supervises the NECs within its zone. A number of zones in turn are grouped under a regional directorate. There are nine regional education administrations. Zonal directorates have planning, programming, and budgeting functions, and appoint directors of the NECs.

From a regional development point of view, educational reform was of double importance: not only did it guarantee a considerable improvement in educational facilities and bring adapted curricula to the less-developed areas of the country, but it also brought about a much more adequate presence of the central administration, countrywide. It is, however, much too early to judge the effect reform has had upon the development potential in lagging regions.

National Support System for Social Mobilization (SINAMOS)

The regional effects of SINAMOS were of some importance. This organization came into effect in 1972, working via regional and zonal offices. As observed, SINAMOS absorbed the staff of the departmental development corporations (CDD). In several instances, such as San Martin, the operation of SINAMOS seemed to be merely a continuation of the CDDs, and SINAMOS was criticized for it. Its role was seen by some to be more that of a mobilizer of the poor and unorganized, and not a role involving the construction of small irrigation dams and feeder roads. In other cases—criticized from a different viewpoint—the offices were staffed by young, generally inexperienced but very enthusiastic academics who were seeing the problems of the interior for the first time. Many did not speak the language of the poor peasants (Quechua or Aymara), and they were hardly trained for the task of social mobilization. When they did mobilize the poor, for example in the north, SINAMOS was sharply criticized, not only by the landowners affected, but also by some government ministers.

The reasoning behind the functioning of SINAMOS is not entirely clear. For many, SINAMOS appeared to be the secretariat of a political party, this one paid for by government. This evaluation was based on the observation that SINAMOS was spreading an ideology among *grupos de base* (base groups, groups of peasants in rural areas, slum dwellers in the cities); it was forming groups under local leadership, and these were used to demonstrate their support of government policy—or as sometimes happened, the ideas of the staff of a zonal SINAMOS office. The main organization created in this way was the CNA, which was dissolved in early 1978. One effect of this activity was that political parties felt there was a downwards pressure on their membership, or at least they feared this would result from organization of the non-organized and poor.

In fact, this was indeed one of the functions of SINAMOS. The government knew it could not return to civilian rule unless new and organized forms of political opinion emerged beforehand to represent the interests of the urban and rural poor. If these pressure groups or even political parties did not exist, then in any election preceding or coming soon after a return to civilian rule, the old power structure would be re-established and a number of the new reforms would be quickly undone. SINAMOS was there to prevent this.

There is no doubt that the functioning of SINAMOS in the years 1972–75 has contributed to radicalization of the process which had started to enter calmer waters, after echoes of the downfall of the *oligarquia* had died away. The Cabinet divided over the importance to be attached to SINAMOS, while followers of APRA especially, and others, began criticizing the government more and more strongly. Illegal land occupations in 1973 were said to be organized by SINAMOS, and the Ministry of Agriculture felt its agrarian reform programme threatened. Similarly, regional and zonal offices of INP found a number of their initiatives thwarted, and a split threatened in the unity of the administration at regional levels.

At regional and zonal levels, SINAMOS had a powerful effect in that it brought about a great number of *ligas campesinas* whereby many peasants never before politically organized, entered the national arena. It is noteworthy, however, that here the military used authoritarian methods in attempting to promote participation. SINAMOS organized election of local leaders in assemblies of urban or rural poor, but the profile of such local leaders had been predetermined at the national level.

If the Peruvian 'revolution' was not capitalist, SINAMOS certainly tried to demonstrate the point. In its emphasis on community organization, it tried to work through small-scale projects, stressing mutual-aid methods, either in attempts to improve urban slums, or to create employment—be it the building of rural roads or small irrigation works, or construction of small manufacturing units in urban areas. But in its use of authoritarian means to implement its aims, SINAMOS did not live up to the government's condemnation of the Soviet system.

The elections for the Constituency, that took place on 18 June 1978, made it clear that the policy of the military in this regard had failed. As in the past the illiterate, two million out of seven million, were excluded from the ballot, and traditional or traditionalist parties won more than half the vote: APRA acquired 36 per cent, and Bedoya's right-wing PPC received 27 per cent. Nonetheless, the combined left gained about 28 per cent of the vote as compared to less than 5 per cent in 1964.

Administrative Reform

Administrative reform was successful in that a number of its objectives were reached. The Ministries of Agriculture and Education were reorganized, a planning system was introduced and started to take effect by 1972–73; collection of taxes was improved, and the first autonomous regional development authority was established in 1973. This was ORDEZA (organization for the development of the zone affected by the earthquake of 1970). The model was used again in 1977–78, as will be seen later.

At the same time, in 1973, departmental coordination committees were introduced. The military commander in each department was made chairman of a committee consisting of the directors of all central government agencies in the department, with the director of the pertinent regional or zonal office of the INP functioning as its secretary. These committees became necessary partially because of the need to control activities of SINAMOS, but also to improve the status of the INP offices in the interior. In addition it made it possible to monitor the progress in decentralization of the various Ministries and the preparation of the Law on Decentralization. This law had been announced early on, but its formulation and discussion were regularly postponed. By 1974 it had reached the newspapers for public discussion. Interestingly, this discussion centred around the issue of the regional capitals selected. Tensions ran so high that some observers felt the government would have to abandon the idea. By

November 1977, however, the government had decreed law 21905 by which it created the second autonomous regional development organization, ORDELORETO (*organismo regional de desarrollo de Loreto*). It is similar in its functions to ORDEZA, the regional development organization set up to reconstruct the area around Huaráz and Chimbote destroyed by the earthquake of 1970. Its chief had the rank of Minister of State. Eight months later, similar organizations were created for the subregion *Suroriente*, called ORDESCO (Decree/Law No. 22213; which includes the departments of Cusco, Apurimac and Madre de Dios), and for the department of Puno, called ORDEPUNO (Decree/Law 22214). Both these laws are based on Decree/Law 22208 which regulates departmental development committees and gives them new functions. Space does not permit going into details of these developments, but they imply that the government has finally decided on regions for each of the Ministries, and on capitals for each of the regions. There are to be twelve administrative regions, most of them consisting of two departments.

EFFECTIVENESS OF IMPLEMENTATION

In 1973 the INP published the two-year plan 1973–74 (INP, 1973). The document contained more than plans and programmes: it included an evaluation of the progress made in the various objectives mentioned in the five-year plan 1971–75.

With regard to policies in the field of spatial planning and regional development, the two-year plan was rather sombre. It said: '. . .they acquire singular relevance considering the slowness with which their implementation has advanced' (INP, 1973, pp. 1–53). Although excuses can be found regarding those activities relating to structural reform, some additional reasons are presented for poor performance. Special attention is given to: (a) the decentralization policy, (b) the policy with regard to zones of concentrated action (ZAC), and (c) *politica territorial*, i.e. macrophysical planning.

Instead of decentralizing, the argument goes, the government in fact centralized during the first years of the process. Now the time had come to return power to local and regional authorities. Therefore, more time should be devoted to defining their roles (INP, 1973, pp. 1–55). As we have seen before, the effort was doomed to fail, and local and departmental authorities never acquired the new roles. The only new institutions that emerged were those relating to agrarian and educational reforms, the departmental coordination committees, and three regional development organizations. The importance of these latter organizations cannot be judged at a time when government has announced a cut in its staff of 25 per cent. Lack of deconcentration on the part of the various Ministries is criticized, but as we have seen, to no avail. The policy with regard to the ZACs was not effective because of projects 'in the pipeline'. In other words, a number of commitments had been entered into previously, and ongoing projects could not just be abandoned (INP, 1973, pp. 1–57). In fact what was hidden in this statement was the reality that government found itself saddled with no capacity to elaborate projects and prepare them for implementation. Historically this work had been mainly in the hands of foreign consulting firms, and they had come to be considered hardly eligible for this work. But government could find no substitutes for these firms amongst local professionals. All in all, the difference between what was planned for the ZACs (almost 100 per cent of state investment), and what they received (only 60 per cent) was rather significant.

The document directs sharp criticism to the fact that no policies had been designed for zones with high population pressures; the Ministry of Agriculture especially was found to be at fault for not developing a solution to the problem of the *minifundistas* in the Sierra. Lack of attention to this part of the country by other Ministries was exposed. Although they fall inside the ZACs, the departments of Cajamarca, Ayacucho, Cuzco, Apurimac, and Puno, all in the Sierra, received only 7 per cent of total government investment, the two-year plan notes, while containing 23 per cent of the population (INP, 1973, pp. 1–58).

Remarkably, however, the list of projects that follows this statement is still one that barely recognizes the needs expressed in it. Concerning the *politica territorial*, the plan goes no further than praising the Ministry of Agriculture for starting its *Programmas integrales de desarrollo* (PIDs), and saying that new agricultural land should be made available to the poor. As we have seen, however, the PIDs never became a reality.

It is clear that by the middle of the period under consideration, the government itself was aware of its failure in regional development policies. At times it even contributed to regional disequilibria, as was the case when the zinc refinery that should have gone to Huacho (in order to promote the decentralization of the Greater Lima agglomeration and avoid dangers of pollution) was located close to Lima after a long and arduous battle between the INP and the Ministry of Mining.

In fact, regional planning began to be viewed differently in 1973–74: if it was to serve the development of a given region, a regional office should not be related only to the regional programming division of the INP, but should have direct insight into all matters pertaining to the region. In other words, the directors of regional offices of the INP wanted to be able to influence decision-making in the other divisions of INP. The directors of the regional offices won this battle, overlooking the fact that a number of issues in regional planning can only be studied from a national point of view. However, they won it at a moment when the importance of INP had begun to decline. More and more, the office of the Prime Minister drew planning powers to itself, even staffing itself with professionals who came from INP.

In earlier years, the INP had considered it a good policy to transfer their staff to the various Ministries. The price they had to pay for this was that time and again their best people left, and these could not be 'reproduced' at the same rate. The technical director (who stayed on until the end of 1976 when he was removed for political reasons) found himself, from mid-1974 onwards, with an

Institute overstaffed with zealots, and lacking professionals. The ensuing crisis moved the centre of gravity in economic policy-making firmly to the Ministry of the Economy and Finance.

Meanwhile regional planning offices returned time and again to the making of *diagnosticos*, the formulation of strategies and objectives; but as they had no counterparts in the regions, their work became an echoless cry. Due to unwillingness on the part of staff of various Ministries, decentralization remained (with the important exceptions aforementioned) a dead letter. Therefore there also were virtually no counterparts in the various sectoral Ministries. As far as the private sector was concerned there were the peasants and the urban poor, the agrarian cooperatives and the industrialists, as possible counterparts. The peasants and the poor were that part of the population over which SINAMOS claimed all rights once the agrarian reform was over, while agrarian cooperatives were and are the domain of the Ministry of Agriculture. The private sector of industrialists was either non-existent, as in the case of the Sierra and Oriente, or virtually removed (as was the case of the Costa Norte fishmeal industry that was nationalized). Thus only in the south could planners come into contact with private groups, whose general critical position *vis-à-vis* the government offered no positive sounding-board for the planners' views. In some cases, therefore, planners in the regions turned to urban planning, but in most cases they were reduced to monitoring the actions of the various Ministries.

CONCLUSIONS

How does one show that something is not there when it should be, especially when some say that it is in fact really there? The *Revolución de la Fuerza Armada* has been called variously 'institutional', 'from within', 'an experiment', and 'ambiguous'. The most candid phrase in the context of this chapter seems to be, however, the one given by Scott Palmer (1973), who called it a 'Revolution From Above'. If there was participation, it was by the military and the professionals involved with the reforms, with INP and with SINAMOS. Only a portion of the students were involved. Some sections of the press showed initial support but later developments, such as nationalization and bans on publicity, make observation impossible.

Studies such as those by Scott Palmer (1973), Conlin (1974) and others show that where participation should have occurred, it did not. The present review shows that where more action on the part of the government should have taken place in the overpopulated areas of the Sierra, there was barely enough. Instead, a return to the old pattern of fostering activity on the coast took place.

Efforts to further integrate the country's economy and bring it into contact with neighbouring markets, did take place. Roads and harbours were built or improved for these purposes.

Poverty in Peru is not due mainly to the openness of the economy, but to isolation and the way proceeds from international trade are distributed, both between persons and between regions. Replacement of private enterprise by the state has not changed much in this regard, the only gain being that more investable funds may remain in the country. The formidable indebtedness that Peru has been led into by the military, however, makes one quickly forget this advantage. The continuing economic crisis bears witness to the situation.

Decentralized government organizations like ORDELORETO and ORDESO may expect to become controlled via their consultative councils by the pre-1968 interest groups. 1978 election results and present government policy point in that direction.

The government's efforts to decrease the country's dependence, by acquiring direct control over virtually all production for export and by launching an enormous investment programme, have had adverse results. The country's dependence on foreign decison-makers has increased, because of its indebtedness.

This nationalist effort went on during most of the period under study, and had the effect of increasing centralization. Thus the objectives of regional development could not be asserted, due to lack of political articulation at the regional level.

Regional planning during the 1968–77 period was definitely 'centre-down', even if a number of the objectives were 'bottom-up' in Stöhr's terms. As observed, regional development objectives were sacrificed during most of the period, and this was especially the case with regard to economic policies. A clear illustration was the regional distribution of investment.

A final conclusion should refer to changes in spatial disparities in living standards. No statistical data are available on which firm conclusions can be based, but one could say that in the period 1968–75 standards of living in urban areas, especially those on the Costa, improved in absolute terms. It must be observed, however, that there was at the same time a decline in the ratio of product-per-capita in rural areas relative to urban areas. This means in fact that in over-all terms the economic position of the population in the Sierra underwent a relative decline in the 1968–75 period. And because of the austerity measures adopted in 1975, living levels in urban areas have since then declined rapidly.

REFERENCES

Andrews Hay, G. (1976) *Educational Finance and Educational Reform in Peru* (Paris: UNESCO, International Institute for Educational Planning).

Boisier, S. (1972) *Polos de Desarrollo: Hipothesis y Politicas Estudio de Bolivia, Chile y Peru* (Geneva: UN Research Institute for Social Development).

Conlin, S. (1974) Participation versus expertise, *International Journal of Comparative Sociology*, XV (3–4), 151–66.

Cotler, J. (1975) The new mode of political domination in Peru, *The Peruvian Experiment*, pp. 44–78 (Princeton: Princeton University Press).

Couriel, A. (1978) *Peru: Estrategia de Desarrollo y Grado de Satisfaccion de las Necesidades Basicas* mimeogr. (PREALC).

Einaudi, L. (1973) The military government in Peru. In Thurberand, C. E. and Graham,

L. S., (eds.), *Development Administration in Latin America*, pp. 294–313 (Durham, NC: Duke University Press).

Einaudi, L. (1975) Revolution from within? Military rule in Peru since 1968. In Fidel, K. (ed.) *Militarism in Developing Countries*, pp. 283–300, (New Brunswick, NJ: Transaction Books).

Fitzgerald, E. V. K. (1976a), *The State and Economic Development: Peru since 1968* (Cambridge).

Fitzgerald, E. V. K. (1976b) State capitalism in Peru, *Boletin de Estudios Latinoamericanos y del Caribe*, 20, 17–33.

Giles, Antonio (1974) Planification regional de base agropecuaria: programao integrados de desarrollo (PID), *Revista Interamericana de Planificacion*, VIII (31) (September), 37–59.

Hilhorst, J. G. M. (1976) The port as an instrument of the economic development of small countries, *Caribbean Journal of Shipping*, III (1).

Hilhorst, J. G. M. (forthcoming), *Migration in Peru*.

Instituto National de Estadistica (INE) (1978) *Oferta y Demandal Global 1970–1976* (Lima: February).

Instituto Nacional de Planificatión (INP) (1967) *Boletin de Estadisticas Regionales* (Lima).

INP (1968) *Estrategia del Dessarrollo National a Largo Plazo*.

INP (1971a) *Los Cambios Fundamentales en la Occupacion del Territorio Peruano*.

INP (1971b) *Plan del Peru, 1971–1975*.

INP (1973) *Plan Bienal de Desarrollo para 1973–1974*.

INP (1976) *Analysis de la Inversion Publica (1975–1976) respecto al Desarrollo Regional-Nivel Nacional*.

INP (1978) *Evaluacion Anual de los Programas Departmentales de Inversion Publica 1977*.

Kruyt, D. and M. Vellinga (1977) The political economy of mining enclaves in Peru, *Boletin de Estudios Latinamericanos y del Caribe*, 23, 97–126.

Lateinamerika Nachrichten (1977) Das Scheitern des Reformismus—die 2. Phase von 1975 bis Heute (Berlin West: 3 October), pp. 23–32.

Lowenthal, A. (ed.) (1975) Peru's ambiguous revolution, *The Peruvian Experiment*, pp. 3–43 (Princeton: Princeton University Press).

Oficina Nacional de Estadistica y Census (ONEC) (1975) Perspectiva de crecimiento de la población del Perú 1960–2000, *Boletin de Análisis Demográfico*, 16 (Lima: December).

Slater, D. (1975) Underdevelopment and spatial inequality, *Progress in Planning*, 4 (2), 97–167.

Smith, C., T. (1976) Agrarian reform and regional development in Peru, Chapter IV in Miller, R., C. T., Smith and J. Fisher (eds.), *Social and Economic Change in Peru*. (Liverpool: University of Liverpool) Monograph Series 6, 87–119.

Stöhr, W. and F. Tödling (1977) Spatial equity—some antitheses to current regional development doctrine. Paper presented at 13th European Regional Science Congress, *Papers of the Regional Science Association*, vol. 38, 33–53.

Webb, R. (1975) Government policy and the distribution of income in Peru, 1963–1973. In Lowenthal, A. (ed.), *The Peruvian Experiment*, Chapter 3 (Princeton: Princeton University Press).

PART III

Conclusions

Chapter 18

Development from Above or Below? Some Conclusions

WALTER B. STÖHR and D. R. F. TAYLOR

Consideration of development 'from above' or 'from below' is in essence a consideration of the nature of development itself and everyone, it seems, knows what development is except the experts! This is perhaps not surprising because in the ultimate sense development is a reflection of personal values, conditioned by the societal framework in which one lives. The values a society holds, which themselves change over time, are the ultimate standard by which development or lack of it will be judged. It is perhaps obvious but worth re-stating that an outside view of a society's 'development' may be very different from an assessment made by that society itself.

The scale at which 'development' is analysed is also critical. The nation-state has been used as a framework in this book but in Third World settings some of these are relatively recent creations and many contain several 'societies' with different value-systems. Nation-states also vary enormously in size and in societal complexity. One region in states such as China or India can be several times the size in population terms of a Nepal of a Papua New Guinea. Differences in areal extent are also great but perhaps it is difference in population which is more significant in development terms as development in the final analysis is about people not places, although people seem to require a certain degree of territorial identity in order to live harmoniously in society. Although the nation-state has been used as a convenient frame of analysis, this book has primarily concerned itself with regional development at the sub-national scale. Each case study author was asked to analyse sub-national regional disparities and policies in a national context as they seem to be determined to a great extent by national variables and objectives.

The nation-state is a political reality and over time has proven to be a remarkably persistent institution. The territorial integrity of even the newest nation-states is likely to be defended vigorously, both against external and internal challenges. In many countries these days, in all parts of the world, the internal threat of secession or devolution is as great if not greater than any external threat to territorial integrity, and is a reflection of dissatisfaction with national development strategies which lack sensitivity to regional values. Schumacher (1973) used the phrase *Small is Beautiful* in the title for his book; other authors feel that small may be beautiful but big is bountiful! (Webber, 1979). Development from below argues essentially for a development which is determined at the lowest feasible territorial scale. Much has been written about the economics of scale; relatively little about how small a unit need be, to be viable in a developmental sense. It has been argued elsewhere (Taylor, 1975) that territorial units containing as few as 30,000–80,000 people are perfectly viable development entities. Friedmann and Douglass (1978) argue for 'agropolitan districts' of 100,000–150,000 and there are a variety of units of about this size emerging throughout the world. Even in the large cities of North America and Western Europe community groups representing populations rarely larger than 50,000 are emerging, challenging big city government. Both the Chinese commune and the Indian block are around this scale.

As Stöhr points out in Chapter 2, however, scale of analysis is not the only, or even the most significant, aspect of development from below. Inherent in development from below are certain basic values. First it is a development determined from within by the people of that society themselves, based on their own resources—human, physical and institutional. Each strategy is therefore unique to the society in which it evolves. Secondly it is egalitarian and self-reliant in nature, emphasizing the meeting of the basic needs of all members of society. It is therefore communalist (Friedmann, 1979) in nature. The ultimate aim of such a strategy is an improvement of both a quantitative and qualitative type in the life-style of all members of society (Goulet, 1978a). It involves selective growth, distribution, self-reliance, employment creation, and above all respects human dignity. It is at one and the same time a new development strategy and a new development ideology.

Hansen has argued in Chapter 1 that the theoretical issues are clear, and that what is required is a thorough empirical examination of specific cases—both to make the assumption of theory more realistic and to provide policy guidance. What then have the empirical studies in this book revealed about the development process?

In *Asia*, China is the exceptional case. Statistical evidence is difficult to obtain but the Chinese peasant has indubitably improved both the quality and the quantity of his life-style since 1950. Between 1950 and 1976 China clearly experienced long-term growth and a marked decrease in inequalities (Wu and Ip, Chapter 6). The other Asian case studies, however, reveal increasing regional and interpersonal disparities and in Thailand, Nepal, and India there has been both relative and absolute impoverishment, especially of rural peoples, over time. In India, in at least 75 districts, there have been negative growth rates. In Papua New Guinea disparities have remained about the same, although since independence in 1976 a trend towards their reduction has emerged. The over-all picture in Asia is one of persistent rural poverty despite comparatively high rates of economic growth. There has been an increase in landlessness and a growth in both urban/rural, rural/rural, and interpersonal disparity and although there has been a rapid increase in industrialization this has not created sufficient jobs to absorb the increasing labour force. Exceptions to this, in addition to China which is considered in this book, are South Korea, Taiwan, Hong Kong, and Singapore which are not. Japan is also an exceptional case, as both Hansen (Chapter 1) and Lo and Salih (Chapter 5) point out, but Japan is not usually classified as a developing nation. Because of the vast numbers of people involved it is in Asia where the problems of development and underdevelopment loom largest.

The *African* case studies also reveal persistent rural poverty and increasing disparity. Disparities are widening most rapidly in Nigeria whereas in Tanzania, although regional disparities persist, there seems to have been a decrease in interpersonal disparity. No Tanzanian is said to be starving but there has been very limited measurable growth. In Algeria there has been growth but only a limited developmental impact; whereas in the Ivory Coast there has been considerable growth. Penouil (Chapter 12) argues that the Ivory Coast is exceptional in Africa, as there has been a positive change in the welfare of all people in that country despite the persistence of both regional and interpersonal disparity.

The *Latin American* nations are, both in absolute and relative aggregate terms, much more fortunate than those of Asia and Africa. In all three Latin American countries studied, however, there have been persistent and widening regional and interpersonal disparities and persistent poverty, which is again most acute in rural areas. In Peru the Sierra population, according to Hilhorst (Chapter 17), is marginal in all senses. There has been no lessening of disparity in Chile over time and in Brazil, Haddad (Chapter 15) reports extensive and persistent regional development inequalities in all sectors.

Lee's (Chapter 4) overall summation on a *world scale* for developing nations is that between 1950 and 1975, despite growth rates averaging 3.4 per cent, poverty and underdevelopment still remain. He estimates that, whereas 33 per cent of people may have increased their welfare, this is more than balanced by 40 per cent for whom life has indubitably become worse. The empirical record of almost three development decades reveals more underdevelopment than development, regardless of what indicators are used for measurement. Existing strategies to bring about a broader participation in development have failed.

The explanations given by authors in this book for this lack of success vary. Several authors focus on the inappropriateness of the model of development used, which has often been urban and industrial in nature and has used growth as a proxy for development both in ideological and strategic terms. Others argue that the problem is essentially a structural one. Blaikie (Chapter 9), for example, argues that Nepal is a dependent periphery of both India and the world, and that the underdevelopment of Nepal will continue unless these basic structural relationships are changed. The underdevelopment of Nepal is not in his view a consequence of isolation but of incorporation. The lack of political will to bring about change is often cited. Even when the reduction of disparity is an explicit aim, as in India, the political will is lacking to take the hard decisions necessary. China's relative success, on the other hand, can be attributed to a revolutionary political ideology accompanied by revolutionary political action. Here there was a considerable degree of local self-reliance in resource mobilization, although it can be argued that China is characterized by participatory centralization and that local control functioned within clearly defined central limits. Wu and Ip (Chapter 6) argue that development from below can function effectively only within a clearly defined national framework in which the peasants have seized power. They identify six key elements which help explain China's success (p. 176–77) all of which are essential elements of a development from below strategy. These integrally related and complementary elements from the basis for a new set of interdependent urban and rural relations. In Peru, Brazil, and Chile the strong central control, felt necessary for political reasons, has militated against any substantive reduction in disparity.

In Africa the need to create integrated nations out of disparate ethnic groups has made central control of society almost a *conditio sine qua non*. This, allied to an urban/industrial, growth-oriented strategy is often used to explain the lack of significant broader developmental success as in the case of both Nigeria and Algeria although in both nations growth has been significant. In Tanzania it is argued that the theory and the reality of development are far apart with the bureaucracy being identified as largely responsible for this gap. Penouil (Chapter 12) argues that the success of the Ivory Coast is due to an innovative mix of strategies over time and an empiricism which is independent of ideology. He describes a mix of both polarization and decentralization and a mix of traditional, transitional, and modern activities.

The case studies present particular examples of the general theoretical points raised by Hansen, Stöhr, Weaver, Lee, and Lo and Salih (Chapters 1 to 5) in the first section of the book. Although there is no general agreement on the reasons for persistent underdevelopment, poverty, and inequality, none of the authors is happy with the present state of affairs and all make suggestions for policies to correct the present situation. Needless to say these again vary.

Decentralization and increased popular participation, especially of rural peoples, in the development process is a popular suggestion being made explicitly for Thailand, Papua New Guinea, India, Nigeria, Tanzania, Brazil, and Chile. In the case of Peru, however, Hilhorst (Chapter 17) argues that to remove the power of both the oligarchy and the unions there will first have to be a centralization process accompanied by legislation against inequality. Blaikie (Chapter 9) argues that for Nepal what is required is a change in the political economy and a recognition that until this happens spatial policies will only be palliative at best. Misra and Natraj (Chapter 10) argue that in India what is required is a return to Ghandian philosophy and its application as an ideology as well as a technique, whereas Penouil (Chapter 12) argues that the Ivory Coast must pursue a global and progressive strategy adapted to the realities of the economy and which pays particular attention to the transitional sector. China, after a period of inward-looking development, has embarked on what some have termed a 'great leap outwards'. The new course in China has been interpreted in a variety of different ways (*International Development Review*, 1979) and it is still too early to give a valid assessment of the possible consequences although it has been argued that

> The long term consequences could be extremely significant, for whatever its limitations as a model for other developing countries, Mao's China had the incalculable value of demonstrating in practice an ideological alternative to capitalism. Only time will tell what the demise of the Maoist model may mean for the future political orientation of Third World Countries; but one result could well be a weakening of the position of those strategists who have argued for a period of 'de-linking' from the West as a pre-requisite for real self-reliance, (International Development Review, 1979: 34).

Wu and Ip (Chapter 6) argue that the essential relationships in Chinese society are likely to remain unchanged. The empirical evidence available suggests that Maoist China was remarkably successful in decreasing the inequality of life of the rural masses. New structural relationships between town and countryside have been established and it is on the basis of new strength that China is once more looking outward. A new balance was also established between centralization and decentralization and the Maoist model contained a high degree of territorially based development and local decision-making, albeit within limits set by central government control (Wu and Ip, Chapter 6).

Stöhr (Chapter 2), Lo and Salih (Chapter 5), and Weaver (Chapter 3) all suggest a territorial approach involving 'selective regional closure' building on internally oriented regional economies. Along with other authors they call for changes both in development ideology and development strategy. There theoretical positions receive empirical support from Wu and Ip in the case study on China (Chapter 6). Hansen (Chapter 1), on the other hand, sees the need to improve the implementation of existing strategies and to concentrate on facts rather than ideology. He attaches particular importance to dealing with the structural problems of human resource development. Lee (Chapter 4) calls for changes in institutional structures with a central element being a greater degree of decentralization in planning.

Where does all of this leave us in terms of both theory and practice? In our view a real acceptance of new development paradigms is required, allied with the political will to implement them. Development from below is one such paradigm. There is, however, a considerable time-lag between the articulation of new paradigms and their acceptance and implementation. Without acceptance of new approaches the empirical evidence suggests that progress towards improvements in the quality of life for most people in developing nations will be slow. China's progress is largely due to the effective political implementation of one new development paradigm. That paradigm is specific to China and, as the Chinese have argued, is probably not transferable; but the Chinese case shows what can be done.

One of the great debates of spatial planning is whether the Kuznets hypothesis of convergence (Kuznets, 1955) and the Williamson divergence/convergence model (1965) are valid. The empirical evidence from the case studies in this book lends support to the divergence part of the hypothesis but little evidence of convergence has yet appeared. It may well be as both Hansen (Chapter 1 in this volume) and Richardson (1973) have suggested that there has been insufficient time for such a process to take place and that in any case a polarization strategy has not been properly implemented, but increasingly this hypothesis is being questioned. Even if the transitional view is correct the period of time for the reduction of poverty to occur as a result of polarized growth is clearly much longer than was thought to be the case 15 or 20 years ago.

It is equally, if not more, likely that the assumptions and the strategy and ideology of development on which these hypotheses rest are faulty, as has been argued by Stöhr (Chapter 2), Weaver (Chapter 3), Lo and Salih (Chapter 5) and others in this book. There is little need to repeat these arguments here except to say that the authors are in broad agreement with them. If this is the case then new approaches to both theory and practice are required. Development from below is one such approach.

The validity of development approaches will not be determined as a result of theoretical and ideological debate but in the realm of practice. The peasant families of Africa, Asia, and Latin America are more likely to judge the validity of a strategy from its results rather than its ideological or methodological soundness. The empirical evidence clearly shows that existing approaches have not brought the results predicted for them despite their application over a time period of over 30 years. Even their most ardent defenders admit that substantial modification is required, while some would argue that nothing will change unless radical restructuring is carried out. There is also evidence that in China a strategy containing essential elements of development from below has had beneficial effects in important respects, although apparently not in aggregate growth.

Development from below argues for flexibility and is as much an ideology as a strategy. It is a way of looking at development rather than a rigid set of policies and ideas. In practice there will be many responses to it over both time and space. Penouil (Chapter 12) argues that one of the greatest errors of the last decade has been to portray centralization and decentralization, or polarization and balanced growth, as being contradictory rather than complementary; his argument merits considerable support. The evidence from China (Wu and Ip, Chapter 6) certainly supports this view as both centralization and decentralization have been essential elements in helping bring about China's success. Development-from-below proponents must not fall into the trap of defending the purity of their ideas despite the realities of the situation. Lee (Chapter 4), utilizing some of Khan's arguments suggests that if specialization and exchange lead to an attainment of a higher level of basic need-fulfilment then it is '... misplaced heroism to condemn the nation to autarchy' (Khan, 1977, p. 113), on the other hand, self-reliance and selective regional closure are not to be equated with autarchy, as has been outlined by Stöhr (Chapter 2).

In pragmatic terms development-from-below proponents will probably have to be satisfied with modifications of existing practice to greater or lesser degrees in most Third World countries as a result of their pleas for change. The question is whether such a strategy can be achieved without radical institutional transformation. In political and ideological terms development from below is unacceptable both to the right and the left but for very different reasons. To the Marxist and neo-Marxist, development from below is not posed in class terms and therefore misses the point. In a society which is committed to equality it is inconceivable that central control could continue to permit regional and interpersonal disparities to exist and these will disappear over time as true communism is attained. That central control by the state itself could be a cause of continuing disparity is considered to be simply ideological heresy. Current debate on the issue bears a striking resemblance to the dispute between Marxists and the Anarchists and Russian populists towards the end of the nineteenth century. Regional and interpersonal disparities, however, continue to exist in centrally planned economies. Weaver (Chapter 3) deals with this issue explicitly and extremely well.

Douglass (Chapter 7) is probably right that a small-community approach is no guarantee that existing inequalities will not persist. It is possible that in the Asian and possibly also other contexts complete 'regional closure' might make the control of landlords and moneylenders even stronger. In cases like this building upon traditional institutions may be counterproductive and new approaches may be required as Friedmann (1978) argues. In various chapters of this book the concept of 'selective spatial closure' is therefore introduced. In many communities exploitation is clearly supported by extraregional forces and processes, and the removal or reduction of these may weaken the power of exploitative groups. In many communities, however, there are traditional institutions through which an egalitarian development-from-below approach could operate. In most rural communities in Africa, for example, this is certainly the case and class formation, where it exists, is only in a nascent stage. In most cases there will be a role for central government where conflict-resolving mechanisms are required.

Change in many rural areas demands land reform; in some instances involving the break-up of large landholdings as in Latin America and parts of Asia; in others, as in many parts of Africa, involving the consolidation of minute and widely scattered fragments. It can be argued with some validity that land reform cannot, and will not, be carried out without central government intervention.

The nation state is a political reality and in any development strategy is must play an important role. To deny it such a role is politically naive regardless of how desirable this might be in theoretical terms. Decentralization might be desirable in developmental strategy terms but be totally unacceptable in political terms if it involves the loss of too much power by the group in charge of the nation's destiny. As Haddad (Chapter 15) argues, national and regional interests are often in conflict. There can also be conflict between different regions within a country. Where differences in the physical and human resource base exist the selective closure of regions may in fact favour the existing 'have' regions at the expense of the others. A central control mechanism is required to deal with problems of this type and to facilitate redistribution between regions where required. Here the Chinese experience is particularly instructive with its emphasis on increasing regional self-reliance rather than major welfare redistribution.

It is, however, certain that the domination of the existing peripheries by the centre must be lessened if development is to take place. What is currently a top-down unidirectional process must be replaced by more equitable relationships, and in the short term this may require a degree of 'selective spatial closure' to create the conditions which will allow a restructuring of these relationships. Spatial integration of a nation is often seen as a positive developmental step. But in our view this is only the case when the terms of the various forms of linkages and modes of production are in favour of the poorer areas, otherwise the situation is more likely to get worse as a result of integration rather than better. A key element in China has been a set of national policies which consistently support the rural sector (Wu and Ip, Chapter 6).

Blaikie (Chapter 9) talks of the different spatial logics which exist in Nepal as a result of different modes of production, and the existence of different spatial systems has also been discerned in other contexts (Taylor, 1972). Some of these are clearly more beneficial to poorer regions and their inhabitants than others. In many countries, at least initially, there is perhaps a need to '*recouler pour mieux sauter*' to allow communities to become more self-sufficient and to centralize upwards to the national level on more favourable terms than at present. It is in this context that selective closure may be a most useful strategy.

Such a new balance cannot be achieved in our view within the framework of existing polarization strategies, especially those of a growth pole type, regardless of how well these are implemented or over what time period they are applied. What is often forgotten is that Perroux's (1964) original concept of growth poles included the concept of dominance—an irreversible or only partially reversible economic influence. Even in China as Wu and Ip (Chapter 6) demonstrate policies of this type have had little real impact on inequality.

International structural relationships clearly influence the policy options available at the nation-state level as Lo and Salih (Chapter 5), Blaikie (Chapter 9), and others in this book indicate. A greater degree of regional self-reliance can not be achieved without a change in external relationships and development from below at the regional level clearly has implications at larger scales. Restructuring of internal relationships will be less than effective if not accompanied by changes at the international level. At the national level development from below demands a more inward and self-reliant approach and in most cases, at least initially, a degree of spatial closure from the international system. The degree of self-reliance possible varies from nation to nation and for some small nation-states possibilities are limited. But there are few nations which could not mobilize more of their own human and physical resources than at present, especially if these were used primarily to meet internally oriented and established demands rather than the growth criteria of the international system.

It is, however, true that in terms of natural resources some nations are much better endowed than others. There are also those whose share of world resources is out of all proportion to their population and whose basic problems are more those of overdevelopment than underdevelopment. Just as conflict-resolving mechanisms are required between regions within a nation, so too are such mechanisms necessary at the international level. Meeting the requirements of north–south dialogue and ensuring that action is taken to lessen tensions and reduce inequalities is one of the greatest challenges of our time.

Reduction of disparity between nations, or between regions within individual nations, does not in itself ensure that interpersonal disparity decreases. In some instances establishment of a New International Economic Order as demanded at the UN by the Group of Seventy-Seven might reduce quantifiable disparity between nations but leave the poor in some developing nations in as bad a position as before. Similarly a reduction in interregional disparity within a country may in fact mask an increase in interpersonal disparity. Misra and Natraj (Chapter 10) argue that a policy which concentrates on the reduction of regional disparity may be used to avoid dealing directly with the more difficult problem of interpersonal differences. Development from below, although essentially territorially based, must involve the development of people rather than places.

A study of the Indian experience will rapidly reveal that the debate between development-from-above advocates and those of development from below is by no means a new one, although today it has emerged in new forms and with renewed energy. Gandhi in many ways epitomized many aspects of the ideology of development from below, whereas Nehru was closer to development from above in his ideas. Both were great men with the true interests of their people at heart. Misra and Natraj (Chapter 10) call for a return to Gandhian philosophy and its application as an *ideology* as well as a technique. Development from below has the ideological underpinnings to give spatial planning in developing nations a new direction, and possibly to bring new hope in the struggle against poverty. But the mechanisms for practical implementation at larger scale still have to be developed and tested under conditions of interacting national and international political systems. The new efforts which appear to become evident in China as from 1980 onward seem to be highly interesting in this context.

SOME CONCLUSIONS FROM CASE STUDIES IN A HISTORICAL SEQUENCE

The case study countries exhibit considerable diversity in natural conditions, level and dynamics of development, demographic and geographic size, economic structure, degree of urbanization, magnitude and kind of external functional relations (see Table B, in the Introduction).

In spite of this great diversity, a number of general conclusions seem to emerge from the case studies regarding the major theme of this book:

1. The Imprint of Pre-Colonial Eras

None of the countries analyzed is still in a stage of formal colonial dependence and therefore only occasional explicit reference is made to these historical conditions.

The argument that in most pre-colonial societies of today's Third World, relatively egalitarian social and landholding structures existed (Stöhr, Chapter 2) seems to be confirmed for several case study countries such as Papua New Guinea (Conyers, Chapter 8), Thailand (Douglass, Chapter 7), and by other sources on Latin American Lowland Indians (Clastres, 1974), and Ceylon (Gunaratna, 1979). 'Most families in their rural environments were able to meet basic needs with relatively little output of energy (Conyers, Chapter 8) and there existed a kind of 'subsistence affluence' which, Conyers maintains, almost sounds too ideal to be true. In some of these countries at least access to land and basic needs was not only a communal right but also a communal responsibility (Douglass, Chapter 7). 'Rights of access to goods and services [were] considered necessary to sustain life and livelihood, by mutual assistance through labour exchange for production and through other systems of reciprocity' (Douglass, Chapter 7). This certainly was not true, however, for certain countries such as India which also in pre-colonial periods had rigid caste structures. For the other countries, however, Douglass's assertion that decision-making was at reasonably low levels until not too long ago would hold true, as would the argument that the term 'devolution' is really nothing very novel or revolutionary but rather quite old and 'conservative' (Douglass, Chapter 7).

2. Changes during Colonial Periods

In many, particularly the smaller, countries it was only during the colonial period that decision-making levels were raised drastically, that an urban elite began to emerge (Conyers, Chapter 8) and that social and income disparities started to increase rapidly. Due to a subsequent continuous economic erosion via rural–urban leakages, the peasantary lost 'its organizational, territorial basis for production' ... 'Land became an exchange commodity ... tenancy and landlessness began to appear, and access to money and credit became essential for even subsistence-level production' (Douglass, Chapter 7). This became particularly accentuated once land reserves, the major resource for the rural population, became exhausted. This fact is stressed in the case studies for Thailand (Douglass, Chapter 7) but also for Brazil (Hadded, Chapter 15) and Nepal (Blaikie, Chapter 9). 'Cooperative production and labour exchanges were replaced by hired labour and wage work' (Douglass, Chapter 7).

Power also became withdrawn from the local to the national level and 'villages, even those which still maintained communal production and distribution systems, were left without any countervailing power. ... Local systems of governance atrophied' (Douglass, Chapter 7). At the same time access to resources and to large-scale markets became differentiated and thereby also to the accumulation of capital, 'access to productive and remunerative employment and access to the consumption of goods and services made possible the entry into each of the "two circuits" which formed' (Douglass, Chapter 7).

Unequal external interaction or external dependence becomes an important determinant for internal class and spatial inequalities, whether in a formal colonial setting, or in a quasi-colonial one such as that described for present-day Nepalese dependency on India (Blaikie, Chapter 9). Nepal in this sense can be considered a 'double periphery' depending on India which, on its part, again depended heavily on Great Britain. The consequences on social and also spatial inequality under these conditions appear to be particularly pronounced in a country such as Nepal. Blaikie offers a detailed description of 'leakage' factors occurring in such a double dependent situation over what he describes to be an 'open frontier', across which selected factor transfers in the interest of the ruling classes on both sides can freely take place. Nepal in this sense experiences all the 'disadvantages of peripherality but none of the advantages' which would exist within a national context, such as in the north of India where the Indian government provided hospitals, schools, agricultural extension services, etc. (Blaikie, Chapter 9).

Apart from the economic inequality created through external dependence, inequality in access to political power also increased in the colonial periods. In part this was caused by the fact that most colonial countries, in order to maintain their dominance over the local population and/or against competing colonial powers, in most cases introduced highly centralized administrative systems. The major interest of the colonial powers, particularly in earlier periods, was the extraction of valuable natural resources for which the necessary transport, urban, and administrative infrastructure was provided with a distinct external orientation. The domination and control of the colonies seemed best facilitated by a hierarchical, centralized administrative system.

3. The Persistence of Centralized Decision-Making Structures After Decolonialization

The case studies show that in most of the countries analysed centralized decision-making structures which were introduced during the colonial period persisted even after decolonialization, for a number of different reasons.

(a) Perceived Need for Integration of Culturally Heterogeneous Nations as Left by a Colonial Heritage

The colonial powers in many cases left as their overseas heritage political/ administrative units which aggregated formerly very differentiated cultures. Cases in point are India (Misra and Natraj, Chapter 10), Peru (Hilhorst, Chapter 17), Ivory Coast Penouil, Chapter 12), Papua New Guinea (Conyers, Chapter 8), but also Brazil and other countries where this historical feature is not specifically discussed in the restricted space that was available for individual case studies. National development initially was considered by most Third World countries after independence not so much as an economic process but rather, in an integrative sense, as a

> process by which a state characterized by sectional, or otherwise competing economies, politics, and cultures, within a given territory, is transformed into a society composed of a single, all-pervasive, and in this sense 'national' economy, policy, and culture. (Hechter, 1975, p. 17.)

A strong central power in many cases was considered a pre-requisite for achieving this. There apparently were few countries which like Nigeria (Filani, Chapter 11) and Papua New Guinea (Conyers, Chapter 8) initiated a policy of decentralization of political power as an instrument to secure national unity.

(b)   Defence of these New National Units Against the Outside

The borders of many of these new political–administrative units were unstable and not yet internationally acknowledged. For purposes of external defence in possible border disputes a strong central power was considered important.

(c)   Physical Integration as a Potential Factor for National Development

Many of these new political/administrative units were physically fractioned territories which made communication and therefore also effective government difficult (Boisier, Chapter 16). The great interregional diversity of natural conditions, however, in many cases could also have been considered an important prerequisite for internal complementarity enhancing the internal development potential of many of these national units (Penouil, Chapter 12). Theoretically such internal diversity and potential complementarity in fact in many cases could have been a promising basis for a internally oriented self-sustained development at least at the national (if not at the sub-national) scale. This chance, however, was rarely taken in view of external orientation and dependence.

(d)   Perceived Need for Rapid Transformation towards Western 'Development'

Development in most Third World countries was considered as a process of transforming (if necessary forcefully) existing economic and social structures so that they could respond as rapidly as possible to the rules of economic interchange, initiated by the highly industrialized countries. Such rapid transformation was considered possible only in a centrally initiated way. At the same time the vast magnitude of economic problems faced by most countries after achieving independence nurtured the belief that these could be solved best by a centrally concerted effort based on 'planning' (Misra and Natraj, Chapter 10). These concerted efforts usually concentrated on those population strata, sectors, and regions able to undergo such transformations most quickly.

Under such conditions internal disparities in many countries increased rather than decreased (Misra and Natraj, Chapter 10).

(e) Perceived Need for Concentration of Development Efforts

It was assumed that the necessary economic as well as social and cultural transformation could not take place in the totality of a social system at the same time without damage (Penouil, Chapter 12). It was therefore considered advisable to concentrate this transformation upon a few centres and let it penetrate to the rest of the territorial system in successive stages. For this purpose also a centralized innovation and decision making structure seemed useful. These pragmatic ideas in fact already were predecessors of the growth centre and trickling down concepts which were formulated in theoretical terms later (Hirschman, 1957; Perroux, 1964).

(f) Perception of the National Level as the Optimum Scale for Promoting Development

The new 'national' scale was implicitly considered the optimum level at which such conflicting objectives as a minimum scale for defence against the outside and a maximum scale for internal integration and full resource use could be made compatible (Blaikie, Chapter 9). In various countries it was realized only considerably later that such a centralization led to a considerable underutilization of resources in the respective national peripheries and consequently to economic and social decline in the latter areas. Marked examples are the Interior of Brazil (Haddad, Chapter 15) and the Hill Areas of Nepal (Blaikie, Chapter 9) to mention only two explicit instances in the case studies of this book.

(g) Formidable Magnitude of Interregional Disparities as a Reason for Evasion of the Regional Issue.

The magnitude of interregional disparities faced by many countries after achieving independence often induced them not to make this issue explicit but rather to play it down as much as possible. In several countries this seems to have been among the reasons why the issue of regional development was raised much later than that of national development. Stressing this issue earlier, without the hope of being able to solve it given scarce central resources, frequently was considered to imply the danger of territorial disruption (Misra and Natraj, Chapter 10). Where the regional issue was addressed explicitly, it was often given mere lip-service. Regional issues in general tended to be given attention essentially as a function of national objectives. On the other hand, in cases where—often for lack of central government funds—sub-national territorial units were encouraged to cope with their own problems (e.g. in India regional units along language lines) regional growth rates diverged considerably due to different natural conditions, different cultural characteristics, and differing qualities of administration (Misra and Natraj, Chapter 10).

(h)   Central and / or External Bias of Inherited Urban and Transport System

The externally oriented and centralized transport networks and urban systems introduced during the colonial period in many cases prejudiced the internal operation of their systems after independence. 'Space . . . is determined by, and is given its rationality by the political economy' (Blaikie, Chapter 9) and undoubtedly such spatial structures again have a feedback effect on a country's political economy. Some of the case study countries deliberately attempted to cut this causal chain by explicitly transforming the spatial configuration of the urban system. Cases in point are Brazil, Tanzania, and Nigeria, which transferred their national capital into the Interior. If these attempts in fact were effective spatial policies they could indeed be considered as 'radical policies', contrary to the negation made by Blaikie (Chapter 9).

4.   The Influence of External Aid Agencies and of the 'Old' International Economic Order on the Centralization of National Development Processes in Third World Countries

From the 1950s onward external national and international aid agencies assumed great influence upon the way in which development strategies were drawn up and implemented in most Third World countries. This was done via the *criteria defined for external aid projects* but also by the *postulate that recipient countries establish central planning offices and elaborate national plans*. A number of case studies in this book show that a regional dimension was lacking in these plans for a long time and has usually been introduced only during the last few years. In many cases, however, even this was only lip-service, as has been mentioned before. The guiding objectives of development were to increase GNP, to improve the capital/output ratio, and to improve the balance of payments. Most variables used in the planning process were aggregate national ones, in many countries also the only ones statistically available. Due to the *lack of planning specialists*—most of whom were economists or engineers (Boisier, Chapter 16)— central planning officers could hardly be staffed adequately, let alone regional or local planning offices.

External factors were largely responsible for the introduction of central planning (Douglass, Chapter 7) and facilitated the application of imported technocratic planning methods and of neoclassical economic doctrine (Boisier, Chapter 16). These external planning influences also led to

import-substitution and urban-based industrialization, assembled from both 'traditional' policies of transferring rural surplus for metropolitan development, and new policies promoting and protecting import-substituting infant industries; . . . giving full duty exemption on capital goods and raw material imports for protected industry which almost without exception chose to locate in the metropolitan area. (Douglass, Chapter 7.)

Like the earlier raw material export policies of colonial periods, these new strategies usually also contributed heavily to the *urban bias* which has been found in most of the countries analysed here.

Changes in spatial disparities of living levels and the influence which development policies have had upon them were discussed briefly at the beginning of this chapter. For various reasons it is difficult to make conclusive statements on the concrete effects regional development policies had on these disparities. First, because it is methodologically very difficult to calculate the quantitative effects of such policies even for countries with good data bases (OECD, 1977); secondly, because most of the countries analyzed have insufficient data available to measure spatial disparities in living levels, particularly regarding their development over time. The mere absence of data on spatial disparities in some countries may be considered as a sympton of the lack of real emphasis given to this problem in actual policies, beyond verbal statements in planning documents. This point has already been referred to above.

Apart from these two facts, the time-period within which effects of regional development policies materialize may in fact need to be considerably longer than these rather recent policies have existed so far (Hansen, Chapter 1; Boisier, Chapter 16). When Hansen argues that this is true for many policies incorporating elements of development 'from above', it is probably even more true for those incorporating elements of development 'from below' which—with the exception of China—are much more recent still. There seems to be a fair amount of agreement that in strategies of development 'from above', a relatively small group of people benefit considerably from large-scale economic interactions of an industrial or commercial type. It is less clear whether the majority of the poor population of these countries is not touched by these benefits only in the short run, or whether their absolute living levels actually decline in the long run. Statements for some of the countries which have applied a 'centre-down' strategy, such as Thailand, seem to indicate that such an absolute deterioration has actually taken place: '... basic needs and poverty line estimates ... indicate that for those in the lower circuit, levels of welfare have fallen for many and the proportion of people below these lines have increased (Douglass, Chapter 7). In another case study, for the Ivory Coast, a broadening of developmental impulses from the modern to the traditional sector via 'transition' is reported in connection with a centre-down strategy (Penouil, Chapter 12).

5. Centre-Down Strategies: Background and Some Consequences

The historical background just sketched laid a perfect basis for centre-down strategies trusting in the trickling down of development through worldwide technocratic planning expertise, through worldwide aid organizations and stimulated by worldwide innovation centres and worldwide demand, as outlined by Stöhr in Chapter 2. This trend towards national centre-down development strategies was reinforced by the fact that individual countries now needed not only to defend their borders against neighbouring states, but also their economic competitiveness as the liberalization of trade initiated after World

War II under schemes such as GATT put countries under increasing economic pressure, without protection, in a worldwide context. Even China recently seems to have reacted to such economic motives in order to be able to better defend her political borders against her neighbours.

This pressure for developing countries for transformations in order to compete on increasingly liberalized worldwide markets for commodities and production factors stressed the need for *rapid and effective centralized decision-making at the national level.* National territories were for all practical purposes considered as aggregate units (comparable to the point locations assumed by neoclassical economics), within which full mobility was assumed or should be brought about as rapidly as possible. The objective of increased national economic and political integration was therefore directly linked, and often identified with, other objectives such as maximum resource mobilization and the improvement of income distribution (Blaikie, Chapter 9; Hilhorst, Chapter 17). The major assumption behind linking these two types of objectives was that national integration would permit the easy transfer of surplus generated in one sector or region to other sectors or regions. This was expected to maximize aggregate national efficiency via an optimal allocation of resources and of the benefits from development. As, particularly in the less-developed countries, this could not be achieved by a functioning market mechanism, the latter was substituted by a central resource allocating mechanism called national planning. This central planning mechanism in most developing countries operated mainly on the sectoral allocation of resources (mainly capital), rarely on an explicit regional allocation of resources, and almost never on the allocation of the benefits from development (neither by social strata nor by regions). The latter failure was in part caused by the lack of solidarity between social strata and between regions in most developing countries, and the consequent lack of political support for such redistribution. Without an effective central redistributive mechanism of the benefits of national development, increasing social and regional disparities were bound to arise.

Under such conditions there occurred in fact a *separation between efficiency oriented economic mechanisms and distribution oriented political mechanisms.* Haddad states that this neat separation was underpinned by the manifestation of the 'new economics as a purely positivistic science while normative issues were delegated to politics. The pervasive belief arose that once economic growth had taken place, distribution would automatically take place afterwards' (Haddad, Chapter 15). In reality, however, such an aggregate allocation of resources at the national scale led to a *disintegration of complementary resources at the regional scale:* natural resources, population, and savings/profits were selectively withdrawn from specific (usually less developed) regions to other (usually core) regions where their use seemed more efficient from an aggregate national point of view.

A tendency towards the *disintegration between the modern and the traditional sectors* also exists. It was hoped that via 'transition activities' (Penouil, Chapter 12) the traditional sectors would become smoothly transformed into modern sectors. While Penouil asserts that this process in aggregate terms was actually successful in Ivory Coast, he admits that in disaggregate terms only an extremely small share of traditional activities actually was able to realize this transformation, while for the great majority of them it meant the death of individual traditional activities and, separately from them, the birth of modern activities often in other sectors and in other regions.

This was accompanied by an *urban–rural disintegration* in the sense of a 'selective (national) integration of a few urban sectors and restricted social groups in isolated regional centres into the production system dominated' by the respective national or international metropolises (Goodman, 1975, quoted by Haddad, Chapter 15). In spatial terms this nationwide transfer of surplus meant that since agricultural productivity in microeconomic terms was lagging behind that of other sectors in most countries, the surplus created by agriculture was sucked off to other sectors and to non-agricultural areas, leading to a deterioration of rural living levels. At the same time most of these countries became net importers of food for which many of them had to expend a considerable share of their export earnings (Sutton, Chapter 14). Various countries attempted to rectify this erosion of less-developed peripheral regions by attracting to them what was considered the scarcest factor of these regions, capital—e.g. via tax incentive schemes such as in Brazil (Haddad, Chapter 15). Such policies, however, while they may have been able to quantitatively increase production in these regions, led to a further regional disintegration as the activities attracted were, in their majority, branch plants of core region or foreign enterprises producing for extraregional markets and utilizing extraregional inputs. The newly attracted activities in fact represented another set of 'islands' disintegrated from the surrounding regions and sectors, and integrated mainly with the national or international metropolises (Haddad, Chapter 15).

An excellent and extremely well documented account of how externally influenced national planning and large-scale functional integration can *mobilize rural surplus to promote metropolitan development* (in this case through a 'rice premium') is given for Thailand. The urban-biased extraction of rural surplus there was complemented by the urban bias of the loan and credit system and led to a strong economic erosion even in the richest agricultural areas of Thailand, the Bangkok plains (Douglass, Chapter 7).

A similar phenomenon occurred in countries where *growth centre policies* were applied. The definition of these growth centres was usually based on a projection of urban population growth and/or on national sectoral projections, but not on the development potential or demand of the surrounding rural areas. Such growth centre policies in most cases therefore often further aggravated the territorial disintegration of the respective regions (Douglass, Chapter 7).

In resource frontiers such as the Brazilian Amazon, regional disintegration manifested itself in *environmental disruption* as the 'screening for new investment alternatives' in areas such as the Amazon 'has intensified destruction of local fauna and flora (Haddad, Chapter 15).

Concentration of resources on the most efficient modern sectors functionally related to the world economy, at the same time in many countries led to a *decline of the traditional sector and particularly of rural industry and crafts*, to rural social fragmentation (Douglass, Chapter 7), and to the underemployment of those natural resources less mobile or in less demand on the world market. On the other hand it led to an externally oriented expansion of communications and transport facilities, to a 'cumulatively expanding power of the national capital as regions became more tightly integrated into the national and international economy' and to a two-circuit dualism...in the course of incorporation of once self-reliant communities into the metropolitan-dominated economy (Douglass, Chapter 7). This was accompanied by the takeover of (usually peripheral) natural resources by large-scale entrepreneurs for external markets (Douglass, Chapter 7; Hadded, Chapter 15).

The *'diversion' of resources*, however, took place not only from agriculture and small artisanry but *also from* a broad number of *locally or regionally operating (often informal) basic needs services*, from which innovative personalities, remuneration (often in kind) and capital were withdrawn. Activities substituting imports (and therefore exogenous in kind and technology) ranked high as recipients for national resources, while there was usually no support forthcoming for traditionally endogenous production or basic services (Lee, Chapter 4).

As an excuse for the inadequacies of these centre-down policies it has often been argued that resources were insufficient to solve the problems of all regions of the respective countries, or of all localities of individual regions simultaneously. This was considered to justify the concentration of resources in specific clusters of sectors (Perroux's *'industries motrices'* or *'poles de croissance'*) and in selected localities ('growth centres'). The question therefore was at which territorial scale functional integration should be emphasized and thereby possibly increase disparities within these territorial units. At the regional level growth centres were expected to fulfil such an integrating function between externally oriented (export base) activities and local/region serving activities, thereby transferring external impulses to these regions and localities. These growth centres, however, in most cases turned out to withdraw resources from their hinterland without actually being able to accommodate the magnitude of rural excess population. For Thailand, Douglass (Chapter 7) estimates that the nine growth centres planned for such purposes would need to provide services and act as industrial and administrative centres for double the population size of the present national capital, in order to fulfil this function. He maintains that the upkeep of these centres will be at least as costly as that of the present metropolis—the cost of which was a major reason for initiating the present decentralization policy.

The question therefore is whether by raising the level of territorial integration (e.g. to the national level) or by lowering it to subregional levels (e.g. that of Friedmann's (1978) 'Agropolitan Districts'), the manageability of the problem can be improved.

Different arguments suggest that emphasis be placed on one specific scale of these three: Hilhorst argues that anything below the national level—and possibly not even at that one—is a scale insufficient for internally oriented self-reliant development (Chapter 17). He feels that this is a valid argument at least for coastal countries of such internal diversity as Peru (and probably also Chile) of which each region formerly had a direct overseas orientation on the basis of its specific natural resources (copper, nitrate, etc.). Hilhorst furthermore argues on political grounds that a regional decentralization of decision-making powers in a country like Peru would arouse again the basis for peasant unrest (such as in 1965) and for rural guerrillas. He therefore considers a strong central government as the only means to maintain national cohesion (Hilhorst, Chapter 17). There is therefore evidently a conflicting relation between national integration (often accompanied by regional disintegration) and regional integration (potentially accompanied by national disintegration) as Boisier (Chapter 16) points out.

A theoretically ideal sequence might therefore be that, once national integration has been achieved, priority should then be given to regional (re)-integration. The problem is that this requires also a shift of real decision-making powers from the national to regional levels. National power groups normally, however, strongly resist giving up power once they have acquired it, nor are they usually prepared to facilitate the changes in economic structure which a higher degree of regional self-reliance would necessarily involve. Of the countries analysed in this book, Nigeria (and possibly Papua New Guinea) seem to be the only cases where the national government voluntarily decentralized decision-making to 12 and later on even to 19 states, as in the case of Nigeria, in a recently established federal system, possibly in a major effort to maintain national unity.

National centre-down policies due to the usual uniformity of their criteria and to the economic, social, and environmental disintegration which they usually cause at sub-national levels, often tend to force sub-national levels of government to sacrifice necessary basic needs investment to compensate disintegrating effects of central government policies (Haddad, Chapter 15). In this way, scarce regional (or state) funds are actually diverted from genuine regional objectives by the need to compensate for negative effects of central policies in the respective regions, such as decongesting nationally stimulated metropolitan concentrations, reducing consequent intra-state disparities, etc.

On the whole it can be said that while national centre-down policies are mainly concerned with the allocation of national resources to the most dynamic sectors and to their most efficient use in a national and international context, the remaining sectors and problem areas are often left as the task of lower level governments, and to spontaneous 'bottom-up' initiatives. While the tax income from the most dynamic and efficient activities also accrues mainly to the national government, lower-level governments and spontaneous initiatives have to try and solve their immediate and often most basic needs with hardly any resources remaining. Such basic needs are often very costly, including, in the case of Thailand for example (Douglass, Chapter 7), cooperative rural development, land reform, tambon (village) development, and political reform. This lack of funds at lower levels of government in fact seems to have been one of the major reasons for the faltering of the earlier generation of Community Development Projects in many countries. It is typical that, for instance, in Peru (Hilhorst, Chapter 17) the 'dynamic' projects for intersectoral growth poles were steered by the central government 'from above', whereas mechanisms for a redistribution of income within (poor) rural regions were left to operate 'from below' with practically no resources available to them. This means that while the relatively profitable projects were decided upon centrally, the least profitable ones were left to local self-determination and to 'broad public motivation'.

It is in part these deficiencies which basic-needs strategies (Lee, Chapter 4) and development from below (Stöhr, Chapter 2) want to overcome. Basic-needs services in particular cannot be supplied in a way satisfactory for the consumer and at reasonable cost by a central agency. Most of these services need to be territorially provided (local, regional, etc.) and tailored to the specific needs of territorially organized population groups at different scales. Their efficient and satisfactory provision therefore requires smaller scales of territorial organization than the production of private goods (Machlup, 1977) and a certain amount of selective spatial closure (Stöhr and Tödtling, 1979).

6. Bottom-Up Strategies: Some Experiences and Prospects

Development from below would essentially be based on integrated regional resource utilization at different spatial scales (Stöhr, Chapter 2). Priorities for such resource integration would be sought at the lowest possible scale in a fashion of a subsidiarity. The development of territorially organized social groups would cater to external demand and utilize external resources only to the extent that this does not reduce the satisfaction of their own needs and the mobilization of their territorially available resources. It represents self-reliant development on a territorially subsidiary basis in order to reduce the negative effects of external dependence.

Such a strategy has many parallels, not only with development paths which occurred in Europe at various earlier stages (Stöhr, Chapter 2 particularly Figure 2.1), but also with development patterns that existed in various case study countries before colonization, as described under (1) above.

More recent examples in Third World countries are reported on in individual case studies such as, for India, the original ideas of Gandhi's development concept (Misra and Natraj, Chapter 10). They relate back to Gandhi's concept of 'a new society consisting of small communities organically linked with each other but undivorced from nature and work'. They seem to be in part revived again in today's India's Block Level Planning scheme (Misra and Natraj, Chapter 10). Attempts along similar lines existed in the initial stages of the Ujamaa scheme in Tanzania (Lundqvist, Chapter 13) or of Guinea Bissau (Goulet, 1978) as well as the sub-national efforts listed at the end of Stöhr's Chapter 2 in this book. A frequently cited example under very special conditions of course is China (Chapter 6). The political, social, economic, and environmental conditions ruling in each of these countries differ widely. Yet, among the pervasive characteristics are that development is based on little-developed, small-scale cellular societies which, in order to avoid major external dependence, pursue endogenously motivated development strategies. They are societies which can receive few aggregate growth impulses from an export-based strategy following the 'small open economy' model (Fei and Ranis, 1973).

In centre-down strategies a less-developed country (or region) will attempt to reduce its balance of trade deficit by increasing its exports and attracting foreign investment and thereby increase external debt and dependence but at the same time neglect requirements of regional (or local) development (Hilhorst, Chapter 17). A strategy of development from below, in contrast, would attempt to solve such a problem by giving highest priority to the satisfaction of national, regional, or subregional requirements by mobilizing a maximum of resources within these entities, thus reducing to a minimum possible external debt and dependence.

Empirical observations of the functioning and of the actual results of bottom-up strategies are even more restricted than the possibilities for empirical observation of centre-down strategies. Some of the reasons are:

(a) In developing countries in which bottom-up strategies were initiated this was possible only after decolonization and therefore the observation period in many of them is necessarily very short.
(b) Bottom-up strategies require major transformations of institutional, economic, and political structures and therefore may require a considerable time until their bases are laid.
(c) Very few countries have actually decided and also been able to attempt such major structural transformations. The number of examples to be observed therefore is small.
(d) In countries where policies along similar lines were attempted, they have often been subjected to periodic backlogs so that a continuous experimental period cannot be evaluated. Alternating sequences between strategies from below and from above have been described explicitly for Chile (Boisier, Chapter 16) for Tanzania (Lundqvist, Chapter 13) and historically for European countries (Stöhr, Chapter 2) in this book. They will be dealt with briefly in the next section.
(e) The statement which Hansen (Chapter 1) makes for the centre-down strategy, namely that it is not necessarily wrong but simply has never been tried for a sufficiently long period, could with even more justification be applied to bottom-up strategies. In post-colonial Third World settings they actually have never been applied for sustained periods of time.

Another set of problems for the evaluation of 'alternative' development strategies from below is that their approaches differ considerably, even between the few countries where they have been attempted, and are therefore difficult to compare amongst each other and to delimit against more conventional centre-down strategies.

In fact Penouil (Chapter 12) maintains that in practice coherent and internally consistent development strategies rarely exist and that success can only be achieved by (and therefore also evaluated for) pragmatic policies via a 'grand empiricism'. Another restriction for defining 'alternative' development strategies 'from below' is that there exists as yet no consistent theory on the basis of which they could be evaluated (Stöhr, Chapter 2; Boisier, Chapter 16) and that most of the professional planners who were involved in implementing them were actually trained along the lines of traditional centre-down strategies and in analytical and planning methods useful to these latter strategies (Boisier, Chapter 16). They were therefore actually unqualified to design such alternative strategies. Due to the lack of trained planners in most developing countries, even in the case of drastic changes in political orientation (such as, for example, at the beginning and the end of the Allende government in Chile), essentially the same planning technicians were used, possibly with changed roles amongst them (Boisier, Chapter 16).

In spite of the diversity of empirical manifestations, there seems to be some consensus, however, that spatial equality of living conditions via development strategies 'from below' require not only specific territorially organized socio-political and administrative structures, but also specific territorially organized systems of economic interaction and the use of territorially available resources and technology (respective chapters by Stöhr, Weaver, Douglass, Boisier, and Haddad in this book). This means that a pure lowering of decision-making scales and popular mobilization at the local or regional levels alone is not sufficient if the national economic emphasis remains on a priority for large-scale national and international projects. In Peru, for example, where the National System of Support for Social Mobilization, SINAMOS, was relatively successful in the political and social mobilization of local population groups for 'the construction of small irrigation dams and feeder roads' (Hilhorst, Chapter 17), such projects were lacking support from national economic resources which remained mainly geared to large-scale national projects. As in many other countries, increasing scarcity of national resources in fact led to even further reductions of resource allocations to regional authorities and programmes (Hilhorst, Chapter 17). Locally elaborated projects usually were not spectacular enough to receive sufficient attention from central government or from the banking system.

On the other hand, in cases where funds were set aside for projects in peripherally less-developed areas, projects corresponding to productivity or efficiency criteria defined by central government or financing agencies could not be locally formulated. In part this was due to the lack of technicians able to elaborate projects along such criteria. Foreign consulting firms which had earlier prepared local projects, were often excluded by regulation from such activities (Hilhorst, Chapter 17) and the government had not found substitutes for these firms among local professionals. Another explanation for this lack of local project-formulating capacity may be that the standard national criteria for project evaluation (effects on foreign exchange balance; GNP, capital productivity, etc.) had little visible relation to the concretely felt requirements at the local level. As a result Hilhorst (Chapter 17) reports that for less-developed, high population pressure areas, very often no development projects were formulated and the central government, in what Hilhorst calls 'pipeline effects' continued to concentrate on the project lines already in progress.

In order to achieve the broad mobilization of individual and social organizations, the governments of many countries have given emphasis to large-scale demonstration projects such as, in Latin America, Brasilia or the Guyana project in Venezuela (Boisier, Chapter 16). In many of these cases it will be difficult to say whether the benefit of such large demonstration projects was greater for regional modernization and development or for the direct benefit of the ruling national strata. Some countries' governments went as far as justifying close-to-dictatorial central measures in order to induce local mobilization and participation, such as for the Ujamaa movement in Tanzania (Lundqvist, Chapter 13) or for the National System of Support for Mobilization in Peru (Hilhorst, Chapter 17).

Development 'from below', however, may also require certain external inputs. In order to facilitate reasonably 'equal access of all population strata to the production and consumption of society's goods, services, and welfare' at the local or regional level, development policies 'must involve some assistance from central decision-making units' (Lo and Salih, Chapter 5). This means that an equity-oriented transformation of local and regional social structures is considered by various authors to require extraregional (national or even international) political or legal support (Lo and Salih, Chapter 5; Lundqvist, Chapter 13).

Whereas centre-down development usually operates on the basis of either a centralized (national or international) or an atomistic individual decision-making system (e.g. via the market mechanism), bottom-up development requires territorially organized communal decision-making at various scales (Lundqvist, Chapter 13). Such a system therefore involves certain elements of 'selective spatial closure' (Stöhr and Tödtling, 1978).

Some types of spatial closure, however, can also contribute to a petrification of existing internal inequalities as Blaikie (Chapter 9) has shown for Nepal, where under historical conditions of a double dependence from India/Great Britain this closure has helped the ruling classes of Nepal to preserve their power by manipulating in their own specific interest the penetration across the national frontier (Blaikie, Chapter 9).

It is equally dangerous, however, if a contradiction exists between the strategies of sociopolitical and of economic development. This seems to have been the case in Ujamaa development in Tanzania where sociopolitical development from below was complemented by economic measures of large-scale integration backed by the advice of external consulting firms which normally operate in a centre-down development context (Lundqvist, Chapter 13). In such cases local and regional decision-making is superseded by large-scale economic projects which may then jeopardize not only the country's external trade balance but also substantially weaken the feasibility of an idea such as that of Ujamaa and consequently of Nyerere's entire development philosophy (Lundqvist, Chapter 13). From this experience the conclusion might be derived that self-reliant local and regional decision-making must be combined also with an egalitarian regional system and with a certain degree with self-reliant economic development.

It appears that an incompatibility in the opposite direction, namely central decisions on concrete features of self-reliant social units, can be equally dangerous. In Tanzania, for example, the site selection for Ujamaa villages was essentially made at the central government level. Accessibility to towns was used as a main criterion while vital local questions such as soil quality and water-availability were not sufficiently taken into account (Lundqvist, Chapter 13).

Boisier (Chapter 16) quotes three conditions for a successful bottom-up development strategy:

(1) The creation of new institutions; this would also require the use of new personnel which in most developing countries, however, is not available due to the lack of trained technicians; using the same technicians as with centre-down strategies will essentially perpetuate formerly used methodologies and theoretical approaches;

(2) A broad understanding by the population of the objectives of regional development 'to mobilize the country's creative energy' (Boisier, Chapter 16); this usually requires spectacular demonstration projects which are directly contradictory to development 'from below'.

(3) The generation of visible results in a relatively short period of time; this is normally impossible as bottom-up strategies would require major structural transformations; in fact such structural transformations, e.g. land reform, usually bring a setback in production during the initial period and therefore also in the short-term results of such policies.

These basic dilemmas must be kept in mind also when evaluating the relatively short active periods of development strategies of a country like Algeria which Sutton (Chapter 14) regards as approaching development 'from below'. Examples there along such lines are the '*Plans Communaux*', containing important elements of local basic needs plans; the Agricultural Reform Cooperatives; the Rural Renovation Projects, the employment effects which, per unit of investment, are considered comparable to those of industry. These strategies are considered by Sutton as a diffusion of sociopolitical innovation from the periphery to the centre (Chapter 14).

The same applies to the Tanzanian experience discussed in detail by Lundqvist (Chapter 13) and to some aspects of regional development policy in Chile where, for example, during Allende's government the concept of growth poles was substituted by that of 'integrated areas' as objects of regional development (Boisier, Chapter 16).

It is interesting that development from below was often allowed to take place by national governments in what Waterston (1965) has called 'wooship relations' towards far outlying geographical areas over which the centre had little control except by wooing. Examples are the North Solomons in Papua New Guinea (Conyers, Chapter 8) or the department of Arica in the extreme North of Chile bordering Peru. It seems that scarce power or difficult physical access by the central power is of help for facilitating development 'from below'.

7. Alternating Sequences Between Centre-Down and Bottom-Up Strategies

It might be assumed that certain countries or regions, depending on their characteristics such as size, internal diversification, availability of resources in high worldwide demand, might be predestined for externally oriented development from above as an 'open economy' able to derive developmental impulses mainly from world demand, while others would be typically destined for internally oriented development 'from below'. Such assumptions might also be taken as an indication that it could be very difficult for a country to break out of such a predetermined development path. Arguments of such predetermination might particularly be levied to fend off popular demands to change development strategies which in the past have created strong internal social or spatial disparities in living levels. They may often also be used as a defence against claims for the introduction of development strategies 'from below' emphasizing local regional economic circuits and broader popular participation.

Evidence shown in this book indicates, however, that such a deterministic linear projection of past development strategies can by no means be empirically verified. In a historical perspective on the past 2500 years, Stöhr (Chapter 2) shows that there have in fact been periodic alternations between development from above and from below related to changes in philosophical outlooks and value systems, technological development, social organization, etc. Although there are no clear-cut cause-and-effect relations between such societal conditions and specific development paths, there seem to exist mutual feedback mechanisms between these two groups of variables. Development from above seems to be associated with predominantly rationalistic eras, with periods of rapid economic growth, large-scale societal interaction, and often rapid technological innovation. Development from below, on the other hand, seems to be associated with metaphysically dominated eras, periods of reduced economic growth, with often small-scale societal interaction dominating (Stöhr, Chapter 2). It is hoped that this book will help to stimulate more systematic historical research on related questions.

For more recent times, a number of case studies also show alternations between the process of development from above and from below. In Chile these sequences have been clearly related to changes in political systems between the three ideological periods described by Boisier. But similar changes between alternate development strategies have also taken place within continuing political systems.

For Algeria, for instance, Sutton (Chapter 14) reports a clear centre-down strategy for the period 1961–71 which in the following period, however, changed to include many bottom-up elements. The initial period was symbolized by 'taking steps to the rear in order to ensure future progress' (Nellis, quoted in Sutton, Chapter 14). He reports that in the early phases the introduction of self-management, mainly in agriculture, proved a failure. During an intermediate period, therefore, major emphasis was given to 'national' policies, whereas in the later period after 1971 strong emphasis was given to bottom-up strategies of an agropolitan type (Sutton, Chapter 14).

For Tanzania Lundqvist, in a similar way, reports a recurring sequence of urban/rural/urban policy emphasis (Lundqvist, Chapter 13). For the Ivory Coast, Penouil gives a detailed account of similar alternations between development sequence in that country (Penouil, Chapter 12). In his view the initial polarization was related to the creation of some essential pieces of infrastructure as soon as a minimum of national revenue was available. The subsequent equilibrium-oriented phase was based on the expansion of the plantation economy (coffee and cocoa) with a relatively wide spatial distribution. The income of these staple products permitted the establishment of an exchange economy which—while without doubt exploiting the population—at the same time facilitated their access to the consumption of new products. This maturation phase, based on a broad demand for new products, successively prepared the following phase of polarized development which Penouil symbolizes not so much by an '*industries môtrice*' but rather by a spatial agglomeration of diverse and mutually little-interacting industries based on the growth of agricultural income. Penouil formulates the need that, during the period to follow, this functional diversification of the economic structure would have to be transformed again into an also geographically integrated diversification. Whether the latter happens will depend to a large extent on whether the large central companies established in the core region merely install isolated modern production plants in peripheral areas (Penouil, Chapter 12) thereby primarily extending the spatial scale of dominance of the core region economy over the periphery, or whether they or the government are able to actually incentivate self-sustained development in the rest of the country.

Without the aid of such multi-regional enterprises the spatial diffusion of development seems to be severely handicapped in countries where 'urban systems are rudimentarily developed so that the trickling down of development impulses is unlikely' to take place (Douglass, Chapter 7). The established tools of regional planners based on the assumption of transmission effects through the urban system, the transport network, industrial linkages, and innovation diffusion are therefore of limited value under such conditions prevalent in many Third World countries, unless the state were able to create the necessary broad infrastructure basis and other determinants of development (Douglass, Chapter 7).

One of the key questions in this context is whether regional development is considered as a merely quantitative economic phenomenon (volume of production and number of jobs created in peripheral areas) or whether it is considered as a process of integral resource mobilization in peripheral areas. Whereas the first can evidently be achieved by a spatial expansion of the scale of action of metropolitan enterprises, the second goal can hardly be achieved unless autochthonous development potentials of peripheral areas can at the same time be stimulated. If such development should happen via Penouil's 'transition activities' (Chapter 12) this would require that such 'transition' is not just the net balance between the death of autochthonous peripheral (traditional) activities and the birth of core regional initiated (modern) activities there, but that actually a continuous transformation of autochthonous traditional activities takes place by integrating within this transformation process also local/regional enterpreneurial, human, institutional, and environmental components and their economic, social, and political interaction within the respective peripheral regions. Evidently certain transformations of these relations with particular emphasis towards more equity would have to be brought about. More detailed micro-case studies of such transition processes at the local or regional level as mentioned at the end of Stöhr's Chapter 2 would be necessary to show under which conditions this is feasible or not.

In this sense the 'grand empiricism' advocated by Penouil (Chapter 12) requires that the empirical results of past strategies should heavily influence the design of future strategies. Changes in strategic orientations between development from above and from below should be strongly guided by the inadequacies of the foregoing period—as seems to have happened in the alternating strategy sequences of some of the quoted case study countries.

The final objective may in fact be that after an alternating sequence of bottom-up and centre-down development phases, a juxtaposition of the major elements of both these strategies remains in the sense that Uphoff and Esman (1974) maintain that 'successful local ... and participatory development depends very much on a high frequency of both top-down and from-below development impulses; local autonomy in isolation provides little leverage for development' (Uphoff and Esman, 1974, quoted in Douglass, Chapter 7). On the other hand, outright large-scale integration tends to erode local and regional development potentials. The intense mutual interaction of both these paradigms seems to be required and may in fact only be feasible after extensive periods in which both of these strategies have been practised subsequently or jointly.

REFERENCES

Clastres, P. (1974) *La Societé contre l'Etat*, (Paris: Les Editions de Minuit).
Fei, J. and D. Ranis (1973) *The Transition in Open Dualistic Economies* (New Haven and London: Yale University Press).
Friedmann, J. (1978) The active community: towards a political–territorial framework for rural development in Asia. UNCRD Seminar Paper, Nagoya, Japan.
Friedmann, J. (1979) Surviving in Rural Asia, An Exhibit. *IFDA Dossier* No. 9, pp. 9–15.
Friedmann, J. and M. Douglass (1978) Agropolitan development: towards a new strategy for regional planning in Asia, in Lo, F.C. and K. Salih (eds.) (1978) *Growth Pole Strategy and Regional Development Policy* (Oxford: Pergamon Press).

Goodman, D. E. (1975) O modelo economico brasileiro e os mercados de trabalho: uma perspectiva regional. *Pesquisa e Planejamento Economico*, 5 (1), 89–116

Goulet, D. (1978) The challenge of development economics, *Communications and Development Review*, 2 (1), 18–23.

Goulet, D. (1978) *Looking at Guinea-Bissau: A New Nations's Development Strategy.* Overseas Development Council, Occasional Paper No. 9, Washington DC.

Gunaratna, K. L. (1979) Spatial Planning in Sri Lanka, unpublished draft (mimeo.).

Hechter, M. (1975) *Internal Colonialism, The Celtic Fringe in British National Development, 1536–1966* (London: Routledge & Kegan Paul).

Hirschman, A. (1958) *The Strategy of Economic Development* (New Haven: Yale University Press).

International Development Review (1979) China: the other foot forward, XXI (2), 35–47.

Khan, A. R. (1977) Production planning for basic needs. In Ghai, D. P. *et al.* (eds.), (1979) *The Basic Needs Approach to Development* (Geneva: ILO).

Kuznets, S. (1955) Economic growth and income inequality, *The American Economic Review* XIV (1), 1–28.

Machlup, F. (1977) *A History of Thought on Economic Integration* (London: Macmillan).

OECD (1977) *Report on Methods of Measuring the Effects of Regional Policies.* (Paris: Organization for Economic Cooperation and Development).

Perroux, F. (1964) La notion de pôle de croissance, *L'économie du XXème Siècle*, 2nd edn., pp. 142–54 (Paris: Presses Universitaires de France).

Richardson, H. W. (1973) *Regional Growth Theory* (New York: John Wiley).

Schumacher, F. (1973) *Economics as if People Mattered: Small is Beautiful* (New York: Harper & Row).

Stöhr, W. and F. Tödtling (1978) An evaluation of regional policies—Experiences in market and mixed economies. In: Hansen, N. M. (ed.), *Human Settlement Systems*, (Cambridge, Mass.: Ballinger).

Stöhr, W. and F. Tödtling (1979) Spatial equity: Some anti-theses to current regional development doctrine. In: Folmer H. and J. Oosterhaven (eds.), *Spatial Inequalities and Regional Development*, (Boston: Nijhoff).

Taylor, D. R. F. (1972) The role of the smaller urban place in development: a case study for Kenya, *African Urban Notes*, VI (3), 7–24.

Taylor, D. R. F. (1975) Spatial organization and rural development. In Fry, M. G. (ed.), *Freedom and Change* (Toronto: McClelland & Stewart).

Uphoff, N. and M. J. Esman (1974), *Local Organization for Rural Development in Asia* (Cornell University: Rural Development Committee).

Waterston, A. (1965) *Development Planning, Lessons of Experience* (London: The Johns Hopkins Press).

Webber, M. (1979) Social policy for the exploding metropolis: what roles for the social sciences. In Pimeri le Moraes, M. and J. M. Pimerta, (eds.) (1979) *Urban Networks, Development of Metropolitan Areas* (Rio de Janeiro: EDUCAM).

Williamson, J. G. (1965) Regional inequality and the process of national development: a description of patterns, *Economic Development and Cultural Change*, 13, 3–45.

Contributing Authors

Walter B. Stöhr is Professor of Regional Planning and Director, Inter-disciplinary Institute for Urban and Regional Studies (IIR), University of Economics (Wirtschaftsuniversität), Vienna, Austria. He was previously Professor of Regional Planning at McMaster University, Canada, and for five years acted as Senior Regional Planning Advisor of the Ford Foundation in Santiago, Chile. He has been a consultant to various international agencies of the UN, OECD, etc., and is the author of *Interurban Systems and Regional Economic Development (A.A.G.., Washington, 1974)*, *Regional Development: Experiences and Prospects in Latin America* (Mouton, Paris 1975), and various papers in professional journals.

D.R.F. Taylor is Professor of Geography and International Affairs, Carleton University, Ottawa, Canada. His research interests in rural and regional development in developing nations focus on Africa where he worked for many years. His publications include *The Computer and Africa: Applications, Problems and Potential* (Praeger, 1977) and *The Spatial Structure of Development: A Case Study of Kenya* (Westview, 1979).

Piers Blaikie received his BA in geography from Cambridge in 1964, his MA in 1965, and his PhD in 1971. He is currently Senior Lecturer in Development Studies at the University of East Anglia and has also taught at the University of Reading. He is author of *Family Planning in India: Diffusion and Policy*, (Arnold, London 1975), and co-author of three books on Nepal (with John Cameron and David Seddon): *Nepal in Crisis: Growth and Stagnation at the Periphery* (Oxford University Press, 1980); *Workers and Peasants in Nepal* (Aris and Phillips, Warminster, 1979); *Struggle for Basic Needs: the case of Nepal* (Paris, OECD, 1980). He spent two years in India and one and a half years in Nepal carrying out fieldwork for these books.

Sergio Boisier is a Chilean economist who graduated from the University of Chile and from the Department of Regional Science of the University of Pennsylvania. He served as a regional planner with the Chilean National Planning Office up to 1970. Later he joined the United Nations working in Argentina, Brazil, and Panamà; currently he is Senior Expert in Regional Planning of the United Nation Latin American Institute for Economic and Social Planning.

Mr Boisier is author of the book, *Diseño de planes regionales*, (Design of Regional Plans), (Madrid, 1976) and also of several articles and papers on regional planning published in various books and professional journals.

Diana Conyers worked in Papua New Guinea from 1973 to 1979, initially as a Research Fellow of the Australian National University and later as a Senior Lecturer in the Papua New Guinea Institute of Administration and as a public servant in the Government's Department of Decentralization. Throughout this period she was closely involved in the planning and implementation of the decentralization programme she describes in her chapter. Prior to coming to Papua New Guinea, she spent three years in the Bureau of Resource Assessment and Land Use Planning at the University of Dar es Salaam, Tanzania, and a year in the USA where she taught at Clark University, Massachusetts and Wilmington College, Ohio. She obtained a doctorate from the University of Sussex, based on some of her work in Tanzania. At the time of writing, she was visiting Senior Lecturer in the Department of Anthropology and Sociology, in the University of Papua New Guinea.

Mike Douglass is currently a lecturer in regional development and spatial analysis at the School of Development Studies, University of East Anglia. He is a PhD candidate in the School of Architecture and Urban Planning at the University of California at Los Angeles and from 1976–1978 was a research associate in the policy research programme of the United Nations Research Centre for Regional Development, Nagoya, Japan.

M. O. Filani is Senior Lecturer in Geography at the University of Ibadan, Nigeria. He received an MS and PhD from Pennsylvania State University, USA, both in Geography. His primary research interests are in development planning and economic geography with a special emphasis on transportation.

Paulo Roberto Haddad is a member of the Center of Regional Development and Planning (CEDEPLAR) at the Federal University of Minas Gerais. He was previously Director of Regional Planning at Fundação João Pinheiro and now is Planning Secretary of the State of Minas Gerais. He has written various articles on urban and regional development in Brazil and has written (or edited) the following books: *Planejamento Regional: métodos e aplicação ao caso Brasileiro* (Rio, 1972); *Desequilibrios Regionais e Descentralização Industrial* (Rio, 1976); *Contabilidade Social e Economia Regional* (Rio, 1977); *Dilemas do Planejamento Urbano e Regional no Brasil* (Rio, 1978); *and Dimensões do Desenvolvimento Brasileiro* (Rio, 1978).

Niles Hansen is Professor of Economics at the University of Texas at Austin. From 1975 to 1977 he was on leave to carry out research on human settlement systems at the International Institute for Applied Systems Analysis, Laxenburg, Austria. He has written numerous books and articles on regional development policies among others *Growth Centers in Regional Economic Development* (New York: Free Press 1972), *Public Policy and Regional Economic Development* (Cambridge, Mass. : Ballinger, 1974). *Human Settlement Systems* (Cambridge, Mass.: Ballinger 1978). He has been a consultant to such organizations as the Ford Foundation, the World Bank and the United Nations.

Jos G. M. Hilhorst is Professor of Regional Planning at the Institute of Social Studies in The Hague, Holland. In addition to his teaching and research tasks at that Institute, he has gained experience in regional planning with assignments of varying duration in, among other places. Argentina, Brazil, Greece, Peru, Pakistan, and Turkey. His publications include books such as *Regional Planning, A Systems Approach* (Rotterdam, 1971) and various papers and articles.

David F. Ip received his PhD from the University of British Columbia, Canada, with a dissertation on rural development in China. He is at present Lecturer of the Department of Sociology, University of Queenland, Brisbane, Australia. Previously he took Doctoral studies at the University of British Columbia, specializing in rural development in South China and the sociology of development, and prior to that, he taught at the University of Hong Kong.

Eddy Lee was educated at the University of Malaya and Magdalen College, Oxford: BPhil. and DPhil. in Economics. Previously he was a lecturer at the University of Malaya and consultant to the Asian Development Bank. Currently he is Senior Economist with the World Employment Programme, ILO. He is joint author of *Poverty and Landlessness in Rural Asia* (Geneva: ILO, 1977); *The Basic Needs Approach to Development* (Geneva: ILO, 1977); and *Agrarian Systems and Rural Development* (London: Macmillan, 1979).

Fu-chen Lo has been chief of Comparative Studies at United Nations Centre for Regional Development since 1973 and is currently in charge of a Policy Research Programme to develop and coordinate collaborative research on key issues in regional development and planning. As part of his responsibility, he has frequently served as consultant and guest lecturer to planning agencies and universities in a number of ESCAP countries. He is a native of Taiwan and, migrated to the USA in 1963, obtaining his PhD in Regional Science at University of Pennsylvania. Prior to UNCRD, he has conducted research for Economic Development Administration, Puerto Rico Planning Board, Environmental Protection Agency, etc., on regional development models and pollution control simulation models. His recent publications include *Growth Pole Strategy and Regional Development Policy, Asian Experience and Alternative Approaches* (edited with K. Salih; Pergamon 1978) and a number of writings on regional development issues in Third World countries.

Jan Lundqvist has a Ph.D. (filosofie doktor, Göteborg). At present he is Professor of Linköpping University. Before he was senior lecturer at the Department of Geography, University of Bergen and responsible for 'non-European studies'. He was educated in Sweden and has carried out research work in Tanzania and Sri Lanka. He has also served as a consultant for the Swedish International Development Authority (SIDA) in the planning of rural water supply in Tanzania. Among his publications are: *The Economic Structure of Morogoro Town* (Uppsala: NAI, 1973); *Local and Central Impulses for Change and Development* (Oslo, 1975); and *Trade and Formation of Production* (People's Bank, Colombo, 1978).

R. P. Misra, Director, Institute of Development Studies, University of Mysore, is currently Deputy Director, U.N. Centre for Regional Development,

Nagoya, Japan. He is a specialist in rural and urban planning closely associated with officially and non-officially sponsored regional development activities in India. His numerous publications include *Regional Planning in India* (Delhi: Vikas 1974), *Regional Planning and National Development* (Delhi: Vikas 1978) and *Million Cities of India* (Delhi, Vikas 1978).

V. K. Natraj, Reader at the Institute of Development Studies, University of Mysore, specializes in economic and regional planning. His publications include *Decentralization of Planning in India* (Mysore, Institute of Development Studies, 1974); *New Perspectives in Centre–State Relations in India* (University of Mysore, 1975); and *Regional Planning and National Development* (Delhi, Vikas, 1978).

Marc Penouil is currently Vice President of the University of Bordeaux and is head of the Faculty of Economic Science. From 1960 to 1962 he was Professor of Economic Science at the University of Abidjan. He is a specialist in problems of economic development with a special interest in the developing world in general and Africa in particular. He has published numerous articles on regional development and economic development in Africa. His publications include *Socio-économie du Sous-développement* (Paris, Dalloz, 1979); *Economie du développement* (Paris Dalloz, 1977); and *Démographie* (Paris, Dalloz, 1978).

Kamal Salih is currently Dean of the School of Comparative Social Sciences, University Sains Malaysia, Penang, and is a member of the Development Studies Group of the faculty. He is also associated with the Centre for Policy Research at the university where he has coordinated several research projects including regional planning evaluation and development of a national integrated data system for urban and regional planning purposes. In the field of regional development research he has collaborated extensively with the United Nations Centre for Regional Development in Nagoya, Japan, with whom he has served as consultant since 1974. He has also served on a number of Malaysian government committees and conducted work for the Economic Planning Unit and the Socioeconomic Research and General Planning Unit of the Prime Minister's Department. He obtained his PhD in Regional Science at the University of Pennsylvania in 1973. His publications include *Growth Pole Strategy and Regional Development Policy, Asian Experience and Alternative Approaches* (edited with Fu Chen-Lo, Pergamon, 1978) and various papers on regional development and other development issues in Malaysia and Third World countries in general.

Keith Sutton is a Lecturer in Geography at the University of Manchester. After postgraduate research at the University of London in the historical geography of the Sologne region in France, he developed an interest in the economic development problems of North Africa. He is author of several chapters and articles on nineteenth-century land clearance in France, on population, agrarian reform, rural settlement, the oil and gas industry of Algeria, and on resettlement in Africa and Malaysia.

Clyde Weaver earned his PhD at the University of California, Los Angeles, and is currently teaching at the University of British Columbia. He has worked as a planning consultant in the United States and Africa, and during the 1977–78 academic year was a Visiting Professor at the Universities of Paris and Aix-Marseille in France. He is co-author with John Friedmann of *Territory and Function: The Evolution of Regional Planning*. (Arnold, 1979)

Chung-Tong Wu is with the Department of Town and Country Planning. The University of Sydney, Australia. He previously taught at the University of Singapore and the University of Hawaii. He is a graduate of the University of California at Berkeley and Los Angeles, and Columbia University, and now specializes in regional planning and development theory.

Index

Basic activities 40–41, 77, 79, 82–83, 88, 97–98, 157
Basic needs 65, 107, 112, 113, 115, 116, 148, 176, 193–94, 262, 270, 273–74, 386, 415, 462, 470–72

Cellular economy 1, 44–46, 63, 93, 94, 97, 132–33, 136–37, 146–47, 149, 162–75, 202–7, 427–28, 457, 458–60, 472, 475
Centralization 17–19, 24, 27–28, 74, 76, 78–84, 93, 98, 125, 155–56, 162–64, 188–89, 195–96, 213, 287, 296–99, 300–1, 308–9, 311, 314–15, 333, 352–56, 384–85, 389, 404, 418–24, 431, 457, 458, 463, 465–66, 469, 470
Centre–periphery relations 20–25, 72, 85, 86, 95, 132–33, 170–75, 253–54, 396, 399, 401–3, 411, 421–22
Circular and cumulative consumption 17, 80, 82–83
City size distribution 32, 45, 66, 81, 84, 131, 139, 293, 465–66
Colonialism (external, internal) 18–21, 27–28, 32, 34–35, 41–42, 70–71, 74, 83–87, 143–45, 232–35, 241, 243–44, 247, 249–50, 297, 333, 347–48, 357–59, 431, 448–49, 461–64
Core–periphery relations 20, 24–25, 71–72, 85, 86, 95, 132–33, 170–71, 253–54, 396, 399, 401–3, 411, 421–22
Core regions and areas, *see* Growth centres

Decentralization 43–46, 52, 86, 96, 119–21, 131, 133–34, 144–45, 157–61, 195, 209, 220–27, 261–62, 274–77, 301–2, 317, 319, 341–42, 351, 355–56, 362–64, 366–67, 372–73, 386, 409–10, 415, 418, 439, 440, 445–48, 457, 458, 471
Deconcentration, *see* Decentralization
Dependency, *see* Colonialism
Development poles, *see* Growth centres
Devolution, *see* Decentralization
Diffusion, *see* Spread effects
Dominance, *see* Colonialism
Dual economy 23–26, 35–44, 141, 188–93, 304, 462–63, 470
Dynamic sectors 40–41, 133–34, 319, 321–22, 324, 326–27, 353–54, 468–69, 472

Economic base, *see* Basic activities
Economic circuits 23–26, 32, 35–36, 40–41, 141, 188–93, 304, 462–63, 470
Export base, *see* Basic activities

Growth centres, growth axis, growth poles 1, 15–17, 19, 21, 32–34, 36, 57–58, 65, 79–86, 92, 95, 98, 123, 129–30, 132, 133, 137–39, 144–46, 161, 183, 196–99, 232, 313–14, 319, 329–30, 357, 362–63, 408, 411–12, 416–17, 421, 433, 436–39, 457–58, 460, 469, 470, 472

Imperialism, *see* Colonialism

Leading industries 40–41, 215, 319, 321, 324, 326–27, 353–54, 468–69, 472
Leakages 42, 84, 95–96, 135, 147, 164–65, 196–97, 462

Minimum needs, *see* Basic needs
Modern activities 40–41, 133–34, 319,
    321, 324, 326–27, 353–54, 468–69
Multi-national, multi-regional
    enterprises, firms, etc. 26–29, 30–32,
    35, 82–85, 91, 97, 113–15, 120, 132–36,
    146–47, 149–50, 295
Multiplier effects 77, 84–85, 97, 98,
    147–48, 202, 240

Peripheral areas (country, regions,
    locations, etc.) 63–64, 84–85, 88–89,
    132–33, 144–45, 198, 231, 295, 317,
    365, 389–96, 462–63
Polarization, *see* Centralization
Polarized development, *see*
    Centralization
Polarized growth, *see* Growth centres
Primacy, *see* City size distribution

Redistribution policies (of income,
    wealth, assets, etc.) 42–43, 45–46, 58,
    110–11, 117–18, 164–75, 267, 317,
    363–66, 371, 388–89, 399, 404–5, 468
Residentiary, non-basic activities 77, 95
Resource development, mobilization
    (material, human, integrated, etc.) 1,

29–34, 36, 39, 61, 65, 70–71, 95–98,
    164–75, 231–32, 344, 367–68, 396–98,
    415, 433–56, 468, 472, 473

Self-reliance (collective, etc.) 42, 44–46,
    68, 116, 118–21, 143–44, 146–47,
    164–70, 216–17, 336–37, 346–47,
    379–84, 454, 458–59, 471, 472, 475–76
Spatial closure (selective, etc.) 1, 44–46,
    63, 93, 94, 98, 132–33, 136–37, 146–47,
    176–77, 185, 202–4, 205–7, 427–28,
    457–60, 472, 475
Spread effects 1, 17–18, 20–23, 26, 32–35,
    41, 42, 45–46, 80, 81, 84, 109, 126, 132,
    172–75, 199, 270, 300–1, 357–59, 408,
    467

Traditional activities, sectors 141, 319,
    335, 468
Transitional activities 41, 308, 309, 312,
    319–28, 467–69, 478–79
Transitional enterprises, firms, etc, *see*
    Multi-national enterprises
Trickle-down effects, *see* Spread effects

Urban system, *see* City size distribution